HISTORY AND PHILOSOPHY OF CONSTRUCTIVE TYPE THEORY

SYNTHESE LIBRARY

STUDIES IN EPISTEMOLOGY,

LOGIC, METHODOLOGY, AND PHILOSOPHY OF SCIENCE

VOLUME 290

HISTORY AND PHILOSOPHY OF CONSTRUCTIVE TYPE THEORY

by

GIOVANNI SOMMARUGA
Albert-Ludwigs University, Freiburg, Germany

KLUWER ACADEMIC PUBLISHERS
DORDRECHT / BOSTON / LONDON

Library of Congress Cataloging in Publication Data is available.

ISBN 0-7923-6180-6

Published by Kluwer Academic Publishers,
P.O. Box 17, 3300 AA Dordrecht, The Netherlands.

Sold and distributed in North, Central and South America
by Kluwer Academic Publishers,
101 Philip Drive, Norwell, MA 02018, U.S.A.

In all other countries, sold and distributed
by Kluwer Academic Publishers, Distribution Center,
P.O. Box 322, 3300 AH Dordrecht, The Netherlands

Printed on acid-free paper

Printed and bound in Great Britain by MPG Books, Bodmin, Cornwall.

TABLE OF CONTENTS

PREFACE

My first acquaintance with constructive type theory goes back to a research stay in Siena many years ago. One day, Giovanni Sambin approached me and whispered in my ear to forget about my research on fixed point constructions in various theories of mathematical logic and to rather spend my time reading a delightful and exciting little logic book with the title 'Intuitionistic Type Theory' by Per Martin-Löf. I did not completely ignore my friend's advice and bought the booklet a bit later, put it straightaway into my bookshelf where it gradually started to gather dust. But the suggestion had continued to work subconsciously. When I wondered a few years ago what my next research project was going to be about, the title 'Intuitionistic Type Theory' quite spontaneously sprung to my mind. And after a first reading of this book, I shared Sambin's enthusiasm for it.

In my first enthusiasm, I wanted to approach this theory in my project from all possible points of view: from a computer scientific, historical, linguistic, logical, mathematical and philosophical point of view. I discussed the issue with Göran Sundholm who, as an expert in the field, warned me about the risks, if not impossibility of such an enterprise. He informed me about the current research on Martin-Löf's type theory which is being carried out all over the world. My discussions with him made my enthusiasm more down to earth and gave rise to a project which promised to be not only original, but also realizable.

A first glance at the Contents of this work may seem to flatly deny that the talks with Sundholm had any effect whatsoever on my first enthusiasm. It looks indeed very comprehensive. To correct this possible false impression, a few explanatory remarks will be made.

First, the subset theory of constructive type theory, i.e. the theory which allows for the introduction and elimination of subsets, does not feature in this work. (There is a neat presentation of it in Nordström *et al.* (1990), part II.)

Second, there is nothing about non-standard constructive type theory, i.e. a non-standard extension of (standard) constructive type theory. The former is obtained by adding appropriately defined choice sequences, or alternatively by adding an indexed fixed point operator to (standard) constructive type theory. Martin-Löf has dealt with this theory and related topics in his (1983B, C), esp. (1988) and (1990–91). (Palmgren's Ph.D. thesis (1991), as well as other papers, are about this theory.)

Third, rather little mention is made of metamathematical investigations of constructive type theory by Coquand, Diller, Schwichtenberg, Troelstra and others.

Moreover, there are no applications of constructive type theory to linguistics (but there are two fine introductions or surveys in Ranta (1991) and (1994) resp.) and hardly any applications of constructive type theory to computer science (as for the latter, cf. Nordström et al. (1990), part IV, and Thompson (1991), Chapt. 6).

Finally, it is regrettable that the comparison of constructive type theory with other type theories or related theories fell a prey to sensibleness and practical realizability, e.g. Coquand and Huet's calculus of constructions, Girard's or Reynold's second order or polymorphic λ-calculus, or de Bruijn's AUTOMATH project or Constable et al.'s Nuprl system. One can only quote Thompson in this respect: 'to do justice to them would require another book of this size'.[1]

It may now be time to counteract the opposite impression which might have formed by reading about all the topics which are missing in the present work. The bulk of my work is historical and philosophical. Considering the widespread and great interest of and in constructive type theory,[2] any lengthy justification of this work would be dwelling on the obvious. It may be pointed out that apart from a brief sketchy history of constructive type theory in Nordström et al. (1990),[3] some scattered (almost confusing) comparisons of various versions in Beeson (1985),[4] and a few highly systematized comparisons of various (constructive type) theories (rather than versions of one theory) in Troelstra and van Dalen (1988),[5] nothing exists on this topic. What transpires, however, is that such a historical reconstruction of the various stages of development of constructive type theory is a desideratum. And as far as the philosophical side is concerned, one can merely point out that whereas on the one hand it seems that since 1987 a majority of Martin-Löf's writings have been dealing with philosophical questions with regards to constructive type theory[6] and, as Beeson dixit, constructive type theory is chiefly philosophy (of logic and mathematics),[7] on the other hand no comprehensive presentation or discussion of the philosophy of constructive type theory exists.

This work proceeds in a reversed chronological order. Chapter 1 starts in the present or what used to be the present in 1995: it presents the last version of constructive type theory in a way which closely follows certain of its subjacent philosophical guidelines (explained in the preliminaries of this chapter). This presentation is honey-combed with historical and philosophical C, n & o's (comments, notes and observations), the bulk of which are taken from Martin-Löf. Chapter 2 elucidates philosophically and metatheoretically various aspects of the version of constructive type theory presented in the preceding chapter. Chapter 3 describes and explains the development of constructive type theory from 1970/71 (version (1)) until 1994/95 (version (8)) by means of almost a dozen, parameters'. Chapter 4 jumps even further back than 1970, namely to 1880, where the preparation of the philosophical and technical ground for constructive type theory all began: in the first part, constructive type theory is philosophically rooted in the development of logic of roughly the last 150 years; the second part presents the most important technical means used in the build-up of constructive type theory, and it does so not in a systematic, but in a historical way. This work could be regarded as a quite complete history of modern logic (with a lot of philosophy) from a rather restricted point of view, namely constructive type theory.

Since Per Martin-Löf has called his theory 'intuitionistic type theory', it may be appropriate to explain why it has here been renamed 'constructive type theory'. The reason is simply that in constructive type theory the Brouwer–Heyting–Kolmogorov interpretation has become the formal theory in which intuitionistic logic or mathematics can be developed. And the Brouwer–Heyting–Kolmogorov interpretation is *the* characteristic of constructive logic or mathematics in general. On the other hand, essential ingredients of intuitionism such as choice sequences, spreads, or the creative subject are missing in (standard) constructive type theory.

Acknowledgements

My sincere thanks goes to the Swiss National Science Foundation (SNSF) as well as to a SNSF/British Academy researchers' exchange program both of which provided me with a grant to work abroad.

I would like to express my gratitude to Thierry Coquand, Per Martin-Löf and Göran Sundholm who ever so often generously spent time answering questions, solving problems, providing clarifications of points I did not understand, questioning my procedure or giving advice. I would like to extend this gratitude to those people who read larger or smaller portions of my manuscript and made many stimulating and valuable comments not only on my handwriting, but also on gaps, obscurities, mistakes, etc. in what I had written. Needless to say, these people, namely Gerhard Jaeger, Marco Swaen, Jaap van der Does and Joost Visser, are by no means to be blamed for all the weaknesses that can still be found in this work.

I am very grateful to those persons who struggled with or bravely mastered the kind of 'writing' which I used next, i.e. LATEX: Joost Visser not only put together a Proof-tree program for me, but whenever I plunged into a crisis because of some major LATEX problem, he also produced as if by magic some tricky solution. And whenever Mrs. Wolf, who typed a considerable part of presumably one of the worst manuscripts she has ever undertaken to type, felt like she was about to have a crisis, I met her with my most sincere sympathy. Finally, half the staff of the Fribourg Math Institute graciously endured being worried ever so often about certain typing or compiling difficulties I had.

I would especially like to thank Evandro Agazzi, Ruedi Imbach, Dan Isaacson, Per Martin-Löf and Göran Sundholm for their great support and encouragement and for making it possible to spend as a visiting scholar a short but interesting time at Oxford University, a long and gratifying time at Leiden University as well as a medium and exciting time at Stockholm University. I am particularly happy that Per Martin-Löf allowed me to use numerous of his unpublished papers and lecture notes – since without them this work would have been a mere farce compared to what it is now.

I wish to thank my relatives and friends who have never stopped letting me know that they believed in what I was doing, even though they have hardly had a faint idea of what it was. Stefano Baratella, Dick de Jongh, Giovanni Sambin and Anne Troelstra – who could have had a very sharp idea, had they chosen to – kept track with relaxed curiosity of what I was working on, and thus kept me going.

Furthermore, I am very thankful to Per Martin-Löf, Alvaro Tasistro and Anne Troelstra for their very valuable and substantial contribution to the revision of an earlier version of this book. Needless to add, they are not to be blamed for not having pointed out all weaknesses, flaws and imperfections which are still part of this book.

Last and most I say thanks to Maria since she made my work much sweeter grapes.

NOTES

[1] Thompson [1991], p. 339.
[2] Manifested e.g. in the fact that both 'bibles' of constructive logic and mathematics, Beeson (1985) and Troelstra and van Dalen (1988), devote a whole chapter to it.
[3] Nordström et al. (1990), pp. 7–9.
[4] Beeson (1985), pp. 250–255, 261, 262.
[5] Troelstra and van Dalen (1988), Chapt. 11.
[6] The caution of this formulation is due to the fact that I am making a statement about Martin-Löf's writings available to me.
[7] Beeson (1985), p. XVI. This remark of Beeson's can readily be challenged for many reasons.

CHAPTER 1
THE PRESENT VERSION OF CONSTRUCTIVE TYPE THEORY (1995)

What is here being called the present version of constructive type theory is a synthesis and elaboration of the versions (7) and (8) of constructive type theory. This chapter sets out with preliminaries which are philosophical methodological guidelines underlying the organization and presentation of the present version of constructive type theory. The following two sections discuss the non-dependent types and objects as well as the dependent types and objects resp. The third section moreover explains, how to construct new types and objects from given ones, where the former will be called function types and the latter function objects. These sections define a whole type structure, but nothing has been said so far about what types there are. The sixth and seventh sections introduce two ground types, namely the ground type of sets and the ground type of propositions resp. And the 'application' of the just mentioned type structure to the ground type of sets yields in the ninth section constructive set theory, while the 'application' of the same type structure to the ground type of propositions yields in the tenth section constructive logic. The eleventh section reveals that the theories of these two ground types, i.e. constructive set theory and constructive logic, not only share the same (type-) structural features, but they also have other essential features in common which allow to establish a precise correspondence (often called isomorphism) between sets and propositions. The twelfth and last section introduces the sequents and shows that there is a precise correspondence (that can also be called isomorphism) between types and sequents. Those sections which have not been mentioned yet, all form digressions on topics related to those in the preceding or following sections.

METHODOLOGICAL PHILOSOPHICAL PRELIMINARIES

The first subsection of the methodological philosophical preliminaries deals with definitions. Not only is the distinction between real and nominal defs. discussed, but also the fundamentals of Curry's theory of definition will be sketched. Both these points are of great importance in constructive type theory as some pertinent hints and examples will show. Since in this subsection it will equally turn out, that the notion of real def. or genuine meaning explanation is not entirely clear, the next subsection is concerned with a method of real def. developed and called 'the syntactic–semantical method' by Martin-Löf. After having introduced the type structure of constructive type theory in the following three sections, a special section (namely Sect. v) will be dedicated to the application of the notion of nominal def. in constructive type theory. At the end of the first subsection of the preliminaries, it is mentioned by the way that the real defs. in constructive type theory give rise to a definitional order or order of conceptual priority. This order is what the third and last subsection is all about. Now, the syntactic–semantical method has two parts, namely a syntactical and a semantical one, each part generates an order of its own. The syntactic part generates the order of conceptual priority, the semantic part the order of semantical explanation. These two orders coincide only partially. The Appendix to the last subsection, which illustrates a segment of the order of conceptual priority, motivates the organization of the material of the rest of this whole chapter.

1 *On Definitions*

According to Aristotle a definition is a phrase signifying a thing's essence. In other words, the answer to the (Socratic) question 'what is ...?' gives a definition of the essence (nature, quiddity) of a thing. Dubislav notes that Aristotle did not explain very clearly what he intended by the term 'essence', and Dubislav himself discusses a metaphysical and a logical intension of this term. Since a definition in this sense is supposed to determine the essence of a thing, it follows that a definition can attain as well as fail to attain its goal and hence can be a true or false statement according to whether the first or second case holds.[1]

But Aristotle also uses 'definition' for a stipulation through which the well-determined meaning of an expression is attributed to an expression which has not any meaning yet.

The definition of the essence of something has later on been called a real definition, and the definition of the meaning of a so far meaningless expression has been called a nominal definition. In modern logic, the real definition has sometimes been interpreted as concept analysis where a certain expression already has a (well-determined) meaning of which the definition claims that it coincides with the meaning of another expression; such a concept analysis is either linguistic or empirical. Occasionally, the real definition has been interpreted as concept explication where a certain ordinary language expression with a somewhat vague meaning is linked with another expression whose meaning is well-determined and precise and where the former expression takes on the well-determined and precise meaning of the latter expression (the meanings of the two expressions are supposed to coincide to a certain extent). Finally, a real definition is sometimes taken to be a nominal definition together with an existence claim. However, this last interpretation of real definition has already been banned by Pascal and Leibniz because it is an obvious mix-up of two different things, namely a linguistic stipulation and a statement. It is possible to conceive of the real definitions of constructive type theory as concept explications. But concept explication as well as all the other attempts at making precise sense of real definitions seem rather unsatisfactory. It may well be that a modern more satisfactory answer to the question how a real definition ought to be understood, can be provided by phenomenology.

Now, Martin-Löf calls a real definition a direct or genuine meaning explanation, and he calls a nominal definition definitional equality or identity. A defined expression of a formal system gets its meaning by a nominal definition, and a prim. expression gets its meaning by a real definition or a genuine meaning explanation.

Martin-Löf's notion of definitional identity is inspired by Curry's theory of definition:[2] Call S_0 a basic system and let $A, B, C, \ldots$ be unspecified entities of S_0. Call S_1 an extended system w.r.t. S_0 and let $X, Y, Z, \ldots$ be arbitrary entities of S_0 or S_1. Entities, operations, axioms, etc. which occur in S_0 are called basic, and entities, operations, axioms, etc. which occur in an extension S_1 of S_0 but not in S_0 are called new. Curry then defines a definitional (def.) extension of S_0 to be a system S_1 satisfying the conditions:

(i) The entities of S_1 are formed by adjoining to those of S_0 certain new operations (the new entities of S_1 will be considered as new operations of degree 0).
(ii) A new binary predicate expressing def. identity occurs in S_1. It is symbolized by $=$.
(iii) The new axioms or introduction rules consist of a certain set E, called the set of defining axioms of S_1, each of which is of the form $\varphi(A_1, \ldots, A_m) = Z$ where φ is a new m-ary operation ($m \geq 0$). Moreover, the new axioms are also propositions of the form $Z = Z$.
(iv) There is a rule of elimination or of def. reduction, which is denoted by Rd, of the form

$$\frac{X = Y \quad \varphi(A_1, \ldots, A_m) = Z}{X = Y'}$$

where φ is a new operation and Y' is the result of replacing an occurrence of $\varphi(A_1, \ldots, A_m)$ in Y by Z.

Call a deduction from the new axioms by Rd a def. reduction. If the conclusion of such a def. reduction D is $X = Y$, X is called the definiendum and Y the definiens of D and of $X = Y$. D is also said to be a def. reduction from X to Y. If Y is a basic entity A, A is called the ultimate definiens of D, and X is said to have A as ultimate definiens or value. If the possibility that there may be more than one defining axiom for each $\varphi(A_1, \ldots, A_m)$ is excluded, E is called a proper set of defining axioms and S_1 a proper def. extension of S_0. In such an extension the def. reduction of any X is in principle an algorithm. At last, Curry calls a def. extension S_1 of S_0 complete if for every new entity X there is at least one def. reduction to an ultimate definiens. And a def. extension S_1 is called univalent if for every new entity X there is at most one def. reduction to an ultimate definiens. It can then be proved that a proper def. extension is univalent but not conversely. (Curry also defines a relation, which he calls definitional identity, as the monotone equivalence relation generated by the defining axioms. In the case of a proper def. extension – and that is the case which is important for constructive type theory – the two notions of def. reduction and def. identity coincide; and that is why they have been amalgamated in the exposition of Curry's theory of definition just given. Also in other respects this exposition has been slightly adapted to certain aspects of constructive type theory.)[3]

Now, Church calls a basic entity an entity in normal form, and a basic entity could be said to constitute the meaning of a basic or prim. expression. If every basic or prim. expression has a real definition, one can easily understand Church insisting that an expression in a formal system has no meaning unless it has a normal form.

For constructive type theory it is of crucial importance that the algorithm mentioned above be effective and the def. extension(s) complete, i.e. that it is always possible to effectively definitionally reduce a new entity to a basic one. It follows from this requirement that all new or, as they are going to be called, non-canonical entities are effectively computable or definitionally reducible to basic or canonical entities, and that all fcts. are effectively computable. It turns out that this principle that definitional identities be effectively eliminable is what essentially makes the difference between constructive and classical logic. (For one of the most prominent examples, cf. C, n & o i) to vii.2).) This principle ensures the fulfillment of the two traditional requirements on nominal defs., namely 1) the requirement of eliminability just mentioned, and 2) the requirement of non-creativity in the sense that whatever is deducible in a (new) theory on the basis of the axioms and rules of inference with the defs. added to them has also to be deducible without those defs. (i.e. in the basic theory).

All of Martin-Löf's meaning explanations can be categorized either into meaning explanations of prim. terms, i.e. direct meaning explanations or real definitions, or into meaning explanations of defined terms or def. identities.

Examples of real definitions are the definitions of: type, object, family of types, fct. in the oldfashioned sense, fct. type, fct. in the modern sense, set, element, proposition, proof-object, natural number, empty set, singleton set, cartesian product of a family of sets, etc.

Examples of nominal definitions or def. identities are definitions of: class, relation, connective, quantifier, $\sim$, $\supset\subset$, Z, Q, R, class union, class intersection, class complement, Bool, true, false, etc.

Note: The real definitions in constructive type theory give rise to a definitional order or order of conceptual priority.

2 *The Syntactic–Semantical Method*

A real definition or genuine meaning explanation can be obtained by means of what Martin-Löf calls the syntactic–semantical method. The application of the syntactic–semantical method thus provides answers to questions of the following form: What does it mean for a . . . to be a . . .?

This method consists of two parts. It serves to determine a basic or prim. mathematical entity

(i) by paying attention to the expression, and in particular to the formation (rules) of the expression by means of which it is expressed (in other terms, by paying attention to the formal structure or composition of the expression by means of which it is expressed). This first part, namely the syntactic part of the method, is also referred to as the process of formalization or divesting of sense;
(ii) by carefully explaining the meaning of the expression obtained as a result of the application of the first part of the method (in other terms, by carefully explaining the constituents or rules of composition occurring or applied in the expression obtained as a result of the application of the first part of the method). This second part, namely the semantical part of the method, is also referred to as the process of contentualization or endowing with sense.

This is how Martin-Löf describes the method himself. If this description of the syntactic–semantical method is compared with the way it is in actual fact applied in the sequel, one could interpret and thus alternatively describe it in the following way:

(i) consists of typing the expression expressing the mathematical entity to be determined or really defined (syntactical part)
(ii) consists of giving a precise semantical explanation of this typed expression, often by means of one or more rules (semantical part)

Martin-Löf purports with this method to investigate senses and essences in a way similar to the one pursued by Frege and Husserl resp. In a paper on 'Sense and Essence: Frege and Husserl' (1970) Solomon draws attention to the fact that for Frege the three concepts of 'concept', 'function' and 'truth-value', and for later analytic philosophers also 'proposition', all share the negative characterization of not being worldly objects, not mere images in someone's mind nor linguistic entities. Although they are sometimes referred to as objects of reference, they mostly appear as ontological relatives of Frege's central notions of 'thought' and 'sense'. In his more informal writings Frege's primary interest was to use these basic, for him 'primitive' notions in an analysis of language and the necessary truths of mathematics. Husserl (as well as e.g. Bolzano or Meinong) accepted similar notions such as 'eidetic judging', 'eidetic judgement' or 'eidetic proposition', 'eidetic truth' and 'eidetic fact'. But unlike Frege, he (they) was (were) not primarily interested in language and its uses, but rather in the nature of mental acts. Despite this different emphasis of interest, Frege's notion of 'sense' and especially of 'thought' can be seen to be remarkably similar to Husserl's notion of 'essence'. Both clearly refer to 'meaning' in some sense and both use these notions of 'meaning' as an approach to the analysis of necessary truths.

Whereas Frege, Russell and Church grappled with inaccessible 'abstract' linguistic-related entities, and ordinary language philosophers turned away from them and turned to how people in fact use language, Husserl actually presented a method of 'eidetic reduction' whose function it is to display 'pure essences' or senses (in Frege's sense). But Husserl's presentation of that method does not meet the standards of clarity of analytic philosophy.[4] Martin-Löf's syntactic–semantical method can be regarded as an attempt to provide an analytical method for extracting essential structures from intuition.

3 *The Order of Conceptual Priority*

Aristotle believed in a 'natural' order of priority of all concepts. He postulated that the conceptual material necessary for the (real) definition of a concept be 'naturally' prior and 'naturally' more intelligible than the concept to be defined by it. One of the consequences of Aristotle's belief just mentioned is his claim that for each domain of objects there is only one correct (real) definition.[5] This order of priority of all concepts has in high scholastics been called the order of definition or the order of essence.

Martin-Löf shares Aristotle's belief in such a 'natural' order of priority of all concepts, which he simply calls 'the order of conceptual priority'.[6] He defines the order relation of conceptual priority as follows:

a concept is conceptually prior to another concept if it is used (and hence presupposed) for the formation (the typing) of the other concept

or alternatively, if the (real) definition (that is here the formation rule) of the second concept refers back to the first concept

Thus, the order of conceptual priority is induced by an order of real defs., and this order is said to be absolutely rigid. The conceptual presuppositions are what is presupposed in the application of the syntactical part of the syntactic–semantical method. Martin-Löf points out that in actual practice the first concepts in the order of conceptual priority are explained last and that the order of conceptual priority is sometimes the reverse order of the order of semantical explanation: the order of conceptual priority is the 'natural' order of priority of the concepts and the order of semantical explanation is the order in which the order of conceptual priority is explained or presented in practice. The presuppositions of the order of semantical explanation are what is presupposed in the application of the semantical part of the syntactic–semantic method.

Examples: Part of the order of conceptual priority of constructive type theory (as expounded in Sect. ii. of Chapt. 2) is the following order between concepts; let the order relation of conceptual priority be denoted by ▷; then

(i)

$$\begin{array}{ccccccc} \begin{array}{c}\text{judgement}\\ \text{(type)}\end{array} & \rhd & \text{proposition} & \rhd & \left.\begin{array}{c}\text{proof (object)}\\ +\\ \text{existence}\end{array}\right\} & \rhd & \text{truth}\\ & & \triangledown & & & & \\ & & \left.\begin{array}{c}\text{disproof (object)}\\ +\\ \text{existence}\end{array}\right\} & \rhd & \text{falsity} & & \end{array}$$

(ii) simplicity ▷ actuality ▷ potentiality

(iii)

$$\left.\begin{array}{c}\text{judgement}\\ \text{demonstration}\\ +\\ \text{existence}\\ +\\ \text{validity}\end{array}\right\} \begin{array}{l} \rhd\ \text{incorrectness}\\ \rhd\ \left.\begin{array}{c}\text{correctness}\\ +\\ \text{actuality}\end{array}\right\} \rhd\ \text{evidence}\end{array}$$

$$\text{(iv)}\quad \text{type} \rhd \left.\begin{array}{c}\text{object}\\ +\\ \text{existence}\end{array}\right\} \rhd \text{type existence}$$

(v)

(non-dependent) set	$\triangleright$	dependent set	$\triangleright$	fct.set
$\triangle$		$\triangle$		$\triangle$
(non-dependent) type	$\triangleright$	dependent type	$\triangleright$	fct.type
$\triangledown$		$\triangledown$		$\triangledown$
(non-dependent) prop	$\triangleright$	dependent prop	$\triangleright$	fct.prop

(vi) The most basic notions of constructive type theory , i.e. the notions s.t. there is no notion conceptually prior to them, are: judgement (type), demonstration, existence, validity (and simplicity).

Some sketchy remarks on the historical development of the notion of order of conceptual priority from Aristotle to Husserl can be found in version (8) of constructive type theory.

Appendix

The order of treatment of all forms of judgement of the following sections i.–iii., vi.–vii. and ix.–x. is determined to a considerable extent by the order of conceptual priority. This order will be explained in some detail in the sequel.

The order of conceptual priority of the following forms of judgement

i. *Form of judgement*	*Conceptual presuppositions*
1) $\alpha : type$	———
2) $\alpha = \beta : type$	$\alpha : type, \beta : type$
3) $a : \alpha$	$\alpha : type$
4) $a = b : \alpha$	$\alpha : type, a : \alpha, b : \alpha$
ii. *Form of judgement*	*Conceptual presuppositions*
1) $x_1 : \alpha_1, \ldots, x_n : \alpha_n \vdash \alpha : type$	$\alpha_1 : type, \ldots,$ $x_1 : \alpha_1, \ldots, x_{n-1} : \alpha_{n-1} \vdash \alpha_n : type$
2) $x_1 : \alpha_1, \ldots, x_n : \alpha_n \vdash \alpha = \beta : type$	$\alpha_1 : type, \ldots,$ $x_1 : \alpha_1, \ldots, x_{n-1} : \alpha_{n-1} \vdash \alpha_n : type,$ $x_1 : \alpha_1, \ldots, x_n : \alpha_n \vdash \alpha : type,$ $x_1 : \alpha_1, \ldots, x_n : \alpha_n \vdash \beta : type$
3) $x_1 : \alpha_1, \ldots, x_n : \alpha_n \vdash a : \alpha$	$\alpha_1 : type, \ldots,$ $x_1 : \alpha_1, \ldots, x_{n-1} : \alpha_{n-1} \vdash \alpha_n : type,$ $x_1 : \alpha_1, \ldots, x_n : \alpha_n \vdash \alpha : type$
4) $x_1 : \alpha_1, \ldots, x_n : \alpha_n \vdash a = b : \alpha$	$\alpha_1 : type, \ldots,$ $x_1 : \alpha_1, \ldots, x_{n-1} : \alpha_{n-1} \vdash \alpha_n : type,$ $x_1 : \alpha_1, \ldots, x_n : \alpha_n \vdash \alpha : type,$ $x_1 : \alpha_1, \ldots, x_n : \alpha_n \vdash a : \alpha,$ $x_1 : \alpha_1, \ldots, x_n : \alpha_n \vdash b : \alpha$
iii. *Form of judgement*	*Conceptual presuppositions*
1) $(x : \alpha)\beta : type$	$\alpha : type, x : \alpha \vdash \beta : type$
2) $(x : \alpha)\beta = (x : \gamma)\beta : type$	$\alpha : type, \gamma : type, \alpha = \gamma : type,$ $x : \alpha \vdash \beta : type, (x : \alpha)\beta : type,$ $(x : \gamma)\beta : type$
$(x : \alpha)\beta = (x : \alpha)\delta : type$	$\alpha : type, x : \alpha \vdash \beta : type,$ $x : \alpha \vdash \delta : type, x : \alpha \vdash \beta = \delta : type,$ $(x : \alpha)\beta : type, (x : \alpha)\delta : type$

Form of judgement	Conceptual presuppositions
3) $f : (x : \alpha)\beta$	$\alpha : type,\ x : \alpha \vdash \beta : type,$ $(x : \alpha)\beta : type$
4) $f = g : (x : \alpha)\beta$	$\alpha : type,\ x : \alpha \vdash \beta : type,$ $(x : \alpha)\beta : type,\ f : (x : \alpha)\beta,$ $g : (x : \alpha)\beta$
vi. *Form of judgement*	*Conceptual presuppositions*
0) $set : type$	——
1) $A : set$	$set : type$
2) $A = B : set$	$set : type,\ A : set,\ B : set$
3) $a : A$	$set : type,\ A : set$
4) $a = b : A$	$set : type,\ A : set,\ a : A,\ b : A$
5) $X : set \vdash X : type$	$set : type$
vii. *Form of judgement*	*Conceptual presuppositions*
0) $prop : type$	——
1) $A : prop$	$prop : type$
2) $A = B : prop$	$prop : type,\ A : prop,\ B : prop$
3) $a : A$	$prop : type,\ A : prop$
4) $a = b : A$	$prop : type,\ A : prop,\ a : A,\ b : A$
5) $X : prop \vdash X : type$	$prop : type$
viii. *Form of judgement*	*Conceptual presuppositions*
1) $N_0,\ N_1, \ldots, N_n : set$	$set : type$
2) $N : set$	$set : type$
3) $+,\ \times,\ \rightarrow\ : (set)(set)set$	$set : type$ $(set)set : type,\ (set)(set)set : type$
4) $\Pi,\ \Sigma : (X : set)(Y : (X)set)set$	$set : type,\ X : set \vdash X : type,$ $X : set \vdash (X)set : type,$ $X : set \vdash ((X)set)set : type,\ (X : set)((X)set)set : type$
5) $I : (X : set)(X)(X)set$	$set : type,\ X : set \vdash set : type,$ $X : set \vdash (X)set : type,\ (X : set)(X)(X)set : type$
6) $U : set$	$set : type$
ix. *Form of judgement*	*Conceptual presuppositions*
1) $\bot,\ \top : prop$	$prop : type$
2) $\vee,\ \&,\ \supset\ : (prop)(prop)prop$	$prop : type,$ $(prop)prop : type,\ (prop)(prop)prop : type$
3) $\forall,\ \exists : (X : set)(Y : (X)prop)prop$	$prop : type,\ set : type,\ X : set \vdash prop : type,$ $X : set \vdash (X)prop : type,$ $X : set \vdash ((X)prop)prop : type,$ $(X : set)((X)prop)prop : type$
4) $I : (X : set)(X)(X)prop$	$set : type,\ prop : type,\ X : set \vdash prop : type,$ $X : set \vdash (X)prop : type,\ (X : set)(X)(X)prop : type$

The order of conceptual priority is fairly strict in the sense that for Sects. i.–iii., vi. and vii.

1. 2), 3) and 4) always presuppose 1)
2. 4) always presupposes 3)

the order of 2), 3) and 4) is interchangeable (as long as condition 2. is respected).

Moreover, there is even more order among the conceptual presuppositions:

3. ii. presupposes i.
4. iii., vi. and vii. presuppose i. and ii.

5. viii. presupposes iii. and vi.
6. ix. presupposes iii. and vii.

and the order between iii. and vi. and vii. is interchangeable as is the order between ix. and x.

The order of semantical explanation
The order of semantical explanation partially deviates from the order of conceptual priority, as mentioned above. It is different also from the latter one in the sense that it is with one exception totally strict:

(1) 1) precedes 3) which precedes 4) which precedes 2) in the case of i., ii. and iii.
(2) 1) precedes 3) which precedes 4) which precedes 2) which precedes 0) which precedes 5) in the case of vi. and vii.

i TYPES AND OBJECTS (NON-DEPENDENT AND NON-FUNCTION)

i.1) What does a judgement of the form '$\alpha : type$' mean? or, What does it mean to be a type?
To make a judgement of the form '$\alpha : type$' one has to know (i) what an object of type α is, and (ii) when two objects of type α are identical.

The identity criterion associated with the objects of a given type has to be defined in such a way that the identity relation is an equivalence relation, i.e. that it satisfies the rules

$$\frac{a : \alpha}{a = a : \alpha}\ \text{(reflex.1)} \qquad \frac{a = b : \alpha}{b = a : \alpha}\ \text{(symmetr.1)} \qquad \frac{a = b : \alpha \quad b = c : \alpha}{a = c : \alpha}\ \text{(trans.1)}$$

These three rules are thus justified for purely conceptual reasons, i.e. by the real definition of what it means to be a type.

Comments, notes and observations (C, n & o)
(1) (Terminological) Dummett calls the criterion implicit in the clause (i) of the real def. above the 'criterion of application',[7] and since Frege's *Begriffsschrift* (1879) the criterion implicit in clause (ii) is called the 'criterion of identity'. Note, however, that Frege did not ask when two objects of a given type were identical, as he did not distinguish any types in his universe of discourse. Everything for him was a thing. (For more on this point, cf. below.)

(2) What comes into play here is the doctrine of types or categories.
Doctrine of types: No entity without type.

Some terminological remarks:
Russell introduced in his *Principles of Mathematics* (1903) the term 'type' for kind or sort. This was an unfortunate choice since 'type' had already a traditional meaning, namely the one explained e.g. by Peirce in the type – token distinction. The traditional philosophical term for what Russell called type is 'category', and the corresponding doctrine was called the doctrine of categories (Kategorienlehre). But if one nowadays wanted to use 'category' instead of 'type' and 'category theory' instead of 'type

theory' the confusion would be perfect, since what is being called category theory today is a well-established part of algebra. (More on Russell's confusing term 'type', cf. note (3) of C, n & o to vi.1).)

Some philosophical–historical remarks:
Aristotle was the first philosopher to have treated categories in a more systematic way. With his distinction of different categories he intended to eliminate certain ambiguities in the use of the predicate 'to be' for which he criticized the platonics. 'κατηγορια' originally means proposition or predication, but Aristotle restricts this meaning to the one of his technical term 'γενος της κατηγοριας' which is kind or way of predication.[8] The typical Aristotelian proposition *S is P* must not be analyzed as if '*is*' were a 2-place predicate applied to '*S*' and '*P*', but may rather be analyzed in the way of constructive type theory as *S* is an object of type/category *P*, i.e. '*S*' applied to '*to be of category P*'. This of course means that something is never tout court, but that it is always something of a certain category, in other terms, the notion of being is being categorized. This was the purpose of Aristotle's doctrine of categories to show that '*is*' can have a whole series of essentially different functions. He was interested in the plurality of the categories and it sufficed for him to find at least ten of them. (It is rather irrelevant that he thought later on that these ten categories actually sufficed to capture all the different functions of '*is*'.)[9]

In his transcendental aesthetics Kant endeavors to show that the concepts of pure mathematics are *a priori* and how they can nevertheless relate to objects. In his transcendental logic he endeavors to show that some non-mathematical concepts are *a priori* and how they can nevertheless relate to objects. Kant calls all elementary non-mathematical and yet applicable concepts *a priori* 'categories'. (According to Körner, Aristotle calls categories all concepts which can be applied to any object whatsoever – and this explains Kant's use of the same term.) The categories are not abstracted from intuition. In order to discover them, one has, according to Kant, to inquire into thinking or judging, which for Kant is the same thing as applying concepts or unifying or compactifying representations (into concepts).

Kant distinguishes judgements of perception (i.e. subjective empirical judgements) and judgements of experience (i.e. objective empirical judgements). The latter differ from the former insofar as they not only contain concepts *a posteriori*, but also a non-mathematical pure concept of understanding, i.e. a category. The idea behind it is the following: in every judgement one can distinguish a) the application of specific concepts, and b) the way these concepts are combined in the judgement, i.e. the logical form of the judgement. Kant reasons that if what conveys objectivity and generality to a judgement of experience does not coincide with any of its specific concepts, then it has to be a concept *a priori* embodied in the form of the judgement. And then there are different categories for each form of a judgement of experience, i.e. for each way in which a judgement of experience conveys objectivity and generality onto the corresponding judgement of perception. Thus, if one can enumerate all the possible forms of judgements of experience, then one can establish a complete list of all the categories.[10]

In Husserl's philosophy the categories play the role of setting limits to objective meaning. The primary task of a 'pure philosophical grammar' is to determine primitive forms of pure categories of meaning the fundamental form of which is that of independent propositional meaning, whereas other primitive forms, e.g. the nominal or adjectival form, are dependent forms of meaning which belong to constituents of a proposition. After determining these pure categories of meaning, pure logical grammar deals with the primitive forms or categories of the composition and modification of propositional meaning. This means that the meaning categories are nothing else than the objective counterparts of the grammatical categories regarded standard in Husserl's time. This pure logical grammar enriched by other parts of pure logic (i.e. the logic of non-contradiction and the logic of truth) may give rise to a formal ontology and develop into a transcendental logic.[11]

Martin-Löf agrees with Aristotle's doctrine of categories that something is never an entity *tout court*, but that it is always of a certain kind, sort or type. This holds in particular for mathematical entities. He endorses the view that the mathematical being is categorized or typed, and this provides half the name of his theory, namely 'type theory'.

In a paper on Kant's philosophy of logic and mathematics (1992A) Martin-Löf presents what he calls a modern analogue of Kant's metaphysical deduction of the categories or pure concepts of understanding: just like Kant he sits down and looks at the forms of judgement of contemporary logic, and by making a complete list of the forms of judgement currently used, he arrives at his categories or types (which are identical, as they were for Kant, with the forms of judgement). Finally, by conceiving of categories or types as meanings of category or type expressions one could recognize some influence of Husserl's philosophy of categories on Martin-Löf's constructive type theory.

(3) What is also endorsed here, is

Quine's slogan: No entity without identity.

Quine motivates his 'precept' by three arguments: He first discusses the argument of a linguist studying the language of a newly discovered tribe. Suppose this language is without known affinities, and the linguist has to carry out the job of lexicography forwards and backwards. 'It was only after somehow deciding what heathen locution to construe as identity and the like that our linguist could begin to say which heathen words serve as terms and what objects they refer to.'[12] Quine argues quite convincingly that after the linguist had decided, 'however arbitrarily, how to accomodate English idioms of identity and quantification in native translation',[13] he had projected his linguistic ontological point of view into the language of this tribe. Second, he speculates about the various phases of a child's development in his learning of words and compares especially the first phase when a child babbles, mutters or utters bulk terms or preindividuative terms with the second phase of the advent of individuative terms. Quine answers his question how one can tell 'whether the child has really got the trick of individuation' in the following way: 'Only by engaging him in sophisticated discourse of "that apple", "not that apple", "an apple", "same apple", "another apple", "these apples" ',[14] i.e. only when the child gets into the swing of correctly using these peculiar words 'same', 'another', 'an', 'that', etc. looking back from the endpoint of this development. Quine comments that 'it was only after getting the knack of identity and similar devices that our child could reasonably be said to be talking in terms and to be speaking of objects.'[15] As a third argument Quine briefly writes about the reference or lack thereof of terms occurring in the context of propositional attitudes. The moral he draws from this third argument is that 'it is to the demands of identity still, that the adult speaker of our language remains answerable as long as he may be said to be using terms to refer.'[16] Quine now observes that there still remains a bulk of bulk terms which adults continue to use as well as certain abstract terms referring to abstract entities such as attributes or propositions which fail to satisfy any identity criteria. Whereas Quine's stance seems largely to have been a descriptive one (describing various situations and phases where the learning and use of terms is strictly associated with identity criteria), it eventually turns into a normative one which makes identity a necessary condition for the possibility of an object. This becomes clear when he writes that positing attributes without any identity criteria 'is perverse, considering that the very use of terms and the very positing of objects are unrecognizable to begin with except as keyed in with idioms of sameness and difference'.[17] From here it is only a small step to his device 'No entity without identity'.[18]

Quine has no doubt demonstrated that identity criteria play a key role in the learning and use of concrete and many singular and general terms; but, as Gottlieb writes, 'the rationale for the requirement and its precise content are not easily made out in the writings of' the widespread adherents to the view that reference requires a criterion of identity.[19] He illustrates his claim with P. Geach and P. Strawson and states that even Quine's position calls for clarification. Gottlieb presents an 'epistemological interpretation of Quine's behavioral link between identity and reference'.[20] After an elaborate construction of his epistemological interpretation he summarizes the rationale for Quine's device as follows: 'Sentences of the form, "This A is F", "That A is G", will claim their rights at the bar of knowledge. We will ask for intersubjective checking, but when we do, we will find that we seldom come across the same description identifying the A whose claim to F-ness is being investigated. To use your belief that that A is F to corroborate my belief that this A is F, we must somehow discover that that A *is* this A. But we are imagining that there are no predicates coextensive with "is the same A as" that are epistemically determinate. The result is that there is no way in general of ascertaining identity for A's; thus is lost the sine qua non of objective knowledge: the interplay of beliefs which variously describe the same object.'[21] However, it seems highly questionable whether Quine had such a constructed and somewhat artificial rationale in mind when he put forward his device.

Furthermore, whereas Quine is inclined to reject attributes or propositions because they have no identity (criteria), Martin-Löf who endorses Quine's device accepts these entities providing them with identity (criteria). (More about this, cf. Subsubsect. i.4.2, Chapt. 2.)

(4) Applying Quine's slogan to types yields the motto: No type without identity.

A strong argument of Martin-Löf's for this motto is indicated by the following slogan (of version (7) of constructive type theory): No synonymy without (type) identity. This slogan is explained in Sect. viii. (There the relation of synonymy is defined as follows: two object expressions 'a' and 'b' of some type expression 'α' are synonymous if their resp. meanings a and b are identical of type α, i.e. if $a = b : \alpha$.

And in order to know whether or not $a = b$ one has to know the type α and its identity.) No arguments are required for the importance of having a synonymy relation in a theory.

Now, Geach has held the converse view expressed by the motto: No identity without type. He has made a case of relativizing identity to types and he thereby speaks of relative identity or sortal identity.[22] Geach's views/claims on relative identity are summarized by Griffin as follows:[23]

(I) 'x is identical with y' is an incomplete expression which either abbreviates 'x is the same A as y' where 'A' represents some sortal or mass noun understood from the context of use of the expression, or else is merely a vague expression of a half-formed thought.

(II) 'Being the same water' cannot be analyzed as 'being the same (something-or-other) and being water'.

(III) The thesis that different As are one and the same B is not objectionable.

Griffin comments on these views as follows:[24] The motivation for the claims (I) and (II) are a) that in ordinary language the expression 'x is the same such-and-such as y' is used very frequently and at any rate more frequently than the expression 'x is identical with y'; b) that identity is interpreted as a 'polygamous predicate', i.e. a predicate which requires the addition of a general term in order to complete its sense and which may therefore be regarded as having different forms depending on the general term which is added. The motivation for claim (III) stems again from ordinary language: As long as the adherent to absolute identity can uphold a distinction between numerical and qualitative identity, (III) does not challenge Leibniz law. But again ordinary language provides examples where the distinction of numerical and qualitative identity is not very obvious and where it is at least desirable to have a theory of identity, such as the one of relative identity, where the two are treated even-handedly.

Geach and the theories of relative identity[25]

If Geach is right in (I), he is right in (II). Geach's claim (I) has been called (D) and is generally stated as follows; let 'A', 'B', 'C' be general terms which are nouns or noun phrases, in short, general nouns.

(D) Absolute identity statements require completion to give a statement of the form 'a is the same A as b'.

But there are different kinds of general nouns and thus different variants of (D):

(D_1) any general noun will serve as a completion

(D_2) all and only sortal general nouns will serve as a completion

(D_3) all sortal general nouns as well as other nouns, but not any general noun will serve as a completion

Geach's position may be the one denoted by (D_3) (cf. his mention of sortal and mass terms). Geach and others also require that the general nouns which serve as a completion in (D) should provide criteria of identity. Since, as Griffin puts it, it is not obvious that general nouns providing criteria of identity are coextensive with any of (D_1), (D_2) or (D_3), there is scope for

(D_4) any general noun which provides a criterion of identity will serve as a completion

Geach's claim (III) has been called (R) in the literature and is the thesis

(R) *a* may be the same *A* as *b* and not the same *B*

Needless to say, there are of course at least as many variants of (R) as there are variants of (D), as each variant of (D) will give a different method of completing '*a* is the same... as *b*' and thus a different variant of (R).

Now, many philosophers have accepted (D) and rejected (R). Geach and others have maintained (D) as well as (R), while some have rejected (D) but accepted (R). It thus turns out, as pointed out by Griffin, that the relations between the absolute and the relative theories of identity are quite complex, for it is possible to construct consistent absolute, (D)-relative, (D)/(R)-relative and (R)-relative identity theories if the basic principles are chosen with sufficient care.

Griffin rejects the principle (D), but he believes that (D)-relativists have an important insight which it is misleading to express by (D).[26] This insight is that the notion of an individual is incoherent without individuation and that the individuative resources required for individuals are provided by (certain kinds of) general nouns in natural language. Thus, Griffin seems in a certain way to accept the slogan: No entity (no object) without type, but he also seems to reject Geach's slogan: No identity without type. It appears that the rationale for a theory of relative identity à la Geach cannot come from mathematics since according to Griffin it is the success of the notion of absolute identity and Leibniz's law in particular in mathematics which has provided it its classical stature with regards to the identity issue and which has made it seem worthwhile to exercise a lot of ingenuity to equip this law for use elsewhere.

i.2) What does a judgement of the form '$\alpha = \beta : type$' mean? or What does it mean for two types to be identical?
A judgement of the form '$\alpha = \beta : type$' means that (i) if a is an object of type α, then a is also an object of type β and vice versa, and (ii) if a and b are identical objects of type α, then they are also identical objects of type β and vice versa.

This real def. of the identity of two types can be expressed by the following two rules

$$\frac{a : \alpha}{\overline{a : \beta}} \quad \text{and} \quad \frac{a = b : \alpha}{\overline{a = b : \beta}}$$

where ═══ means that one can infer in both directions, i.e. downwards and upwards.

The following rules of inference are readily semantically explained or justified by means of the semantical explanation just given

$$\frac{\alpha : type}{\alpha = \alpha : type}\ (\text{reflex.2}) \qquad \frac{\alpha = \beta : type}{\beta = \alpha : type}\ (\text{symmetr.2})$$

$$\frac{\alpha = \beta : type \quad \beta = \gamma : type}{\alpha = \gamma : type} \text{ (trans.2)}$$

Justification

(reflex.2) Obviously, if α : *type* then $\frac{a : \alpha}{a : \alpha}$ and $\frac{a = b : \alpha}{a = b : \alpha}$; this however means, by the explanation above, that $\alpha = \alpha$: *type*.

(symmetr.2) Let $\alpha = \beta$: *type*; this means $\frac{a : \alpha}{a : \beta}$ which is the same as $\frac{a : \beta}{a : \alpha}$, and it also means that $\frac{a = b : \alpha}{a = b : \beta}$ which again is the same as $\frac{a = b : \beta}{a = b : \alpha}$. But the rewriting of the two rules given by $\alpha = \beta$: *type* is just the definition of $\beta = \alpha$: *type*.

(trans.2) Let $\alpha = \beta$: *type* and $\beta = \gamma$: *type*. This means that $\begin{array}{c} a : \alpha \\ \hline a : \beta \\ \hline a : \gamma \end{array}$ but also that $\begin{array}{c} a : \gamma \\ \hline a : \beta \\ \hline a : \alpha \end{array}$ which together yields $\frac{a : \alpha}{a : \gamma}$. The same holds for $a = b : \alpha$ and $a = b : \gamma$ resp. Altogether one thus obtains $\frac{a = b : \alpha}{a = b : \gamma}$ which of course is just the real definition of $\alpha = \gamma$: *type*. □

The following identity of types rules are equally readily justified by the meaning of the judgement '$\alpha = \beta$: *type*':

$$\frac{a : \alpha \quad \alpha = \beta : type}{a : \beta} \text{ (id. of types 1)} \qquad \frac{a = b : \alpha \quad \alpha = \beta : type}{a = b : \beta} \text{ (id. of types 2)}$$

Justification

(id. of types 1) Straightforward, since '$\alpha = \beta$: *type*' means $\frac{a : \alpha}{a : \beta}$.

(id. of types 2) Analogously, '$\alpha = \beta$: *type*' also means $\frac{a = b : \alpha}{a = b : \beta}$. □

i.3) What does a judgement of the form '$a : \alpha$' mean? or What does it mean to be an object of some type?
To explain what a judgement of the form '$a : \alpha$' means one has to know that α : *type*, since knowledge of the latter judgement is a presupposition of the knowledge of the former. The direct meaning explanation of the latter constitutes a meaning explanation of the former, since the first clause of a real def. of 'α : *type*' says exactly what an object of type α is.

C, n & o

In two papers on the problem of universals (1956–57) Stegmüller endorses Quine's claim according to which the ontology which the user of a (formal) language presupposes is expressed precisely by the bound variables. This is so, because a range of values has to be associated with each variable used in the language and the truth conditions of general propositions are formulated by referring to the values of such ranges. This range of values either contains only concrete entities, or it contains concrete as well as abstract entities.

- A platonist can be defined to be a person who admits variables for abstract entities (classes, predicates, numbers, fcts., propositions, etc.) in her (formal) language.
- A nominalist can be defined to be a person who admits variables only for concrete entities in her (formal) language (or alternatively, who does not admit variables for abstract entities in her (formal) language).
- Platonism can be defined to be a 'system' consisting of a platonist's (formal) language together with its semantics.
- Nominalism can be defined analogously.

Two consequences can be drawn from these defs.: 1) As indicated by the alternative def. of a nominalist, platonism and nominalism form a complete disjunction. Tertium non datur. 2) Since platonism admits all that nominalism admits but not conversely, platonism is clearly the richer system than nominalism.

Now, n-ary predicate expressions can be part of the (formal) language of a platonist or a nominalist, but in each case they are interpreted in a different way: For a nominalist whole or closed sentences (closed formulas) are meaningful. And predicate expressions are meaningful only in an indirect or derived way, namely because as open sentences (formulas) they can be supplemented to form whole or closed sentences (formulas). (They can be called synsemantical expressions.) For a platonist predicate expressions are directly meaningful; their meanings are abstract entities (i.e. they are names of abstract entities and thus downright semantical expressions.) According to whether these abstract entities are taken to be extensional or intensional entities one has

- extensional platonism: the predicate expressions are names of their extensions (i.e. of classes (of n-tuples) of objects)
- intensional platonism: the predicate expressions are names of their intensions (i.e. of properties of or n-ary relations between objects)

Hence at least three forms of platonism are possible:

- extensional platonism: the values of whose predicate variables are only extensions
- intensional platonism: the values of whose predicate variables are only intensions
- extensional–intensional platonism: the values of whose predicate variables are extensions as well as intensions

The fundamental nominalist thesis is: Whatever can be expressed in platonism can equally be expressed in nominalism. (*Note*: it is wrong to express the thesis as: Whatever can be expressed in a platonist language can equally be expressed in a nominalist language, since the two languages (in the syntactical sense) can be the same which would make the thesis trivial.) This thesis has not been proved or verified up to now.

Until today only a limited number of platonist contexts have been translated into nominalism.

The converse platonist thesis: Whatever can be expressed in nominalism can be expressed in platonism, is obviously correct. Any nominalist sentence (formula) of the kind $P(a_1, \ldots, a_n)$ can be interpreted e.g. in extensional platonism as expressing $(a_1, \ldots, a_n) \in P$ (and the translation of $P(a_1, \ldots, a_n)$ into intensional platonism could be the very same expression).

Stegmüller also distinguishes two kinds of conceptualism, namely a psychological and a constructive kind. Conceptualism, i.e. the acceptance and interpretation of abstract entities as mental entities or products of the activity of the human mind, originated in the Middle Ages as an attempt to overcome platonism. Later on it developed in two directions: in the direction of psychological conceptualism, i.e. a psychological theory which is supposed to describe and explain the formation of general concepts (universals, abstract entities) in the human mind, and in the direction of constructive conceptualism, i.e. a logical-mathematical theory to describe and prescribe the formation of abstract entities by certain principles of construction. According to Stegmüller, psychological conceptualism is doomed to failure for reasons that will not be discussed here; constructive conceptualism is, due to its acceptance of abstract entities (described and prescribed above), a variant of platonism.

This yields – always following Stegmüller – the following options concerning the question whether or not there exist non-concrete (abstract) entities:

I platonism in general

- I.1 strict platonism
- I.2 constructive conceptualism (constructivism)

II nominalism

Finally, the three conceptions of nominalism, constructive conceptualism, strict platonism have exact quantitative correlates with the following three notions: finite collection, countable infinite collection, uncountable infinite collection resp. The constructive conceptualist recognizes infinity, rejects however the notion of uncountable infinity. The nominalist rejects the notion of infinity altogether. Nominalism is essentially also finitism. Note that strict platonism and constructive conceptualism can also precisely be distinguished by the question whether or not they admit non-constructive (impredicative) methods of concept formation.

Stegmüller's framework which has here been presented allows to assess the position of constructive type theory w.r.t. the problem of universals. W.r.t. the opposition nominalism vs. platonism Martin-Löf clearly takes a platonist stance: in constructive type theory there exist types and not merely type expressions (which would be the opposite nominalist claim). For this stance, he puts forward the following argument: as soon as one knows, understands or grasps e.g. what a set (or a proposition) is and what it means for two sets (or two propositions) to be identical, one knows the type of sets (or the type of propositions resp.) and not only how to use the type expression '*set*' (or '*proposition*' resp.) (which would be the nominalist interpretation of type expressions). Implicit in Martin-Löf's argument seems to be a Husserlian conception of the understanding of the meaning of a predicate or type expression: namely to understand or grasp the meaning of a type expression presupposes the carrying out of a specific 'intentional act' which is not directed at something individual, but at a universal essence (e.g. the essence of a set).

Stegmüller also rejects nominalism but for another reason: it suffers from a great weakness which however does not disqualify it once and for all on *a priori* grounds (in Stegmüller's terms, this weakness is not lethal); this weakness manifests itself in the fact that so far nobody seems to have the slightest idea how to construct mathematics in a strictly nominalist way (cf. also the note following the fundamental nominalist thesis); and from today's point of view this appears to be an unrealizable fiction. At any rate, nominalism today is very much weaker than the weakest platonism in the form of constructive conceptualism.

Furthermore, constructive type theory's platonism is an intensional platonism. Each set and thus each type 'derived' from a set has an intensional identity (cf. the second formation rules (... F2) of all the set-theoretical operations of Sect. ix., which rules are instances of the set type identity $set = set : type$, cf. C, n & o to vi.0).) Hence, the meaning of a type expression (or of most type expressions) is also its intension. Moreover, one has decidability of judgements and of definitional identity in particular in constructive type theory, and the latter requires an intensional interpretation of identity.

Finally, as is readily guessed because of its name and largely confirmed by its radically constructive and predicative character, constructive type theory is a variant of constructive conceptualism. If constructive type theory is compared with Hao Wang's system Σ (which according to Stegmüller is the best known system of constructive conceptualism), the former can be said to have all but one the advantages of the latter, namely

- the axiom of choice is derivable in the theory
- the axiom of infinity and the axiom of reducibility are superfluous
- the operations of the theory are close to intuitive mathematical reasoning
- there are no formally undecidable propositions in the theory
- the theory can be shown to be consistent
- the theory is open (in the sense that it allows to transgress every limit which has been reached concerning the knowledge of possibilities to form new sets from given ones, and that no ultimate limit puts a final end to these possibilities of construction)

as well as some additional advantages:

- constructive type theory is richer than Σ because a) it contains intensions (whereas Σ is extensional) and because b) proofs can appear as parts of propositions.
- constructive type theory is more radically constructive since it contains a formalization of intuitionistic mathematics which is not the case with Σ.

The only advantage of the system Σ not shared by constructive type theory is, that Σ is better known than CTT – but that advantage is actually in the process of being made to disappear. At last, an order of conceptual priority pervades the whole of constructive type theory. The fact that one has $a : \alpha$ provided $\alpha : type$ shows that the type has to be there before its objects (or the type precedes its objects) in the order of conceptual priority. Martin-Löf points out that the scholastic principle 'universalia ante res' holds in his theory. This may be a second point where Stegmüller's judgement that the opposition between 'universalia ante res' and 'universalia in rebus' is fictitious, has to be revised.[27]

i.4) What does a judgement of the form '$a = b : \alpha$' mean? or What does it mean for two objects of some type to be identical?
To explain what a judgement of the form '$a = b : \alpha$' means one has to know that $a : \alpha$, $b : \alpha$ and especially that $\alpha : type$. The direct meaning explanation of the last mentioned judgement constitutes a meaning explanation of '$a = b : \alpha$', since the second clause of a real def. of '$\alpha : type$' says exactly what two identical objects of type α are.

ii DEPENDENT TYPES AND DEPENDENT OBJECTS (OR FAMILIES OF TYPES AND FCTS. IN THE OLD-FASHIONED SENSE)

Dependent types i.e. types depending on variables ranging over previously defined types (types depending on assumptions in progressive dependence).

Let $\alpha_1 : type$. Then $x_1 : \alpha_1 \vdash \alpha : type$ expresses symbolically that α is a type depending on the variable $x_1 : \alpha_1$. A list of assumptions $x_1 : \alpha_1, \ldots, x_n : \alpha_n$ is said to be in progressive dependence if the following conditions are met:[28]

i. x_i may only appear free in assumptions $x_j : \alpha_j$ for $j > i$
ii. $\alpha_{j+1} : type$ only if $x_1 : \alpha_1, \ldots, x_j : \alpha_j$, for each $0 < j < n$
iii. the variables x_j are distinct

Such a list of assumptions $x_1 : \alpha_1, \ldots, x_n : \alpha_n$ is often denoted by Γ and called a context (this is de Bruijn's AUTOMATH terminology). Then $x_1 : \alpha_1, \ldots, x_n : \alpha_n \vdash \alpha : type$ expresses that α is a type depending on the variables $x_1 : \alpha_1, \ldots, x_n : \alpha_n$, or more precisely, depending on the assumptions $x_1 : \alpha_1, \ldots, x_n : \alpha_n$ in progressive dependence. (This notation is Gentzen's sequent notation.)

Dependent objects i.e. objects depending on variables ranging over previously defined types (or objects depending on assumptions in progressive dependence)

In a judgement of the form $x_1 : \alpha_1, \ldots, x_n : \alpha_n \vdash a : \alpha$ not only the type α but also its object a depends on the assumptions $x_1 : \alpha_1, x_2 : \alpha_2, \ldots, x_n : \alpha_n$ in progressive dependence.

ii.1) What does a judgement of the form '$x_1 : \alpha_1, \ldots, x_n : \alpha_n \vdash \alpha : type$' mean? or What does it mean to be a dependent type (or a family of types)?
Presuppositions in the order of explanation: Let the forms of judgement $\alpha_1 : type, \ldots, x_1 : \alpha_1, \ldots, x_{n-1} : \alpha_{n-1} \vdash \alpha_n : type$ be known.
To make a judgement of the form '$x_1 : \alpha_1, \ldots, x_n : \alpha_n \vdash \alpha : type$' one has to know that (i) $\alpha(a_1/x_1, \ldots, a_n/x_n) : type$ provided $a_1 : \alpha_1, \ldots, a_n :$

$\alpha_n(a_1/x_1, \ldots, a_{n-1}/x_{n-1})$, and (ii) $\alpha(a_1/x_1, \ldots, a_n/x_n)$ and $\alpha(b_1/x_1, \ldots, b_n/x_n)$ are identical types provided $a_1 = b_1 : \alpha_1, \ldots, a_n = b_n : \alpha_n(a_1/x_1, \ldots, a_{n-1}/x_{n-1})$.

Alternatively, '$x_1 : \alpha_1, \ldots, x_n : \alpha_n \vdash \alpha : type$' is really defined by the following two rules:

$$\frac{\begin{array}{l} a_1 : \alpha_1 \\ \vdots \\ a_n : \alpha_n(a_1/x_1, \ldots, a_{n-1}/x_{n-1}) \quad x_1 : \alpha_1, \ldots, x_n : \alpha_n \vdash \alpha : type \end{array}}{\alpha(a_1/x_1, \ldots, a_n/x_n) : type} \text{ (gen. subst 1)}$$

$$\frac{\begin{array}{l} a_1 = b_1 : \alpha_1 \\ \vdots \\ a_n = b_n : \alpha_n(a_1/x_1, \ldots, a_{n-1}/x_{n-1}) \quad x_1 : \alpha_1, \ldots, x_n : \alpha_n \vdash \alpha : type \end{array}}{\alpha(a_1/x_1, \ldots, a_n/x_n) = \alpha(b_1/x_1, \ldots, b_n/x_n) : type} \text{ (gen. subst 2)}$$

ii.2) What does a judgement of the form '$x_1 : \alpha_1, \ldots, x_n : \alpha_n \vdash \alpha = \beta : type$' mean? or What does it mean for two dependent types (two families of types) to be identical?
Presuppositions in the order of explanation: Let the judgements $\alpha_1 : type, \ldots, x_1 : \alpha_1, \ldots, x_{n-1} : \alpha_{n-1} \vdash \alpha_n : type$, $x_1 : \alpha_1, \ldots, x_n : \alpha_n \vdash \alpha : type$, and $x_1 : \alpha_1, \ldots, x_n : \alpha_n \vdash \beta : type$ be known.
A judgement of the form '$x_1 : \alpha_1, \ldots, x_n : \alpha_n \vdash \alpha = \beta : type$' means that $\alpha(a_1/x_1, \ldots, a_n/x_n)$ and $\beta(a_1/x_1, \ldots, a_n/x_n)$ are identical types provided a_1 is an object of type α_1, ..., a_n is an object of type $\alpha_n(a_1/x_1, \ldots, a_{n-1}/x_{n-1})$.

Alternatively, '$x_1 : \alpha_1, \ldots, x_n : \alpha_n \vdash \alpha = \beta : type$' is really defined by the rule:

$$\frac{\begin{array}{l} a_1 : \alpha_1 \\ \vdots \\ a_n : \alpha_n(a_1/x_1, \ldots, a_{n-1}/x_{n-1}) \quad x_1 : \alpha_1, \ldots, x_n : \alpha_n \vdash \alpha = \beta : type \end{array}}{\alpha(a_1/x_1, \ldots, a_n/x_n) = \beta(a_1/x_1, \ldots, a_n/x_n) : type} \text{ (gen. subst 3)}$$

ii.3) What does a judgement of the form '$x_1 : \alpha_1, \ldots, x_n : \alpha_n \vdash a : \alpha$' mean? or What does it mean to be a dependent object (or a fct. in the oldfashioned sense)?
Presuppositions in the order of explanation: Let the judgements $\alpha_1 : type, \ldots, x_1 : \alpha_1, \ldots, x_{n-1} : \alpha_{n-1} \vdash \alpha_n : type$ and $x_1 : \alpha_1, \ldots, x_n : \alpha_n \vdash \alpha : type$ be known.
To make a judgement of the form '$x_1 : \alpha_1, \ldots, x_n : \alpha_n \vdash a : \alpha$' one has to know that (i) $a(a_1/x_1, \ldots, a_n/x_n)$ is an object of type $\alpha(a_1/x_1, \ldots, a_n/x_n)$ provided a_1

is an object of type $\alpha_1, \ldots, a_n$ is an object of type $\alpha_n(a_1/x_1, \ldots, a_{n-1}/x_{n-1})$, and (ii) $a(a_1/x_1, \ldots, a_n/x_n)$ and $a(b_1/x_1, \ldots, b_n/x_n)$ are identical objects of type $\alpha(a_1/x_1, \ldots, a_n/x_n)$ provided a_1 and b_1 are identical objects of type $\alpha_1, \ldots, a_n$ and b_n are identical objects of type $\alpha_n(a_1/x_1, \ldots, a_{n-1}/x_{n-1})$.

Alternatively, '$x_1 : \alpha_1, \ldots, x_n : \alpha_n \vdash a : \alpha$' is really defined by the following two rules:

$$\frac{\begin{array}{l} a_1 : \alpha_1 \\ \vdots \\ a_n : \alpha_n(a_1/x_1, \ldots, a_{n-1}/x_{n-1}) \quad x_1 : \alpha_1, \ldots, x_n : \alpha_n \vdash a : \alpha \end{array}}{a(a_1/x_1, \ldots, a_n/x_n) : \alpha(a_1/x_1, \ldots, a_n/x_n)} \text{(gen. subst 4)}$$

$$\frac{\begin{array}{l} a_1 = b_1 : \alpha_1 \\ \vdots \\ a_n = b_n : \alpha_n(a_1/x_1, \ldots, a_{n-1}/x_{n-1}) \quad x_1 : \alpha_1, \ldots, x_n : \alpha_n \vdash a : \alpha \end{array}}{a(a_1/x_1, \ldots, a_n/x_n) = a(b_1/x_1, \ldots, b_n/x_n) : \alpha(a_1/x_1, \ldots, a_n/x_n)} \text{(gen. subst 5)}$$

ii.4) What does a judgement of the form '$x_1 : \alpha_1, \ldots, x_n : \alpha_n \vdash a = b : \alpha$' mean? or What does it mean for two dependent objects (two fcts. in the oldfashioned sense) to be identical?
Presuppositions in the order of explanation: Let the judgements $\alpha_1 : type, \ldots, x_1 : \alpha_1, \ldots, x_{n-1} : \alpha_{n-1} \vdash \alpha_n : type$, $x_1 : \alpha_1, \ldots, x_n : \alpha_n \vdash \alpha : type$, $x_1 : \alpha_1, \ldots, x_n : \alpha_n \vdash a : \alpha$, $x_1 : \alpha_1, \ldots, x_n : \alpha_n \vdash b : \alpha$ be known.
To explain what a judgement of the form '$x_1 : \alpha_1, \ldots, x_n : \alpha_n \vdash a = b : \alpha$' means one has to know that $a(a_1/x_1, \ldots, a_n/x_n)$ and $b(a_1/x_1, \ldots, a_n/x_n)$ are identical objects of type $\alpha(a_1/x_1, \ldots, a_n/x_n)$ provided a_1 is an object of type $\alpha_1, \ldots, a_n$ is an object of type $\alpha_n(a_1/x_1, \ldots a_{n-1}/x_{n-1})$.

Alternatively, '$x_1 : \alpha_1, \ldots, x_n : \alpha_n \vdash a = b : \alpha$' is really defined by the following rule:

$$\frac{\begin{array}{l} a_1 : \alpha_1 \\ \vdots \\ a_n : \alpha_n(a_1/x_1, \ldots, a_{n-1}/x_{n-1}) \quad x_1 : \alpha_1, \ldots, x_n : \alpha_n \vdash a = b : \alpha \end{array}}{a(a_1/x_1, \ldots, a_n/x_n) = b(a_1/x_1, \ldots, a_n/x_n) : \alpha(a_1/x_1, \ldots, a_n/x_n)} \text{(gen. subst 6)}$$

C, n & o

(1) Throughout this section the notion of (multiple or simultaneous) substitution has played a crucial role. Dependent types (dependent objects) are namely really defined by means of (non-dependent) types ((non-dependent) objects resp.) and substitution. Notations

in logic (e.g. in Curry)	$(a_1/x_1, \ldots, a_n/x_n)$ (prefixed)
in ordinary mathematical notation	$(x_1 = a_1, \ldots, x_n = a_n)$
in computer science (e.g. ALGOL)	$(x := a_1, \ldots, x := a_n)$

(2) The notion of fct. in the oldfashioned sense
Euler defined a fct. to be what would now be called a functional expression. The notion of fct. in the oldfashioned sense is older though and stems roughly from the 1690s. A fct. in the oldfashioned sense is always an expression containing free variables and it is always a fct. of its variables. *Examples*: $x^2 + 3y$, $\sin x$, $\log x$, etc.

(3) The history of the concept of fct. in the oldfashioned sense[29]
The word 'function' appears for the first time in 1673 in an unpublished manuscript of Leibniz with the title 'The inverse method of tangents, or about functions'. There Leibniz writes:

I denote as functions all pieces of straight lines which one obtains by letting lines pass through a fixed point and points on a curve; for example abscissa, ordinate, chord, secant, tangent, normal, subtangent, subnormal, etc. In addition there are infinitely many others whose construction is more involved and which cannot be represented easily.

Leibniz called any part of a straight line (the length of which depends on the position of a point of the given curve) which satisfies certain relations in the largest sense concerning the curve (i.e. a tangent, a normal or a subtangent) a function. This conception can also be found in some of his articles from 1692 and 1694.[30]

But in a letter of Joh. Bernoulli to Leibniz from 1694 one can already read:

by n I understand a quantity somehow formed from indeterminate and constant [quantities]

and in an article of the same year Joh. Bernoulli proposes to denote by

$\overset{1}{X}, \overset{2}{X} \cdots$ diverse quantities given somehow by an indeterminate [quantity] x and by constants ... [either] algebraically or transcendentally.

Finally, the first explicit def. of function as an analytic expression appears in an article by Joh. Bernoulli in 1718:

One calls here Function of a variable a quantity composed in any manner whatever of this variable and of constants.

All these defs. precede and some of them anticipate the famous Euler def. of fct. from 1748 in his *Introductio in analysin infinitorum*:

1. A constant quantity is a determinate quantity keeping the same value permanently ...
2. A variable quantity is an indeterminate or universal quantity which comprises in itself all determinate values ...
3. A function of a variable quantity is an analytical expression composed in any manner from that variable quantity and numbers or constant quantities.

But in his *Institutiones calculi differentialis* from 1755 Euler generalized his notion of function in the following way:

If, however, some quantities depend on others in such a way that if the latter are changed the former undergo changes themselves then the former quantities are called functions of the latter quantities. This is a very comprehensive notion and comprises in itself all the modes through which one quantity can be determined by others. If, therefore, x denotes a variable quantity then all the quantities which depend on x in any manner whatsoever or are determined by it are called its functions; ...

In 1822 Fourier gave a similar def.:

In general, the function $f(x)$ represents a succession of values or ordinates each of which is arbitrary. An infinity of values being given to the abscissa x, there are an equal number of ordinates $f(x)$. All have actual numerical values, either positive or negative or nul. We do not suppose these ordinates to be subject to a common law; they succeed each other in any manner whatsoever, and each of them is given as it were a single quantity.

Thus, Euler's second def. and Fourier's def. can be seen to anticipate the famous Dirichlet def. of fct. from 1837 in his *Über die Darstellung ganz willkürlicher Functionen durch Sinus- und Cosinusreihen*:

Let us suppose that a and b are two definite values and x is a variable quantity which is to assume, gradually, all values located between a and b. Now, if to each x there corresponds a unique, finite y in such a way that, as x continuously passes through the interval from a to b, $y = f(x)$ varies likewise gradually then y is called a continuous ... function of x for this interval. It is, moreover, not at all necessary, that y depends on x in this whole interval according to the same law; indeed it is not necessary to think of only relations that can be expressed by mathematical operations. Geometrically represented, i.e. x and y imagined as abscissa and ordinate, a continuous function appears as a connected curve, for which only one point corresponds to each abscissa between a and b.

The last step in the def. of fct. in the oldfashioned sense was taken by Frege when he refrained from linking functions to numbers. In his *Begriffsschrift* (1879) Frege writes

If in an expression, whose content need not be capable of investigating, a simple or a compound sign has one or more occurrences and if we regard that sign as replaceable in all or some of these occurrences by something else (but everywhere by the same thing), then we call the part that remains invariant in the expression a function, and the replaceable part the argument of the function.

By calling 'the part that remains invariant in the expression' a function, Frege comes pretty close to the modern notion of function.

But when fcts. of fcts. are considered (i.e. fcts. taking fcts. as arguments), the oldfashioned fcts. are not suitable anymore. That is why in the late 19th century Dedekind and probably Frege introduced the modern notion of fct. (For more on this point, cf. also Sect. iii. below).

(4) Some history of the concept of function in the modern sense

In his *Was sind und was sollen die Zahlen?* from 1887 Dedekind defines the modern concept of fct. as follows:[31]

By a mapping of a system S a law is understood, in accordance with which to each determinate element s of S there is associated a determinate object, which is called the image of s and is denoted by $\varphi(s)$; we say, too, that $\varphi(s)$ corresponds to the element s, that $\varphi(s)$ is caused or generated by the mapping φ out of s, that s is transformed by the mapping φ into $\varphi(s)$.

It is not easy to decide whether or not Frege had the modern notion of fct. According to his spirit, i.e. his conception of a fct. as an unsaturated expression, one would answer affirmatively, according to his notation, one would answer negatively. It may well be that Church introduced the idea of functional abstraction and the corresponding λ-notation in order to analyze with full formal rigor Frege's not very clear and unsatisfactory functional notation. With Church's notation a sharp and precise distinction is introduced between function expressions (fcts. in the oldfashioned sense) such as $x^2 + 3y, \sin x, \log z$, and functions (fcts. in the modern sense) e.g. $(\lambda xy)x^2 + 3y$, $(\lambda x)\ \sin x$, $(\lambda z)\ \log z$ (or, in the notation of constructive type theory, $(x)(y)x^2 + 3y$, $(x) \sin x$, $(z) \log z$).

iii FUNCTION TYPES AND FUNCTION OBJECTS (OR FUNCTION TYPES AND FCTS. IN THE MODERN SENSE)

In the preceding Sects. i. and ii. only the notion of type in the sense of non-dependent type (and object) as well as the notion of type depending on variables ranging over previously defined types (objects depending on variables ranging over previously defined types resp.) have been introduced. But no ground types have been mentioned so far nor has any way been indicated how to form new types out of given ones. Before tackling the first point the second will be dealt with in the sequel. Those new types generated from given ones will be called function types and their objects will be called function objects or briefly functions (in the modern sense). Since those new types are also types and the new objects also objects, the direct meaning explanations of forms of judgement involving fct. types or fct. objects are essentially the same as those given in i.1) to i.4) for non-function non-dependent types or objects.

iii.1) What does a judgement of the form '$(x : \alpha)\beta$: *type*' mean? or What does it mean to be a fct. type?
To make a judgement of the form '$(x : \alpha)\beta$: *type*' one has to know (i) what an object f of type $(x : \alpha)\beta$ is, and (ii) when two objects f and g of type $(x : \alpha)\beta$ are identical (in other terms, by explaining (i) what a fct. in the modern sense is, and (ii) what it means for two fcts. in the modern sense to be identical).

The formation rule for fct. types is

$$\frac{\alpha : type \quad x : \alpha \vdash \beta : type}{(x : \alpha)\beta : type}\ \text{(fct. type)}$$

C, n & o

The development of type structures from Frege to Martin-Löf/de Bruijn[32]

(1) *Frege's type structure*

In his *Grundgesetze der Arithmetik* (1893–1903) Frege implicitly had a type structure of the following kind:

$$\alpha_1, \ldots, \alpha_n \text{ are types} \Rightarrow (\alpha_1, \ldots, \alpha_n) \text{ is a type} \qquad n = 0, 1, \ldots$$

In particular, this def. yields the following types:

$()$	the type of objects
$\underbrace{((), \ldots, ())}_{n}$	the type of n-ary fcts.
$((()))$	the type of fcts. which take a 1-ary fct. of objects into an object, e.g. the course-of-values operator

etc.

Frege actually used the term ‘level’ (Stufe) and not type. He defined the levels l as follows:

i. $l(()) = 0$
ii. $l((\alpha_1, \ldots, \alpha_n)) = \max_{1 \leq i \leq n}(l(\alpha_i) + 1)$

At first Frege tried to work with one huge all-encompassing universal type of things, but he did not succeed, since he also needed the various fct. types $(\alpha_1, \ldots, \alpha_n)$ for $n > 0$. The original notion of thing (Gegenstand) was then restricted to the notion of object of type ( ).

(2) *Russell's simple type structure*
In his *Principles of Mathematics* (1903) Russell enriched Frege's type structure in the following way:

i. $\alpha_1, \ldots, \alpha_n$ are types $\Rightarrow (\alpha_1, \ldots, \alpha_n)$ is a type $\quad n = 0, 1, \ldots$
ii. $\alpha_1, \ldots, \alpha_n$ are types $\Rightarrow [\alpha_1, \ldots, \alpha_n]$ is a type $\quad n = 0, 1, \ldots$

Clause *i*. generates the type of individual-valued functionals and clause *ii.* the type of proposition-valued functions. Alternatively, this inductive def. can be expressed by the rules

$$\frac{\alpha_1 : type \ \ldots \ \alpha_n : type}{(\alpha_1, \ldots, \alpha_n) : type} \qquad \frac{\alpha_1 : type \ \ldots \ \alpha_n : type}{[\alpha_1, \ldots, \alpha_n] : type} \qquad (n = 0, 1, \ldots)$$

(this is Dummett's formulation).

Another way of formulating Russell's simple type structure is the following: *i.* ι is a type, *ii.* o is a type, *iii.* α and β are types $\Rightarrow (\alpha)\beta$ is a type, and this def. expressed by rules is

$$\iota : type \qquad o : type \qquad \frac{\alpha : type \quad \beta : \text{type}}{(\alpha)\beta : type}$$

(this is the Church–Schütte formulation).

Note that there is a 1–1 correspondence between the $(\alpha_1, \ldots, \alpha_n)$- and the $(\alpha_1) \ldots (\alpha_n)\iota$-notation as well as between the $[\alpha_1, \ldots, \alpha_n]$- and the $(\alpha_1) \ldots (\alpha_n)o$-notation.

Frege's concept of level can be adapted to this new type structure:

i. $l([\,]) = l(()) = 0$
ii. $l([\alpha_1, \ldots, \alpha_n]) = l((\alpha_1, \ldots, \alpha_n)) = \max_{1 \leq i \leq n}(l(\alpha_i) + 1)$

Frege's objects of type ( ) are called individuals by Russell rather than objects. In particular, Russell's def. yields the following types:

$[\,] = o$	the type of e.g. $\bot$, $\top$
$[[\,]] = (o)o$	the type of $\sim$
$[[\,], [\,]] = (o)(o)o$	the type of &, $\vee$, $\supset$
$[[()]] = ((\iota)o)o$	the type of $\forall$, $\exists$
$\underbrace{((), \ldots, ())}_{n} = \underbrace{(\iota) \ldots (\iota)}_{n}\,\iota$	the type of n-ary fcts.
$\underbrace{[(), \ldots, ()]}_{n} = \underbrace{(\iota) \ldots (\iota)}_{n}\,o$	the type of n-ary relations

(3) *Russell's ramified type structure*
In his and Whitehead's *Principia Mathematica* (1910–13) Russell changed his simple type structure by defining his ramified type structure (except for the type of individuals without the types of fcts. and without the type of propositions) in the following way; let the level fct. l be defined as in Russell's simple type structure; then

i. $()/0$ is a type
ii. $\alpha_1/m_1, \ldots, \alpha_n/m_n$ are types $\Rightarrow$ $[\alpha_1, \ldots, \alpha_n]/k/(m_1, \ldots, m_n)$ is a type provided $m_i \geq l(\alpha_i)$, $k \geq \max_{1 \leq i \leq n}(m_i) + 1$ and if $\alpha_i = ()$ then $m_i = 0$

Clause *i.* introduces the type of individuals, clause *ii.* generates the types of proposition-valued fcts. of order k.Note that the numerals after the first stroke / are called the orders of the resp. types. The concept of order of a type is a generalization of the one of level and is defined as follows:

i. $ord(()) = 0$
ii. $ord([\alpha_1, \ldots, \alpha_n]) = l([\alpha_1, \ldots, \alpha_n]) + j \quad j = 0, 1, \ldots$

The numerals in parenthesis after the second stroke / of a type indicate the orders of the resp. component types of the given type (in the very same order) if there are any.[33]

(4) *Problems with Frege's or Russell's type structures*
One can assume that one has a fairly clear notion of proposition as well as of fct. from one type to another. The main problem with Frege's or Russell's type structure(s) lies in the understanding of Frege's notion of object or Russell's notion of individual resp. In first order logic this problem is solved by limiting oneself to a specific interpretation, i.e. by interpreting an object or an individual resp. as an element of an arbitrarily given set. This in turn presupposes that the notions of set and element of a set are well enough understood, which at this stage can be taken for granted. If D is a set and P is a propositional fct. depending on D, i.e. in the notation of constructive type theory, $D : set$, $\iota = elem(D) : type$ and $P : (D)o$, then a universally or existentially quantified proposition can be written as follows:

$$\begin{array}{ccc} (\forall x \in D)P(x) & = & \forall(D, (x)P) \\ \exists & & \exists \end{array}$$

Now, another problem arises: even if the type of sets and the type of propositions are introduced as new ground types, it is impossible to type a binary operation which takes a set D and a propositional fct. P defined on that set into a proposition by means of the simple (or ramified) type structure, since the type of the second argument of such an operation depends on the type of the first argument. Thus, to type the universal or existential quantifier provided the quantifier's domain is being varied, one needs a type structure which is richer than the simple (or ramified) type structure.

(5) *Martin-Löf's and de Bruijn's dependent type structure*
Martin-Löf and de Bruijn (AUTOMATH)[34] independently from each other enriched Russell's simple type structure by defining a dependent type structure in the

following way:

i. *set* is a type

ii. *prop* is a type

iii. $\dfrac{A : set}{elem(A) : type}$ or simply $\dfrac{A : set}{A : type}$

iv. $\dfrac{A : prop}{proof(A) : type}$ or simply $\dfrac{A : prop}{A : type}$

v. α is a type and β is a type depending on the variable $x : \alpha \Rightarrow (x : \alpha)\beta$ is a type

and this def. expressed slightly more precisely by rules is

$$set : type \quad prop : type \quad \frac{\alpha : type \quad x : \alpha \vdash \beta : type}{(x : \alpha)\beta : type} \quad X : set \vdash X : type$$

$X : prop \vdash X : type$
(this is Martin-Löf's formulation)

Note that the fourth and fifth rule of the dependent type structure are refinements of the first and second rule resp. of the simple type structure (in the Church–Schütte formulation). Note also that the third rule of the simple type structure is just a special case of the third rule of the dependent type structure. If β is independent from α, then the two rules coincide. The independent fct. types, as they will be called, can be introduced into the dependent type structure by the following nominal def.

$$\frac{\alpha : type \quad \beta : type}{(\alpha)\beta : type}\,(*) \quad \text{and} \quad \frac{\alpha : type \quad \beta : type}{(\alpha)\beta = (x : \alpha)\beta : type}\,(**)$$

Note also that the problem of the simple type structure can be solved by the dependent type structure:

(1)	$set : type$	
(2)	$X : set$	assumption
(3)	$X : type$	by the third rule of the dep. type structure
(4)	$prop : type$	
(5)	$(X)prop : type$	by (*)
(6)	$((X)prop)prop : type$	by (*)
(7)	$(X : set)((X)prop)prop : type$	discharging the assumption by the third rule of the dep. type structure

It is easy to see that $\forall : (X : set)((X)prop)prop$ and $\exists : (X : set)((X)prop)prop$ which is the solution of the problem mentioned above.

iii.2) What does a judgement of the form '$(x : \alpha)\beta = (x : \gamma)\beta : type$' or of the form '$(x : \alpha)\beta = (x : \alpha)\delta : type$' mean? or What does it mean for two function types to be identical?
Presuppositions in the order of explanation: Let the forms of judgement '$(x : \alpha)\beta : type$', '$f : (x : \alpha)\beta$' and '$f = g : (x : \alpha)\beta$' be known.

A judgement of the form '$(x : \alpha)\beta = (x : \gamma)\beta : type$' (or '$(x : \alpha)\beta = (x : \alpha)\delta : type$') means that (i) if f is an object of type $(x : \alpha)\beta$, then f is also an object of type $(x : \gamma)\beta$ (or $(x : \alpha)\delta$ resp.) and vice versa, and (ii) if f and g are two identical objects of type $(x : \alpha)\beta$, then they are also identical objects of type $(x : \gamma)\beta$ (or of type $(x : \alpha)\delta$ resp.) and vice versa.

This real def. of the identity of two function types can be expressed by the following two rules

$$\frac{f : (x : \alpha)\beta}{\overline{f : (x : \gamma)\beta}} \qquad \frac{f = g : (x : \alpha)\beta}{\overline{f = g : (x : \gamma)\beta}}$$

(or by the two rules

$$\frac{f : (x : \alpha)\beta}{\overline{f : (x : \alpha)\delta}} \qquad \frac{f = g : (x : \alpha)\beta}{\overline{f = g : (x : \alpha)\delta}}$$

resp.)

iii.3) What does a judgement of the form '$f : (x : \alpha)\beta$' mean? or What does it mean to be a fct. object (or a fct. in the modern sense)?
Presuppositions in the order of explanation: Let the following forms of judgement $\alpha : type$, $x : \alpha \vdash \beta : type$ be known.
To explain what a judgement of the form '$f : (x : \alpha)\beta$' means one has to know (i) that $f(a)$ is an object of type $\beta(a/x)$ provided a is an object of type α, and (ii) that $f(a)$ and $f(b)$ are identical objects of type $\beta(a/x)$ provided a and b are identical objects of type α.

Alternatively, '$f : (x : \alpha)\beta$' is really defined by the following two rules

$$\frac{f : (x : \alpha)\beta \quad a : \alpha}{f(a) : \beta(a/x)}\ \text{(app.1)} \qquad \frac{f : (x : \alpha)\beta \quad a = c : \alpha}{f(a) = f(c) : \beta(a/x)}\ \text{(app.2)}$$

iii.4) What does a judgement of the form '$f = g : (x : \alpha)\beta$' mean? or What does it mean for two fcts. in the modern sense to be identical?
Presuppositions in the order of explanation: Let the forms of judgement $\alpha : type$, $x : \alpha \vdash \beta : type$ and $f : (x : \alpha)\beta$ be known.
To explain what a judgement of the form '$f = g : (x : \alpha)\beta$' means one has to know that $f(a)$ and $g(a)$ are identical objects of type $\beta(a/x)$ provided a is an object of type α.

Alternatively, '$f = g : (x : \alpha)\beta$' is really defined by the following rule

$$\frac{f = g : (x : \alpha)\beta \quad a : \alpha}{f(a) = g(a) : \beta(a/x)}\ \text{(app.3)}$$

C, n & o
(app. 3) and (app.2) can be condensed into one rule

$$\frac{f = g : (x : \alpha)\beta \quad a = c : \alpha}{f(a) = g(c) : \beta(a/x)}\ \text{(app.4)}$$

The semantical explanations of the preceding forms of judgements serve to justify the two identity rules associated with the fct. type formation rule, i.e. the following two rules:

$$\frac{\alpha = \gamma : type \quad x : \alpha \vdash \beta : type}{(x : \alpha)\beta = (x : \gamma)\beta : type} \text{ (fct. type id. 1)}$$

$$\frac{\alpha : type \quad x : \alpha \vdash \beta = \delta : type}{(x : \alpha)\beta = (x : \alpha)\delta : type} \text{ (fct. type id. 2)}$$

Justification

(fct. type id. 1) By the semantic explanations above '$(x : \alpha)\beta = (x : \gamma)\beta : type$' means that $\dfrac{f : (x : \alpha)\beta}{f : (x : \gamma)\beta}$ and $\dfrac{f = g : (x : \alpha)\beta}{f = g : (x : \gamma)\beta}$. The first part of the rule is shown by the following two derivations:

$$\cfrac{\cfrac{a : \gamma \quad \cfrac{\alpha = \gamma : type}{\gamma = \alpha : type}}{a : \alpha} \quad f : (x : \alpha)\beta}{f(a) : \beta(a/x)} \quad \text{and} \quad \cfrac{\cfrac{a : \alpha \quad \alpha = \gamma : type}{a : \gamma} \quad f : (x : \gamma)\beta}{f(a) : \beta(a/x)}$$

and the second part by the following derivations

$$\cfrac{\cfrac{a : \gamma \quad \cfrac{\alpha = \gamma : type}{\gamma = \alpha : type}}{a : \alpha} \quad f = g : (x : \alpha)\beta}{f(a) = g(a) : \beta(a/x)} \quad \text{and}$$

$$\cfrac{\cfrac{a : \alpha \quad \alpha = \gamma : type}{a : \gamma} \quad f = g : (x : \gamma)\beta}{f(a) = g(a) : \beta(a/x)}$$

(fct. type id. 2) Similarly. □

C, n & o

The meaning explanation of the most general form of judgement '$(x : \alpha)\beta = (x : \gamma)\delta : type$' concerning the identity of fct. types is then straightforward.

Until now the formation of new types out of given ones has been presented; and now it will be shown how to form new objects out of given objects. Application is one way of constructing a new object out of a given one. There are two other ways of doing the same thing, namely argument removal and argument move; but before presenting them, a rule will be introduced for forming objects at all, namely the assumption rule:

$$\frac{\alpha : type}{x : \alpha \vdash x : \alpha} \text{ (ass) (in an empty context)}$$

$$\frac{x_1 : \alpha_1, \ldots, x_n : \alpha_n \vdash \alpha : type}{x_1 : \alpha_1, \ldots, x_n : \alpha_n, x : \alpha \vdash x : \alpha} \text{ (gen. ass) (in a non-empty context)}$$

Justification
(gen. ass) Suppose the meaning of the judgement which forms the premiss to be given. Then the following two rules are obviously valid:

$$\frac{\begin{array}{l} a_1 : \alpha_1 \\ \vdots \\ a_n : \alpha_n(a_1/x_1, \ldots, a_{n-1}/x_{n-1}) \\ a : \alpha(a_1/x_1, \ldots, a_n/x_n) \end{array}}{x(a_1/x_1, \ldots, a_n/x_n, a/x) = a : \alpha(a_1/x_1, \ldots, a_n/x_n, a/x) = \alpha(a_1/x_1, \ldots, a_n/x_n)}$$

where a/x has no effect on $\alpha(a_1/x_1, \ldots, a_n/x_n, a/x)$, x being a completely new variable.

$$\frac{\begin{array}{l} a_1 = c_1 : \alpha_1 \\ \vdots \\ a_n = c_n : \alpha_n(a_1/x_1, \ldots, a_{n-1}/x_{n-1}) \\ a = c : \alpha(a_1/x_1, \ldots, a_n/x_n) \end{array}}{a = c : \alpha(a_1/x_1, \ldots, a_n/x_n, a/x) = \alpha(a_1/x_1, \ldots, a_n/x_n)}$$

for the same reason as above. These two rules, however, constitute the meaning of the conclusion of (gen. ass). □

The two ways of constructing a new object out of a given one – other than application – are called argument removal and argument move.

First the argument removal rule and its corresponding identity rule

$$\frac{x : \alpha \vdash f(x) : \beta}{f : (x : \alpha)\beta} \text{ (arg. removal 1)} \quad \text{if } f \text{ does not depend on } x$$

$$\frac{x : \alpha \vdash f(x) = g(x) : \beta}{f = g : (x : \alpha)\beta} \text{ (arg. removal 2)} \quad \text{same proviso}$$

This is the ζ- or extensionality rule in combinatory logic.

Justification
(arg. removal 1) The judgement '$f : (x : \alpha)\beta$' was semantically explained as meaning $f(x) : \beta$ provided $x : \alpha$ (cf. iii.3)).

(arg. removal 2) And the judgement '$f = g : (x : \alpha)\beta$' was semantically explained as meaning $f(x) = g(x) : \beta$ provided $x : \alpha$ (cf. iii.4)). □

The argument move rules are

$$\frac{a : \alpha \quad x : \alpha \vdash b : \beta}{((x)b)(a) : \beta(a/x)} \text{ (arg. move 1)} \qquad \frac{a : \alpha \quad x : \alpha \vdash b : \beta}{((x)b)(a) = b(a/x) : \beta(a/x)} \text{ (arg. move 2)}$$

which is β-conversion in combinatory logic.

Justification
The argument move rules constitute a nominal def. of the definiendum $((x)b)(a)$ by first typing the new expression (first rule) and then introducing its definiens (second rule). □

C, n & o
(1) The rules of argument move are here considered to be more basic than (functional) abstraction which is the following rule given together with its corresponding identity rule:

$$\frac{x:\alpha \vdash b:\beta}{(x)b:(x:\alpha)\beta}\ \text{(abs 1)} \qquad \frac{x:\alpha \vdash b=d:\beta}{(x)b=(x)d:(x:\alpha)\beta}\ \text{(abs 2)}$$

The latter rule is the ξ-rule in combinatory logic. Finally, there is the rule of η-conversion in combinatory logic, i.e.

$$\frac{x:\alpha \vdash b(x):\beta}{(x)b(x)=b:(x:\alpha)\beta}\ \text{(abs 3)}$$

if b does not depend on x. (abs 1), (abs 2) and (abs 3) can be derived by means of the argument removal and argument move rules:

i. (functional) abstraction (abs 1)

$$\frac{\dfrac{(x:\alpha) \quad x:\alpha \vdash b:\beta}{((x)b)(x):\beta\ (=\beta(x/x))}\ \text{arg. move}}{(x)b:(x:\alpha)\beta}\ \text{arg. removal} \qquad \text{and} \qquad \frac{x:\alpha \quad x:\alpha \vdash b:\beta}{((x)b)(x)=b:\beta}$$

where $((x)b)(x)$ does not depend on x

ii. ξ-rule (abs 2)

$$\frac{\dfrac{\dfrac{(x:\alpha) \quad x:\alpha \vdash b:\beta}{((x)b)(x)=b:\beta} \qquad \dfrac{(x:\alpha) \quad x:\alpha \vdash d:\beta}{((x)d)(x)=d:\beta}}{((x)b)(x)=((x)d)(x):\beta}\ \text{since } x:\alpha \vdash b=d:\beta}{(x)b=(x)d:(x:\alpha)\beta}$$

iii. η-rule (abs 3)

$$\frac{\dfrac{(x:\alpha) \quad x:\alpha \vdash b(x):\beta}{((x)b(x))(x)=b(x):\beta}\ \text{arg. move}}{(x)b(x)=b:(x:\alpha)\beta}\ \zeta\text{-rule}$$

Note that the inference of η-conversion actually goes in both directions, i.e. downwards and upwards:

$$\frac{x:\alpha \vdash b(x):\beta}{b:(x:\alpha)\beta}\ \downarrow \quad \text{arg. removal (b does not depend on x)}$$

$$\frac{b:(x:\alpha)\beta \quad x:\alpha}{b(x):\beta}\ \uparrow \quad \text{application}$$

Lastly it may be observed that the rule of argument move can be derived from (functional) abstraction:

$$\dfrac{a:\alpha \qquad \dfrac{x:\alpha \vdash b:\beta}{(x)b:(x:\alpha)\beta}\ \text{(funct.) abstraction}}{((x)b)(a):\beta(a/x)}\ \text{application}$$

however, the derived rule in constructive type theory is (funct.) abstraction.

(2) The argument removal is meant to operate in the sense of turning a fct. in the oldfashioned sense into a fct. in the modern sense. This operation was carried out before (in versions earlier than version (8) of constructive type theory) by (functional) abstraction. Now, the single step of abstraction is split into two steps: first, the fct. in the oldfashioned sense is transformed by argument move into the application of the corresponding fct. in the modern sense to an object; second, from this application of the fct. in the modern sense to an object one obtains the fct. in the modern sense itself by argument removal. The following diagram will illustrate this:

fct. in the oldfashioned sense	b, a	$\xrightarrow{\text{substitution}}$	$b(a/x)$	value substituted for a variable in a fct. in the oldfashioned sense
	$\downarrow$ abstraction		$\searrow$	
fct. in the modern sense	$(x)b, a$	$\overset{\text{application}}{\underset{\longleftarrow}{\longrightarrow}}$	$((x)b)(a)$	fct. in the modern sense applied to an argument

Substitution is the process of providing a fct. in the oldfashioned sense with an input, and application is the process of providing a fct. in the modern sense with an input. Abstraction is the process of passing from a fct. in the oldfashioned sense to the corresponding fct. in the modern sense. The fct. in the oldfashioned sense can be compared to the fct. in the modern sense as follows

Oldfashioned fct.	Modern fct.
$x:\alpha \vdash b:\beta$ means that	$f:(\alpha)\beta$ means that (where β is non-dependent)
1) $b(a/x):\beta$ provided $a:\alpha$	1) $f(a):\beta$ provided $a:\alpha$
2) $b(a/x)=b(c/x):\beta$ provided $a=c:\alpha$	2) $f(a)=f(c):\beta$ provided $a=c:\alpha$ a and c are called the arguments or inputs of f, and $f(a)$ and $f(c)$ are called the values or outputs resp. of f

Note: if '$(\alpha)\beta$' is replaced by '$(x:\alpha)\beta$' and every occurrence of 'β' in 1) and 2) is replaced by '$\beta(a/x)$', one obtains the comparison for dependent types.

(3) What has been explained in (2) makes explicit what has been implicit so far, namely: in the order of conceptual priority the notion of fct. in the oldfashioned sense

(or dependent object) is prior/precedes the notion of fct. in the modern sense (or fct. object). A fct. in the modern sense is obtained from (presupposes) its application to a variable which in turn is obtained from (presupposes) a shifting of variables in a fct. in the oldfashioned sense.

(4) Curry started working in combinatory logic on the convertibility relation (i.e. formal identity in a theory of combinators) and worked on typing later.[35] According to Martin-Löf this was not a good choice. To avoid meaningless expressions one has to type the convertibility or identity relation as well, i.e. one has to work on typing first and the convertibility relation afterwards. Martin-Löf put this into the following slogan: application comes before identity. (This is not entirely correct. For an exact statement of Martin-Löf's position, cf. Subsect. i.2, Chapt. 2.)

(5) The rules of (gen.) substitution in the theory expounded on in Sects. i.–iii. (esp. in Sect. ii.) are not fully explained, i.e. the operations of simultaneously assigning in α (or in a of α) objects of appropriate types to the variables on which α (or a of α resp.) depends, so that the result of this operation is a type (an object of a type resp.), are left unspecified. (This fact is also manifested by the constant use of the ellipsis $- \cdots -$ in the formulation of the rules mentioned above.) The following section on explicit substitutions is intended to change this state of affairs, i.e. to fully explain (gen.) substitution.

iv DIGRESSION. EXPLICIT SUBSTITUTIONS

iv.1 *New Notation*

The calculus of explicit substitutions requires the following changes in notation:

'$\beta : \alpha \vdash type$' is used instead of '$x : \alpha \vdash \beta : type$' for dependent types
'$(\alpha)\beta : type$' instead of '$(x : \alpha)\beta : type$' for dependent (!) fct. types
'$\beta a : type$' instead of '$\beta(a/x) : type$' for $\beta : \alpha \vdash type$ and $a : \alpha$ (application w.r.t. families of types, also called substitution).[36]

iv.2 *Contexts, Environments and Context Extensions*

New forms of judgement are going to be introduced and semantically explained.

iv.2.1) What does a judgement of the form '$\Gamma : context$' mean? or What does it mean for Γ to be a list of variable declarations?
'$\Gamma : context$' is really defined by the following two rules

$$\frac{}{(\,) : context}\ (\text{contexF 1}) \qquad \frac{\Gamma : context \quad \alpha : type\,[\Gamma]}{(\Gamma, x : \alpha) : context}\ \text{provided } x\ \Gamma\text{-fresh}\ (\text{contexF 2})$$

where $\alpha : type\,[(\,)]$ is written $\alpha : type\,[\,]$ and $((\,), x : \alpha)$ is written $(x : \alpha)$. 'x Γ-fresh' means that x is a variable not occurring in Γ.

iv.2.2) What do judgements of the form '$\gamma : \Gamma$' and '$\gamma = \delta : \Gamma$' (where $\Gamma : context$) mean? or What does it mean for γ to be an environment for the context Γ? And what does it mean for the environments γ and δ for Γ to be identical?

'$\gamma : \Gamma$' means that for every variable declaration $x : \alpha$ in Γ (i) $\alpha\gamma$: *type*, and (ii) $x\gamma : \alpha\gamma$.

'$\gamma = \delta : \Gamma$' means that for every variable declaration $x : \alpha$ in Γ (i) $\alpha\gamma = \alpha\delta$: *type*, and (ii) $x\gamma = x\delta : \alpha\gamma$.

The judgements '$\gamma : \Gamma$' and '$\gamma = \delta : \Gamma$' can alternatively be really defined as follows:

If Γ is empty, i.e. Γ is $(\,)$, then $\gamma : (\,)$ is vacuously satisfied. And if $\gamma : (\,)$ and $\delta : (\,)$, then $\gamma = \delta : (\,)$ is vacuously satisfied. If Γ is non-empty, i.e. Γ is $(\Delta, x : \alpha)$, then '$\gamma : (\Delta, x : \alpha)$' means that (i) $\gamma : \Delta$ and (ii) $x\gamma : \alpha\gamma$. And if $\gamma : (\Delta, x : \alpha), \delta : (\Delta, x : \alpha)$, then '$\gamma = \delta : (\Delta, x : \alpha)$' means that (i) $\gamma = \delta : \Delta$, and (ii) $x\gamma = x\delta : \alpha\gamma$. '$x\gamma$' is read as the closure of x with γ.

iv.2.3) What does a judgement of the form '$\Gamma \preceq \Delta$' (where Γ : *context*, Δ : *context*) mean? or What does it mean for a context Δ to be an extension of a context Γ?
'$\Gamma \preceq \Delta$' is really defined by the following two rules

$$\frac{\Delta : \mathit{context}}{(\,) \preceq \Delta}\ \text{(contextF 1)}$$

$$\frac{\Gamma : \mathit{context} \quad \Delta : \mathit{context} \quad \Gamma \preceq \Delta}{(\Gamma, x : \alpha) \preceq \Delta}\ \text{provided } x : \alpha \text{ in } \Delta \text{ and } x\ \Gamma\text{-fresh} \qquad \text{(contextF 2)}$$

$\preceq$ expresses a partial ordering on the contexts, i.e. one can easily justify the following two rules by means of the rules given above:

$$\frac{\Gamma : \mathit{context}}{\Gamma \preceq \Gamma}\ \text{(contextreflex)}$$

$$\frac{\Gamma_1 : \mathit{context} \quad \Gamma_2 : \mathit{context} \quad \Gamma_3 : \mathit{context} \quad \Gamma_1 \preceq \Gamma_2 \quad \Gamma_2 \preceq \Gamma_3}{\Gamma_1 \preceq \Gamma_3}\ \text{(contexttrans)}$$

iv.3 *Relative Types and Relative Objects*

iv.3.1) What does a judgement of the form 'α : *type* $[\Gamma]$' mean? or What does it mean to be a type under a context, i.e. a relative type?
What it means to be a relative type α under Γ is explained by explaining (i) that for any $\gamma : \Gamma$, $\alpha\gamma$: *type*, and (ii) that for any $\gamma : \Gamma$ and $\delta : \Gamma$, if $\gamma = \delta : \Gamma$, then $\alpha\gamma = \alpha\delta$: *type*. $\alpha\gamma$ is read as the closure of α with γ.

C, n & o

This presentation of the semantical explanations of 'Γ : *context*', '$\gamma : \Gamma$' and 'α : *type* $[\Gamma]$' may hide the fact that these three forms of judgement are actually defined simultaneously.

iv.3.2) What does a judgement of the form '$\alpha = \beta$: *type* $[\Gamma]$' mean? or What does it mean for two relative types to be identical?
Presuppositions in the order of conceptual priority: α : *type* $[\Gamma]$, β : *type* $[\Gamma]$
'$\alpha = \beta$: *type* $[\Gamma]$' means that for any $\gamma : \Gamma$, $\alpha\gamma = \beta\gamma$: *type*.

iv.3.3) What does a judgement of the form '$a : \alpha\,[\Gamma]$' mean? or What does it mean to be an object under a context of a type under a context, i.e. a relative object of a relative type?
Presuppositions in the order of conceptual priority: $\alpha : type\,[\Gamma]$
'$a : \alpha\,[\Gamma]$' means that (i) for any $\gamma : \Gamma, a\gamma : \alpha\gamma$, and (ii) for any $\gamma : \Gamma$ and $\delta : \Gamma$, if $\gamma = \delta : \Gamma$ then $a\gamma = a\delta : \alpha\gamma$.

iv.3.4) What does a judgement of the form '$a = b : \alpha\,[\Gamma]$' mean? or What does it mean for two relative objects of a relative type to be identical?
Presuppositions in the order of conceptual priority: $\alpha : type\,[\Gamma], a : \alpha\,[\Gamma], b : \alpha\,[\Gamma]$
'$a = b : \alpha\,[\Gamma]$' means that for any $\gamma : \Gamma, a\gamma = b\gamma : \alpha\gamma$.

A digression on absolute types or simply types (non-relative) and absolute objects or simply objects
If the types and objects occurring in the categorical judgements are called absolute, and the types and objects occurring in hypothetical judgements are called relative, then it is possible to regard the absolute notions as special cases of the relative ones. Thus, it is possible to formulate constructive type theory in such a way that no use is made of categorical judgements and hence of absolute notions. Just adopt the following conventions:

(1) The judgement '$\alpha : type\,[\,]$' is explained by explaining that (i) $\alpha : type$, (ii) for any $\gamma : (\,),\ \alpha\gamma : type$, and (iii) for any $\gamma : (\,),\ \alpha\gamma = \alpha : type$.
(2) The judgement '$\alpha = \beta : type\,[\,]$' means $\alpha = \beta : type$.
(3) The judgement '$a : \alpha\,[\,]$' means that (i) $a : \alpha$, (ii) for any $\gamma : (\,), a\gamma : \alpha$, and (iii) for any $\gamma : (\,),\ a\gamma = a : \alpha$.
(4) The judgement '$a = b : \alpha\,[\,]$' means that $a = b : \alpha$.

iv.4 *Relative Families of Types (or Relative Dependent Types) and Relative Fct. Types*

iv.4.1) What does a judgement of the form '$\beta : \alpha \vdash type\,[\Gamma]$' mean? or What does it mean to be a family of types β over an index type α under a context Γ?
'$\beta : \alpha \vdash type\,[\Gamma]$' is explained by explaining that (i) for any $\gamma : \Gamma, \beta\gamma : \alpha\gamma \vdash type$, and (ii) for any $\gamma : \Gamma$ and $\delta : \Gamma$, if $\gamma = \delta : \Gamma$ then $\beta\gamma = \beta\delta : \alpha\gamma \vdash type$.

iv.4.2) What does a judgement of the form '$\beta_1 = \beta_2 : \alpha \vdash type\,[\Gamma]$' mean? or What does it mean for two families of types under a context to be identical?
'$\beta_1 = \beta_2 : \alpha \vdash type\,[\Gamma]$' means that for any $\gamma : \Gamma, \beta_1\gamma = \beta_2\gamma : \alpha\gamma \vdash type$.

iv.4.3) What does a judgement of the form '$(\alpha)\beta : type\,[\Gamma]$' mean? or What does it mean to be a fct. type under a context Γ?
'$(\alpha)\beta : type\,[\Gamma]$' is explained by explaining (i) that for any $\gamma : \Gamma$, $((\alpha)\beta)\gamma : type$, and (ii) that for any $\gamma : \Gamma$ and $\delta : \Gamma$, if $\gamma = \delta : \Gamma$, then $((\alpha)\beta)\gamma = ((\alpha)\beta)\delta : type$.

iv.4.4) What does a judgement of the form '$(\alpha_1)\beta = (\alpha_2)\beta$: *type* $[\Gamma]$' (or '$(\alpha)\beta_1 = (\alpha)\beta_2$: *type* $[\Gamma]$') mean? or What does it mean for two fct. types under a context to be identical?

'$(\alpha_1)\beta = (\alpha_2)\beta$: *type* $[\Gamma]$' (or '$(\alpha)\beta_1 = (\alpha)\beta_2$: *type* $[\Gamma]$') means that for any $\gamma : \Gamma$, $((\alpha_1)\beta)\gamma = ((\alpha_2)\beta)\gamma$: *type* (or $((\alpha)\beta_1)\gamma = ((\alpha)\beta_2)\gamma$: *type* resp.)

Digression continued
The 'absolute families of types' are then defined as follows:

(5) The judgement '$\beta : \alpha \vdash$ *type* $[\,]$' means that (i) $\beta : \alpha \vdash$ *type*, (ii) for any $\gamma : (\,)$, $\beta\gamma : \alpha \vdash$ *type*, and (iii) for any $\gamma : (\,)$, $\beta\gamma = \beta : \alpha \vdash$ *type*.
(6) The judgement '$\beta_1 = \beta_2 : \alpha \vdash$ *type* $[\,]$' means that $\beta_1 = \beta_2 : \alpha \vdash$ *type*.

And the 'absolute fct. types' are defined as follows:

(7) The judgement '$(\alpha)\beta$: *type* $[\,]$' means that (i) $(\alpha)\beta$: *type*, (ii) for any $\gamma : (\,)$, $((\alpha)\beta)\gamma$: *type*, and (iii) for any $\gamma : (\,)$, $((\alpha)\beta)\gamma = (\alpha)\beta$: *type*.
(8) The judgement '$(\alpha_1)\beta = (\alpha_2)\beta$: *type* $[\,]$' (or '$(\alpha)\beta_1 = (\alpha)\beta_2$: *type* $[\,]$') means that $(\alpha_1)\beta = (\alpha_2)\beta$: *type* (or $(\alpha)\beta_1 = (\alpha)\beta_2$: *type* resp.)

Thus, on the one hand every categorical or absolute construction is made a construction under the empty context, which means just assuming that it is introduced as such, and that its closure with any environment for the empty context is defined to be identical with the original construction itself. On the other hand every relative entity or entity under a context is ultimately always semantically explained by absolute entities. E.g. a relative judgement on types is a judgement on (non-dependent, dependent or fct.) types depending on variables $x_1, \ldots, x_n$ of corresponding types. And a relative judgement on a (non-dependent, dependent or fct.) type α is semantically explained by explaining that by simultaneously assigning objects of appropriate types to the variables $x_1, \ldots, x_n$ in α one gets an absolute (non-dependent, dependent or fct.) type (resp.). The same holds *mutatis mutandis* for objects of some type.

iv.5 *Substitutions*

iv.5.1) What does a judgement of the form '$\delta : \Gamma \to \Delta$' mean? or What does it mean for δ to be a substitution for Δ over Γ?

Presuppositions in the order of conceptual priority: Γ : *context*, Δ : *context*

'$\delta : \Gamma \to \Delta$' is explained by explaining that (i) for any $\gamma : \Gamma$, $\delta\gamma : \Delta$, and (ii) for any $\gamma_1 : \Gamma$ and $\gamma_2 : \Gamma$, if $\gamma_1 = \gamma_2 : \Gamma$ then $\delta\gamma_1 = \delta\gamma_2 : \Delta$.

In the special case where Γ is $(\,)$, the judgement '$\delta : (\,) \to \Delta$' means that (i) $\delta : \Delta$, (ii) for any $\gamma : (\,)$, $\delta\gamma : \Delta$, and (iii) for any $\gamma : (\,)$, $\delta\gamma = \delta : \Delta$.

iv.5.2) What does a judgement of the form '$\delta_1 = \delta_2 : \Gamma \to \Delta$' mean? or What does it mean for two substitutions for Δ over Γ to be identical?

Presuppositions in the order of conceptual priority: Γ : *context*, Δ : *context*, $\delta_1 : \Gamma \to \Delta$, $\delta_2 : \Gamma \to \Delta$

'$\delta_1 = \delta_2 : \Gamma \to \Delta$' means that for any $\gamma : \Gamma$, $\delta_1\gamma = \delta_2\gamma : \Delta$.

C, n & o

(1) Suppose $\alpha : type\,[\Delta]$ and $\delta : \Gamma \to \Delta$, where $\Gamma : context$, $\Delta : context$. Then, by the above semantical explanations, for any $\gamma : \Gamma$, $\delta\gamma : \Delta$ and so also $\alpha(\delta\gamma) : type$. What is here given is a construction taking us from each environment $\gamma : \Gamma$ to a type, i.e. what is given is a type under the context Γ. This construction is what is usually called the result of operating the substitution δ on α, or simply of substituting δ in α.

(2) If the type under the context Γ mentioned in note (1) is written as $\alpha\delta : type\,[\Gamma]$ (where $\alpha\delta$ is read as the closure of α with δ), it can be justified by defining: for any $\gamma : \Gamma$, $(\alpha\delta)\gamma$ is the same type as $\alpha(\delta\gamma)$. Then substitutions become components of expressions and there is no reason on principle why such an expression would be simplifiable into one not containing substitutions. This is to be contrasted with the usual idea of substitution as an operation defined over expressions. Expressions containing substitutions are called expressions with substitutions.

(3) In $\delta : \Gamma \to \Delta$, Γ is called the source context and Δ the target context.

iv.6 *The Calculus of Explicit Substitutions*

The calculus of explicit substitutions contains just about all the rules of the previous Sects. i.–iii., but now in a relativized form. According to the digressions of the preceding subsects. iii. and iv. of this section, those rules which have just been referred to are nothing but the special absolute forms (i.e. the context Γ is empty) of the new general relative forms. Where a former rule is being relativized, this will be indicated by prefexing 'rel.' to the name of that rule. The presentation of these relativized rules will roughly follow the order in which they appear in Sects. i.–iii.

According to the enriched language of explicit substitutions there are quite a number of new rules concerning or involving contexts, environments or substitutions. In the sequel the rules of the calculus will be divided into 3 groups:

1) general rules
2) special rules for substitutions
3) special rules for relative families of types, for relative fct. types, and for two particular relative types

1) *General rules*

1a) General rules concerning identity of relative objects, relative types and relative families of types

$$\frac{a : \alpha\,[\Gamma]}{a = a : \alpha\,[\Gamma]}\ \text{(rel. reflex. 1)} \qquad \frac{a = b : \alpha\,[\Gamma]}{b = a : \alpha\,[\Gamma]}\ \text{(rel. symmetr. 1)}$$

$$\frac{a = b : \alpha\,[\Gamma] \quad b = c : \alpha\,[\Gamma]}{a = c : \alpha\,[\Gamma]}\ \text{(rel. trans. 1)}$$

$$\frac{\alpha : type\,[\Gamma]}{\alpha = \alpha : type\,[\Gamma]}\ \text{(rel. reflex. 2)} \qquad \frac{\alpha = \beta : type\,[\Gamma]}{\beta = \alpha : type\,[\Gamma]}\ \text{(rel. symmetr. 2)}$$

$$\frac{\beta_1 = \beta_2 : type\,[\Gamma] \quad \beta_2 = \beta_3 : type\,[\Gamma]}{\beta_1 = \beta_3 : type\,[\Gamma]} \text{ (rel. trans. 2)}$$

$$\frac{\beta : \alpha \vdash type\,[\Gamma]}{\beta = \beta : \alpha \vdash type\,[\Gamma]} \text{ (reflex. 3)} \qquad \frac{\beta_1 = \beta_2 : \alpha \vdash type\,[\Gamma]}{\beta_2 = \beta_1 : \alpha \vdash type\,[\Gamma]} \text{ (symmetr. 3)}$$

$$\frac{\beta_1 = \beta_2 : \alpha \vdash type\,[\Gamma] \quad \beta_2 = \beta_3 : \alpha \vdash type\,[\Gamma]}{\beta_1 = \beta_3 : \alpha \vdash type\,[\Gamma]} \text{ (trans. 3)}$$

identity of relative types and relative families of types rules

$$\frac{\alpha = \beta : type\,[\Gamma] \quad a : \alpha\,[\Gamma]}{a : \beta\,[\Gamma]} \text{ (rel. id. of types 1)}$$

$$\frac{\alpha = \beta : type\,[\Gamma] \quad a = b : \alpha\,[\Gamma]}{a = b : \beta\,[\Gamma]} \text{ (rel. id. of types 2)}$$

$$\frac{\alpha_1 = \alpha_2 : type\,[\Gamma] \quad \beta : \alpha_1 \vdash type\,[\Gamma]}{\beta : \alpha_2 \vdash type\,[\Gamma]} \text{ (id. of rel. types 3)}$$

$$\frac{\alpha_1 = \alpha_2 : type\,[\Gamma] \quad \beta_1 = \beta_2 : \alpha_1 \vdash type\,[\Gamma]}{\beta_1 = \beta_2 : \alpha_2 \vdash type\,[\Gamma]} \text{ (id. of rel. types 4)}$$

1b) Assumption rule

$$\frac{}{x : \alpha\,[\Gamma]} \quad provided\ x : \alpha\ in\ \Gamma \quad (rel.\ ass.)$$

1c) Thinning rules

$$\frac{\Gamma : context \quad \Delta : context \quad \Gamma \preceq \Delta \quad \alpha : type\,[\Gamma]}{\alpha : type\,[\Delta]} \text{ (thin 1)}$$

$$\frac{\Gamma : context \quad \Delta : context \quad \Gamma \preceq \Delta \quad \alpha = \beta : type\,[\Gamma]}{\alpha = \beta : type\,[\Delta]} \text{ (thin 2)}$$

$$\frac{\Gamma : context \quad \Delta : context \quad \Gamma \preceq \Delta \quad a : \alpha\,[\Gamma]}{a : \alpha\,[\Delta]} \text{ (thin 3)}$$

$$\frac{\Gamma : context \quad \Delta : context \quad \Gamma \preceq \Delta \quad a = b : \alpha\,[\Gamma]}{a = b : \alpha\,[\Delta]} \text{ (thin 4)}$$

$$\frac{\Gamma : context \quad \Delta : context \quad \Gamma \preceq \Delta \quad \beta : \alpha \vdash type\,[\Gamma]}{\beta : \alpha \vdash type\,[\Delta]} \text{ (thin 5)}$$

$$\frac{\Gamma : context \quad \Delta : context \quad \Gamma \preceq \Delta \quad \beta_1 = \beta_2 : \alpha \vdash type\,[\Gamma]}{\beta_1 = \beta_2 : \alpha \vdash type\,[\Delta]} \text{ (thin 6)}$$

1d) General substitution rules

$$\frac{\alpha : type\,[\Delta] \quad \delta : \Gamma \to \Delta}{\alpha\delta : type\,[\Gamma]} \text{ (type subst 1)} \qquad \frac{\alpha : type\,[\Delta] \quad \delta_1 = \delta_2 : \Gamma \to \Delta}{\alpha\delta_1 = \alpha\delta_2 : type\,[\Gamma]} \text{ (type subst 2)}$$

$$\frac{\alpha = \beta : type\,[\Delta] \quad \delta : \Gamma \to \Delta}{\alpha\delta = \beta\delta : type\,[\Gamma]} \text{ (type subst 3)}$$

$$\frac{a : \alpha\,[\Delta] \quad \delta : \Gamma \to \Delta}{a\delta : \alpha\delta\,[\Gamma]} \text{ (object subst 1)} \qquad \frac{a : \alpha\,[\Delta] \quad \delta_1 = \delta_2 : \Gamma \to \Delta}{a\delta_1 = a\delta_2 : \alpha\delta_1\,[\Gamma]} \text{ (object subst 2)}$$

$$\frac{a = b : \alpha\,[\Delta] \quad \delta : \Gamma \to \Delta}{a\delta = b\delta : \alpha\delta\,[\Gamma]} \text{ (object subst 3)}$$

$$\frac{\beta : \alpha \vdash type\,[\Delta] \quad \delta : \Gamma \to \Delta}{\beta\delta : \alpha\delta \vdash type\,[\Gamma]} \text{ (fam. of types subst 1)}$$

$$\frac{\beta : \alpha \vdash type\,[\Delta] \quad \delta_1 = \delta_2 : \Gamma \to \Delta}{\beta\delta_1 = \beta\delta_2 : \alpha\delta_1 \vdash type\,[\Gamma]} \text{ (fam. of types subst 2)}$$

$$\frac{\beta_1 = \beta_2 : \alpha \vdash type\,[\Delta] \quad \delta : \Gamma \to \Delta}{\beta_1\delta = \beta_2\delta : \alpha\delta \vdash type\,[\Gamma]} \text{ (fam. of types subst 3)}$$

C, n & o

(1) Tasistro calls the general rules concerning identity of relative objects, relative types and relative families of types 'equality rules' and the identity of relative types and relative families of types 'rules of type conversion'.[37] Using the meaning explanations of the judgements '$\alpha : type\,[\Gamma]$', '$\alpha = \beta : type\,[\Gamma]$', '$a : \alpha\,[\Gamma]$', '$a = b : \alpha\,[\Gamma]$', '$\beta : \alpha \vdash type\,[\Gamma]$' and '$\beta_1 = \beta_2 : \alpha \vdash type\,[\Gamma]$' the justification of these rules is straightforward. Likewise, the justification of the rule (rel. ass.) is based on the meaning explanations of '$\gamma : \Gamma$', for non-empty context Γ, and '$a : \alpha\,[\Gamma]$'.

(2) The thinning rules express that if the meaning of a hypothetical judgement under a given context is explained, then the meaning of the judgement under any extension of the context is equally explained. Tasistro justifies the rules by distinguishing two cases: namely Δ being empty and Δ being non-empty.[38]

(3) The general substitution rules constitute formation rules of judgements with substitutions. Tasistro contracts each pair of rules (type subst 2) and (type subst 3), (object subst 2) and (object subst 3), (fam. of types subst 2) and (fam. of types subst 3) into a single rule resp.[39] He justifies these rules by first defining $(\alpha\delta)\gamma$ to be the same type as $\alpha(\delta\gamma)$, $\gamma : \Gamma$, in the case of the (type subst)-rules (and by defining $(a\delta)\gamma$ to be the same object as $a(\delta\gamma)$ of type $\alpha(\delta\gamma)$ in the case of the (object subst)-rules, etc.), and by then distinguishing the two cases of context Γ being empty and non-empty.[40]

2) *Special substitution rules*

2a) Substitution formation rules

$$\frac{}{() : \Gamma \to \Gamma} \text{ where } \Gamma : context \quad \text{(substF 1)}$$

$$\frac{\alpha : type\,[\Delta] \quad \delta : \Gamma \to \Delta \quad a : \alpha\delta\,[\Gamma]}{(\delta, a/x) : \Gamma \to (\Delta, x : \alpha)} \text{ provided } x\ \Delta\text{-fresh} \quad \text{(substF 2)}$$

$$\frac{\vartheta : \Delta \to \Theta \quad \delta : \Gamma \to \Delta}{\vartheta\delta : \Gamma \to \Theta} \text{ (substF 3)} \qquad \frac{\delta : \Psi \to \Delta \quad \Psi \preceq \Gamma}{\delta : \Gamma \to \Delta} \text{ (substF 4)}$$

$$\frac{\delta : \Gamma \to \Theta \quad \Delta \preceq \Theta}{\delta : \Gamma \to \Delta} \text{(substF 5)}$$

2b) The associativity substitution rules

$$\frac{\alpha : type\,[\Delta] \quad \delta : \Gamma \to \Delta \quad \gamma : \Psi \to \Gamma}{(\alpha\delta)\gamma = \alpha(\delta\gamma) : type\,[\Psi]} \text{(assoc. subst 1)}$$

$$\frac{a : \alpha\,[\Delta] \quad \delta : \Gamma \to \Delta \quad \gamma : \Psi \to \Gamma}{(a\delta)\gamma = a(\delta\gamma) : \alpha(\delta\gamma)\,[\Psi]} \text{(assoc. subst 2)}$$

$$\frac{\beta : \alpha \vdash type\,[\Delta] \quad \delta : \Gamma \to \Delta \quad \gamma : \Psi \to \Gamma}{(\beta\delta)\gamma = \beta(\delta\gamma) : \alpha(\delta\gamma) \vdash type\,[\Psi]} \text{(assoc. subst 3)}$$

2c) Identity substitution rules

$$\frac{\alpha : type\,[\Gamma] \quad (\,) : \Gamma \to \Gamma}{\alpha(\,) = \alpha : type\,[\Gamma]} \text{(id. subst 1)} \qquad \frac{a : \alpha\,[\Gamma] \quad (\,) : \Gamma \to \Gamma}{a(\,) = a : \alpha(\,)\,[\Gamma]} \text{(id. subst 2)}$$

$$\frac{\beta : \alpha \vdash type\,[\Gamma] \quad (\,) : \Gamma \to \Gamma}{\beta(\,) = \beta : \alpha(\,) \vdash type\,[\Gamma]} \text{(id. subst 3)}$$

2d) Identity of substitution rules

$$\frac{\delta_1 : \Gamma \to (\,) \quad \delta_2 : \Gamma \to (\,)}{\delta_1 = \delta_2 : \Gamma \to (\,)} \text{(id. of subst 1)}$$

$$\frac{\delta_1 = \delta_2 : \Gamma \to \Delta \quad x\delta_1 = x\delta_2 : \alpha\delta_1\,[\Gamma]}{\delta_1 = \delta_2 : \Gamma \to (\Delta, x : \alpha)} \text{(id. of subst 2)}$$

2e) General rules concerning identity of substitutions

$$\frac{\delta : \Gamma \to \Delta}{\delta = \delta : \Gamma \to \Delta} \text{(reflex. subst)} \qquad \frac{\delta = \gamma : \Gamma \to \Delta}{\gamma = \delta : \Gamma \to \Delta} \text{(symmetr. subst)}$$

$$\frac{\delta_1 = \delta_2 : \Gamma \to \Delta \quad \delta_2 = \delta_3 : \Gamma \to \Delta}{\delta_1 = \delta_3 : \Gamma \to \Delta} \text{(trans. subst)}$$

C, n & o

(1) Tasistro calls (substF 1) **id** since $(\,) : \Gamma \to \Gamma$ is the identity substitution, (substF 2) **upd** for updating, (substF 3) **com** for composition of two substitutions, (substF 4) **s-th** for source context thinning, and (substF 5) **t-th** for target context thinning.[41] The justifications of the first three of these rules start off with a definition. Let $\gamma : \Gamma$; then

- (substF 1) $(\,)\gamma$ is the same environment for Γ as γ (Def.)
- (substF 2) $(\delta, a/x)\gamma$ is the same environment for $(\Delta, x : \alpha)$ as $(\delta\gamma, a\gamma/x)$. (Def.)
- (substF 3) $(\vartheta\delta)\gamma$ is the same environment for Θ as $\vartheta(\delta\gamma)$. (Def.)

Then they proceed by the usual distinction of cases: Γ is empty and Γ is non-empty. The justifications of (substF 4) and (substF 5) require no such definitions and proceed also by distinguishing the cases of Γ being empty and Γ not being empty. But they do presuppose a particular system of environments set up and analyzed by Tasistro.[42]

(2) The justifications of the associativity substitution rules are implicitly contained in the justification of rule (substF 3).

(3) The justifications of the identity substitution rules first make use of the definition introduced in the justification of the (type subst)-rules and second of the definition introduced in the justification of (substF 1).

(4) Tasistro calls the identity of substitution rules ζ-rules.[43] Their justification relies on the particular system of environments mentioned in note (1) above. The criterion for the identity of substitutions is extensional: roughly speaking, two substitutions are identical if they assign the same objects to the same variables.

(5) Justification of the general rules concerning identity of substitutions is straightforward.

(6) Tasistro defines the notion of normal substitution as follows:

A substitution is said to be normal if it is generated from $() : \Gamma \rightarrow ()$ by means of (substF 2) only.[44] Note that $() : \Gamma \rightarrow ()$ is obtained from $() : \Gamma \rightarrow \Gamma$ by (substF 5): $\dfrac{() : \Gamma \rightarrow \Gamma \quad () \preceq \Gamma}{() : \Gamma \rightarrow ()}$. In order to introduce a specific identity notion between normal substitutions, he observes that the following two rules are derivable:

$$\frac{}{() = () : \Gamma \rightarrow ()}\ (\xi_0)$$

$$\frac{\delta_1 = \delta_2 : \Gamma \rightarrow \Delta \quad a_1 = a_2 : \alpha\delta_1\ [\Gamma]}{(\delta_1, a_1/x) = (\delta_2, a_2/x) : \Gamma \rightarrow (\Delta, x : \alpha)}\ (\xi_1)$$

Two normal substitutions are said to be identical as normal substitutions if this identity is generated by means of (ξ_0) and (ξ_1) only. Tasistro then proves two interesting properties of normal substitutions:[45]

I Every substitution is equal to a normal substitution.
II If two normal substitutions are identical, then they are identical as normal substitutions.

3) *Special rules for relative families of types, relative fct. types and for two particular relative types*

3a) Family of types formation and identity rules

$$\frac{\beta x : type\ [\Gamma, x : \alpha]}{\beta : \alpha \vdash type\ [\Gamma]}\ \text{provided } x\ \Gamma\text{-fresh and } \beta \text{ non-dependent on } x \qquad \text{(fam. of types 1)}$$

$$\frac{\beta_1 : \alpha \vdash type\ [\Gamma] \quad \beta_2 : \alpha \vdash type\ [\Gamma] \quad \beta_1 x = \beta_2 x : type\ [\Gamma, x : \alpha]}{\beta_1 = \beta_2 : \alpha \vdash type\ [\Gamma]}\ \text{(fam. of types 2)}$$

provided x Γ-fresh

$$\frac{\beta : type\,[\Gamma, x : \alpha] \quad \gamma : \Psi \to \Gamma \quad a : \alpha\gamma\,[\Psi]}{(((x)\beta)\gamma)a : type\,[\Psi]} \text{ (fam. of types 3)*}$$

$$\frac{\Gamma \preceq \Delta \quad \beta : type\,[\Gamma, x : \alpha] \quad a : \alpha\,[\Delta]}{((x)\beta)a : type\,[\Delta]} \text{ (fam. of types 3)}$$

$$\frac{\beta : type\,[\Gamma, x : \alpha] \quad \gamma : \Psi \to \Gamma \quad a : \alpha\gamma\,[\Psi]}{(((x)\beta)\gamma)a = \beta(\gamma, a/x) : type\,[\Psi]} \text{ (fam. of types 4)*}$$

$$\frac{\Gamma \preceq \Delta \quad \beta : type\,[\Gamma, x : \alpha] \quad a : \alpha\,[\Delta]}{((x)\beta)a = \beta(a/x) : type\,[\Delta]} \text{ (fam. of types 4)}$$

The rules (fam. of types 1)–(fam. of types 4) serve to derive the following rules

$$\frac{\beta : type\,[\Gamma, x : \alpha]}{(x)\beta : \alpha \vdash type\,[\Gamma]} \text{ (fam. of types 5)} \qquad \frac{\beta_1 = \beta_2 : \alpha \vdash type\,[\Gamma, x : \alpha]}{(x)\beta_1 = (x)\beta_2 : \alpha \vdash type\,[\Gamma]} \text{ (fam. of types 6)}$$

$$\frac{\beta : \alpha \vdash type\,[\Gamma]}{\beta = (x)(\beta x) : \alpha \vdash type\,[\Gamma]} \text{ provided } x \text{ } \Gamma\text{-fresh} \qquad \text{(fam. of types 7)}$$

3b) Family of types application rules

$$\frac{\beta : \alpha \vdash type\,[\Gamma] \quad a : \alpha\,[\Gamma]}{\beta a : type\,[\Gamma]} \text{ (fam. of types app. 1)}$$

$$\frac{\beta : \alpha \vdash type\,[\Gamma] \quad a = b : \alpha\,[\Gamma]}{\beta a = \beta b : type\,[\Gamma]} \text{ (fam. of types app. 2)}$$

$$\frac{\beta_1 = \beta_2 : \alpha \vdash type\,[\Gamma] \quad a : \alpha\,[\Gamma]}{\beta_1 a = \beta_2 a : type\,[\Gamma]} \text{ (fam. of types app. 3)}$$

$$\frac{\beta : \alpha \vdash type\,[\Gamma] \quad a : \alpha\,[\Gamma] \quad \gamma : \Psi \to \Gamma}{(\beta a)\gamma = (\beta\gamma)(a\gamma) : type\,[\Psi]} \text{ (fam. of types app. 4)}$$

C, n & o

According to the convention above, the rules (fam. of types app. 1)–(fam. of types app. 3) should actually have been called (rel. gen. subst 1)–(rel. gen. subst 3) resp., but since the term ‘substitution’ has assumed a new meaning in this section the names of those rules have been changed accordingly.

3c) Fct. type formation and identity rules

$$\frac{\alpha : type\,[\Gamma] \quad \beta : \alpha \vdash type\,[\Gamma]}{(\alpha)\beta : type\,[\Gamma]} \text{ (rel. fct. type)}$$

$$\frac{\alpha_1 = \alpha_2 : type\,[\Gamma] \quad \beta : \alpha_1 \vdash type\,[\Gamma]}{(\alpha_1)\beta = (\alpha_2)\beta : type\,[\Gamma]} \text{ (rel. fct. type id. 1)}$$

$$\frac{\alpha : type\,[\Gamma] \quad \beta_1 = \beta_2 : \alpha \vdash type\,[\Gamma]}{(\alpha)\beta_1 = (\alpha)\beta_2 : type\,[\Gamma]} \text{ (rel. fct. type id. 2)}$$

$$\frac{\alpha : \mathit{type}\,[\Gamma] \quad \beta : \alpha \vdash \mathit{type}\,[\Gamma] \quad \gamma : \Psi \to \Gamma}{((\alpha)\beta)\gamma = (\alpha\gamma)(\beta\gamma) : \mathit{type}\,[\Psi]} \text{ (fct. type subst)}$$

3d) Fct. type application rules

$$\frac{f : (\alpha)\beta\,[\Gamma] \quad a : \alpha\,[\Gamma]}{fa : \beta a\,[\Gamma]} \text{ (rel. app. 1)} \qquad \frac{f : (\alpha)\beta\,[\Gamma] \quad a = b : \alpha\,[\Gamma]}{fa = fb : \beta a\,[\Gamma]} \text{ (rel. app. 2)}$$

$$\frac{f = g : (\alpha)\beta\,[\Gamma] \quad a : \alpha\,[\Gamma]}{fa = ga : \beta a\,[\Gamma]} \text{ (rel. app. 3)}$$

$$\frac{f : (\alpha)\beta\,[\Gamma] \quad a : \alpha\,[\Gamma] \quad \gamma : \Psi \to \Gamma}{(fa)\gamma = (f\gamma)(a\gamma) : (\beta a)\gamma\,[\Psi]} \text{ (fct. type app. 4)}$$

3e) Fct. object formation and identity rules

$$\frac{fx : \beta x\,[\Gamma, x : \alpha]}{f : (\alpha)\beta\,[\Gamma]} \text{ provided } x \ \Gamma\text{-fresh and } \beta \text{ non-dependent on } x \qquad \text{(rel. arg. removal 1)}$$

$$\frac{fx = gx : \beta x\,[\Gamma, x : \alpha]}{f = g : (\alpha)\beta\,[\Gamma]} \text{ provided } x \ \Gamma\text{-fresh and } \beta \text{ non-dependent on } x \qquad \text{(rel. arg. removal 2)}$$

$$\frac{b : \beta\,[\Gamma, x : \alpha] \quad \gamma : \Psi \to \Gamma \quad a : \alpha\gamma\,[\Psi]}{(((x)b)\gamma)a : \beta(\gamma, a/x)\,[\Psi]} \text{ (rel. arg. move 1)*}$$

$$\frac{\Gamma \preceq \Delta \quad b : \beta\,[\Gamma, x : \alpha] \quad a : \alpha\,[\Delta]}{((x)b)a : \beta(a/x)\,[\Delta]} \text{ (rel. arg.move 1)}$$

$$\frac{b : \beta\,[\Gamma, x : \alpha] \quad \gamma : \Psi \to \Gamma \quad a : \alpha\gamma\,[\Psi]}{(((x)b)\gamma)a = b(\gamma, a/x) : \beta(\gamma, a/x)\,[\Psi]} \text{ (rel. arg. move 2)*}$$

$$\frac{\Gamma \preceq \Delta \quad b : \beta\,[\Gamma, x : \alpha] \quad a : \alpha\gamma\,[\Psi]}{((x)b)a = b(a/x) : \beta(a/x)\,[\Delta]} \text{ (rel. arg. move 2)}$$

From these fct. object formation rules can be derived

$$\frac{b : \beta\,[\Gamma, x : \alpha]}{(x)b : (\alpha)((x)\beta)\,[\Gamma]} \text{ (rel. abs. 1)} \qquad \frac{b_1 = b_2 : \beta\,[\Gamma, x : \alpha]}{(x)b_1 = (x)b_2 : (\alpha)((x)\beta)\,[\Gamma]} \text{ (rel. abs. 2)}$$

$$\frac{b : (\alpha)\beta\,[\Gamma]}{b = (x)(bx) : (\alpha)\beta\,[\Gamma]} \text{ provided } x \ \Gamma\text{-fresh} \qquad \text{(rel. abs. 3)}$$

3f) The particular relative type of sets

$$\frac{}{\mathit{set} : \mathit{type}\,[\Gamma]} \text{ (rel. set)}$$

$$\frac{}{\mathit{elem} : (\mathit{set})\mathit{type}\,[\Gamma]} \text{ which is somehow the relativization of (dep. set type)}$$

$$\frac{\gamma : \Psi \to \Gamma}{(\mathit{set})\gamma = \mathit{set} : \mathit{type}\,[\Psi]} \text{ (set subst)} \qquad \frac{\gamma : \Psi \to \Gamma}{(\mathit{elem})\gamma = \mathit{elem} : (\mathit{set})\mathit{type}\,[\Psi]} \text{ (elem subst)}$$

3g) The particular relative type of propositions

$$\frac{}{prop : type\,[\Gamma]}\ \text{(rel. prop)}$$

$$\frac{}{proof : (prop)type\,[\Gamma]}\ \text{which is somehow the relativization of (dep. prop type)}$$

$$\frac{\gamma : \Psi \to \Gamma}{(prop)\gamma = prop : type\,[\Psi]}\ \text{(prop subst)}$$

$$\frac{\gamma : \Psi \to \Gamma}{(proof)\gamma = proof : (prop)type\,[\Psi]}\ \text{(proof subst)}$$

C, n & o

(1) Following Martin-Löf's idea of replacing the abstraction rule (abs) of Sect. iii. by more basic rules, namely the (arg. removal)- and the (arg. move)- rules, Tasistro's abstraction rule for families of types, here denoted by (fam. of types 5), has been replaced by the rules (fam. of types 1) and (fam. of types 3), from which it is derivable.

Tasistro calls (fam. of types 2) ζ, (fam. of types 4)* β^*, (fam. of types 4) β, (fam. of types 6) ξ and (fam. of types 7) η.

Note: (fam. of types 4)* is just a preliminary form of the β- or (fam. of types 4)-rule, i.e. the latter follows readily from the former.[46]

As far as the justifications of (fam. of types 2), (fam. of types 4) and (fam. of types 5) are concerned, cf. Tasistro[47].

(2) Tasistro contracts the pair of rules (fam. of types app. 2) and (fam. of types app. 3) into one single rule.[48] The justifications of these rules rest on defining $(\beta\alpha)\gamma$ to be the same type as $(\beta\gamma)(\alpha\gamma)$, $\gamma : \Gamma$, and on distinguishing the usual two cases: Γ is empty and Γ is non-empty.[49]

(3) Tasistro contracts the pair of rules (rel. fct. type id. 1) and (rel. fct. type id. 2) into one single rule and likewise with the pair of rules (rel. app. 2) and (rel. app. 3).[50] As far as the justification of these rules is concerned, Tasistro may be quoted once again: this 'justification is now just routine'.[51]

(4) For the same reason that was mentioned in the previous note (1), the abstraction rule (rel. abs. 1), called 'abs' by Tasistro,[52] has been replaced by the rules (rel. arg. removal 1) and (rel. arg. move 1), from which it is derivable.

Tasistro calls (rel. arg. removal 2) ζ, (rel. arg. move 2)* β^*, (rel. arg. move 2) β, (rel. abs. 2) ξ and (rel. abs. 3) η.[53] *Note*: (rel. arg. move 1)* is just a preliminary form of (rel. arg. move 1), and the same applies to (rel. arg. move 2)* w.r.t. (rel. arg. move 2).

(5) Since the relative types and relative objects now play the role previously played by the dependent types (or families of types) and dependent objects (or fcts. in the oldfashioned sense), the rules (gen. subst 1)–(gen. subst 6) of Sect. ii. can now be considered relativized by the following rules resp.

$$\frac{\alpha : type\,[\Delta] \quad \delta : \Delta}{\alpha\delta : type} \qquad \frac{\alpha : type\,[\Delta] \quad \delta_1 = \delta_2 : \Delta}{\alpha\delta_1 = \alpha\delta_2 : type}$$

$$\frac{\alpha = \beta : type\,[\Delta] \quad \delta : \Delta}{\alpha\delta = \beta\delta : type}$$

$$\frac{a : \alpha\,[\Delta] \quad \delta : \Delta}{a\delta : \alpha\delta} \qquad \frac{a : \alpha\,[\Delta] \quad \delta_1 = \delta_2 : \Delta}{a\delta_1 = a\delta_2 : \alpha\delta_1}$$

$$\frac{a = b : \alpha\,[\Delta] \quad \delta : \Delta}{a\delta = b\delta : \alpha\delta}$$

which are exactly the general substitution rules for types and objects in the special case where Γ is empty, i.e. $\delta : (\,) \to \Delta$.

(6) The two collections of rules 3f) and 3g) actually presuppose acquaintance with the following two sections vi. and vii. For a proper understanding of these rules the reader is thus referred to those two sections.

V DIGRESSION. EXPLICIT (ABBREVIATORY)AND PARTIAL (CONDITIONAL) DEFINITIONS IN CONSTRUCTIVE TYPE THEORY

The previous four sections focused on the real definition of what it was to be: a (non-dependent) type, a dependent type, a (non-dependent) object, a dependent object, a context, an environment, identical types, identical objects, etc. This section, however, focuses on the other kind of definition, namely nominal definition in general, and on explicit and partial def. in constructive type theory in particular. Explicit defs. are the most simple form of nominal defs. In constructive type theory either types are identified, in which case types are abbreviated, or objects are identified, in which case objects are abbreviated. Hence two levels will be distinguished:

(i) *The type level*

The general rules for making explicit (abbreviatory) defs. on this level are

$$\frac{\begin{array}{l} x_1 : \alpha_1, \ldots, x_n : \alpha_n \vdash \beta : type \\ a_1 : \alpha_1 \\ \vdots \\ a_n : \alpha_n(a_1/x_1, \ldots, a_{n-1}/x_{n-1}) \end{array}}{\varphi(a_1)\ldots(a_n) : type} \qquad \frac{\begin{array}{l} x_1 : \alpha_1, \ldots, x_n : \alpha_n \vdash \beta : type \\ a_1 : \alpha_1 \\ \vdots \\ a_n : \alpha_n(a_1/x_1, \ldots, a_{n-1}/x_{n-1}) \end{array}}{\varphi(a_1)\ldots(a_n) = \beta(a_1/x_1, \ldots, a_n/x_n) : type}$$

which can be abbreviated in the following way:

$$\frac{\begin{array}{l} x_1 : \alpha_1, \ldots, x_n : \alpha_n \vdash \beta : type \\ a_1 : \alpha_1 \\ \vdots \\ a_n : \alpha_n(a_1/x_1, \ldots, a_{n-1}/x_{n-1}) \end{array}}{\begin{cases} \varphi(a_1)\ldots(a_n) : type \\ \varphi(a_1)\ldots(a_n) = \beta(a_1/x_1, \ldots, a_n/x_n) : type \end{cases}}$$

C, n & o

(1) The second rule means that one is licensed to infer

$$\frac{a : \beta(a_1/x_1, \ldots, a_n/x_n)}{a : \varphi(a_1) \ldots (a_n)} \quad \text{and} \quad \frac{a = c : \beta(a_1/x_1, \ldots, a_n/x_n)}{a = c : \varphi(a_1) \ldots (a_n)}$$

(2) An explicit (abbreviatory) def. of a type always consists of two rules, i. the typing rule, and ii. the (definitional) identity rule.

Examples of explicit defs. on the type level

a.

$$\frac{A : set \quad X : set \vdash (X)prop : type}{\begin{cases} class(A) : type \\ class(A) = (A)prop : type \end{cases}}$$

b.

$$\frac{B : set \qquad \dfrac{A : set \quad X : set, Y : set \vdash (X)(Y)prop : type}{Y : set \vdash (A)(Y)prop : type}}{\begin{cases} rel(A, B) : type \\ rel(A, B) = (A)(B)prop : type \end{cases}}$$

c.

$$\frac{(prop)(prop)prop : type}{\begin{cases} connective : type \\ connective = (prop)(prop)prop : type \end{cases}}$$

d.

$$\frac{(X : set) \quad ((X)prop)prop : type}{\begin{cases} quantifier : type \\ quantifier = (X : set)((X)prop)prop : type \end{cases}}$$

(3) The following two rules

$$\frac{\alpha : type \quad \beta : type}{(\alpha)\beta : type} \text{(indep. fct. type)} \qquad \frac{\alpha : type \quad \beta : type}{(\alpha)\beta = (x : \alpha)\beta : type}$$

are no example or instance of the general scheme, since in the scheme only objects and not types can occur as arguments of φ.

(ii) *The object level*

Here the general rules are

$$\frac{\begin{array}{l} x_1 : \alpha_1, \ldots, x_n : \alpha_n \vdash b : \beta \\ a_1 : \alpha_1 \\ \vdots \\ a_n : \alpha_n(a_1/x_1, \ldots, a_{n-1}/x_{n-1}) \end{array}}{f(a_1) \ldots (a_n) : \beta(a_1/x_1, \ldots, a_n/x_n)}$$

$$\frac{\begin{array}{l} x_1 : \alpha_1, \ldots, x_n : \alpha_n \vdash b : \beta \\ a_1 : \alpha_1 \\ \vdots \\ a_n : \alpha_n(a_1/x_1, \ldots, a_{n-1}/x_{n-1}) \end{array}}{f(a_1)\ldots(a_n) = b(a_1/x_1, \ldots, a_n/x_n) : \beta(a_1/x_1, \ldots, a_n/x_n)}$$

which can be abbreviated by

$$\frac{\begin{array}{l} x_1 : \alpha_1, \ldots, x_n : \alpha_n \vdash b : \beta \\ a_1 : \alpha_1 \\ \vdots \\ a_n : \alpha_n(a_1/x_1, \ldots, a_{n-1}/x_{n-1}) \end{array}}{\begin{cases} f(a_1)\ldots(a_n) : \beta(a_1/x_1, \ldots, a_n/x_n) \\ f(a_1)\ldots(a_n) = b(a_1/x_1, \ldots, a_n/x_n) : \beta(a_1/x_1, \ldots, a_n/x_n) \end{cases}}$$

C, n & o

(1) *Examples* of explicit defs. on the object level

a.

$$\frac{0 : N \quad x : N \vdash s : N}{\begin{cases} 1 : N \\ 1 = s(0/x) : N \end{cases}}$$

b.

$$\frac{1 : N \quad x : N \vdash s : N}{\begin{cases} 2 : N \\ 2 = s(1/x) : N \end{cases}}$$

c.

$$\frac{\begin{array}{l} X : prop,\ Y : prop \vdash \supset : prop \\ A : prop \\ \bot : prop \end{array}}{\begin{cases} \sim A : prop \\ \sim A = \supset (A/X, \bot/Y) : prop \end{cases}}$$

d.

$$\frac{\begin{array}{l} X : prop,\ Y : prop \vdash \& : prop \\ A \supset B : prop \\ B \supset A : prop \end{array}}{\begin{cases} A \supset\subset B : prop \\ A \supset\subset B = \&(A \supset B/X, B \supset A/Y) : prop \end{cases}}$$

(2) (Terminological) f is called a defined fct. constant or, in Curry's terminology, a combinator.

(3) When $x_1 : \alpha_1, \ldots, x_n : \alpha_n \vdash b : \beta$, then $x_1 : \alpha_1, \ldots, x_n : \alpha_n \vdash f(x_1)\ldots(x_n) = b : \beta$, i.e. all the variables on which the definiens b depends have to be shown or explicitly indicated in the definiendum f but not conversely. This requirement for

variables in explicit defs. was correctly stated in Russell and Whitehead's *Principia Mathematica* (1910–13). Frege in his *Grundgesetze der Arithmetik* (1893–1903) also required that all the variables in the definiendum should be shown or should appear in the definiens. But Martin-Löf observes that this requirement is wrong, since it prevents the def. of constant fcts.

(4) $x_1 : \alpha_1, \ldots, x_n : \alpha_n \vdash f(x_1)\ldots(x_n) : \beta$ and $f : (x_1 : \alpha_1)\ldots(x_n : \alpha_n)\beta$ are interderivable. Pass from $x_1 : \alpha_1, \ldots, x_n : \alpha_n \vdash f(x_1)\ldots(x_n) : \beta$ to $f : (x_1 : \alpha_1)\ldots(x_n : \alpha_n)\beta$ by n argument removals and from $f : (x_1 : \alpha_1)\ldots(x_n : \alpha_n)\beta$ to $x_1 : \alpha_1, \ldots, x_n : \alpha_n \vdash f(x_1)\ldots(x_n) : \beta$ by n applications. $f : (x_1 : \alpha_1)\ldots(x_n : \alpha_n)\beta$ and $f = (x_1)\ldots(x_n)b : (x_1 : \alpha_1)\ldots(x_n : \alpha_n)\beta$ will be called the linear way and

$$\frac{\begin{array}{l} x_1 : \alpha_1, \ldots, x_n : \alpha_n \vdash f(x_1)\ldots(x_n) : \beta \\ a_1 : \alpha_1 \\ \vdots \\ a_n : \alpha_n(a_1/x_1, \ldots, a_{n-1}/x_{n-1}) \end{array}}{\begin{cases} f(a_1)\ldots(a_n) : \beta(a_1/x_1, \ldots, a_n/x_n) \\ f(a_1)\ldots(a_n) = b(a_1/x_1, \ldots, a_n/x_n) : \beta(a_1/x_1, \ldots, a_n/x_n) \end{cases}}$$

the natural way of defining f.

Example: The natural way of defining $\sim$ is

$$\frac{A : prop}{\sim A : prop} \qquad \text{and} \qquad \frac{A : prop}{\sim A = A \supset \bot : prop}$$

which can also be written as $X : prop \vdash \sim (X) : prop$ and $X : prop \vdash \sim (X) = X \supset \bot : prop$. The linear way of defining $\sim$ is

$$\sim : (prop)prop \qquad \text{and} \qquad \sim = (X)(X \supset \bot) : (prop)prop$$

One passes from the linear to the natural way of defining $\sim$ by means of a single application. As for passing from the natural to the linear way of defining $\sim$, apply argument removal to $X : prop \vdash \sim (X) : prop$ to obtain $\sim : (prop)prop$. As for the second rule one has

$$\dfrac{\dfrac{\dfrac{(X : prop)}{\sim (X) = X \supset \bot : prop} \qquad \dfrac{(X : prop) \quad X : prop \vdash X \supset \bot : prop}{X \supset \bot = ((X)X \supset \bot)(X) : prop}\text{ arg. move}}{\sim (X) = ((X)X \supset \bot)(X) : prop}}{\sim = (X)(X \supset \bot) : (prop)prop}\ \zeta\text{-rule}$$

(5) One can raise the question of whether a nominal (explicit) def. defines a term or a type expression, or whether it defines an object or a type resp. The term 'nominal def.' suggests that it is the name of an object or type which is being defined. However, the examples of the def. of *connective* or $\sim A$ indicate that it is rather

a type (e.g. the type *connective* or the type *quantifier*) or an object of a type (e.g. the proposition $\sim A$ or the number 1) which are being defined. Of course, one thereby also defines a linguistic expression but, as it appears, only in a secondary sense.

Partial or conditional defs. do not fully satisfy the requirement i. (of eliminability) on nominal defs. They are called partial or conditional because they allow the replacement of the definiendum by the definiens (or they are explicit) only if a certain condition is met. The rules for partial or conditional defs. on the object level are like the rules for making explicit defs. on that level except that $f(a_1)\ldots(a_n) = b(a_1/x_1,\ldots,a_n/x_n) : \beta(a_1/x_1,\ldots,a_n/x_n)$ holds only on condition that at least one of the a_i's ($1 \leq i \leq n$) is of a certain form c.

C, n & o

Examples of partial or conditional defs. on the object level are the defs. of N_nE, NE, $+E$, ΠE, $\rightarrow E$, ΣE, etc. which can be seen from the N_n-equality axioms (N_n-equ 1) – (N_n-equ n), the N-equality axioms (N-equ 1) and (N-equ 2), the $+$-equality axioms ($+$-equ 1), ($+$-equ 2), the Π-equality axiom (Π-equ), etc. (cf. Sect. ix below)

Other forms of partial or conditional defs. are the rec. and transfinite rec. defs.

vi SETS AND ELEMENTS

In Sects. i.–iii. the notions of different kinds of type and their resp. objects have been presented and explicated: the non-dependent non-function type, the dependent type and the function type. Nothing has been said so far about what types there are. One ground type which will be introduced here is the type of sets. It will turn out in the sequel that the elements of a set form themselves a ground type. Thus, there will be as many new ground types as there are sets. Applying the real def. of 'type' to the type of sets one has to explain (i) what it means to be a set, and (ii) when two sets are identical. This constitutes the answer to the following questions:

vi.0) What does the judgement '*set* : *type*' mean? or What does it mean for the sets to form a type?

Presuppositions in the order of explanation: Let the judgements A : *set* and $A = B$: *set* be known.

To make the judgement '*set* : *type*' one has to know (i) what a set A is, and (ii) when two sets A and B are identical.

The formation rule for the ground type *set* is the rule with 0 premisses

$$\textit{set} : \textit{type} \quad (\text{set})$$

C, n & o

If the type *set* is to be an entity it ought, by comment (3) on the real def. of 'type', to have an identity (criterion). This set type identity is provided by the following judgement: *set* = *set* : *type*, which is just an instance of the general type identity reflexivity rule (reflex. 2).

vi.1) What does the judgement '$A : set$' mean? or What does it mean to be a set (an object of type *set*)?
Presuppositions in the order of explanation: Let the judgements $a : A$ and $a = b : A$ be known, for a, b canonical elements of A.
To explain what a judgement of the form '$A : set$' means one has to know (i) what a canonical element a of set A is, and (ii) when two canonical elements a, b of set A are identical.

C, n & o

(1) *Three concepts of sets*

i) *the concept of set as class*

A class is a collection of individuals of a universe of discourse. Class theory was founded in the 40s of the 19th century by de Morgan and Boole and was further developed by Peirce and Schröder in the 60s and 70s of the same century.[54] Hence class theory existed when Cantor founded set theory in the 1880s. Class theory can be quite easily developed in constructive type theory; first, 'class' is explicitly defined by the two rules

$$\frac{A : set}{class(A) : type} \qquad \frac{A : set}{class(A) = (A)prop : type}$$

second, the class operations of union, intersection and complement are defined as follows: let $D : class(A)$ and $E : class(A)$; then

$$\frac{D : class(A) \quad E : class(A)}{\begin{cases} D \cup E : class(A) \\ D \cup E = (x)(D(x) \vee E(x)) : class(A) \end{cases}} \qquad \frac{D : class(A) \quad E : class(A)}{\begin{cases} D \cap E : class(A) \\ D \cap E = (x)(D(x) \& E(x)) : class(A) \end{cases}}$$

$$\frac{D : class(A)}{\begin{cases} \bar{D} : class(A) \\ \bar{D} = (x)(D(x) \supset \bot) : class(A) \end{cases}}$$

where the proposition forming operations $\bot, \vee, \supset$ and & will be introduced later; third, the relation of class inclusion is defined as follows, for $D : class(A)$ and $E : class(A)$:

$$(D \subseteq E) = (\forall x : A)(D(x) \supset E(x)) : prop$$

The definitions of the class operations can be motivated as follows:

$$\cfrac{\cfrac{D : class(A) \quad class(A) = (A)prop : type}{D : (A)prop} \quad (x : A)}{D(x) : prop}$$

$$\cfrac{\cfrac{E : class(A) \quad class(A) = (A)prop : type}{E : (A)prop} \quad (x : A)}{E(x) : prop}$$

$$\dfrac{\dfrac{\dfrac{\begin{array}{c}\vdots\\ D(x) : prop\end{array} \quad \begin{array}{c}\vdots\\ E(x) : prop\end{array}}{D(x) \vee E(x) : prop}}{(x)(D(x) \vee E(x)) : (A)prop} \quad (A)prop = class(A) : type}{(x)(D(x) \vee E(x)) : class(A)}$$

Similarly for $(x)(D(x)\&E(x))$.

$$\dfrac{\dfrac{D : class(A) \quad class(A) = (A)prop : type}{D : (A)prop} \qquad (x : A)}{D(x) : prop}$$

$$\dfrac{(A)prop = class(A) : type \quad \dfrac{\dfrac{\begin{array}{c}\vdots\\ D(x) : prop\end{array} \quad \bot : prop}{D(x) \supset \bot : prop}}{(x)(D(x) \supset \bot) : (A)prop}}{(x)(D(x) \supset \bot) : class(A)}$$

The typing of the class operations yields

$$\begin{array}{lcl} \cup & : & (X : set)(Y : class(X))(Z : class(X))class(X) \\ \cap & : & (X : set)(Y : class(X))(Z : class(X))class(X) \\ ^{-} & : & (X : set)(Y : class(X))class(X) \end{array}$$

Now, the question arises naturally as to what the universe of discourse is? Since classes always presuppose a universe of discourse, a universe of discourse ultimately cannot be a class (infinite regresses and vicious circles being excluded). Thus, there is a natural way from class theory to Cantor's set theory.

ii) *The concept of set in Cantor's sense*

Cantor used the terms 'Mannigfaltigkeit', 'Inbegriff' or 'Menge' for set and it is the last one which eventually prevailed. Cantor's original intention was to clarify or make precise what a universe of discourse was. His first def. of set in 1882 was:[55]

> I call a manifold (a totality, a set) of elements which belong to some conceptual sphere well-defined, if on the basis of its definition and as a consequence of the logical principle of excluded middle it must be seen as *internally determined both* whether some object belonging to the same conceptual sphere belongs to the imagined manifold as an object or not, *as well as* whether two objects belonging to the set are equal to one another or not, despite formal differences in the way they are given.

This means that a set is defined by a criterion of application and a criterion of identity. This def. is obviously fairly similar to the real def. of set in constructive type theory. (Cantor's third def. of set in 1895 was:

> By a 'set' we understand every collection to a whole M of definite, well-differentiated objects m of our intuition or our thought. (We call these objects the 'elements' of M.)

If one interprets 'definite, well-differentiated' the way Parsons does, namely as meaning that it must be determinate what the elements are, that identity and difference have to be well-defined for the elements, and that a set must be determined by its elements, then one has the characteristics of the def. of set in constructive type theory again: a criterion of application, a criterion of identity, as well as an extensional interpretation of sets.[56])

Moreover, Cantor's basic set-theoretical operations are quite different from the basic class operations; they are the disjoint union of two sets $A + B$, the cartesian product of two sets $A \times B$ and the formation of the set of all fcts. from a set A to a set B, i.e. B^A or $A \rightarrow B$.

These can be expressed in the framework of constructive type theory as follows:

$$\frac{A : set \quad B : set}{A + B : set} \text{ disjoint union}$$

the set $A + B$ is defined by a criterion of application

$$\frac{a : A}{i(a) : A + B} \quad \frac{b : B}{j(b) : A + B}$$

and a criterion of identity

$$\frac{a = c : A}{i(a) = i(c) : A + B} \quad \frac{b = d : B}{j(b) = j(d) : A + B}$$

$$\frac{A : set \quad B : set}{A \times B : set} \text{ cartesian product}$$

the set $A \times B$ is defined by a criterion of application and by a criterion of identity, i.e. by

$$\frac{a : A \quad b : B}{(a, b) : A \times B} \qquad \frac{a = c : A \quad b = d : B}{(a, b) = (c, d) : A \times B}$$

resp.

$$\frac{A : set \quad B : set}{A \rightarrow B : set} \text{ formation of all fcts. from set } A \text{ to set } B$$

the set $A \rightarrow B$ is defined by a criterion of application and a criterion of identity, i.e. by

$$\frac{b : (A)B}{\lambda(b) : A \rightarrow B} \quad \frac{b = d : (A)B}{\lambda(b) = \lambda(d) : A \rightarrow B}$$

resp. (for explanations of the defs. of the sets $A + B$, $A \times B$, $A \rightarrow B$, cf. Sect. ix.)

Finally, it should be pointed out that the most basic sets in Cantor's set theory are the finite sets and the set of natural numbers. (This latter point is of interest because it illustrates, as does the indication of Cantor's basic set-theoretic operations, how close set theory of constructive type theory is to Cantor's original theory.)

iii) *The iterative or cumulative hierarchy concept of set*

Around the turn of last century the so-called logical or set-theoretical paradoxes were discovered. This led logicians like Zermelo, Fraenkel, von Neumann, Bernays, Gödel to formalize and axiomatize Cantor's set theory. The result was a concept of set which clearly differs from Cantor's resp. concept. According to the iterative or cumulative hierarchy notion of set, a set is defined as follows:[57] Let $\bar{\in}$ be the transitive closure of $\in$, i.e. for sets A, B, '$A\bar{\in}B$' means that if A is an element of an element of... of an element of B, then A is an element of B, i.e. $A \in B$. Let ϱ be the unique rank operation satisfying the rec. equation $\varrho(A) = \{\varrho(B)|B\bar{\in}A\}$. Let *OR* be the collection of ordinals and V the collection of all sets. Finally, define the operation υ on *OR* by means of the rec. equation $\upsilon_\alpha = \bigcup\{\mathcal{P}\upsilon_\beta|\beta < \alpha\}$. One readily proves that $\upsilon_0 = \emptyset$, $\upsilon_{\alpha+1} = \mathcal{P}\upsilon_\alpha$ and $\upsilon_\gamma = \bigcup_{\xi<\gamma} \upsilon_\xi$ if $\lim \gamma$. Moreover, one then proves the theorem: $\upsilon_\alpha = \{A|\varrho(A) < \alpha\}$. It follows from this theorem that $V = \bigcup_{\alpha \in OR} \upsilon_\alpha$ (*Note*: $\alpha, \beta, \gamma \in OR$).

That is, A is a set according to the iterative or cumulative hierarchy notion of set if A is an element of υ_α, for some $\alpha \in OR$. ϱ measures the rank or the height of a set in the so-called cumulative hierarchy $(\upsilon_\alpha)_{\alpha \in OR}$, i.e. it indicates how high up one has to go in the cumulative hierarchy until one reaches the height at which the resp. set can be or is being formed. (E.g. the rank ϱ of $\emptyset$ is 0, the rank $\varrho(\{\emptyset\}) = 1$, the rank $\varrho(\{\{\emptyset\}\}) = 2$, and $\varrho(\emptyset \cup \mathcal{P}\emptyset \cup \mathcal{PP}\emptyset \cup \cdots) = \omega$).

Moreover, the fundamental operations in ZF set theory are quite different from Cantor's basic operations, namely they are:[58] pairing $\{a, b\}$, sum $\bigcup a$, separation $\{x \in a|P(x)\}$, power set $\mathcal{P}a$, and replacement $\{f(x)|x \in a\}$.

Now, the foundational problem with ZF set theory is the following: ZF set theory is based on

1) the set-theoretic universe V
2) the iterative sets i.e. the elements of V
3) classes of iterative sets

ad 3) class theory within the framework of constructive type theory is unproblematic (cf. i) the concept of set as class)

ad 2) the concept of iterative set in the framework of constructive set theory is also relatively unproblematic. In order to give an interpretation of a constructive version of the classical conception of the cumulative hierarchy of sets (in the form of constructive ZF set theory) in constructive type theory, Aczel had to assume a new type or set (in the sense of constructive type theory) V of sets (in the sense of the cumulative hierarchy conception).[59] (*Note*: for the understanding of the rules (VF 1)–(V-equ) below, cf. Sect. ix.)

The formation rules of V are

$$V : set \quad (\text{VF 1}) \qquad\qquad V = V : set \quad (\text{VF 2})$$

The introduction rules for V are

$$\frac{A : U \vdash f : (x : A)V}{\{f(x)|x : A\} : V}\ (\text{VI 1}) \qquad \frac{A = B : U \vdash f = g : (x : A)V}{\{f(x)|x : A\} = \{g(x)|x : B\} : V}\ (\text{VI 2})$$

V's elimination rules, called set recursion by Aczel, are

$$\frac{\begin{array}{l} C : (V)set \\ b : (X : U)(f : (X)V)(g : (x : X)C(f(x)))C(\{f(x)|x : X\}) \\ \delta : V \end{array}}{VE(C, b, \delta) : C(\delta)}\ (\text{VE 1})$$

$$\frac{\begin{array}{l} C : (V)set \\ b = c : (X : U)(f : (X)V)(g : (x : X)C(f(x)))C(\{f(x)|x : X\}) \\ \delta = \delta' : V \end{array}}{VE(C, b, \delta) = VE(C, c, \delta') : C(\delta)} \text{ (VE 2)}$$

and its equality rule is

$$\frac{\begin{array}{l} C : (V)set \\ b : (X : U)(f : (X)V)(g : (x : X)C(f(x)))C(\{f(x)|x : X\}) \\ A : U \\ \{f(x)|x : A\} : V \end{array}}{VE(C, b, \{f(x)|x : A\}) = b(A, f, (x)VE(C, b, f(x))) : C(\{f(x)|x : A\})} \text{ (V-equ)}$$

Aczel illustrates how familiar sets from the cumulative hierarchy of sets are explicitly defined in constructive type theory by means of the rule (VI 1):

– the hereditarily finite sets $\{\delta_1, \ldots, \delta_k\}$, $k \in \omega$, given the sets $\delta_1, \ldots, \delta_k$

$$\frac{N_k : U \vdash f : (x : N_k)V}{\begin{cases} \{\delta_1, \ldots, \delta_k\} : V \\ \{\delta_1, \ldots, \delta_k\} = \{f(x)|x : N_k\} : V \end{cases}}$$

where $f(1) = \delta_1, \ldots, f(k) = \delta_k$. In particular, for $N_0 : U$, $\{f(x)|x : N_0\} = \emptyset$.

– the disjoint union of two sets $\delta \cup \delta'$

Let δ be $\{f(x)|x : A\} : V$ and δ' be $\{g(y)|y : B\} : V$. Then

$$\frac{A + B : U \vdash h : (z : A + B)V}{\begin{cases} \delta \cup \delta' : V \\ \delta \cup \delta' = \{h(z)|z : A + B\} : V \end{cases}}$$

where

$$h(+I1(a)) = f(a) : V \quad \text{for } a : A \qquad \text{and}$$
$$h(+I2(b)) = g(b) : V \quad \text{for } b : B$$

– the disjoint union of a family of sets $\bigcup \delta$

Let δ be the set $\{f(x)|x : A\}$ where $f(a)$ is a set $\{g(a, y)|y : B(a)\}$ for $a : A$, $A : U$. Then

$$\frac{\Sigma(A, B) : U \vdash h : (z : \Sigma(A, B))V}{\begin{cases} \bigcup \delta : V \\ \bigcup \delta = \{h(z)|z : \Sigma(A, B)\} : V \end{cases}}$$

where $h((a, b)) = g(a, b) : V$ for $a : A$ and $b : B(a)$.

– the infinite set ω

$$\frac{N : U \vdash f : (x : N)V}{\begin{cases} \omega : V \\ \omega = \{f(x)|x : N\} : V \end{cases}}$$

where $f(NI1) = \emptyset$ and $f(NI2(x)) = f(x) \cup \{f(x)\}$, for $x : N$.

The whole cumulative hierarchy of sets can be constructed with these operations, and thus the concept of iterative set is interpreted in constructive type theory.[60]

(*Note*: (1) Sometimes, the term 'set' is used in the sense of the iterative conception of set and sometimes in the sense of constructive type theory (the latter is always the case when 'set' occurs in a judgement of the form A : *set*). The reader will easily distinguish from the context which sense is intended in a certain occurrence of the term. (2) Aczel uses 'V' for 'U' and 'U' for 'V'.)

ad 1) the set-theoretic universe V is the real problem; because one can ask oneself the question: what exactly do the quantifiers range over? or What is the quantificational domain D going to be?

$$\frac{D\ domain \qquad x \in D \vdash P\ prop}{\begin{array}{c}(\forall x \in D)P\ prop\\(\exists x \in D)P\ prop\end{array}}$$

The quantificational domain D can be

a) a domain of individuals or a class
b) a set in the sense of constructive type theory
c) the set-theoretic universe

ad a) poses no problem, because if one has quantification over classes one can define

$$\begin{aligned}(\forall x \in \{x \in D|P\})Q &= (\forall x \in D)(P \supset Q)\\(\exists x \in \{x \in D|P\})Q &= (\exists x \in D)(P \,\&\, Q)\end{aligned}$$

ad b) poses no problem, because if D is a set in the sense of constructive type theory, then D is constructively well-defined.

ad c) poses the problem, because if $D = V$ then the constructive meaning of D is most unclear and problematic.

(2) Observe also that predicate logic presupposes set theory because there is always at least one domain of quantification involved. Thus, if logicism means that set theory (and mathematics) is (are) reducible to logic, then logicism is plainly wrong.

(3) In comment (2) on i.1) it was said that Russell introduced the term 'type' for kind or sort or for what had traditionally been called a category. In his *Principles of Mathematics* (1903) he mentions the type of propositions, the type of individuals, the type of properties of individuals and relations between individuals etc. which obviously suggest that types and categories are the same. However, in the same place Russell defines types also as ranges of significance of propositional fcts., i.e. as domains of quantification in which case they are clearly conceived as sets. Martin-Löf has pointed out that it is exactly the confusion of the two concepts of type (the concept of type as category and the concept of type as set) which has given rise to the impredicativity of simple type theory. Actually, it is not so much this confusion as the fact of basing type theory on the conception of type as category which makes simple type theory impredicative. And he also drew the attention to the fact that the ramified

types of properties, of relations, etc. of some finite order differ from the simple types of all properties, of all relations, etc. in the important sense that the ramified types can be conceived as sets and thus as domains of quantification, whereas the simple types cannot.

To put it more explicitly: (Simple, ramified) Type theory is intended as a language which serves to describe the (simple, ramified) type structure presented in C, n & o to iii.1), Sect. iii. The types are defined by propositional fcts., which will later on be called open wffs, and all the variables (and constants) of the language will have types (or type indices) and will be conceived of as ranging only over relations of the indicated type. If the approach of type theory is constructive, all higher types have to be defined from lower types which implies that the lower types are defined first. One way of formalizing this constructive procedure is by means of the concept of predicativity which is defined below.[61]

Def. of the terms and wffs of simple (ramified) type theory

i. A variable $x, y, z, \ldots$ or constant $a, b, \ldots$ of a given type is a term of that type
ii. If t is a term of type $[\alpha_1, \ldots, \alpha_n]$ ($[\alpha_1, \ldots, \alpha_n]/k/(m_1, \ldots, m_n)$) and $t_1, \ldots, t_n$ are terms, each t_i of type α_i (α_i/m_i', where $m_i' \leq m_i$) resp., $1 \leq i \leq n$, then $t(t_1, \ldots, t_n)$ is a wff. The wffs are closed as usual under $\sim, \&, \vee, \supset, \forall$ and $\exists$.
iii. If A is a wff in which the list of variables $y_1, \ldots, y_n, n > 0$, is possibly a not exhaustive list of distinct free variables (in canonical order relative to A [62]) of A and α_i (α_i/m_i) is the type of y_i, $1 \leq i \leq n$, then the expression – called an abstract – obtained from A by placing a circumflex over all occurrences of the free variables in the list $y_1, \ldots, y_n$ and enclosing the result in parenthesis is a term of type $[\alpha_1, \ldots, \alpha_n]$ ($[\alpha_1, \ldots, \alpha_n]/k/(m_1, \ldots, m_n)$), the variables being bound in the term so formed.

Def. of the order *ord* of a term and of an open wff of (simple, ramified) type theory

i. (simple) The order ord of a variable or a constant of type α is $l(\alpha)$.
i. (ramified) The order ord of a variable or constant of type α/k is k.
ii. Let t be an abstract and let n be the highest order of all the variables (bound or free) and all the constants of t. If there is a bound variable y s.t. $ord(y) = n$, then $ord(t) = n + 1$, o.w. $ord(t) = n$.
iii. Let A be an open wff and let $y_1, \ldots, y_n$ be all the free variables of A (in canonical order relative to A), and let $(A(\hat{y}_1, \ldots, \hat{y}_n))$ be the abstract formed from A by circumflexing all the free variables of A. Then $ord(A) = ord((A(\hat{y}_1, \ldots, \hat{y}_n)))$.

Def. of predicative and impredicative abstracts and wffs

(simple) An abstract $(A(\hat{y}_1, \ldots, \hat{y}_n))$ of type α is predicative iff $ord((A(\hat{y}_1, \ldots, \hat{y}_n))) = l(\alpha)$; o.w. it is impredicative. An open wff A is predicative iff $ord(A) = l(\alpha)$, where α is the type of $(A(\hat{y}_1, \ldots, \hat{y}_n))$ (where $y_1, \ldots, y_n$ are all the free variables of A). O.w. A is impredicative (i.e. $ord(A) > l(\alpha)$).

(ramified) An open wff A and its associate abstract $(A(\hat{y}_1, \ldots, \hat{y}_n))$ of type $[\alpha_1, \ldots, \alpha_n]/k/(m_1, \ldots, m_n)$ are predicative iff $k = \max_{1 \leq i \leq n}(m_i) + 1$; o.w. impredicative (i.e. if $k > \max_{1 \leq i \leq n}(m_i) + 1$).

If the following predicativity restriction is introduced in simple type theory, namely (1) that terms substituted for variables be predicative, and (2) that the abstract in the axiom of abstraction $(A(\hat{y}_1, \ldots, \hat{y}_n))(z_1, \ldots, z_n) \supset\subset A(z_1, \ldots, z_n)$ be predicative, then simple type theory is predicative but not strong enough to develop either analysis in its fullness or the general theory of cardinal numbers. If, however, the predicativity restriction is dropped, then simple type theory is not predicative anymore, i.e. the definitions of sets become definitions of categories, but the theory is now strong enough to do all the things it could not do before.

In order to obtain a theory which is at the same time as strong as (impredicative) simple type theory and yet still predicative, Russell introduced the ramified type structure with order-types. Although ramified type theory is stronger than predicative simple type theory as well as constructive (i.e. the types can be thought of as sets), it is not strong enough to develop the usual analysis of real numbers in a predicative way. Russell dealt with this problem by including the axiom of reducibility (asserting roughly that, whatever can be done in ramified type theory in an impredicative way, can also be done in a predicative way) in his ramified type theory.

However, Chwistek and Ramsey showed that if the theory based on the ramified type structure contains the axiom of reducibility (which PM does and which it somehow has to in order to develop the usual analysis of real numbers), then it is nothing but a roundabout way of formulating (impredicative) simple type theory, i.e. the theory based on the simple type structure.

Hence ramified type theory intended to eliminate this confusion of the two concepts of type. What makes it possible to conceive of the ramified types as sets is the constructive character of the former, i.e. types of higher orders have to be defined from types of lower orders which implies that the lower orders are defined first. This is not the case with simple type theory where lower types can be defined by reference to higher types. (This is what characterizes impredicative defs. of simple types.) In constructive type theory, sets have to satisfy a more radical constructive condition, namely they have to be given by an inductive def. As far as the types as categories are concerned, there are in Russell's simple type structure two ground categories, namely the one of individuals and the one of propositions. In Martin-Löf's dependent type structure there are two basic ground categories, the one of sets and the one of propositions (which are eventually identified (cf. Curry–Howard isomorphism, Sect. xi.)) each of which gives rise to infinitely many other ground categories (each non-dependent non-fct. set and proposition constituting a ground category). It seems more difficult to pinpoint the differences between Russell's simple categories and Martin-Löf's categories. (More about the difference between sets and types (in the sense of categories), cf. C, n & o to vi.5)).[63]

vi.2) What does a judgement of the form '$a : A$' mean? or What does it mean to be an element of some set?

To explain what a judgement of the form '$a : A$' means one has to know that (i) a has the form of an object formed by one of the element-formation or -construction rules of A (a has the form of a canonical element) or (ii) a can be evaluated to one of those forms (a is a non-canonical element). (Concerning the concepts of canonical element and evaluation, cf. Sect. viii.)

vi.3) What does a judgement of the form '$a = b : A$' mean? or What does it mean for two elements of a set to be identical?
To explain what a judgement of the form '$a = b : A$' means one has to know that (i) a and b have the form of two objects formed by one of the identity rules associated with the element-formation rules of A (a and b are identical canonical elements of A) or (ii) a and b can be evaluated to one of those forms (a or b are identical non-canonical elements of A).

C, n & o

(Terminological) If the rules mentioned in (i) of vi.2) and vi.3) are called introduction rules, then a set A can be said to be really defined by its introduction rules. Alternatively, a set A is defined by its constructors, i.e. by the prim. operations whose values are canonical elements of the set.

vi.4) What does a judgement of the form '$A = B : set$' mean? or What does it mean for two sets to be identical?
A judgement of the form '$A = B : set$' means that (i) if a is an element of set A then a is an element of set B and vice versa, and (ii) if a and b are identical elements of set A, then a and b are identical elements of set B and vice versa.

This real def. of the identity of two sets can be expressed by the following two rules:

$$\frac{a : A}{\overline{a : B}} \quad \text{and} \quad \frac{a = b : A}{\overline{a = b : B}}$$

After having explained what the type of sets is, one can explain what it means for a set to be a type. Applying the real def. of 'dependent type' to the judgement $X : set \vdash X : type$ one obtains:

vi.5) What does a judgement of the form '$X : set \vdash X : type$' mean? or What does it mean for a set to be a type?
Presuppositions in the order of explanation: Let the judgements $A : set$ and $A = B : set$ be known.
To make a judgement of the form '$X : set \vdash X : type$' one has to know that (i) A is a type provided A is a set, and (ii) A and B are identical types provided A and B are identical sets.

Alternatively, '$X : set \vdash X : type$' is really defined by the following two rules:

$$\frac{A : set}{A : type} \quad \text{and} \quad \frac{A = B : set}{A = B : type}$$

Justification

$$\frac{A : set}{A : type}$$

$A : type$: Straightforward – for if one knows that $A : set$, one knows what an element of A is and hence what an object of A is (analogously for two identical elements/objects of A); whence one knows that $A : type$.

$$\frac{A = B : set}{A = B : type}$$

$A = B : type$: if one knows that $A = B : set$ then $\dfrac{a : A}{a : B}$ and $\dfrac{a = b : A}{a = b : B}$ and this is exactly the meaning of '$A = B : type$', if A and B are considered to be the types corresponding to the resp. sets. □

The formation rule for dependent set types is the rule with 0 premisses

$$X : set \vdash X : type \quad (\text{dep. set type})$$

C, n & o

The differences between sets and types in constructive type theory:

On the one hand one knows or understands the judgement '$\alpha : type$' if one knows (i) what an object of type α is, and (ii) when two objects of type α are identical. One grasps or has a certain intuition about what is to count as an α and what is not, and the same applies to identical objects of type α. On the other hand one knows or understands '$A : set$' if one knows (i) how to generate the canonical elements of A by means of certain rules (namely the element-formation or introduction rules) and (ii) how to generate identical canonical elements of A by means of identity rules associated with those rules. In other terms, a set A is always (by real def.) inductively defined or generated, and to know it means to know its inductive def.

Jacobs claims that the difference between types and sets is epistemological: sets are better known than types.[64] A judgement of the form $A : set$ presupposes that A is very well known – which indeed it is if it is inductively defined – so that sufficient information about its elements is given to make sense of quantification over A by means of Π, Σ, $\forall$ and $\exists$. In general, knowledge of types is insufficient – no inductive generation, but only a certain intuition of their objects is required – to make sense of quantification over them. Thus, the differences between types and sets originate from the difference in knowledge about the formation of their objects or elements resp. To have a better understanding of what it means to be a type one ought to have a better understanding of the sort of knowledge which above has been called 'grasping' or 'having a certain intuition'.

vii PROPOSITIONS AND PROOF-OBJECTS

In Sect. vi. the first ground types were introduced, namely the type of sets and the types generated from sets. In the sequel other ground types will be introduced and explained, e.g. the type of propositions. Furthermore, it will be explained why the propositions form a type rather than a set and why the proofs of a proposition form a type. The special relationship between the (type of) sets and the (type of) propositions will only be discussed later on (cf. Sect. xi).

vii.0) What does a judgement of the form '*prop* : *type*' mean? or What does it mean for the propositions to form a type?
Presuppositions in the order of explanation: Let the judgements A : *prop* and $A = B$: *prop* be known.
To make the judgement '*prop* : *type*' one has to know (i) what a proposition A is, and (ii) when two propositions A and B are identical.

The formation rule for the ground type *prop* is the rule with 0 premisses

$$prop : type \;\; (\text{prop})$$

C, n & o

As already observed in the case of the type *set*, the type *prop* is an entity only if it has an identity (criterion), which, again as in the case of type *set*, is expressed by $prop = prop : type$ and constitutes a simple instance of the type identity reflexivity (reflex. 2).

vii.1) What does a judgement of the form 'A : *prop*' mean? or What does it mean to be a proposition (an object of the type *prop*)?
Presuppositions in the order of explanation: Let the judgements $a : A$ and $a = b : A$ be known, for a, b canonical proof-objects of A.
To explain what a judgement of the form 'A : *prop*' means one has to know (i) what a canonical proof-object a of proposition A is, and (ii) when two canonical proof-objects a and b of proposition A are identical.

C, n & o

Two concepts of proposition

The concept of proposition has evoked a lot of philosophical discussion, and at least three different concepts have emerged from this discussion; the third one will be discussed in C, n & o to vii.3).

i) *The Boolean concept of proposition*

According to the Boolean concept of proposition, a proposition is just a truth value; i.e. if A is a proposition, then A is conceived as

$$\begin{array}{rll} \text{either} & A = \mathit{true} & (\text{alternatively, } A = 1) \\ \text{or} & A = \mathit{false} & (\text{alternatively, } A = 0) \end{array}$$

This is the conception which Frege tried to build his *Grundgesetze der Arithmetik* (1893–1903) on. (In his *Begriffsschrift* (1879) the propositions were still conceived as 'beurteilbare Inhalte'.) It works well for the logical connectives which are defined by the well-known truth tables. But there is also quantificational logic in which one has at least:

$$\frac{D \; domain \qquad x \in D \vdash P(x) : prop}{\begin{array}{rcl} (\forall x \in D)P(x) & : & prop \\ (\exists x \in D)P(x) & : & prop \end{array}}$$

Thus, there is a domain D and a Boolean fct. $P(x)$ where $P(d) = \mathit{true}$ or *false* for $d \in D$. The quantifiers (or rather the quantified propositions) are then defined as follows:

$$(\forall x \in D)P(x) = \begin{cases} \mathit{true} & \text{if } P(d/x) = \mathit{true} \text{ for every } d \in D \\ \mathit{false} & \text{o.w.} \end{cases}$$
$$(\exists x \in D)P(x) = \begin{cases} \mathit{true} & \text{if } P(d/x) = \mathit{true} \text{ for some } d \in D \\ \mathit{false} & \text{o.w.} \end{cases}$$

Now, the following difficulty arises: the identity above is a definition or definitional identity and the difficulty is that, unless domain D is finite, one cannot effectively pass over from the definiendum to the definiens anymore. This is the real crux of the two-valued or Boolean semantics: definitions are not eliminable anymore. In constructive type theory the principle or requirement of eliminability (cf. Subsect. 1 on definitions above) of nominal defs. is valued higher than two-valued semantics which is hence abandoned. More practically, if mathematical calculations cannot be carried out because certain steps are reached in those calculations where one cannot proceed any further, then mathematics cannot serve the purpose of providing calculable models for the sciences anymore. For these reasons the Boolean concept of proposition is unsatisfactory.

ii) *The Frege/Wittgenstein concept of proposition*

According to the Frege/Wittgenstein concept of proposition, a proposition is not defined by its truth value anymore, but rather by its truth condition. This means that to know the truth or falsity of a proposition A one has to know under what conditions A is true. It is called the Frege/Wittgenstein concept because Frege already wrote about truth conditions of propositions in his *Grundgesetze der Arithmetik* (1893–1903) and the term becomes of course very important in Wittgenstein's *Tractatus logico-philosophicus* (1922). The standard truth conditions for the logical operations are:

$$\frac{A\ \mathit{true} \quad B\ \mathit{true}}{A \,\&\, B\ \mathit{true}} \qquad \frac{A\ \mathit{true}}{A \vee B\ \mathit{true}} \qquad \frac{B\ \mathit{true}}{A \vee B\ \mathit{true}} \qquad \frac{A\ \mathit{true} \vdash B\ \mathit{true}}{A \supset B\ \mathit{true}}$$

The number of clauses in the def. of $\bot$ is 0;

$$\frac{x \in D \vdash P(x)\ \mathit{true}}{(\forall x \in D)P(x)\ \mathit{true}} \qquad \frac{d \in D \quad P(d)\ \mathit{true}}{(\exists x \in D)P(x)\ \mathit{true}}$$

where D is a domain. The Frege/Wittgenstein concept of proposition clearly succeeds in avoiding the difficulty of the Boolean concept of proposition, but it runs into another difficulty: when one tries to define when two propositions A and B are identical, then the only and obvious answer is: A and B are identical iff the truth conditions of A are the same as the truth conditions of B, or expressed alternatively by a rule

$$\frac{A\ \mathit{true}}{B\ \mathit{true}} \quad \text{if } A = B : \mathit{prop}$$

This of course means that identity between two propositions is material equivalence between them. And this again means that e.g. Euclid's theorem: $(\forall x : N)(\exists y : N)$

$(x \leq y \,\&\, Prim(y))$ and a trivial, say logically valid proposition $\top$ are both identical propositions, i.e.

$$(\forall x : N)(\exists y : N)(x \leq y \,\&\, Prim(y)) = \top : prop$$

This, i.e. essentially the identification of all the true and the identification of all the false propositions, is of course counterintuitive and absurd. Whereas the Boolean concept of proposition runs into trouble when trying to semantically explain the judgement 'A : *prop*', the Frege/Wittgenstein concept of proposition runs into trouble when trying to explain semantically the judgement '$A = B$: *prop*'. A third and viable concept of proposition will be expounded on below.

vii.2) What does a judgement of the form '$a : A$' mean? or What does it mean to be a proof-object of some proposition?
To explain what a judgement of the form '$a : A$' means one has to know that (i) a has the form of an object formed by one of the proof-object construction rules of A (a has the form of a canonical proof object) or (ii) a can be evaluated to one of those forms (a is a non-canonical proof-object).
(Concerning the concept of evaluation and canonical proof-object, cf. Sect. viii.)

vii.3) What does a judgement of the form '$a = b : A$' mean? or What does it mean for two proof-objects of some proposition to be identical?
To explain what a judgement of the form '$a = b : A$' means one has to know that (i) a and b have the form of two objects formed by one of the identity rules associated with the proof-object construction rules of A (a and b are identical canonical proof-objects of A) or (ii) a and b can be evaluated to one of those forms (a or b are identical non-canonical proof-objects of A). (Concerning the concepts of evaluation and canonical proof-object, cf. Sect. viii.)

C, n & o

(1) *A third concept of proposition: the intuitionistic concept of proposition*
In intuitionism, truth of a proposition is defined in terms of more basic concepts, namely the concept of proof and the concept of existence. According to the intuitionists a judgement of the form A *true* where A : *prop* is an incomplete judgement saying that there exists a proof of the proposition A, i.e. one has A *true* = there exists a proof of A. A judgement of the form A *true* is thus an incomplete or abbreviated way of saying: a is a proof of A, or symbolically, $a : A$, whereby the abbreviation consists in skipping the proof of the proposition and merely retaining the information that there is a proof of it. It should be noted that the concept of existence mentioned here is not the kind of existence expressed by the existential quantifier. If one tried to write 'There exists a proof of A' with an existential quantifier as $\exists x : A$, the result would not be a proposition, and if one wrote $(\exists x : A)\top$, the result would be a proposition, but one would get into an infinite regress (because one could then judge $(\exists x : A)\top$ *true* which is an incomplete way of saying $(\exists x : (\exists x : A)\top)\top$, etc.). (More on the intuitionistic concepts of truth and existence, cf. Subsubsect. ii.3.4, Chapter 2.) Since one may want to prove judgements of the form 'a is a proof of A', i.e. $a : A$ where A : *prop*, the word 'proof' has become ambiguous. It can mean 1) the proof of a judgement (i.e. the proof of a mathematical theorem) or 2) the proof of a proposition (i.e. a mathematical

object). In the first case one could call a proof a demonstration and in the second case a verification.

a is a proof of A } proof = demonstration of a judgement

↗ proof = verification of a proposition

judgement demonstrated (= theorem)

Moreover, the term 'proof' can refer to the act of proving a judgement or to the object/product of such an act. One thus gets at least 4 concepts of proof

		Proof	
		Act	Object
Proof of a proposition	non-canonical	—	(d)
	canonical	—	(c)
Proof of a judgement		(a)	(b)

meaning (a): when one proves a mathematical theorem one carries out a certain (mental) act of proof (an act of judgement) which is an act of mathematical construction

meaning (b): the result or product of this act of proving a judgement/theorem is the judgement made or the mathematical theorem (saying that a proposition A is true or that c is a proof-object of A)

meaning (c): the act of proof (mentioned in (a)) yields a construction object d that satisfies certain conditions depending on the proposition A which serves to give content to the theorem in question. These conditions are given by the meaning explanation of A.

meaning (d): proof as a proof trace or proof 'blue print'. Proofs in this sense can be considered as indications or means to find corresponding proofs in the sense of (c). Thus, they can be called non-canonical (indirect or defined) proof-objects, whereas their corresponding proof-objects in the sense of (c) are called canonical (direct or prim.). This is Brouwer's and Dummett's terminology. The non-canonical proof-objects are like shorthand forms of canonical proof-objects; they can be transformed/evaluated into canonical proof objects. (This will be made more precise in Sect. x.) The proofs in mathematical journals and textbooks, etc. are in a certain way such proof traces.

(One could even introduce another meaning of 'proof', namely proof as act/process of proving (in the sense of (a)) regarded as an object, but the above distinctions suffice for the present discussion.)

Proofs in the sense (a) are acts or processes of getting to know the resp. judgement or theorem. A proof in this sense is acquisition of knowledge, and this judgement proved/demonstrated is the piece of knowledge acquired. Now, one cannot know something except as the result of having gotten to know it. Hence if proof in the sense of (a) is the very process or act of getting to know a judgement, one cannot know a judgement except as the result of having proved or demonstrated it. Knowledge is thus by its very nature demonstrative knowledge (provided whatever one knows has the form of a judgement). (More on this topic, cf. Subsect. ii.5, Chapter 2)

The idea of introducing proof-objects as mathematical objects seems to be completely new with intuitionism. If proofs of a proposition are to be treated as mathematical objects or entities, they must have a type (according to the slogan: No entity without type (cf. C, n & o (2) to i.1)). Now, the proposition itself will serve as the type of its proofs. Hence one has $\frac{A : \mathit{prop}}{A : \mathit{type}}$. If $A : \mathit{prop}$, then an object of type A will be called a proof (object) of A. If the proofs of a proposition are to form a type, one has to lay down the identity criterion for that type, i.e. the meaning explanation of a proposition need not only include its proof conditions, but also the conditions under which two proofs of the proposition are identical. These conditions always stipulate that two proofs of a proposition are identical if they have identical form and identical parts. If identity is always to be preserved under substitution, one must in particular be allowed to infer by the rule $\frac{A = B : \mathit{prop}}{A = B : \mathit{type}}$. This provides the key for the identity criterion for the type of propositions: two propositions are defined to be identical, i.e. $A = B : \mathit{prop}$, if the types they give rise to are identical, i.e. $A = B : \mathit{type}$. This is the weakest identity criterion for the type *prop* validating the rule $\frac{A = B : \mathit{prop}}{A = B : \mathit{type}}$.

According to the intuitionistic concept of proposition, a proposition is not defined by its truth conditions anymore, but by its proof conditions. This means that to know the truth or falsehood of a proposition A one has to know under what conditions A is proved or disproved resp.

The following table shows how the standard truth conditions for the logical operations change if they are taken as proof conditions (under the intuitionistic interpretation $A\ \mathit{true}$ = there exists a proof (object) of A, say $a : A$ (where $A : \mathit{prop}$)).

Log. op.	Truth conditions	Proof conditions
$\bot$	———	———
&	$\frac{A\ \mathit{true} \quad B\ \mathit{true}}{A \,\&\, B\ \mathit{true}}$	$\frac{a : A \quad b : B}{(a, b) : A \,\&\, B}$ or $\mathit{pair}(a, b) : A \& B$ $\frac{a = c : A \quad b = d : B}{(a, b) = (c, d) : A \,\&\, B}$

$\vee$	$\dfrac{A\ \mathit{true}}{A \vee B\ \mathit{true}} \quad \dfrac{B\ \mathit{true}}{A \vee B\ \mathit{true}}$	$\dfrac{a : A}{i(a) : A \vee B} \quad \dfrac{b : B}{j(b) : A \vee B}$
		$\dfrac{a = c : A}{i(a) = i(c) : A \vee B} \quad \dfrac{b = d : B}{j(b) = j(d) : A \vee B}$
$\supset$	$\dfrac{A\ \mathit{true} \vdash B\ \mathit{true}}{A \supset B\ \mathit{true}}$	$\dfrac{x : A \vdash b : B}{\lambda((x)b) : A \supset B}$
		$\dfrac{x : A \vdash b = d : B}{\lambda((x)b) = \lambda((x)d) : A \supset B}$
$\forall$	$\dfrac{x : A \vdash B(x)\ \mathit{true}}{(\forall x : A)B(x)\ \mathit{true}}$	$\dfrac{x : A \vdash b : B(x)}{\lambda((x)b) : (\forall x : A)B(x)}$
		$\dfrac{x : A \vdash b = d : B(x)}{\lambda((x)b) = \lambda((x)d) : (\forall x : A)B(x)}$
$\exists$	$\dfrac{a : A \quad B(a)\ \mathit{true}}{(\exists x : A)B(x)\ \mathit{true}}$	$\dfrac{a : A \quad b : B(a)}{(a, b) : (\exists x : A)B(x)}$
		$\dfrac{a = c : A \quad b = d : B(a)}{(a, b) = (c, d) : (\exists x : A)B(x)}$

Note: the column of the standard proof conditions for the logical operations can also be conceived as filling in the proof-objects into the rules for standard truth conditions.

The intuitionistic concept of proposition succeeds in avoiding the difficulty of the Frege/Wittgenstein concept of proposition by proposing a reasonable criterion of identity for the type *prop* which is obviously stricter than material equivalence.[65]

(2) Martin-Löf agrees with Quine's criticism of propositions according to which no propositions should be assumed (to exist) unless there is an identity criterion for them. He, however, disagrees with Quine on the conclusion to be drawn from this criticism. Quine concludes that one should dispense with propositions altogether.[66] Martin-Löf concludes that one should provide propositions with an identity such as to turn them into proper entities or objects. And, as was shown above, he indeed does provide an identity criterion for propositions (cf. also Subsubsect. i.4.2, Chapter 2).

vii.4) What does a judgement of the form '$A = B : \mathit{prop}$' mean? or What does it mean for two propositions to be identical?

A judgement of the form '$A = B : prop$' means that (i) if a is a proof-object of proposition A, a is a proof-object of proposition B and vice versa, and (ii) if a and b are identical proof-objects of proposition A, a and b are identical proof-objects of proposition B and vice versa.

This real def. of the identity of two propositions can be expressed by the following two rules:

$$\frac{a : A}{\overline{a : B}} \qquad \frac{a = b : A}{\overline{a = b : B}}$$

After it has been explained what the type of propositions is , one can explain what it means for a proposition to be a type. Applying the real def. of 'dependent type' to the judgement $X : prop \vdash X : type$, one obtains

vii.5) What does a judgement of the form '$X : prop \vdash X : type$' mean? or What does it mean for a proposition to be a type?
Presuppositions in the order of explanation: Let the judgements $A : prop$ and $A = B : prop$ be known.
To make a judgement of the form '$X : prop \vdash X : type$' one has to know that (i) A is a type provided A is a proposition, and (ii) A and B are identical types provided A and B are identical propositions.

Alternatively, '$X : prop \vdash X : type$' is really defined by the following two rules

$$\frac{A : prop}{A : type} \qquad \frac{A = B : prop}{A = B : type}$$

Justification

$\dfrac{A : prop}{A : type}$ Straightforward.

$\dfrac{A = B : prop}{A = B : type}$ '$A = B : prop$' means that $\dfrac{a : A}{\overline{a : B}}$ and $\dfrac{a = b : A}{\overline{a = b : B}}$, and this is exactly the meaning of '$A = B : type$' if A and B are considered to be the types corresponding to the resp. propositions. □

The formation rule for dependent proposition types is the rule with 0 premisses

$$X : prop \vdash X : type \quad (\text{dep.prop.type})$$

C, n & o

This completes the real def. of the type of propositions. In order to define the type *prop* it was necessary to define what it means for two propositions (objects of the type *prop*) to be identical. And in order to define the meaning of '$A = B : prop$' for $A : prop$ and $B : prop$, it was necessary to introduce proofs of a proposition as mathematical objects. But then the question of the type of these new mathematical objects immediately arose, and it was answered by taking the proposition itself as the type of the proofs of the proposition and by taking the identity between the two propositions as the criterion of identity for the corresponding two types.

viii DIGRESSION. SYNTAX AND SEMANTICS/THE THEORY OF INDIRECT REFERENCE IN CONSTRUCTIVE TYPE THEORY

viii.1 *Syntax and Semantics*

If one has a name-like linguistic expression '*a*', then '*a*' is the expression of *a* and conversely, *a* is the meaning of '*a*'. It should be noted that in the whole of this digression '*a*' as well as *a* are here supposed to be linguistic entities. Moreover, what is assumed here is an object-oriented (gegenständlich gerichtet) or natural attitude (natürliche Einstellung)(Husserl), i.e. when speaking about *a* it is assumed that one does not speak about the expression '*a*', but about the meaning of the expression '*a*'.

Syntax deals with the process of formalization (divesting of sense) i.e. with finding or determining the formal structure of an expression; this process is directed from the meaning of '*a*' to the expression '*a*' itself or from *a* to '*a*'.

Syntax does not pose any particular problems in the case of formal languages, the expressions of which are already highly structured. It does pose enormous problems though in the case of natural languages.

Semantics deals with the process of contentualization (endowing with sense) i.e. with the interpretation of a formalized expression; this process reverses the direction of syntax, it is directed from the expression '*a*' to its meaning, i.e. from '*a*' to *a*.

All of this can be summarized as follows:

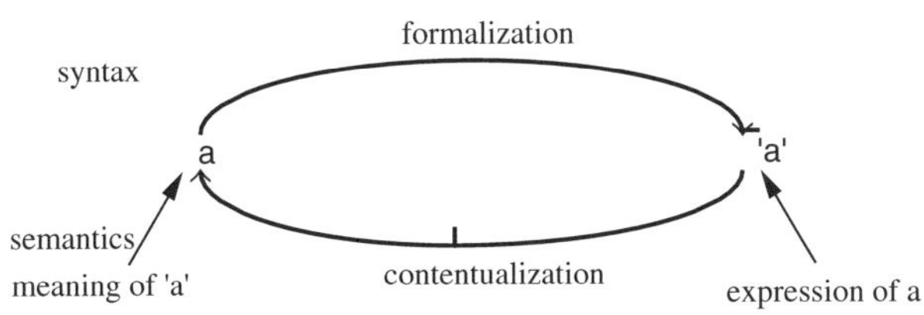

Martin-Löf's original problem/point of departure was: Mathematics or intuitionistic mathematics is not entirely clear; what is needed is a clarification (e.g. of what a natural number, what a fct., what a set, what a proof is, etc.). This clarification involves at least two tasks:

1. a formalization or structuring of the mathematical language in order to find all the atoms, rules of composition and rules of transformation (syntactical task).
2. an interpretation (contentualization) of all the atoms, rules of composition and transformation s.t. all that can be expressed in the new mathematical language has a meaning (semantical task).

Note: From the point of view to be developed here, there is a clear priority implicitly given, namely objects, etc. are prior to their expressions, or *a* is prior to '*a*'. This is in striking contrast to Frege and the whole analytic tradition which has moved the focus of attention from objects, etc. to their expressions. However, it is completely conform to mathematical practice where mathematicians in the natural attitude do not speak about numerals or expressions for fcts., expressions for vectors or for spaces, etc. but about numbers, fcts., vectors, etc. which are actually the meanings of certain mathematical expressions.

The above schema can be presented in a slightly more general form and the objectual and semantic categories are the same since the object is the meaning of its expression. It appears that the only way of getting to mathematical objects is by making sense of the mathematical expressions. Once one has grasped the meaning of a mathematical expression one knows the object which it stands for. Hence the only way into the mathematical world seems to be the one through making sense of the mathematical language – contrary to what Brouwer believed.

Note: Henceforth in this digression the term 'entity' will be used instead of 'object' if reference will be made indiscriminately to objects as well as to types. What has been said so far (in this section) holds for all entities and their expressions (and not only for objects).

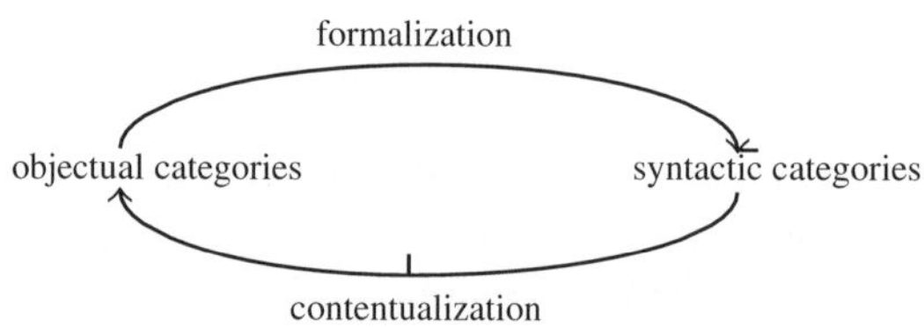

This view concerning objectual and semantic categories and what underlies it, deviates from the standard view inherent in the semantic triangle (from Aristotle to de Saussure and Ogden and Richards):

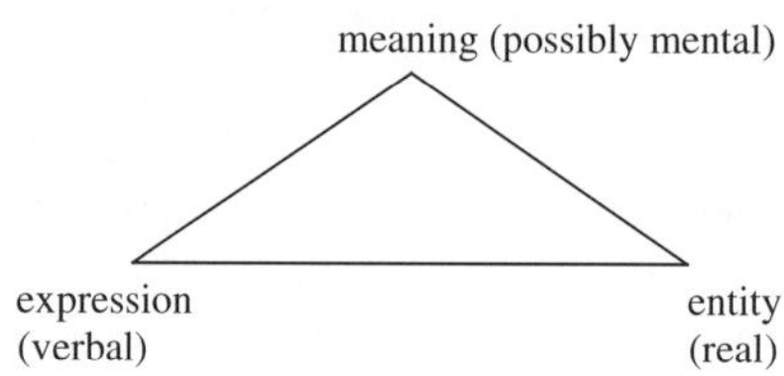

where meaning was often conceived of as something mental (psychical) such as a passion in the soul or a thought (Aristotle, scholastics, Ogden & Richards and others). But sometimes meaning was also conceived as something external, objective, not mental (Stoics, late scholastics, Bolzano, Frege).[67] If meanings are nothing mental or psychical and nothing physical either, and if they are not the expressions themselves, then a third realm has to be accepted, namely the realm of the ideal, objective and non-mental meanings. (e.g. Frege's 'Reich der Sinne').

According to Martin-Löf the realm of concepts or meanings (in the objective sense) and the realm of entities, i.e. the third and the second realm resp. are the same. From the standpoint of constructive type theory there is no fundamental difference between them. In other words, there is no categorical difference between what appears at the top and at the bottom right hand side of the semantic triangle. An example may illustrate this:

Example

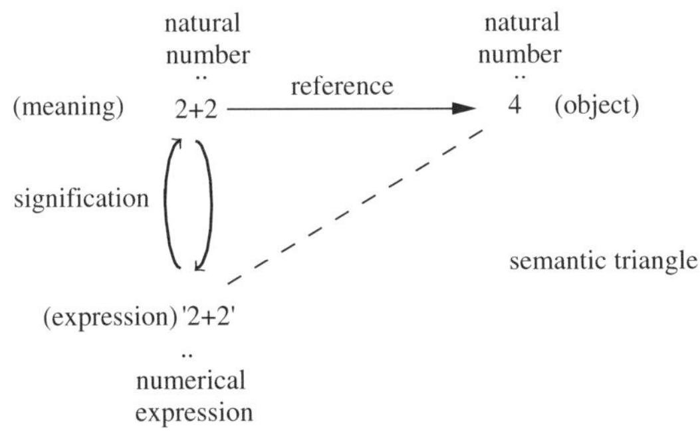

There is no need for a third realm since the inhabitants of the third realm already inhabit the second one or vice versa. The senses (in Frege's sense) or meanings and the entities are meanings of expressions and there are different ways of presenting the same thing. The meanings and entities are all human constructions.

An expression 'a' becomes (or is taken to be) an object or 'receives a meaning' a, when it is sorted into a type or into a semantical category. Hence an object is never an object tout court, it is always an object of a certain category (cf. doctrine of types, Sect. i).

In the natural attitude an act of turning an expression into an object is an act of meaning (an object by means of the expression) or an act of understanding (the expression as an object), in either case it is an act of endowing an expression with meaning. Metaphorically speaking, the meaning of an expression is the way the expression is used (this is of course reminiscent of Wittgenstein's dictum that meaning is use). And the type or semantic category (or semantic function or semantic role) indicates the way in which or the purpose for which the expression is used: object a is an α (or is of type α), object a is used for the purpose α, object a fulfills the function α, object a plays the role α, etc.[68] Three components are then required for the making of an object: a type or semantical category, an expression and an act of understanding it or meaning something with it.

Example completed and some terminology

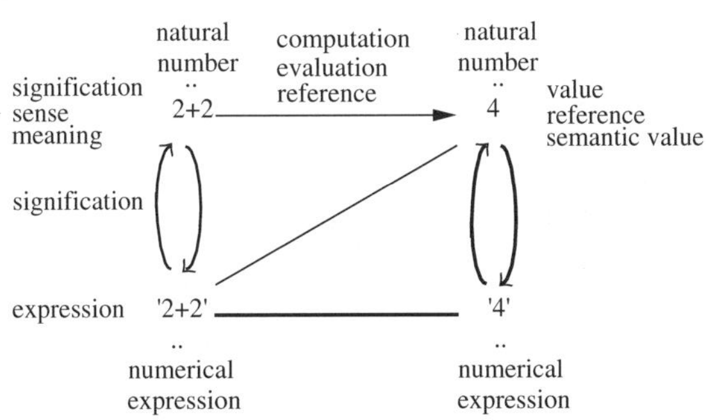

Note: An expression 'a' is also an object as much as the object a which it signifies or stands for, but they are objects of different categories: whereas 'a' is an object of a syntactical category, a is an object of a type or semantical category.

viii.2 *The Theory of Indirect Reference in Constructive Type Theory*

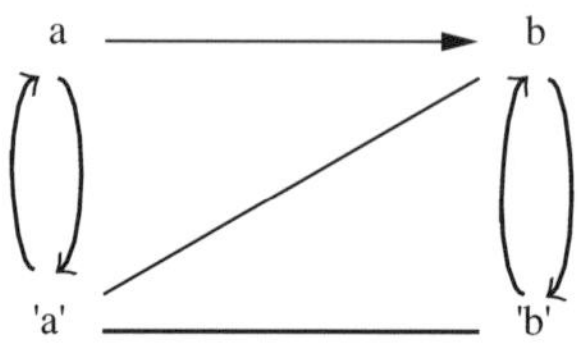

The diagonal line from 'a' to b in the figure above purports to show that the theory of meaning or signification presented here is not a theory of direct reference, but rather a theory of indirect reference (similar to Frege's theory of meaning). This means that an expression 'a' always refers to entity b via its meaning a and possibly via a whole chain of meanings $a,\ a_1,\ a_2, \ldots,\ a_n$.

Some terminology:

It is said	that	'a'	refers to b
			has value b
	that	'a'	signifies a
			stands for a
			means a
			expresses (ausdrücken) a (Frege)
	that	a	evaluates to b
			has value b
			refers to (bedeuten, bezeichnen) b (Frege)

A crucial role is played in this theory of indirect meaning by the relation of evaluation. The basic assumption is: Every entity is either prim. or (nominally) defined. Evaluation comes in, in a very natural way:

i. A prim. entity a evaluates in one step to itself.
ii. A defined entity a_1 evaluates in one step to its definiens a_2.

By this def. the evaluation of an entity always terminates in a prim. entity in a finite number of steps (if it terminates at all). A (nominally) defined entity cannot have a reference unless the process of evaluation, i.e. of successively replacing a definiendum (entity) by its corresponding definiens (entity) terminates with a prim. entity. That is why Church insisted that an expression cannot be meaningful (or have a reference) unless it possesses a normal (or prim.) form.

With each non-dependent entity is thus implicitly associated a definitional chain

$$\underbrace{a = a_1 = a_2 \cdots = a_n}_{\text{definitional chain}}$$

where $n = 0$ if a is prim., $n > 0$ if a is defined. a_n is the ultimate definiens (Curry) or the (semantic) value of a.

Let 'a has value b' be symbolically expressed by $a \Longrightarrow b$, and let 'definiendum a has definiens b' be expressed by $a = b$. The evaluation relation can be inductively defined as follows:

(i) $a \Longrightarrow a$ if a is prim.

(ii) $$\frac{a = b \quad b \Longrightarrow c}{a \Longrightarrow c} \quad \text{o.w.}$$

Evaluation is thus the process leading from an entity to its value (semantic value).

An object is always of a certain type and to each type belongs an identity criterion, i.e. an identity relation between its objects. If b is the value of a and $a : \alpha$ and $b : \alpha$, then a and b are identical objects of type α but differently expressed. If b is the value of a and a : *type* and b : *type*, then a and b are identical types but differently expressed (for notational convenience or abbreviatory purposes). This identity is called definitional or semantical identity (in version (8), and also intensional identity in version (7)). But there is a second identity relation on the semantical level whose def. though presupposes the concept of syntactical identity. Two expressions are said to be syntactically identical if they have identical form and syntactically identical parts. The second identity relation between objects of some type or between types is induced by this syntactical identity: entity a is said to be identical with entity b, if expression 'a' is syntactically identical with expression 'b'. But it is then possible to define a second identity relation on the level of expressions (syntactical level): two expressions 'a' and 'b' are identical if their meanings a and b are definitionally identical.

If definitional or semantical identity and the semantically induced identity are both denoted by $=$ and the syntactical as well as syntactically induced identity are both denoted by $\equiv$, one gets the following table

	$\equiv$	$=$
entities	syntactically induced identity	semantical identity
expressions	syntactical identity	semantically induced identity

and the following scheme

$$\begin{array}{lllll} \text{sense} & a \equiv a_0 = a_1 =^{\text{evaluation}}_{\cdots} = a_n \equiv & \overset{\text{type}}{\ddot{b}} : \alpha & & \text{reference} \\ \text{signification} \ \uparrow\downarrow & & & & \\ \text{expression} & `a' \equiv `a_0' = `a_1' = \cdots = `a_n' \equiv & \overset{\text{type expr.}}{\ddot{`b'}} & :`\alpha' & \text{syntactic value} \end{array}$$

where on the level of entities a,

$$\begin{array}{lll} & \text{sense} = \text{reference} & \text{(in the sense of value)} \\ \text{and} & \text{sense} \not\equiv \text{reference} & \text{unless } a \text{ is prim.} \end{array}$$

Reference in the sense of evaluation comes out as computation and this is what makes constructive type theory interesting for computer science. Now, in a computation there is something which changes all the time, and this is here the expression of a, i.e. 'a' $\equiv$ 'a_0' $\not\equiv$ 'a_1' $\not\equiv \cdots \not\equiv$ 'a_n' $\equiv$ 'b' ('b' is the result of the computation) but there is something else which is invariant under the computation, and that is here the meaning of 'a' or the entity a which 'a' stands for, i.e. $a \equiv a_0 = a_1 = \cdots = a_n \equiv b$. Thus, computation certainly has a semantic significance in constructive type theory.

The semantically induced identity is as a matter of fact identity of meaning or the relation of synonymy: two expressions 'a' and 'b' are synonymous if their resp. meanings are identical, i.e. $a = b$. This conception of synonymy is not in agreement with Frege's conception: 'a' and 'b' are synonymous if a and b are identically expressed, i.e. if 'a' $\equiv$ 'b'. Frege's relation of synonymy interpreted this way thus coincides with syntactical identity. Clearly then, the question whether in a nominal def. definiendum and definiens are synonymous is answered affirmatively according to the theory of indirect reference expounded on above, and it is answered negatively according to Frege's theory of meaning.

The sense–reference distinction applied

So far several ground types have been introduced, namely the type of sets, the types of elements of sets, the type of propositions as well as the types of proof-objects of propositions. Accordingly, one can distinguish between defined sets and prim. sets as well as between defined propositions and prim. propositions, etc. (examples of defined propositions are $\sim A$ or $A \supset\subset B$, for $A : prop$, $B : prop$, cf. Sect. v). Furthermore, one can distinguish between defined elements and prim. elements of a given set, or using the terminology of Sect. vi, between non-canonical and canonical elements of a set, and one can distinguish between defined proof-objects and prim. proof-objects, or in the terminology of Sect. vii., between non-canonical and canonical proof-objects of a given proposition. Another disagreement with Frege is that the references of the propositions are not truth values (as in classical semantics in general), but prim. propositions. Moreover, one can ask what the reference of dependent objects or fct. objects is or what the sense/meaning of dependent object expressions or fct. object expressions is. The former ones are open expressions and the latter ones closed higher level expressions, and together they are called functional expressions as opposed to the non-functional expressions, i.e. expressions of (simple) objects. If one asks what value (or reference) a dependent object (a fct. in the oldfashioned sense) has, the natural response is to ask: For what argument do you want to have the value of the dependent object? Similarly, if one asks what value (or reference) a fct. object (a fct. in the modern sense) has, the natural thing to do is to ask back: For what argument do you want to have the value of the fct. object? This shows that dependent objects or fct. objects have no value or reference in themselves; they only have a value after they have been supplied with an argument. In other terms, functional expressions have no reference (unless or until they are supplied with an argument). This is again in disagreement with Frege who believed functional expressions to have a reference.

The reason for Frege's belief is the fact that one and the same fct. can be expressed in many different ways, and Frege accounts for this fact by saying that a functional expression may have different senses but it has the same reference. The theory of indirect reference which is being explained here accounts for the same fact simply by saying that one and the same fct. can be expressed by several different functional expressions; i.e. f and g are equal as fcts. although 'f'≢'g', and reference has no place in this account.

This digression can very roughly be summarized as follows:
There is a double distinction which comes into play:

– the syntax–semantics distinction between expressions and their meanings/objects

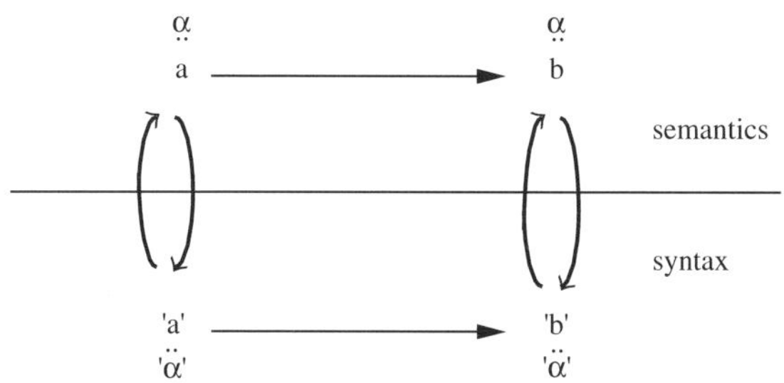

– the distinction between prim. and defined entities (expressions)

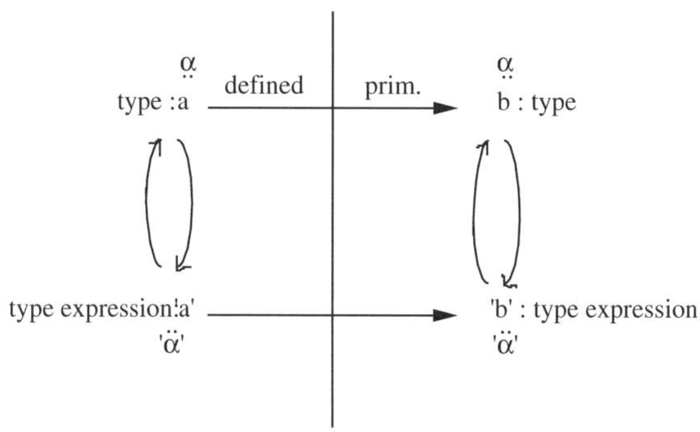

where

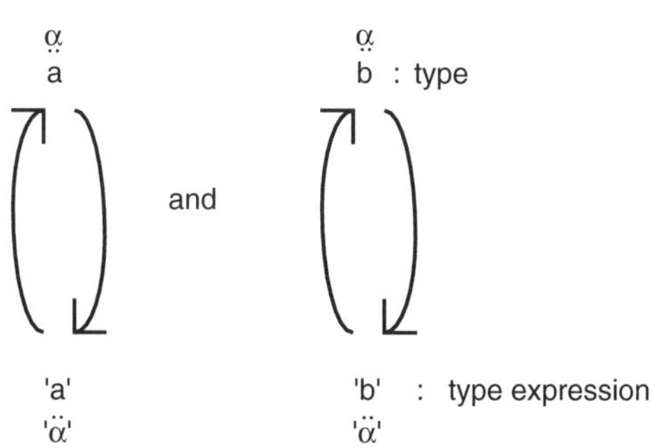

collapse or coincide if a is prim.

Terminology: Instead of 'prim./defined' the terms 'direct/indirect' or 'canonical/non-canonical' (Dummett w.r.t. proof-objects[69]) or finally 'normal/non-normal' (combinatory logic, proof theory) have been used in this context.

ix CONSTRUCTIVE SET THEORY: GROUND SETS, DEPENDENT SETS AND SETS OF DIFFERENT SIZES

In Sect. vi. it was only really defined what it means to be a set, but nothing has been said about what sets there are. In the following a denumerable infinity of ground sets will be introduced, then some set operations will be indicated with which a finite set structure can be generated, and finally it will be shown how to pass from a finite to a transfinite set structure.

With each of the following sets or set operations 4 kinds of axioms or rules will be given: (i) the formation axioms or rules which say under what conditions a set and under what conditions two identical sets can be formed or constructed; (ii) the introduction axioms or rules which give the meaning of the set by indicating how the canonical and how the identical canonical elements of the set are formed; (iii) the elimination axioms and rules which introduce non-canonical elements of the set as well as a method or fct., called a selector. They are a kind of structural induction rules and they tell how fcts. are defined on that set. The principle of structural induction says that to prove that a property $B(a)$ holds for an arbitrary element a of set A, prove that the property holds for each of the canonical elements of A. (*Note*: this only holds for the main (standard) selectors and will not apply to what will in Sect. x. be called a Gentzen or a Schroeder–Heister selector.); (iv) the equality rules which determine the meaning of these methods or fcts. (i.e. selectors) associated with the set by indicating how the selectors operate on canonical elements of the set.

C, n & o

(1) Note that the introduction rules state that all elements of a certain form are canonical elements of the corresponding set (and that all identical elements of a certain form are identical canonical elements of the set), but they leave open the possibility of there being elements of other forms which are also canonical elements of that set. The elimination rules, however, can be seen to rule out any elements (or identical elements) of forms other than those indicated by the introduction rules as canonical elements (or identical canonical elements) of the resp. set. The elimination rules can thus be regarded as 'closure rules' for the corresponding introduction rules.

(2) From the programming point of view the formation rules express the syntax of the set or set operation; the introduction and elimination rules assert the existence of certain (canonical and non-canonical) objects of certain types, thus specifying the syntax of the programming language; and the computation or equality rules explain how to evaluate programs written in that programming language.[70]

To start with, an infinity of ground sets will be introduced, namely the finite sets N_n, $n \geq 0$, with exactly n elements as well as the set N of natural numbers. Next will be explained in the following order: the disjoint union $+$ of two such sets, the cartesian product Π of a family of such sets, the fct. set formation $\rightarrow$ of two sets, the disjoint union Σ of a family of such sets, the cartesian product $\times$ of two sets. Finally, the intensional identity set formation I of a set and the set formation U of the small sets will be introduced which alone allow for the formation of dependent sets or families of sets.

The explanation of all these sets will not only comprise the four kinds of axioms as well as the four kinds of corresponding rules mentioned above, but also a justification of the various elimination- and equality rules and a justification of certain formation rules where this is indicated. Finally, each introduction of a new set forming operator will be supplied with a smaller or greater number of comments, notes and observations.

ix.1.1) What does a judgement of the form 'N_n : *set*' mean? or What does it mean to be a finite set?
To explain what it means to be a finite set N_n one has to know (i) what a canonical element of N_n is, and (ii) when two canonical elements of N_n are identical.

The formation rules for the ground set N_n is the axiom or rule with 0 premisses

$$N_n : set \quad n = 0, 1, \ldots \quad (N_n\text{F 1})$$

As usual, if the set N_n is to be a respectable entity at all, it has to have an identity. This N_n set identity is given by the judgement

$$N_n = N_n : set \quad (N_n\text{F 2})$$

ix.1.2) What does a judgement of the form '$c : N_n$' mean? or What does it mean to be an element of some finite set $N_n, n \geq 1$?
To explain what a judgement of the form '$c : N_n$' means one has to know that (i) c has the form of an object formed by one of the element-formation rules of N_n (i.e. c has the form of a canonical element of N_n), or (ii) c can be evaluated to one of those forms (c is a non-canonical element).

The element-formation rules, i.e. introduction rules/axioms of N_n are

$$N_n I0 : N_n \quad \ldots \quad N_n In - 1 : N_n \quad (N_n\text{I 1}) - (N_n\text{I n})$$

For $n = 0$ there is no element in N_0 and hence no canonical element c. (Cf. Note (2) in C, n & o to ix.1.3))

ix.1.3) What does a judgement of the form '$c = d : N_n$' mean? or What does it mean for two elements c and d of a finite set $N_n, n \geq 1$, to be identical?
To explain what a judgement of the form '$c = d : N_n$' means one has to know that (i) c and d have the form of two objects formed by one of the identity rules associated with the element-formation rules of N_n (c and d are identical canonical elements of N_n) or (ii) c and d can be evaluated to one of those forms (i.e. c or d are identical non-canonical elements of N_n).

The identity rules associated with the element-formation rules i.e. the introduction axioms/rules of N_n are

$$N_nI0 = N_nI0 : N_n \quad \ldots \quad N_nIn-1 = N_nIn-1 : N_n \quad (N_n\text{I n+1}) - (N_n\text{I n+n})$$

The following is a type

$$(Z : (N_n)set)(Z(N_nI0))\ldots(Z(N_nIn-1))(x : N_n)Z(x) : type$$

as can be seen from the derivation

$$\dfrac{\dfrac{\begin{array}{c}N_nI0 : N_n\\(Z : (N_n)set)\end{array}}{Z(N_nI0) : set}}{Z(N_nI0) : type} \qquad \dfrac{\dfrac{\begin{array}{c}N_nIn-1 : N_n\\(Z : (N_n)set)\end{array}}{Z(N_nIn-1) : set}}{Z(N_nIn-1) : type} \qquad \dfrac{\dfrac{N_n : set}{N_n : type} \quad \dfrac{\dfrac{\begin{array}{c}(x : N_n)\\(Z : (N_n)set)\end{array}}{Z(x) : set}}{Z(x) : type}}{(x : N_n)Z(x) : type}$$

$$\dfrac{\dfrac{\dfrac{N_n : set}{N_n : type} \quad set : type}{(N_n)set : type} \quad \dfrac{\begin{array}{c}\vert\\Z(N_nI0) : type\end{array} \quad \ldots \quad \dfrac{\begin{array}{c}\vert\\Z(N_nIn-1) : type\end{array} \quad \begin{array}{c}\vert\\(x : N_n)Z(x) : type\end{array}}{(Z(N_nIn-1))(x : N_n)Z(x) : type}}{(Z(N_nI0))\ldots(Z(N_nIn-1))(x : N_n)Z(x) : type}}{(Z : (N_n)set)(Z(N_nI0))\ldots(Z(N_nIn-1))(x : N_n)Z(x) : type}$$

Now, the selector or eliminatory operator N_nE of the set N_n is an object of this type, i.e.

$$N_nE : (Z : (N_n)set)(Z(N_nI0))\ldots(Z(N_nIn-1))(x : N_n)Z(x)$$

This N_n-elimination axiom can alternatively be expressed by the following two rules

$$\dfrac{C : (N_n)set \quad c_0 : C(N_nI0) \quad \ldots \quad c_{n-1} : C(N_nIn-1) \quad c : N_n}{N_nE(C, c_0, \ldots, c_{n-1}, c) : C(c)} \quad (N_n\text{E 1})$$

$$\dfrac{\begin{array}{l}C : (N_n)set\\c = c' : N_n \quad c_0 = c_0' : C(N_nI0) \quad \ldots \quad c_{n-1} = c_{n-1}' : C(N_nIn-1)\end{array}}{N_nE(C, c_0, \ldots, c_{n-1}, c) = N_nE(C, c_0', \ldots, c_{n-1}', c') : C(c)} \quad (N_n\text{E 2})$$

The meaning of the selector N_nE is explained as follows

1. First evaluate c to canonical form.
2. If $c \Longrightarrow N_nIk \quad (0 \leq k \leq n-1)$ then $N_nE(C, c_0, \ldots, c_{n-1}, c) \Longrightarrow c_k$ and $N_nE(C, c_0, \ldots, c_{n-1}, c) = c_k$. O.w. evaluation of the selector is aborted.

Justification

of (N_nE 1) Assume the premisses of the rule. One has then to show that the value of $N_nE(C, c_0, \ldots, c_{n-1}, c)$ is a canonical element of $C(c)$.

From the last premiss one knows that $c \Longrightarrow N_n Ik$ for some $0 \leq k \leq n-1$. According to the meaning of $N_n E$, the value $N_n E(C, c_0, \ldots, c_{n-1}, c)$ is then identical with the value of c_k. From one of the premisses one knows that the value of c_k is a canonical element of $C(N_n Ik)$. Thus, the value of $N_n E(C, c_0, \ldots, c_{n-1}, c)$ is a canonical element of $C(N_n Ik)$. But since one has $c = N_n Ik : N_n$, it follows from the first premiss that the value of $N_n E(C, c_0, \ldots, c_{n-1}, c)$ is a canonical element of $C(c)$.

of (N_nE 2) Analogously to (N_nE 1). □

The N_n-equality axioms are

$$N_n E(C, c_0, \ldots, c_{n-1}, N_n I0) = c_0 : C(N_n I0)$$
$$\vdots$$
$$N_n E(C, c_0, \ldots, c_{n-1}, N_n In-1) = c_{n-1} : C(N_n In-1)$$

where $C : (N_n)set$ and $c_k : C(N_n Ik)$ for $0 \leq k \leq n-1$.

They can be expressed by the following n rules

$$\frac{C : (N_n)set \quad c_0 : C(N_n I0) \quad \ldots \quad c_{n-1} : C(N_n In-1)}{N_n E(C, c_0, \ldots, c_{n-1}, N_n I0) = c_0 : C(N_n I0)} \; (N_n\text{-equ 1})$$
$$\vdots$$
$$\frac{C : (N_n)set \quad c_0 : C(N_n I0) \quad \ldots \quad c_{n-1} : C(N_n In-1)}{N_n E(C, c_0, \ldots, c_{n-1}, N_n In-1) = c_{n-1} : C(N_n In-1)} \; (N_n\text{-equ n})$$

Justification

of (N_n-equ 1) – (N_n-equ n) : straightforward by the meaning explanation of $N_n E$. □

C, n & o

(1) $N_n E$ is recursion over the finite set N_n, it is some kind of a definition by cases – which explains why Martin-Löf uses in version (6) 'R_n' instead of '$N_n E$' and why Nordström et al. use '*case*'.

(2) The empty set

The formation rules for the empty set N_0 or { } or Ø are the axioms or rules with 0 premisses

$$N_0 : set \quad \text{and} \quad N_0 = N_0 : set$$

There are no introduction axioms or rules (i.e. no element-formation rules and associated identity rules for N_0) since N_0 has no elements.

The elimination axiom for the empty set N_0 is

$$N_0 E : (Z : (N_0)set)(x : N_0)Z(x)$$

and since there are no introduction axioms or rules for N_0, there also are no equality-axioms for N_0.

(3) The one-element set

The formation rules for the one-element set N_1 are the axioms or rules with 0 premisses

$$N_1 : set \quad \text{and} \quad N_1 = N_1 : set$$

The introduction axioms or rules for N_1 are

$$N_1 I : N_1 \quad \text{and} \quad N_1 I = N_1 I : N_1$$

The elimination axiom for the one-element set N_1 is

$$N_1 E : (Z : (N_1)set)(Z(N_1 I))(x : N_1)Z(x)$$

The equality axiom for N_1 is

$$N_1 E(C, c_0, N_1 I) = c_0 : C(N_1 I)$$

where $C : (N_1)set$ and $c_0 : C(N_1 I)$.

(4) The set *Bool*

The set *Bool* is the set of boolean values, i.e. $Bool = \{true, false\}$. One can explicitly define

$$\frac{N_2 : set}{\begin{cases} Bool : set \\ Bool = N_2 : set \end{cases}} \qquad \frac{N_2 I0 : Bool}{\begin{cases} true : Bool \\ true = N_2 I0 : Bool \end{cases}} \qquad \frac{N_2 I1 : Bool}{\begin{cases} false : Bool \\ false = N_2 I1 : Bool \end{cases}}$$

The formation rules for the set *Bool* or the two-element set N_2 are obviously the axioms or rules with 0 premisses

$$Bool : set \quad \text{and} \quad Bool = Bool : set$$

The introduction axioms or rules for *Bool* are

$$true : Bool \quad false : Bool \quad \text{and}$$

$$true = true : Bool \quad false = false : Bool$$

The elimination axiom for the set *Bool* is

$$N_2 E : (Z : (Bool)set)(Z(true))(Z(false))(x : Bool)Z(x)$$

The equality axioms for *Bool* are

$$N_2 E(C, c_0, c_1, true) = c_0 : C(true)$$
$$N_2 E(C, c_0, c_1, false) = c_1 : C(false)$$

where $C : (Bool)set,\ c_0 : C(true)$ and $c_1 : C(false)$.

Note that *true* as one of the two elements of *Bool* has nothing to do with the truth of a proposition. *true* is merely a convenient name in the programming practice for one of the two canonical elements of a certain set, namely the two-element set *Bool*, whereas the judgement C *true* for $C : prop$ intuitionistically means that there exists a proof of the proposition C. (cf. comment C, n & o to vii.4) above)

(5) Alternative formulation of the finite or enumeration sets by means of a set forming operator S [71]

ix.1.4) What does a judgement of the form '$S(A)$: *set*' mean, where A : *set*? What does it mean to be a finite set?
To explain what it means to be a finite set $S(A)$, A : *set*, one has to know (i) what a canonical element of $S(A)$ is, and (ii) when two canonical elements of $S(A)$ are identical.

The formation axiom for the composite set $S(A)$ is

$$S : (X : set)set$$

Equivalently, the formation axiom can be given by the two rules

$$\frac{A : set}{S(A) : set}\ (\text{SF 1}) \qquad \frac{A = C : set}{S(A) = S(C) : set}\ (\text{SF 2})$$

If the set $S(A)$ is to be a respectable entity, it has to have an identity. This $S(A)$ set identity is given by the judgement $S(A) = S(A)$: *set* provided A : *set* and is merely an instance of the set type identity $(set)set = (set)set : type$.

ix.1.5) What does a judgement of the form '$c : S(A)$', A : *set*, mean? or What does it mean to be an element of the finite set $S(A)$?
To explain what a judgement of the form '$c : S(A)$' means one has to know that (i) c has the form of an object formed by one of the element-formation rules of $S(A)$ (i.e. c has the form of a canonical element of $S(A)$), or (ii) c can be evaluated to one of those forms (c is a non-canonical element of $S(A)$).

The element-formation or introduction axioms of $S(A)$ are

$$SI1 : (X : set)S(X) \quad \text{and} \quad SI2 : (X : set)(X)S(X)$$

which can be expressed alternatively by the introduction rules

$$\frac{A : set}{SI1(A) : S(A)}\ (\text{SI 1}) \qquad \frac{A : set \quad a : A}{SI2(A, a) : S(A)}\ (\text{SI 2})$$

ix.1.6) What does a judgement of the form '$c = d : S(A)$', A : *set*, mean? or What does it mean for two elements c and d of $S(A)$ to be identical?
To explain what a judgement of the form '$c = d : S(A)$', A : *set*, means one has to know that (i) c and d have the form of two objects formed by one of the identity rules associated with the element-formation rules of $S(A)$ (c and d are identical canonical elements of $S(A)$) or (ii) c or d can be evaluated to one of those forms (i.e. c or d are identical non-canonical elements of $S(A)$).

The identity rules associated with the element-formation rules of $S(A)$ i.e. the introduction rules of $S(A)$ are

$$\frac{A = B : set}{SI1(A) = SI1(B) : S(A)}\ (\text{SI 3}) \qquad \frac{A = B : set \quad a = b : A}{SI2(A, a) = SI2(B, b) : S(A)}\ (\text{SI 4})$$

The following is a type

$$(X : set)(Z : (S(X))set)(Z(SI1(X)))((x : X)Z(SI2(X, x)))(z : S(X))Z(z) : type$$

as can be seen from the derivation

$$\frac{\dfrac{(X : set)}{S(X) : set}}{S(X) : type} \qquad \frac{\dfrac{\dfrac{\begin{array}{l}(X : set)\\ SI1 : (X : set)S(X)\end{array}}{SI1(X) : S(X)} \quad (Z : (S(X))set)}{Z(SI1(X)) : set}}{Z(SI1(X)) : type}$$

$$\frac{\dfrac{\dfrac{\begin{array}{l}(X : set)\\ (x : X)\\ SI2 : (X : set)(X)S(X)\end{array}}{SI2(X, x) : S(X)} \quad (Z : (S(X))set)}{Z(SI2(X, x)) : set}}{Z(SI2(X, x)) : type} \qquad \frac{\dfrac{\dfrac{(X : set)}{S(X) : set}}{S(X) : type} \quad \dfrac{\dfrac{\begin{array}{l}(Z : (S(X))set)\\ (z : S(X))\end{array}}{Z(z) : set}}{Z(z) : type}}{(z : Set(X))Z(z) : type}$$

$$\frac{\dfrac{\dfrac{(X : set)}{X : type} \quad \begin{array}{c}\mid\\ Z(SI2(X, x)) : type\end{array}}{(x : X)Z(SI2(X, x)) : type} \quad \begin{array}{c}\mid\\ (z : Set(X))Z(z) : type\end{array}}{((x : X)Z(SI2(X, x)))(z : Set(X))Z(z) : type}$$

$$\frac{set : type \quad \dfrac{\dfrac{\begin{array}{c}\mid\\ S(X) : type\end{array} \quad set : type}{(S(X))set : type} \quad \dfrac{\begin{array}{c}\mid\\ Z(SI1(X)) : type\end{array} \quad \begin{array}{c}\mid\\ ((x : X)Z(SI2(X, x)))(z : Set(X))Z(z) : type\end{array}}{(Z(SI1(X)))((x : X)Z(SI2(X, x)))(z : Set(X))Z(z) : type}}{(Z : (S(X))set)(Z(SI1(X)))((x : X)Z(SI2(X, x)))(z : Set(X))Z(z) : type}}{(X : set)(Z : (S(X))set)(Z(SI1(X)))((x : X)Z(SI2(X, x)))(z : Set(X))Z(z) : type}$$

The selector or eliminatory operator SE is an object of this type, i.e.

$$SE : (X : set)(Z : (S(X))set)(Z(SI1(X)))((x : X)Z(SI2(X, x)))(z : S(X))Z(z)$$

This S-elimination axiom can alternatively be expressed by the following two rules

$$\frac{\begin{array}{l}A : set\\ C : (S(A))set\\ d : C(SI1(A))\\ e : (x : A)C(SI2(A, x))\\ c : S(A)\end{array}}{SE(A, C, d, e, c) : C(c)} \; \text{(SE 1)}$$

$$\frac{\begin{array}{l}A : set\\ C : (S(A))set\\ d = g : C(SI1(A))\\ e = h : (x : A)C(SI2(A, x))\\ c = f : S(A)\end{array}}{SE(A, C, d, e, c) = SE(A, C, g, h, f) : C(c)} \; \text{(SE 2)}$$

The meaning of the selector SE is explained as follows

1. First evaluate c to canonical form.
2a. If $c \Longrightarrow SI1(A)$ then $SE(A, C, d, e, c) \Longrightarrow d$ and the value of $SE(A, C, d, e, c)$ is identical with the value of d, i.e. $SE(A, C, d, e, c) = d$.
2b. If $c \Longrightarrow SI2(A, a)$ then $SE(A, C, d, e, c) \Longrightarrow e(a)$ and $SE(A, C, d, e, c) = e(a)$.

Justification

of (SE 1) Assume the premisses of the rule. Then one has to show that the value of $SE(A, C, d, e, c)$ is a canonical element of $C(c)$.

From the last premiss one knows that c evaluates either to $SI1(A)$ or to $SI2(A, a)$.

If $c \Longrightarrow SI1(A)$, then by the meaning explanation of SE, the value of $SE(A, C, d, e, c)$ is identical with the value of d, whereby the value of d is, by the third premiss, a canonical element of $C(SI1(A))$. By the second premiss and the fact that $SI1(A) = c : S(A)$ one gets $C(SI1(A)) = C(c) : set$. Since the value of $SE(A, C, d, e, c)$ is identical with the value of d, the value of $SE(A, C, d, e, c)$ is a canonical element of $C(SI1(A))$ and hence a canonical element of $C(c)$.

If, however, $c \Longrightarrow SI2(A, a)$, then by the meaning explanation of SE, the value of $SE(A, C, d, e, c)$ is identical with the value of $e(a)$, and the value of $e(a)$ is, by the fourth premiss, a canonical element of $C(SI2(A, a))$. By the second premiss and the fact that $SI2(A, a) = c : S(A)$, one gets $C(SI2(A, a)) = C(c) : set$. Since the value of $SE(A, C, d, e, c)$ is identical with the value of $e(a)$, the value of $SE(A, C, d, e, c)$ is a canonical element of $C(SI2(A, a))$ and hence of $C(c)$.

Hence in either case is the value of $SE(A, C, d, e, c)$ a canonical element of $C(c)$.

of (SE 2) Analogous to (SE 1). □

The S-equality axioms are

$$SE(A, C, d, e, SI1(A)) = d : C(SI1(A))$$
$$SE(A, C, d, e, SI2(A, a)) = e(a) : C(SI2(A, a))$$

where $A : set$, $C : (S(A))set$, $d : C(SI1(A))$, $e : (x : A)C(SI2(A, x))$ and $a : A$, which can be expressed by the following two rules

$$\frac{\begin{array}{l} A : set \\ C : (S(A))set \\ d : C(SI1(A)) \\ e : (x : A)C(SI2(A, x)) \end{array}}{SE(A, C, d, e, SI1(A)) = d : C(SI1(A))} \text{ (S-equ 1)}$$

$$\frac{\begin{array}{l} A : set \\ C : (S(A))set \\ d : C(SI1(A)) \\ e : (x : A)C(SI2(A, x)) \\ a : A \end{array}}{SE(A, C, d, e, SI2(A, a)) = e(a) : C(SI2(A, a))} \text{ (S-equ 2)}$$

Justification
of (S-equ 1) and (S-equ 2): straightforward by the meaning explanation of SE. □

ix.2.1) What does a judgement of the form 'N : *set*' mean? or What does it mean to be the set of natural numbers?
To make the judgement 'N : *set*' one has to know (i) what a canonical element of N is, and (ii) when two canonical elements of N are identical.

The formation rules for the ground set N are the axioms or rules with 0 premisses

$$N : set \quad \text{(NF 1)} \quad \text{and} \quad N = N : set \quad \text{(NF 2)}$$

($N = N$: *set* is, as usual, just an instance of set type identity.)

ix.2.2) What does a judgement of the form '$a : N$' mean? or What does it mean to be a natural number (an element of the set of natural numbers)?
To explain what a judgement of the form '$a : N$' means, one has to know that (i) a has the form of an object formed by one of the element-formation rules of N (i.e. a has the form of a canonical element of N) or (ii) a can be evaluated to one of those forms (i.e. a is a non-canonical element of N).

The element-formation or introduction axioms of N are

$$NI1 : N \qquad NI2 : (N)N$$

which can be expressed alternatively by the introduction rules

$$NI1 : N \quad \text{(NI 1)} \qquad \frac{NI2 : (N)N \quad a : N}{NI2(a) : N} \text{(NI 2)}$$

ix.2.3) What does a judgement of the form '$a = b : N$' mean? or What does it mean for two natural numbers to be identical?
To explain what a judgement of the form '$a = b : N$' means one has to know that (i) a and b have the form of two objects formed by one of the identity rules associated with the element-formation rules of N (a and b are identical canonical elements of N) or (ii) a or b can be evaluated to one of those forms (a or b are identical non-canonical elements of N).

The identity rules associated with the element-formation rules, i.e. the introduction rules of N are

$$NI1 = NI1 : N \quad \text{(NI 3)} \qquad \frac{NI2 : (N)N \quad a = b : N}{NI2(a) = NI2(b) : N} \text{(NI 4)}$$

The following is a type

$$(Z : (N)set)(Z(NI1))((x : N)(Z(x))Z(NI2(x)))(z : N)Z(z) : type$$

as can be seen from the derivation

$$\frac{\dfrac{\begin{array}{l}0 : N\\(Z : (N)set)\end{array}}{Z(0) : set}}{Z(0) : type} \qquad \frac{\dfrac{\begin{array}{l}(x : N)\\(Z : (N)set)\end{array}}{Z(x) : set}}{Z(x) : type} \qquad \frac{\dfrac{\dfrac{\begin{array}{l}s : (N)N\\(x : N)\end{array}}{s(x) : N} \quad (Z : (N)set)}{Z(s(x)) : set}}{Z(s(x)) : type}$$

$$\frac{\dfrac{N : set}{N : type} \qquad \dfrac{\begin{array}{c}|\\Z(x) : type\end{array} \quad \begin{array}{c}|\\Z(s(x)) : type\end{array}}{(Z(x))Z(s(x)) : type}}{(x : N)(Z(x))Z(s(x)) : type} \qquad \frac{\dfrac{N : set}{N : type} \qquad \dfrac{\dfrac{\begin{array}{l}(z : N)\\(Z : (N)set)\end{array}}{Z(z) : set}}{Z(z) : type}}{(z : N)Z(z) : type}$$

$$\frac{\dfrac{\dfrac{N : set}{N : type} \quad set : type}{(N)set : type} \qquad \dfrac{\begin{array}{c}|\\Z(0) : type\end{array} \quad \dfrac{\begin{array}{c}|\\(x : N)(Z(x))Z(s(x)) : type\end{array} \quad \begin{array}{c}|\\(z : N)Z(z) : type\end{array}}{((x : N)(Z(x))Z(s(x)))(z : N)Z(z) : type}}{(Z(0))((x : N)(Z(x))Z(s(x)))(z : N)Z(z) : type}}{(Z : (N)set)(Z(0))((x : N)(Z(x))Z(s(x)))(z : N)Z(z) : type}$$

where 0 is used instead of $NI1$ and the successor fct. symbol s is used instead of $NI2$.

The selector or eliminatory operator NE of set N is an object of this type and the N-elimination axiom states just this fact

$$NE : (Z : (N)set)(Z(NI1))((x : N)(Z(x))Z(NI2(x)))(z : N)Z(z)$$

The N-elimination axiom can be expressed alternatively by the two rules

$$\frac{\begin{array}{l}C : (N)set\\d : C(NI1)\\e : (x : N)(C(x))C(NI2(x))\\a : N\end{array}}{NE(C, d, e, a) : C(a)} \ \text{(NE 1)}$$

$$\frac{\begin{array}{l}C : (N)set\\d = f : C(NI1)\\e = g : (x : N)(C(x))C(NI2(x))\\a = b : N\end{array}}{NE(C, d, e, a) = NE(C, f, g, b) : C(a)} \ \text{(NE 2)}$$

The meaning of the selector NE is explained as follows:

1. Evaluate a to canonical form.

2a. If $a \Longrightarrow NI1$ then $NE(C, d, e, a) \Longrightarrow d$ and the value of $NE(C, d, e, a)$ is identical with the value of d.

2b. If $a \Longrightarrow NI2(b)$ then $NE(C, d, e, a) \Longrightarrow e(b, NE(C, d, e, b))$ and $NE(C, d, e, a) = e(b, NE(C, d, e, b))$.

Justification

of (NE 1) Assuming the premisses one has to show that the value of $NE(C, d, e, a)$ is a canonical element of $C(a)$.

If $a \Longrightarrow NI1$ then $NE(C, d, e, a) = d$, by the meaning explanation of NE, and the value of $NE(C, d, e, a)$ is identical with the value of d which, by the second premiss, is a canonical element of $C(NI1)$. From the first premiss and the fact that $a = NI1 : N$ follows that $C(NI1) = C(a) : set$ and hence that the value of $NE(C, d, e, a)$ is a canonical element of $C(a)$.

If $a \Longrightarrow NI2(b)$ where $b : N$, then $NE(C, d, e, a) = e(b, NE(C, d, e, b))$, by the meaning explanation of NE, and the value of $NE(C, d, e, a)$ is identical with the value of $e(b, NE(C, d, e, b))$. It follows from the third premiss that the value of $e(b, NE(C, d, e, b))$ is a canonical element of $C(NI2(b))$ and thus, by the first premiss and the fact that $a = NI2(b) : N$, that it is a canonical element of $C(a)$.

By a similar reasoning one shows that $NE(C, d, e, b) : C(b)$, and if $b \Longrightarrow NI2(c)$, that $NE(C, d, e, c) : C(c)$, etc. This method terminates, since every natural number is generated by applying the successor fct. $NI2$ to $NI1$ a finite number of times, and by the last premiss, a is a natural number.

of (NE 2) Analogous to (NE 1). □

The N-equality axioms are

$$NE(C, d, e, NI1) = d : C(NI1)$$
$$NE(C, d, e, NI2(a)) = e(a, NE(C, d, e, a)) : C(NI2(a))$$

where $C : (N)set,\ d : C(NI1),\ e : (x : N)(C(x))C(NI2(x))$ and $a : N$.

These axioms can be expressed by the following two rules

$$\frac{\begin{array}{l} C : (N)set \\ d : C(NI1) \\ e : (x : N)(C(x))C(NI2(x)) \end{array}}{NE(C, d, e, NI1) = d : C(NI1)} \text{ (N-equ 1)}$$

$$\frac{\begin{array}{l} C : (N)set \\ d : C(NI1) \\ e : (x : N)(C(x))C(NI2(x)) \\ a : N \end{array}}{NE(C, d, e, NI2(a)) = e(a, NE(C, d, e, a)) : C(NI2(a))} \text{ (N-equ 2)}$$

Justification

(N-equ 1), (N-equ 2) are readily justified by the meaning explanation of NE. □

C, n & o

(1) NE is recursion over the set of natural numbers N, and that is why Nordström et al. use '*natrec*' and Martin-Löf in version (6) uses 'R' rather than 'NE'. In version (7) 'R', 'rec', '*natrec*' as well as 'NE' occur.

(2) NE makes it possible to define fcts. by prim. recursion. Say, defining a fct. f by prim. recursion

$$\begin{aligned} f(0) &= d \\ f(n+1) &= e(n, f(n)) \end{aligned}$$

is expressed in constructive type theory by the def.

$$f = (n)\, NE(N, d, e, n)$$

This allows us to define e.g. the following fct. constants pd, $+$, $\cdot$ by the explicit defs.

$$\begin{aligned} pd(x) &= NE(N, 0, (u)(v)u, x) \\ +(x, y) &= NE(N, y, (u)(v)s(v), x) \\ \cdot(x, y) &= NE(N, 0, (u)(v) + (y, v), x) \end{aligned}$$

and even to define the bounded μ-operator by prim. recursion but on a higher type, namely $(N \to N_2) \to N$.[72]

(3) Peano's third axiom formulated as the rule

$$\frac{m : N \quad n : N \quad NI2(m) = NI2(n) : N}{m = n : N}$$

can be derived in constructive type theory by means of the predecessor fct. pd explicitly defined in note (2) above.[73]

(4) Mathematical induction

If one thinks of C in the rule (NE 1) in terms of a propositional fct. over N, i.e. $C : (N)prop$, one obtains the following rule

$$\frac{\begin{array}{l} C : (N)prop \\ d : C(NI1) \\ x : N, y : C(x) \vdash e(x)(y) : C(NI2(x)) \\ a : N \end{array}}{NE(C, d, (x)(y)e, a) : C(a)}$$

which, if one suppresses the proof-objects of the occurring propositions and writes instead '*true*' behind the resp. propositions, is turned into the rule

$$\frac{\begin{array}{l} C : (N)prop \\ C(NI1)\ true \\ x : N, C(x)\ true \vdash C(NI2(x))\ true \\ a : N \end{array}}{C(a)\ true}$$

which is nothing else but mathematical induction i.e. Peano's fifth axiom.

In note (1) NE was said to be recursion over the set N of natural numbers. The above transformations show that recursion and induction essentially express the same concept when sets are interpreted as propositions. (This note presupposes the Curry–Howard isomorphism, cf. Sect. xi)

ix.3.1) What does a judgement of the form '$A+B$: *set*' mean? or What does it mean to be the disjoint union of two sets?
To make a judgement of the form '$A + B$: *set*' one has to know (i) what a canonical element of $A+B$ is, and (ii) when two canonical elements of $A+B$ are identical.

The formation axiom for the composite set $A + B$ is

$$+ : (set)(set)set \quad \text{where clearly } (set)(set)set : type$$

Equivalently, the formation axiom can be given by the two rules

$$\frac{A : set \quad B : set}{A + B : set}\,(+\text{F } 1) \qquad \frac{A = C : set \quad B = D : set}{A + B = C + D : set}\,(+\text{F } 2)$$

ix.3.2) What does a judgement of the form '$c : A + B$' mean? or What does it mean to be an element of the set $A + B$?
To explain what a judgement of the form '$c : A+B$' means, one has to know that (i) c has the form of an object formed by one of the element-formation rules of $A + B$ (c has the form of a canonical element of $A + B$) or (ii) c can be evaluated to one of those forms (c is a non-canonical element of $A + B$).

The element-formation or introduction axioms of $A + B$ are

$$+I1 : (X : set)(Y : set)(X)X + Y \qquad + I2 : (X : set)(Y : set)(Y)X + Y$$

which can be expressed alternatively by the introduction rules

$$\frac{A = C : set \quad B = D : set \quad a : A}{+I1(A, B, a) : C + D}\,(+\text{I } 1)$$

$$\frac{A = C : set \quad B = D : set \quad b : B}{+I2(A, B, b) : C + D}\,(+\text{I } 2)$$

Justification

of (+F 2):
$$\frac{\begin{array}{l}A = C : set\\ B = D : set\\ a : A\end{array}}{+I1(A, B, a) : C + D}\,\text{by } (+\text{I } 1)$$

$$\frac{\dfrac{A = C : set \quad C = E : set}{A = E : set} \quad \dfrac{B = D : set \quad D = F : set}{B = F : set} \quad a : A}{+I1(A, B, a) : E + F}\,\text{by } (+\text{I } 1)$$

$$\begin{array}{l} A = E : set \\ B = F : set \\ a : A \\ \hline \end{array}$$

but also $+I1(A, B, a) : E + F$ and

$$\frac{\dfrac{A = E : set \quad E = C : set}{A = C : set} \qquad \dfrac{B = F : set \quad F = D : set}{B = D : set} \qquad a : A}{+I1(A, B, a) : C + D}\ \text{by } (+\text{I } 1)$$

The same holds w.r.t. (+I 2) if in the derivations above $a : A$ is replaced by $b : B$ and $+I1(A, B, a)$ by $+I2(A, B, b)$ resp.

This then yields a justification of $\dfrac{C = E : set \quad D = F : set}{C + D = E + F : set}$ □

ix.3.3) What does a judgement of the form '$c = d : A + B$' mean? or What does it mean for two elements of $A + B$ to be identical?
To explain what a judgement of the form '$c = d : A + B$' means, one has to know that (i) c and d have the form of two objects formed by one of the identity rules associated with the element-formation rules of $A + B$ (c and d are identical canonical elements of $A + B$) or (ii) c and d can be evaluated to one of those forms (c or d are identical non-canonical elements of $A + B$).

The identity rules associated with the element-formation rules are

$$\frac{A = C : set \quad B = D : set \quad a = c : A}{+I1(A, B, a) = +I1(C, D, c) : A + B}\ (+\text{I } 3)$$

$$\frac{A = C : set \quad B = D : set \quad b = d : B}{+I2(A, B, b) = +I2(C, D, d) : A + B}\ (+\text{I } 4)$$

C, n & o

That $(X : set)(Y : set)(X)X + Y : type$ can be shown as follows

$$\frac{\dfrac{(X : set)}{X : type} \qquad \dfrac{\dfrac{(X : set) \quad (Y : set)}{X + Y : set}}{X + Y : type}}{(X)X + Y : type} \qquad \frac{set : type \qquad \dfrac{set : type \qquad \begin{array}{c}\vert \\ (X)X + Y : type\end{array}}{(Y : set)(X)X + Y : type}}{(X : set)(Y : set)(X)X + Y : type}$$

analogously $(X : set)(Y : set)(Y)X + Y : type$

Moreover, the following judgement can be derived

$$\begin{array}{l}(X : set)(Y : set)(Z : (X + Y)set)((x : X)Z(+I1(X, Y, x))) \\ \quad ((y : Y)Z(+I2(X, Y, y)))(z : X + Y)Z(z) : type\end{array}$$

$$\frac{\dfrac{\dfrac{\begin{array}{l}(X : set)\\(Y : set)\end{array}}{X + Y : set}}{X + Y : type} \qquad set : type}{(X + Y)set : type} \qquad \frac{\dfrac{\dfrac{\begin{array}{l}(X : set)\\(Y : set)\\(x : X)\\+I1 : (X : set)(Y : set)(X)X + Y\end{array}}{+I1(X, Y, x) : X + Y} \qquad (Z : (X + Y)set)}{Z(+I1(X, Y, x)) : set}}{Z(+I1(X, Y, x)) : type}$$

$$\dfrac{\dfrac{\dfrac{\begin{array}{l}(X:set)\\(Y:set)\\(y:Y)\\+I2:(X:set)(Y:set)(Y)X+Y\end{array}}{+I2(X,Y,y):X+Y}\qquad (Z:(X+Y)set)}{Z(+I2(X,Y,y)):set}}{Z(+I2(X,Y,y)):type}$$

$$\dfrac{\dfrac{\dfrac{(X:set)\quad (Y:set)}{X+Y:set}}{X+Y:type}\qquad \dfrac{\dfrac{(z:X+Y)\quad (Z:(X+Y)set)}{Z(z):set}}{Z(z):type}}{(z:X+Y)Z(z):type}$$

$$\dfrac{\dfrac{\dfrac{(X:set)}{X:type}\quad \begin{array}{c}\vdots\\ Z(+I1(X,Y,x)):type\end{array}}{(x:X)Z(+I1(X,Y,x)):type}\qquad \dfrac{\dfrac{\dfrac{(Y:set)}{Y:type}\quad \begin{array}{c}\vdots\\ Z(+I2(X,Y,y)):type\end{array}}{(y:Y)Z(+I2(X,Y,y)):type}\quad \begin{array}{c}\vdots\\ (z:X+Y)Z(z):type\end{array}}{((y:Y)Z(+I2(X,Y,y)))(z:X+Y)Z(z):type}}{((x:X)Z(+I1(X,Y,x)))((y:Y)Z(+I2(X,Y,y)))(z:X+Y)Z(z):type}$$

$$\dfrac{set:type\quad \dfrac{set:type\quad \dfrac{\begin{array}{c}\vdots\\ (X+Y)set:type\end{array}\quad \begin{array}{c}\vdots\\ ((x:X)Z(+I1(X,Y,x)))((y:Y)Z(+I2(X,Y,y)))(z:X+Y)Z(z):type\end{array}}{(Z:(X+Y)set)((x:X)Z(+I1(X,Y,x)))((y:Y)Z(+I2(X,Y,y)))(z:X+Y)Z(z):type}}{(Y:set)(Z:(X+Y)set)((x:X)Z(+I1(X,Y,x)))((y:Y)Z(+I2(X,Y,y)))(z:X+Y)Z(z):type}}{(X:set)(Y:set)(Z:(X+Y)set)((x:X)Z(+I1(X,Y,x)))((y:Y)Z(+I2(X,Y,y)))(z:X+Y)Z(z):type}$$

The $+$-elimination axiom states that the selector or eliminatory operator $+E$ for the set $A+B$ is an object of the type just derived, i.e.

$$\begin{array}{l}+E:(X:set)(Y:set)(Z:(X+Y)set)((x:X)Z(+I1(X,Y,x)))\\ \qquad((y:Y)Z(+I2(X,Y,y)))(z:X+Y)Z(z)\end{array}$$

The $+$-elimination axiom can be expressed alternatively by the two rules

$$\dfrac{\begin{array}{l}A:set\\ B:set\\ C:(A+B)set\\ d:(x:A)C(+I1(A,B,x))\\ e:(y:B)C(+I2(A,B,y))\\ c:A+B\end{array}}{+E(A,B,C,d,e,c):C(c)}\ (+\text{E } 1)$$

$$\dfrac{\begin{array}{l}A:set\\ B:set\\ C:(A+B)set\\ d=g:(x:A)C(+I1(A,B,x))\\ e=h:(y:B)C(+I2(A,B,y))\\ c=f:A+B\end{array}}{+E(A,B,C,d,e,c)=+E(A,B,C,g,h,f):C(c)}\ (+\text{E } 2)$$

The meaning of the selector $+E$ is explained as follows

1. Evaluate c to canonical form.
2a. If $c \Longrightarrow +I1(A, B, a)$ then $+E(A, B, C, d, e, c) \Longrightarrow d(a)$ and $+E(A, B, C, d, e, c) = d(a)$.
2b. If $c \Longrightarrow +I2(A, B, b)$ then $+E(A, B, C, d, e, c) \Longrightarrow e(b)$ and $+E(A, B, C, d, e, c) = e(b)$.

Justification

of (+E 1) Assuming the premisses ...

If $c \Longrightarrow +I1(A, B, a)$ then $+E(A, B, C, d, e, c) = d(a)$ by the meaning explanation of $+E$ and the value of $+E(A, B, C, d, e, c)$ is identical with the value of $d(a)$ which, by the fourth premiss, is a canonical element of $C(+I1(A, B, a))$. From the third premiss and $c = +I1(A, B, a) : A + B$ follows that $C(+I1(A, B, a)) = C(c) : set$ and hence the value of $+E(A, B, C, d, e, c)$ is a canonical element of $C(c)$.

If, however, $c \Longrightarrow +I2(A, B, b)$, then $+E(A, B, C, d, e, c) = e(b)$, by the meaning explanation of $+E$, and the value of $+E(A, B, C, d, e, c)$ is identical with the value of $e(b)$ which, by the fifth premiss, is a canonical element of $C(+I2(A, B, b))$. From the third premiss and $c = +I2(A, B, b) : A+B$ follows that $C(+I2(A, B, b)) = C(c) : set$ and hence the value of $+E(A, B, C, d, e, c)$ is a canonical element of $C(c)$.

of (+E 2) Analogous to (+E 1). □

The +-equality axioms are

$$+E(A, B, C, d, e, +I1(A, B, a)) = d(a) : C(+I1(A, B, a))$$
$$+E(A, B, C, d, e, +I2(A, B, b)) = e(b) : C(+I2(A, B, b))$$

where $A : set$, $B : set$, $C : (A + B)set$, $d : (x : A)C(+I1(A, B, x))$, $e : (y : B)C(+I2(A, B, y))$, $a : A$, $b : B$.

These axioms can be expressed by the following two rules

$$\frac{\begin{array}{l} A : set \\ B : set \\ C : (A + B)set \\ d : (x : A)C(+I1(A, B, x)) \\ e : (y : B)C(+I2(A, B, y)) \\ a : A \end{array}}{+E(A, B, C, d, e, +I1(A, B, a)) = d(a) : C(+I1(A, B, a))}\ (\text{+-equ 1})$$

$$\frac{\begin{array}{l} A : set \\ B : set \\ C : (A + B)set \\ d : (x : A)C(+I1(A, B, x)) \\ e : (y : B)C(+I2(A, B, y)) \\ b : B \end{array}}{+E(A, B, C, d, e, +I2(A, B, b)) = e(b) : C(+I2(A, B, b))}\ (\text{+-equ 2})$$

Justification
of (+-equ 1) and (+-equ 2): straightforward by the meaning explanation of $+E$. □

C, n & o
Note that $+I1$ and $+I2$ are the left and right injections from A and B resp. to $A+B$. That is why Nordström et al. choose to use '*inl*' and '*inr*' instead of '$+I1$' and '$+I2$' resp. Martin-Löf uses 'i' and 'j' in version (6) and 'i' and 'j' as well as '$+I1$' and '$+I2$' in version (7). Nordström et al. use '*when*' instead of '$+E$' and Martin-Löf uses 'D' in version (6) and '$+E$', 'D' and '$case$' in version (7).

ix.4.1) What does a judgement of the form '$\Pi(A, B)$: *set*' mean? or What does it mean to be the cartesian product of a family of sets?
To make a judgement of the form '$\Pi(A, B)$: *set*' one has to know (i) what a canonical element of $\Pi(A, B)$ is, and (ii) when two canonical elements of $\Pi(A, B)$ are identical.

The formation axiom for the composite set $\Pi(A, B)$ is

$$\Pi : (X : set)(Y : (X)set)set$$

The formation axiom is tantamount to the following two rules

$$\frac{A : set \quad B : (A)set}{\Pi(A, B) : set} \ (\Pi F\ 1) \qquad \frac{A = C : set \quad B = D : (A)set}{\Pi(A, B) = \Pi(C, D) : set} \ (\Pi F\ 2)$$

ix.4.2) What does a judgement of the form '$c : \Pi(A, B)$' mean? or What does it mean to be an element of the cartesian product of a family of sets B over a set A?
To explain what a judgement of the form '$c : \Pi(A, B)$' means, one has to know that (i) c has the form of an object formed by the element-formation rule of $\Pi(A, B)$ (c has the form of a canonical element of $\Pi(A, B)$) or (ii) c can be evaluated to that form (c is a non-canonical element of $\Pi(A, B)$).

The element-formation or introduction axiom of $\Pi(A, B)$ is

$$\Pi I : (X : set)(Y : (X)set)((x : X)Y(x))\Pi(X, Y)$$

which can be expressed alternatively by the introduction rule

$$\frac{\begin{array}{l} A = C : set \\ B = D : (A)set \\ b : (x : A)B(x) \end{array}}{\Pi I(A, B, b) : \Pi(C, D)} \ (\Pi I\ 1)$$

C, n & o
The first two premisses are necessary in this form to secure that $\Pi(C, D) = \Pi(E, F)$: *set*, i.e. that given the premisses $C = E$: *set*, $D = F$: (A)*set*, if $\Pi I(A, B, b) : \Pi(C, D)$, then also $\Pi I(A, B, b) : \Pi(E, F)$ and vice versa. This would obviously not be the case if the Π-introduction rule above were formulated without the just mentioned premisses in their resp. form. (This is the justification of (ΠF 2).)

ix.4.3) What does a judgement of the form '$c = d : \Pi(A, B)$' mean? or What does it mean for two elements of $\Pi(A, B)$ to be identical?
To explain what a judgement of the form '$c = d : \Pi(A, B)$' means, one has to know that (i) c and d have the form of two objects formed by the identity rule associated with the element-formation rule of $\Pi(A, B)$ (c and d are identical canonical elements of $\Pi(A, B)$) or (ii) c and d can be evaluated to that form (c or d are identical non-canonical elements of $\Pi(A, B)$).

The identity rule associated with the element-formation rule or second introduction rule of $\Pi(A, B)$ is

$$\frac{\substack{A = C : \mathit{set} \\ B = D : (A)\mathit{set} \\ b = d : (x : A)B(x)}}{\Pi I(A, B, b) = \Pi I(C, D, d) : \Pi(A, B)}\ (\Pi\text{I } 2)$$

C, n & o

The type $(X : \mathit{set})(Y : (X)\mathit{set})((x : X)Y(x))\Pi(X, Y)$ is derived as follows

$$\dfrac{\dfrac{(X:\mathit{set})}{X:\mathit{type}}\quad \mathit{set}:\mathit{type}}{(X)\mathit{set}:\mathit{type}} \qquad \dfrac{\dfrac{(X:\mathit{set})}{X:\mathit{type}}\quad \dfrac{\dfrac{\substack{(x:X)\\(Y:(X)\mathit{set})}}{Y(x):\mathit{set}}}{Y(x):\mathit{type}}}{(x:X)Y(x):\mathit{type}} \qquad \dfrac{\dfrac{\substack{(X:\mathit{set})\\(Y:(X)\mathit{set})}}{\Pi(X,Y):\mathit{set}}}{\Pi(X,Y):\mathit{type}}$$

$$\dfrac{\mathit{set}:\mathit{type}\quad \dfrac{\substack{| \\ (X)\mathit{set}:\mathit{type}}\quad \dfrac{\substack{| \\ (x:X)Y(x):\mathit{type}}\quad \substack{| \\ \Pi(X,Y):\mathit{type}}}{((x:X)Y(x))\Pi(X,Y):\mathit{type}}}{(Y:(X)\mathit{set})((x:X)Y(x))\Pi(X,Y):\mathit{type}}}{(X:\mathit{set})(Y:(X)\mathit{set})((x:X)Y(x))\Pi(X,Y):\mathit{type}}$$

Moreover, the following type can be derived

$$(X : \mathit{set})(Y : (X)\mathit{set})(Z : (\Pi(X, Y))\mathit{set})((y : (x : X)Y(x)) \\ Z(\Pi I(X, Y, y)))(z : \Pi(X, Y))Z(z) : \mathit{type}$$

$$\dfrac{\dfrac{\dfrac{\substack{(X:\mathit{set})\\(Y:(X)\mathit{set})}}{\Pi(X,Y):\mathit{set}}}{\Pi(X,Y):\mathit{type}}\quad \mathit{set}:\mathit{type}}{(\Pi(X,Y))\mathit{set}:\mathit{type}} \qquad \dfrac{\dfrac{(X:\mathit{set})}{X:\mathit{type}}\quad \dfrac{\dfrac{\substack{(x:X)\\(Y:(X)\mathit{set})}}{Y(x):\mathit{set}}}{Y(x):\mathit{type}}}{(x:X)Y(x):\mathit{type}}$$

$$\dfrac{\dfrac{\dfrac{\substack{(X:\mathit{set})\\(Y:(X)\mathit{set})\\(y:(x:X)Y(x))\\\Pi I:(X:\mathit{set})(Y:(X)\mathit{set})((x:X)Y(x))\Pi(X,Y)}}{\Pi I(X,Y,y):\Pi(X,Y)}\quad (Z:(\Pi(X,Y))\mathit{set})}{Z(\Pi I(X,Y,y)):\mathit{set}}}{Z(\Pi I(X,Y,y)):\mathit{type}}$$

$$\dfrac{\dfrac{\dfrac{(X : set) \quad (Y : (X)set)}{\Pi(X, Y) : set}}{\Pi(X, Y) : type} \qquad \dfrac{\dfrac{(z : \Pi(X, Y)) \quad (Z : (\Pi(X, Y))set)}{Z(z) : set}}{Z(z) : type}}{(z : \Pi(X, Y))Z(z) : type}$$

$$\dfrac{\dfrac{(x : X)Y(x) : type \quad Z(\Pi I(X, Y, y)) : type}{(y : (x : X)Y(x))Z(\Pi I(X, Y, y)) : type} \qquad (z : \Pi(X, Y))Z(z) : type}{((y : (x : X)Y(x))Z(\Pi I(X, Y, y)))(z : \Pi(X, Y))Z(z) : type}$$

$$\dfrac{set : type \qquad \dfrac{\dfrac{\dfrac{(X : set)}{X : type} \quad set : type}{(X)set : type} \qquad \dfrac{(\Pi(X, Y))set : type \quad ((y : (x : X)Y(x))Z(\Pi I(X, Y, y)))(z : \Pi(X, Y))Z(z) : type}{(Z : (\Pi(X, Y))set)((y : (x : X)Y(x))Z(\Pi I(X, Y, y)))(z : \Pi(X, Y))Z(z) : type}}{(Y : (X)set)(Z : (\Pi(X, Y))set)((y : (x : X)Y(x))Z(\Pi I(X, Y, y)))(z : \Pi(X, Y))Z(z) : type}}{(X : set)(Y : (X)set)(Z : (\Pi(X, Y))set)((y : (x : X)Y(x))Z(\Pi I(X, Y, y)))(z : \Pi(X, Y))Z(z) : type}$$

The Π-elimination axiom states that the selector or eliminatory operator ΠE for the set $\Pi(A, B)$ is an object of the type just derived, i.e.

$$\begin{aligned}\Pi E : (X : set)(Y : (X)set)(Z : (\Pi(X, Y))set)((y : (x : X)Y(x))\\ Z(\Pi I(X, Y, y)))(z : \Pi(X, Y))Z(z)\end{aligned}$$

The Π-elimination axiom can be expressed alternatively by the rules

$$\frac{\begin{array}{l} A : set \\ B : (A)set \\ C : (\Pi(A, B))set \\ d : (y : (x : A)B(x))C(\Pi I(A, B, y)) \\ c : \Pi(A, B) \end{array}}{\Pi E(A, B, C, d, c) : C(c)} \; (\Pi E\ 1)$$

$$\frac{\begin{array}{l} A : set \\ B : (A)set \\ C : (\Pi(A, B))set \\ d = f : (y : (x : A)B(x))C(\Pi I(A, B, y)) \\ c = e : \Pi(A, B) \end{array}}{\Pi E(A, B, C, d, c) = \Pi E(A, B, C, f, e) : C(c)} \; (\Pi E\ 2)$$

The meaning of the selector ΠE is explained as follows:

1. Evaluate c to canonical form.
2. If $c \Longrightarrow \Pi I(A, B, b)$ then $\Pi E(A, B, C, d, c) \Longrightarrow d(b)$ and $\Pi E(A, B, C, d, c) = d(b)$.

Justification
of (ΠE 1) Assuming the premisses ...
From the meaning of the last premiss it follows that c must evaluate to canonical form. If $c \Longrightarrow \Pi I(A, B, b)$, then, by the meaning explanation of ΠE, the value of $\Pi E(A, B, C, d, c)$ is identical with the value of $d(b)$ which, by the fourth

premiss, is a canonical element of set $C(\Pi I(A, B, b))$. Because of this and because $\Pi I(A, B, b) = c : \Pi(A, B)$ and by the third premiss, C is a fct. set over $\Pi(A, B)$, $C(\Pi I(A, B, b)) = C(c) : set$ and the value of $\Pi E(A, B, C, d, c)$ is a canonical value of $C(c)$.

of (ΠE 2) Analogous. □

The Π-equality axiom is

$$\Pi E(A, B, C, d, \Pi I(A, B, b)) = d(b) : C(\Pi I(A, B, b))$$

where $A : set$, $B : (A)set$, $C : (\Pi(A, B))set$, $d : (y : (x : A)B(x))\, C(\Pi I(A, B, y))$ and $b : (x : A)B(x)$.

This Π-equality axiom can also be expressed by the following rule

$$\frac{\begin{array}{l} A : set \\ B : (A)set \\ C : (\Pi(A, B))set \\ d : (y : (x : A)B(x))C(\Pi I(A, B, y)) \\ b : (x : A)B(x) \end{array}}{\Pi E(A, B, C, d, \Pi I(A, B, b)) = d(b) : C(\Pi I(A, B, b))} \ (\Pi\text{-equ } 1)$$

Justification
of (Π-equ 1): straightforward by the meaning explanation of ΠE. □

C, n & o

(1) In all the versions preceding version (7), Martin-Löf uses 'λ' instead of 'ΠI'. Only in version (7) does he introduce 'ΠI' as well, especially for monomorphic notation. (In version (8) 'Π*intr*' also appears.)

Notations for $\Pi I(A, B, b)$ in lambda calculus are: $\lambda(b) = \lambda((x)b(x)) = (\lambda x)b(x) : \Pi(A, B)$.

Frege's notation is $\breve{\varepsilon}\phi(\varepsilon)$.

Martin-Löf uses in version (6a) 'F' instead of 'ΠE' and in version (7) 'F' as well as 'ΠE' and in version (8) only 'ΠE'. Nordström et al. use '*funsplit*' for the polymorphic notation of ΠE.

(2) There is an alternative (non-equivalent) Π-elimination rule which is the well-known β-reduction rule from λ-calculus. This rule though does not follow the principle of structural induction. (Cf. beginning of this section.) Until version (6a), this alternative Π-elimination rule was the only one of versions (1) – (6).

The alternative Π-elimination axiom is

$$\Pi E : (X : set)(Y : (X)set)(\Pi(X, Y))(x : X)Y(x)$$

The meaning of this alternative selector ΠE is:

1. Evaluate c to canonical form.
2. If $c \Longrightarrow \Pi I(A, B, b)$ then $\Pi E(A, B, c, a) \Longrightarrow b(a)$ and $\Pi E(A, B, c, a) = b(a)$.

With this meaning explanation of the alternative eliminatory operator ΠE, the Π-elimination rules corresponding to the alternative Π-elimination axiom can easily be justified. The alternative Π-equality axiom

$$\Pi E(A, B, \Pi I(A, B, b), a) = b(a) : B(a)$$

is equally readily justified by the meaning explanation of the alternative ΠE.

To avoid any confusion, the alternative ΠE-operator will henceforth be referred to by '*app*' (as Martin-Löf does in version (7) as well as in version (8)).

Now, *app* can be explicitly defined by ΠE in the following way:

$$\frac{\begin{array}{l} A : set \\ B : (A)set \\ c : \Pi(A, B) \\ a : A \end{array}}{\begin{cases} app(A, B, c, a) : B(a) \\ app(A, B, c, a) = \Pi E(A, B, (z)B(a), (y)y(a), c) : B(a) \end{cases}}$$

That this def. is correct, is quickly verified by the following observation

$$\frac{\begin{array}{l} A : set \\ B : (A)set \\ b : (x : A)B(x) \\ a : A \end{array}}{\begin{array}{ll} app(A, B, \Pi I(A, B, b), a) & = \Pi E(A, B, (z)B(a), (y)y(a), \Pi I(A, B, b)) \\ & = ((y)y(a))(b) = b(a) : B(a) \end{array}}$$

(3) Until now, two notions of fct. have been discussed: the notion of fct. in the oldfashioned sense which is the most fundamental syntactical notion of fct. as an expression with holes in it, i.e. an expression which is not saturated; and the notion of fct. in the modern sense, which is already a notion of fct. derived from or dependent on the first notion of fct. and which refers to the objects of a fct. type. Finally, there is a third notion of fct. in the sense of an element of a cartesian product of a family of sets or of a power set.

If $A : set,\ B : (A)set$ then

$x : A \vdash$	$b : B(x)$	is a fct. in the oldfashioned sense
	$b : (x : A)B(x)$	is a fct. in the modern sense
	$b : \Pi(A, B)$	is a fct. as an element of a Π-set or a fct.set

Example

in $x : N, y : N \vdash x^2 + 3y : N$, $\quad x^2 + 3y$ is a fct. in the oldfashioned sense
in $(x)(y)(x^2 + 3y) : (N)(N)N$, $\quad (x)(y)(x^2 + 3y)$ is a fct. in the modern sense
in $\lambda((x)\lambda((y)(x^2 + 3y))) : \Pi(N, (x)\Pi(N, (y)N))$, $\quad \lambda((x)\lambda((y)(x^2 + 3y)))$ or $\Pi I((x)\Pi I((y)(x^2 + 3y)))$ is a fct. as an element of a Π-set.

Note that it is always possible to pass from a fct. of one kind to the corresponding one of the other kinds:
from the first to the second and to the third notion

$$\cfrac{\cfrac{\cfrac{x : A \vdash b : B(x) \quad (x : A)}{((x)b)(x) : B(x)}}{(x)b : (x : A)B(x)}}{\Pi I(A, B, (x)b) : \Pi(A, B)}$$

from the third to the first and the second notion

$$\cfrac{\cfrac{\cfrac{b : \Pi(A, B) \quad (x : A)}{app(A, B, b, x) : B(x)}}{x : A \vdash app(A, B, b, x) : B(x)}}{(x)app(A, B, b, x) : (x : A)B(x)}$$

from the second to the first notion

$$\cfrac{\cfrac{b : (x : A)B(x) \quad (x : A)}{b(x) : B(x)}}{x : A \vdash b(x) : B(x)}$$

The notion of fct. common in mathematical texts is the third notion of fct., i.e. $b : \Pi(A, B)$. The fct. sets or Π-sets are introduced because one wants to be able to quantify over fcts. Note that the notion of fct. in the oldfashioned sense has almost disappeared from today's math books.

If one wants to quantify over fcts., then the A in $(\exists x : A)B(x)$ or in $(\forall x : A)B(x)$ has to be at least a non-dependent type. Hence if A is a Π-set or a fct. set, then A clearly is also a non-dependent type and quantification over fcts. as elements of this Π-set is possible. If A is a fct. type $(x : \alpha)\beta$, then A is also a non-dependent type and hence quantification over fcts. as objects of this fct. type is possible. However, what is impossible is quantification over fcts. as objects of a family of types, or in other terms, quantification over fcts. in the oldfashioned sense.

Finally, to the three ways of fct. formation correspond three ways of providing the resp. fcts. with arguments: let $A : set$, $B : (A)set$ and $a : A$

$x : A \vdash b : B(x)$	$b(a/x) : B(a/x)$	substitution
$b : (x : A)B(x)$	$b(a) : B(a)$	type-theoretical application
$b : \Pi(A, B)$	$app(A, B, b, a)$	set-theoretical application

(4) A Π-set is more general than the usual set of fcts. $A \rightarrow B$, as the result of applying a fct. to an argument is in a set, which may depend on the argument to which the fct. is applied. A reason for this generality is, as was remarked upon already in the case of the fct. types (cf. C, n & o to iii.1)), that it is needed for the def. of the quantifiers. (Some kind of dependent sets, analogous to dependent types implicit in the fct. types,

are implicitly contained in the notion of Π-set.) Another reason for this generality is, that it is needed when sets are used to specify programs.[74]

(5) There is a second derived Π-equality judgement

$$c = \Pi I(A, B, \Pi E(A, B, C, d, c)) : \Pi(A, B)$$

where $A : set$, $B : (A)set$, $C : (\Pi(A, B))set$, $d : (y : (x : A)B(x))\, C(\Pi I(A, B, y))$, $c : \Pi(A, B)$, which can be expressed by the rule

$$\frac{\begin{array}{l} A : set \\ B : (A)set \\ C : (\Pi(A, B))set \\ d : (y : (x : A)B(x))C(\Pi I(A, B, y)) \\ c : \Pi(A, B) \end{array}}{c = \Pi I(A, B, \Pi E(A, B, C, d, c)) : \Pi(A, B)} \ (\Pi\text{-equ } 2)$$

Justification
of (Π-equ 2) has to be postponed till after the introduction of the set-theoretical extensional identity. □

ix.5.1) What does a judgement of the form '$A \to B : set$' mean? or What does it mean to be a fct. set?
To make a judgement of the form '$A \to B : set$' one has to know (i) what a canonical element of $A \to B$ is, and (ii) when two canonical elements of $A \to B$ are identical.

The formation axiom for the composite set $A \to B$ is

$$\to : (set)(set)set$$

The formation axiom is tantamount to the following two rules

$$\frac{A : set \quad B : set}{A \to B : set} \ (\to\text{F } 1) \qquad \frac{A = C : set \quad B = D : set}{A \to B = C \to D : set} \ (\to\text{F } 2)$$

C, n & o

(1) A fct. set or →-set is nothing but a special case of a Π-set in which B is not a family of sets dependent on A, but B is a set independent of A, i.e. $B : (A)set$ is s.t. $B : set$. $A \to B$ is the set of fcts. from set A to set B and is often also notated by B^A. In version (6) Martin-Löf writes '$(\Pi x \in A)B(x)$' for '$\Pi(A, B)$' and '$(\Pi x \in A)B$' for '$A \to B$'.

(2) However, → can also be explicitly defined by Π as follows:

$$\frac{A : set \quad B : set}{\begin{cases} A \to B : set \\ A \to B = \Pi(A, (x)B) : set \end{cases}}$$

The $\rightarrow$-introduction, $\rightarrow$-elimination and $\rightarrow$-equality axioms and rules will not be given in the sequel, as they can easily be recovered from the corresponding Π-axioms or Π-rules.

C, n & o

There is an alternative $\rightarrow$-elimination axiom which is but a special case of the alternative Π-elimination axiom (cf. C, n & o (2) to Π-equality), namely

$$\rightarrow E : (X : set)(Y : set)((X)Y)(X)Y$$

The meaning of this alternative eliminatory operator $\rightarrow E$ is

1. Evaluate c to canonical form.
2. If $c \Longrightarrow \rightarrow I(A, B, b)$ then $\rightarrow E(A, B, c, a) \Longrightarrow b(a)$ and $\rightarrow E(A, B, c, a) = b(a)$.

With this meaning explanation of the alternative selector $\rightarrow E$, the $\rightarrow$-elimination rules corresponding to the alternative $\rightarrow$-elimination axiom can easily be justified.

The same holds for the alternative $\rightarrow$-equality axiom

$$\rightarrow E(A, B, \rightarrow I(A, B, b), a) = b(a) : B$$

where $A : set,\ B : set,\ b : (A)B$ and $a : A$.

ix.6.1) What does a judgement of the form '$\Sigma(A, B) : set$' mean? or What does it mean to be a disjoint union of a family of sets?
To make a judgement of the form '$\Sigma(A, B) : set$' one has to know (i) what a canonical element of $\Sigma(A, B)$ is, and (ii) when two canonical elements of $\Sigma(A, B)$ are identical.

The formation axiom for the composite set $\Sigma(A, B)$ is

$$\Sigma : (X : set)(Y : (X)set)set$$

Alternatively, the formation axiom can be given by two rules

$$\frac{A : set \quad B : (A)set}{\Sigma(A, B) : set}\ (\Sigma F\ 1) \qquad \frac{A = C : set \quad B = D : (A)set}{\Sigma(A, B) = \Sigma(C, D) : set}\ (\Sigma F\ 2)$$

ix.6.2) What does a judgement of the form '$c : \Sigma(A, B)$' mean? or What does it mean to be an element of the disjoint union of a family of sets B over a set A?
To explain what a judgement of the form '$c : \Sigma(A, B)$' means, one has to know that (i) c has the form of an object formed by the element-formation rule of $\Sigma(A, B)$ (c has the form of a canonical element of $\Sigma(A, B)$) or (ii) c can be evaluated to that form (c is a non-canonical element of $\Sigma(A, B)$).

The element-formation or introduction axiom of $\Sigma(A, B)$ is

$$\Sigma I : (X : set)(Y : (X)set)(x : X)(Y(x))\Sigma(X, Y)$$

which can be expressed alternatively by the introduction rule

$$\frac{\begin{array}{l} A = C : set \\ B = D : (A)set \\ a : A \\ b : B(a) \end{array}}{\Sigma I(A, B, a, b) : \Sigma(C, D)}\ (\Sigma\text{I } 1)$$

C, n & o

(1) The reason for adopting the premisses $A = C : set$ and $B = D : (A)set$ was given in C, n & o to ix.4.2).

(2) The derivation of $(X : set)(Y : (X)set)(x : X)(Y(x))\Sigma(X, Y) : type$ is merely a slight modification of the derivation of $(X : set)(Y : (X)set)((x : X)Y(x))\Pi(X, Y) : type$ in C, n & o to ix.4.3).

ix.6.3) What does a judgement of the form '$c = d : \Sigma(A, B)$' mean? or What does it mean for two elements of $\Sigma(A, B)$ to be identical?

To explain what a judgement of the form '$c = d : \Sigma(A, B)$' means, one has to know that (i) c and d have the form of two objects formed by the identity rule associated with the element-formation rule of $\Sigma(A, B)$ (c and d are identical canonical elements of $\Sigma(A, B)$) or (ii) c and d can be evaluated to that form (c or d are identical non-canonical elements of $\Sigma(A, B)$).

The identity rule associated with the element-formation rule or second introduction rule of $\Sigma(A, B)$ is

$$\frac{\begin{array}{l} A = C : set \\ B = D : (A)set \\ a = c : A \\ b = d : B(a) \end{array}}{\Sigma I(A, B, a, b) = \Sigma I(C, D, c, d) : \Sigma(A, B)}\ (\Sigma\text{I } 2)$$

The following type can be derived as follows

$$(X : set)(Y : (X)set)(Z : (\Sigma(X, Y))set)((x : X)(y : Y(x))Z(\Sigma I(X, Y, x, y)))$$
$$(z : \Sigma(X, Y))Z(z)$$

$$\cfrac{\cfrac{\cfrac{\begin{array}{l}(X : set)\\(Y : (X)set)\end{array}}{\Sigma(X, Y) : set}}{\Sigma(X, Y) : type} \quad set : type}{(\Sigma(X, Y))set : type}
\qquad
\cfrac{\cfrac{\cfrac{\begin{array}{l}(X : set)\\(Y : (X)set)\\(x : X)\\(y : Y(x))\\ \Sigma I : (X : set)(Y : (X)set)(x : X)(Y(x))\Sigma(X, Y)\end{array}}{\Sigma I(X, Y, x, y) : \Sigma(X, Y)} \quad (Z : (\Sigma(X, Y))set)}{Z(\Sigma I(X, Y, x, y)) : set}}{Z(\Sigma I(X, Y, x, y)) : type}$$

$$\cfrac{\cfrac{\cfrac{\begin{array}{l}(X : set)\\(Y : (X)set)\end{array}}{\Sigma(X, Y) : set}}{\Sigma(X, Y) : type} \quad \cfrac{\cfrac{\begin{array}{l}(z : \Sigma(X, Y))\\(Z : (\Sigma(X, Y))set)\end{array}}{Z(z) : set}}{Z(z) : type}}{(z : \Sigma(X, Y))Z(z) : type}
\qquad
\cfrac{\cfrac{\cfrac{\begin{array}{l}(x : X)\\(Y : (X)set)\end{array}}{Y(x) : set}}{Y(x) : type} \quad \begin{array}{c}|\\ Z(\Sigma I(X, Y, x, y)) : type\end{array}}{(y : Y(x))Z(\Sigma I(X, Y, x, y)) : type}$$

$$\cfrac{\cfrac{\cfrac{(X : set)}{X : type} \qquad \overset{\vdots}{(y : Y(x))Z(\Sigma I(X, Y, x, y)) : type}}{(x : X)(y : Y(x))Z(\Sigma I(X, Y, x, y)) : type} \qquad \overset{\vdots}{(z : \Sigma(X, Y))Z(z) : type}}{((x : X)(y : Y(x))Z(\Sigma I(X, Y, x, y)))(z : \Sigma(X, Y))Z(z) : type}$$

$$\cfrac{set : type \qquad \cfrac{\cfrac{\cfrac{(X : set)}{X : type} \quad set : type}{(X)set : type} \qquad \cfrac{\overset{\vdots}{(\Sigma(X, Y))set : type} \quad \overset{\vdots}{((x : X)(y : Y(x))Z(\Sigma I(X, Y, x, y)))(z : \Sigma(X, Y))Z(z) : type}}{(Z : (\Sigma(X, Y))set)((x : X)(y : Y(x))Z(\Sigma I(X, Y, x, y)))(z : \Sigma(X, Y))Z(z) : type}}{(Y : (X)set)(Z : (\Sigma(X, Y))set)((x : X)(y : Y(x))Z(\Sigma I(X, Y, x, y)))(z : \Sigma(X, Y))Z(z) : type}}{(X : set)(Y : (X)set)(Z : (\Sigma(X, Y))set)((x : X)(y : Y(x))Z(\Sigma I(X, Y, x, y)))(z : \Sigma(X, Y))Z(z) : type}$$

The Σ-elimination axiom states that the selector or eliminatory operator ΣE for the set $\Sigma(A, B)$ is an object of the type just derived, i.e.

$$\begin{aligned}\Sigma E : (X : set)(Y : (X)set)(Z : (\Sigma(X, Y))set)((x : X)(y : Y(x))\\ Z(\Sigma I(X, Y, x, y)))(z : \Sigma(X, Y))Z(z)\end{aligned}$$

The Σ-elimination axiom can be expressed alternatively by the two rules

$$\frac{\begin{array}{l} A : set \\ B : (A)set \\ C : (\Sigma(A, B))set \\ d : (x : A)(y : B(x))C(\Sigma I(A, B, x, y,)) \\ c : \Sigma(A, B) \end{array}}{\Sigma E(A, B, C, d, c) : C(c)} \; (\Sigma\text{E } 1)$$

$$\frac{\begin{array}{l} A : set \\ B : (A)set \\ C : (\Sigma(A, B))set \\ d = f : (x : A)(y : B(x))C(\Sigma I(A, B, x, y)) \\ c = e : \Sigma(A, B) \end{array}}{\Sigma E(A, B, C, d, c) = \Sigma E(A, B, C, f, e) : C(c)} \; (\Sigma\text{E } 2)$$

The meaning of the selector ΣE is explained as follows:

1. Evaluate c to canonical form.
2. If $c \Longrightarrow \Sigma I(A, B, a, b)$ then $\Sigma E(A, B, C, d, c) \Longrightarrow d(a, b)$ and $\Sigma E(A, B, C, d, c) = d(a, b)$.

Justification

of (ΣE 1) Assuming the premisses ...

From the last premiss follows that c evaluates to canonical form. If c evaluates to $\Sigma I(A, B, a, b)$ then, by the meaning explanation of ΣE, the value of $\Sigma E(A, B, C, d, c)$ is identical with the value of $d(a, b)$ which, by the fourth premiss, is a canonical element of $C(\Sigma I(A, B, a, b))$. Since $\Sigma I(A, B, a, b) = c : \Sigma(A, B)$, and by the third premiss $C(\Sigma I(A, B, a, b)) = C(c) : set$, the value of $\Sigma E(A, B, C, d, c)$ is a canonical element of $C(c)$.

of (ΣE 2) Analogous. □

The Σ-equality axiom is

$$\Sigma E(A, B, C, d, \Sigma I(A, B, a, b)) = d(a, b) : C(\Sigma I(A, B, a, b))$$

where $A : set$, $B : (A)set$, $C : (\Sigma(A, B))set$, $d : (x : A)(y : B(x))\ C(\Sigma I(A, B, x, y))$, $a : A$ and $b : B(a)$.

This Σ-equality axiom can also be expressed by the rule

$$\frac{\begin{array}{l} A : set \\ B : (A)set \\ C : (\Sigma(A, B))set \\ d : (x : A)(y : B(x))C(\Sigma I(A, B, x, y)) \\ a : A \\ b : B(a) \end{array}}{\Sigma E(A, B, C, d, \Sigma I(A, B, a, b)) = d(a, b) : C(\Sigma I(A, B, a, b))}\ (\Sigma\text{-equ } 1)$$

which is justified by the meaning explanation of ΣE.

C, n & o

(1) In version (7), Martin-Löf uses at the same time 'ΣI' as well as '*pair*'. The latter can also be found in Nordström et al. And in version (7) the following all denote the same selector ΣE: 'ΣE', 'E', and '*split*', where again the last expression is also Nordström et al.'s notation. 'E' refers back to the polymorphic form of ΣE in version (6).

(2) The left and right projections, i.e. the fcts. for obtaining the value of the left and right coordinate resp. of the pair produced by an arbitrary element of $\Sigma(A, B)$, can be explicitly defined by means of the eliminatory operator ΣE of $\Sigma(A, B)$.

The typing rule and the definitional identity rule for the left projection are in the abbreviated form

$$\frac{\begin{array}{l} A : set \\ B : (A)set \\ c : \Sigma(A, B) \end{array}}{\begin{cases} p(A, B, c) : A \\ p(A, B, c) = \Sigma E(A, B, A, (x)(y)x, c) : A \end{cases}}$$

The typing rule and the definitional identity rule for the right projection are in the abbreviated form

$$\frac{\begin{array}{l} A : set \\ B : (A)set \\ c : \Sigma(A, B) \end{array}}{\begin{cases} q(A, B, c) : B(p(c)) \\ q(A, B, c) = \Sigma E(A, B, B, (x)(y)y, c) : B(p(c)) \end{cases}}$$

The typing rules for $p(A, B, c)$ and $q(A, B, c)$ are just special cases of (ΣE 1) where instead of C one takes A or $B(x)$ with $x : A$ resp. and instead of d one takes

$(x)(y)x$ or $(x)(y)y$ resp. The meaning of the defined selectors p and q is given by the meaning explanation of ΣE. These two facts provide at the same time the justification for the left and right projection.

Accordingly, one gets the corresponding special cases of (Σ-equ), namely

$$\frac{\begin{array}{l} A : set \\ B : (A)set \\ a : A \\ b : B(a) \end{array}}{p(A, B, \Sigma I(A, B, a, b)) = a : A}$$

$$\frac{\begin{array}{l} A : set \\ B : (A)set \\ a : A \\ b : B(a) \end{array}}{q(A, B, \Sigma I(A, B, a, b)) = b : B(a)}$$

(3) The projections p and q make it possible to derive a second Σ-equality judgement

$$c = \Sigma I(A, B, \Sigma E(A, B, A, (x)(y)x, c), \Sigma E(A, B, B, (x)(y)y, c)) : \Sigma(A, B)$$

where $A : set$, $B : (A)set$ and $c : \Sigma(A, B)$.

This judgement can be expressed alternatively by the rule

$$\frac{\begin{array}{l} A : set \\ B : (A)set \\ c : \Sigma(A, B) \end{array}}{c = \Sigma I(A, B, \Sigma E(A, B, A, (x)(y)x, c), \Sigma E(A, B, B, (x)(y)y, c)) : \Sigma(A, B)} \ (\Sigma\text{-equ } 2)$$

This derived Σ-equality judgement (rule) can only be justified after the introduction of set-theoretical identity. (Cf. ix.8)

Let c be the name of a pair of $\Sigma(A, B)$. Then the second Σ-equality judgement is needed to obtain a notation for an element of $\Sigma(A, B)$, of which only the name is known.

(4) The typing rules for p and q plus their corresponding equality rules plus the following rule

$$\frac{\begin{array}{l} A : set \\ B : (A)set \\ c : \Sigma(A, B) \end{array}}{c = \Sigma I(A, B, p(A, B, c), q(A, B, c)) : \Sigma(A, B)}$$

provide a formulation of the Σ-sets which is equivalent to the rules (ΣE 1), (ΣE 2) and (Σ-equ) where Σ-formation and Σ-introduction are the same in both formulations. This is shown in version (6).

(5) Apart from being the disjoint union of a family of sets B over A, $\Sigma(A, B)$ has another interpretation, namely the set of all $a : A$ s.t. $B(a)$ holds.

Let A : *set* and $x : A \vdash B(x)$: *prop*. Suppose one wants to define the set of all $a : A$ s.t. $B(a)$ holds, i.e. the set $\{x : A | B(x)\}$. To have an a s.t. $B(a)$ constructively means to have an element $a : A$ and a proof of $B(a)$, i.e. an element (or proof-object) $b : B(a)$. Thus, the elements of the set of all elements of A satisfying $B(x)$ are pairs (a, b) with $b : B(a)$, which in monomorphic notation is written $\Sigma I(A, B, a, b)$ with $b : B(a)$. That is, they are elements of $\Sigma(A, B)$. (This note presupposes the Curry–Howard isomorphism, cf. Sect. xi.)

The Σ-axioms (or rules) then play the role of the comprehension axiom or separation principle of ZF set theory.

Example (the real numbers as Cauchy sequences)
The set of reals R is explicitly defined as follows

$$\frac{\begin{array}{l} N : set \\ Q : set \\ N \to Q : set \\ Cauchy : (N \to Q)prop \end{array}}{\begin{cases} R : set \\ R = \Sigma(N \to Q, Cauchy) : set \end{cases}}$$

which is the def. of the reals as sequences of rational numbers satisfying the Cauchy condition

$$Cauchy(a) = (\forall d : Q)(d > 0 \supset (\exists m : N)(\forall n : N)(|a_{m+n} - a_m|) \leq d)$$

where $a : N \to Q$, i.e. a is a sequence $a_0, a_1, a_2, \ldots$

Thus, a real number is here defined to be a sequence of rationals and a proof that it satisfies the Cauchy condition. If $c : R$, then the projections yield

$$\Sigma E(N \to Q, Cauchy, N \to Q, (x)(y)x, c) : N \to Q \quad \text{and}$$

$$\Sigma E(N \to Q, Cauchy, Cauchy, (x)(y)y, c) : Cauchy(\Sigma E(N \to Q,$$
$$Cauchy, N \to Q, (x)(y)x, c))$$

ix.7.1) What does a judgement of the form '$A \times B$: *set*' mean? or What does it mean to be a cartesian product of a set A and a set B?
To make a judgement of the form '$A \times B$: *set*' one has to know (i) what a canonical element of $A \times B$ is, and (ii) when two canonical elements of $A \times B$ are identical.

The formation axiom for the composite set $A \times B$ is

$$\times : (set)(set)set$$

The formation axiom is tantamount to the following rules

$$\frac{A : set \quad B : set}{A \times B : set}\,(\times F\,1) \qquad \frac{A = C : set \quad B = D : set}{A \times B = C \times D : set}\,(\times F\,2)$$

C, n & o

(1) A cartesian product of two sets is nothing but a special case of a Σ-set in which B is not a family of sets over A, but B is a set independent of A, i.e. $B : (A)set$ is s.t. $B : set$. $A \times B$ is the set of pairs whose first coordinate is an element of A and whose second coordinate is an element of B.

(2) $\times$ can also be explicitly defined by Σ as follows: the typing rule and the definitional identity rule are in the abbreviated form

$$\frac{A : set \quad B : set}{\begin{cases} A \times B : set \\ A \times B = \Sigma(A, (x)B) : set \end{cases}}$$

The $\times$-introduction-, $\times$-elimination- and $\times$-equality rules will not be given, as they can easily be recovered from the corresponding Σ-axioms or Σ-rules.

C, n & o

The set of integers Z and the set of rationals Q can be introduced by explicit def. as follows:

Integers

$$\frac{\begin{array}{l} X : set, Y : set \vdash \ \times : set \\ N : set \end{array}}{\begin{cases} Z : set \\ Z = \ \times(N/X, N/Y) : set \end{cases}}$$

and identity on Z, i.e. $\underbrace{(a, b)}_{a-b} = \underbrace{(c, d)}_{c-d} : Z$ for $a : N,\ b : N,\ c : N$ and $d : N$, is defined by the fact that there is a term t s.t.

$$t : I(Z, (a, b), (c, d)) \supset\subset I(N, a + d, b + c)$$

Note that Z can be introduced alternatively as a primitive set (like N) as follows:

$$Z : set \text{ (ZF 1)} \qquad Z = Z : set \text{ (ZF 2)} \qquad \text{and}$$

$$0 : Z \text{ (ZI 1)} \qquad \frac{n : N}{+n : Z}\text{ (ZI 2)} \qquad \frac{n : N}{-n : Z}\text{ (ZI 3)}$$

$$0 = 0 : Z \text{ (ZI 4)} \qquad \frac{n = m : N}{+n = +m : Z}\text{ (ZI 5)} \qquad \frac{n = m : N}{-n = -m : Z}\text{ (ZI 6)}$$

where N in this case is the set $\{1, 2, 3, \dots\}$ of positive integers.

Rationals

$$\frac{\begin{array}{l} X : set, Y : set \vdash \ \times : set \\ Z : set \\ N \setminus \{0\} : set \end{array}}{\begin{cases} Q : set \\ Q = \ \times(Z/X, N \setminus \{0\}/Y) : set \end{cases}}$$

and identity on Q, i.e. $\underbrace{(a,b)}_{\frac{a}{b}} = \underbrace{(c,d)}_{\frac{c}{d}} : Q$ for $a : Z,\ c : Z,\ b : N \setminus \{0\},\ d : N \setminus \{0\}$ is defined by the fact that there is a term t s.t.

$$t : I(Q, (a,b), (c,d)) \supset\subset I(Z, a \cdot d, b \cdot c)$$

Note that Q can likewise be introduced as a primitive set (like N or Z):

$$Q : set \ \text{(QF 1)} \qquad Q = Q : set \ \text{(QF 2)} \qquad \text{and}$$

$$\frac{m : Z \quad n : N \setminus \{0\}}{\frac{m}{n} : Q} \text{(QI 1)} \qquad \frac{m = m' : Z \quad n = n' : N \setminus \{0\}}{\frac{m}{n} = \frac{m'}{n'} : Q} \text{(QI 2)}$$

ix.8.1) What does a judgement of the form '$I(A, a, b) : set$' mean? or What does it mean to be a set expressing that a and b are (intensionally) identical elements of a set A?
To make a judgement of the form '$I(A, a, b) : set$' one has to know (i) what a canonical element of $I(A, a, b)$ is, and (ii) when two canonical elements of $I(A, a, b)$ are identical.

The formation axiom for the composite set $I(A, a, b)$ is

$$I : (X : set)(X)(X)set$$

Equivalently, the formation axiom can be given by the two rules

$$\frac{A : set \quad a : A \quad b : A}{I(A, a, b) : set} \text{(IF 1)} \qquad \frac{A = C : set \quad a = c : A \quad b = d : A}{I(A, a, b) = I(C, c, d) : set} \text{(IF 2)}$$

ix.8.2) What does a judgement of the form '$c : I(A, a, b)$' mean? or What does it mean to be an element of the intensional identity set $I(A, a, b)$?
To explain what a judgement of the form '$c : I(A, a, b)$' means one has to know that (i) c has the form of an object formed by the element-formation rule of $I(A, a, b)$ (c has the form of a canonical element of $I(A, a, b)$) or (ii) c can be evaluated to that form (c is a non-canonical element of $I(A, a, b)$).

The element-formation or introduction axiom of $I(A, a, b)$ is

$$II : (X : set)(x : X)I(X, x, x)$$

which can be expressed alternatively by the introduction rule

$$\frac{A : set \quad a : A}{II(A, a) : I(A, a, a)} \text{(II 1)}$$

C, n & o

(1) By means of the rule (II 1) it is possible to derive another I-introduction rule, namely

$$\frac{a = b : A}{II(A, a) : I(A, a, b)}$$

as follows:

$$\frac{II(A, a) : I(A, a, a) \qquad \dfrac{a = b : A}{I(A, a, a) = I(A, a, b) : set}}{II(A, a) : I(A, a, b)}$$

(2) Set-theoretical identity I can thus be seen to mimic judgemental identity inside constructive set theory.

ix.8.3) What does a judgement of the form '$c = d : I(A, a, b)$' mean? or What does it mean for two elements of an intensional identity set $I(A, a, b)$ to be identical?
To explain what a judgement of the form '$c = d : I(A, a, b)$' means, one has to know that (i) c and d have the form of two objects formed by the identity rule associated with the element-formation rule of $I(A, a, b)$ (c and d are identical canonical elements of $I(A, a, b)$) or (ii) c and d can be evaluated to that form (c or d are identical non-canonical elements of $I(A, a, b)$).

The identity rule associated with the element-formation rule or introduction rule of $I(A, a, b)$ is

$$\frac{A = B : set \quad a = b : A}{II(A, a) = II(B, b) : I(A, a, b)} \text{ (II 2)}$$

C, n & o

The following type can be derived as follows:

$$(X : set)(x : X)(y : X)(Z : ((x : X)(y : X)I(X, x, y))set)$$
$$((w : X)Z(w, w, II(X, w)))(z : I(X, x, y))Z(x, y, z) : type$$

$$\frac{\dfrac{(X : set)}{X : type} \qquad \dfrac{\dfrac{\begin{array}{l}(X : set)\\(x : X)\\(y : X)\end{array}}{I(X, x, y) : set}}{I(X, x, y) : type}}{(y : X)I(X, x, y) : type} \qquad \frac{\dfrac{\begin{array}{l}(w : X)\\II(X, w) : I(X, w, w)\\(Z : ((x : X)(y : X)I(X, x, y))set)\end{array}}{Z(w, w, II(X, w)) : set}}{Z(w, w, II(X, w)) : type}$$

$$\frac{\dfrac{\begin{array}{l}(X : set)\\(x : X)\\(y : X)\end{array}}{I(X, x, y) : set}}{I(X, x, y) : type} \qquad \frac{\dfrac{\begin{array}{l}(x : X)\\(y : X)\\(z : I(X, x, y))\\(Z : ((x : X)(y : X)I(X, x, y))set)\end{array}}{Z(x, y, z) : set}}{Z(x, y, z) : type}$$

$$
\dfrac{\dfrac{\dfrac{(X:set)}{X:type} \quad \begin{matrix}\vdots\\ (y:X)I(X,x,y):type\end{matrix}}{(x:X)(y:X)I(X,x,y):type} \quad set:type}{((x:X)(y:X)I(X,x,y))set:type}
\quad
\dfrac{\dfrac{\dfrac{(X:set)}{X:type} \quad \begin{matrix}\vdots\\ Z(w,w,II(X,w)):type\end{matrix}}{(w:X)Z(w,w,II(X,w)):type} \quad \dfrac{\begin{matrix}\vdots\\ I(X,x,y):type\end{matrix} \quad \begin{matrix}\vdots\\ Z(x,y,z):type\end{matrix}}{(z:I(X,x,y))Z(x,y,z):type}}{((w:X)Z(w,w,II(X,w)))(z:I(X,x,y))Z(x,y,z):type}
$$
$$
\overline{(Z:((x:X)(y:X)I(X,x,y))set)((w:X)Z(w,w,II(X,w)))(z:I(X,x,y))Z(x,y,z):type}
$$

$$
\dfrac{set:type \quad \dfrac{\dfrac{(X:set)}{X:type} \quad \dfrac{\dfrac{(X:set)}{X:type} \quad \begin{matrix}\vdots\\ (Z:((x:X)(y:X)I(X,x,y))set)((w:X)Z(w,w,II(X,w)))(z:I(X,x,y))Z(x,y,z):type\end{matrix}}{(y:X)(Z:((x:X)(y:X)I(X,x,y))set)((w:X)Z(w,w,II(X,w)))(z:I(X,x,y))Z(x,y,z):type}}{(x:X)(y:X)(Z:((x:X)(y:X)I(X,x,y))set)((w:X)Z(w,w,II(X,w)))(z:I(X,x,y))Z(x,y,z):type}}{(X:set)(x:X)(y:X)(Z:((x:X)(y:X)I(X,x,y))set)((w:X)Z(w,w,II(X,w)))(z:I(X,x,y))Z(x,y,z):type}
$$

The I-elimination axiom states that the selector or eliminatory operator for the set $I(A,a,b)$ is an object of the type just derived, i.e.

$$
\begin{aligned}
IE : {} & (X:set)(x:X)(y:X)(Z:((x:X)(y:X)I(X,x,y))set)((w:X) \\
& Z(w,w,II(X,w)))(z:I(X,x,y))Z(x,y,z)
\end{aligned}
$$

The I-elimination axiom can be expressed alternatively by the two rules

$$
\dfrac{\begin{array}{l} A:set \\ a:A \\ b:A \\ C:((x:A)(y:A)I(A,x,y))set \\ d:(w:A)C(w,w,II(A,w)) \\ c:I(A,a,b) \end{array}}{IE(A,a,b,C,d,c):C(a,b,c)} \; \text{(IE 1)}
$$

$$
\dfrac{\begin{array}{l} A:set \\ a=e:A \\ b=f:A \\ C:((x:A)(y:A)I(A,x,y))set \\ d=g:(w:A)C(w,w,II(A,w)) \\ c=h:I(A,a,b) \end{array}}{IE(A,a,b,C,d,c)=IE(A,e,f,C,g,h):C(a,b,c)} \; \text{(IE 2)}
$$

The meaning of the selector IE is explained as follows:

1. Evaluate c to canonical form.
2. If $c \Longrightarrow II(A,a)$ then $IE(A,a,b,C,d,c) \Longrightarrow d(a)$ and $IE(A,a,b,C,d,c) = d(a)$.

Justification

of (IE 1) Assuming the premisses ...

From the last premiss follows that c evaluates to canonical form. If $c \Longrightarrow II(A,a)$ then, by the meaning explanation of IE, the value of $IE(A,a,b,C,d,c)$ is identical with the value of $d(a)$ which, by the fifth premiss, is a canonical element of $C(a, a, II(A, a))$. Because of this and because $II(A, a) = c : I(A, a, b)$ and by the fourth premiss, $C(a, b, II(A, a)) = C(a, b, c) : set$ and the value of $IE(A, a, b, C, d, c)$ is a canonical element of $C(a, b, c)$.

of (IE 2) Analogous. □

The I-equality axiom is

$$IE(A, a, b, C, d, II(A, a)) = d(a) : C(a, b, II(A, a))$$

where $A : set$, $a : A$, $b : A$, $C : ((x : A)(y : A)I(A, x, y))set$ and $d : (w : A)C(w, w, II(A, w))$.

This I-equality axiom can also be expressed by the rule

$$\frac{\begin{array}{l} A : set \\ a : A \\ b : A \\ C : ((x : A)(y : A)I(A, x, y))set \\ d : (w : A)C(w, w, II(A, w)) \end{array}}{IE(A, a, b, C, d, II(A, a)) = d(a) : C(a, b, II(A, a))} \text{(I-equ)}$$

Justification
of (I-equ): straightforward by the meaning explanation of IE. □

C, n & o

(1) Martin-Löf uses in version (8) for the introductory operator II the symbols 'II' as well as 'r' (for reflexivity), the latter of which he already used in version (6) although in a different sense. Nordström et al. use 'Id' instead of 'I', 'id' instead of 'II' and '*idpeel*' instead of 'IE'.

(2)[75] To prove symmetry of the set-theoretical identity choose $C = (x)(y)(z) I(A, y, x) : set$ and let $e : I(A, a, b)$ where $A : set$, $a : A$ and $b : A$. Then (IE 1) yields

$$\frac{\begin{array}{l} A : set \\ a : A \\ b : A \\ (x)(y)(z)I(A, y, x) : ((x : A)(y : A)I(A, x, y))set \\ II(A) : (x : A)I(A, x, x) \\ e : I(A, a, b) \end{array}}{IE(A, a, b, (x)(y)(z)I(A, y, x), II(A, a), e) : I(A, b, a)}$$

To prove transitivity of set-theoretical identity, let $A : set$, $a : A$, $b : A$ and $c : A$, $e : I(A, a, b)$ and $f : I(A, b, c)$. Choose $C = (x)(y)(z)(I(A, y, c) \to I(A, x, c))$. Then (IE 1) and subsequently the alternative (→E 1) yield

$A : set$
$a : A$
$b : A$
$e : I(A, a, b)$
$(x)(y)(z)(I(A, y, c) \to I(A, x, c)) : ((x : A)(y : A)I(A, x, y))set$

$$\frac{\to I(I(A,x,c),I(A,x,c),(x)x):(w:A)(I(A,w,c)\to I(A,w,c))}{IE(A,a,b,(x)(y)(z)(I(A,y,c)\to I(A,x,c)),\to I(I(A,x,c),I(A,x,c),(x)x),e):I(A,b,c)\to I(A,a,c)}$$

$$\frac{f:I(A,b,c)}{\to E(I(A,b,c),I(A,a,c),IE(A,a,b,(x)(y)(z)(I(A,y,c)\to I(A,x,c)),\to I(I(A,x,c),I(A,x,c),(x)x),e),f):I(A,a,c)}$$

(3) Note that the formation axiom for the composite set $I(A, a, b)$ is the first axiom which allows the formation of dependent types or families of types. If only the formation axioms for the sets N_n, N, $A+B$, $\Pi(A, B)$, $A \to B$, $\Sigma(A, B)$ and $A \times B$ were given, one could only generate independent or constant types. Another axiom which also allows for the formation of dependent types is the introduction axiom for the composite set U (cf. the next point ix.9))

(4) In versions (5) and (6) Martin-Löf uses a different notion of identity from the one introduced above: he then has an extensional rather than an intensional notion:

ix.8.4) What does a judgement of the form '$E(A, a, b)$: *set*' mean? or What does it mean to be a set expressing that a and b are extensionally identical elements of a set A?
To make a judgement of the form '$E(A, a, b)$: *set*' one has to know (i) what a canonical element of $E(A, a, b)$ is, and (ii) when two canonical elements of $E(A, a, b)$ are identical.

The formation axiom for the composite set $E(A, a, b)$ is

$$E : (X : set)(X)(X)set$$

Equivalently, the formation axiom can be given by the two rules

$$\frac{A : set \quad a : A \quad b : A}{E(A, a, b) : set} \text{(EF 1)} \qquad \frac{A = C : set \quad a = c : A \quad b = d : A}{E(A, a, b) = E(C, c, d) : set} \text{(EF 2)}$$

ix.8.5) What does a judgement of the form '$c : E(A, a, b)$' mean? or What does it mean to be an element of the extensional identity set $E(A, a, b)$?
To explain what a judgement of the form '$c : E(A, a, b)$' means, one has to know that (i) c has the form of an object formed by the element-formation rule of $E(A, a, b)$ (c has the form of a canonical element of $E(A, a, b)$) or (ii) c can be evaluated to that form (c is a non-canonical element of $E(A, a, b)$).

The element-formation or introduction axiom of $E(A, a, b)$ is

$$EI : E(X, x, x) \quad \text{provided } X : set \text{ and } x : X$$

which can be expressed alternatively by the introduction rule

$$\frac{A : set \quad a : A}{EI : E(A, a, a)} \text{(EI 1)}$$

C, n & o

By means of the rule (EI 1) it is possible to derive another E-introduction rule, namely

$$\frac{a = b : A}{EI : E(A, a, b)}$$

basically the same way as $\dfrac{a = b : A}{II(A, a) : I(A, a, b)}$ was derived from (II 1).

ix.8.6) What does a judgement of the form '$c = d : E(A, a, b)$' mean? or What does it mean for two elements of an extensional identity set $E(A, a, b)$ to be identical?

To explain what a judgement of the form '$c = d : E(A, a, b)$' means, one has to know that (i) c and d have the form of two objects formed by the identity rule associated with the element-formation rule of $E(A, a, b)$ (c and d are identical canonical elements of $E(A, a, b)$) or (ii) c and d can be evaluated to that form (c or d are identical non-canonical elements of $E(A, a, b)$).

The identity rule associated with the element-formation rule or introduction rule of $E(A, a, b)$ is

$$\frac{A = B : set \quad a = b : A}{EI = EI : E(A, a, b)} \text{ (EI 2)}$$

The E-identity set has a strong elimination rule of a form different from the ones of all the other sets and set-forming operations; its elimination rule has no selector or eliminatory operator and it is not a structural induction principle. The strong E-elimination rules are

$$\frac{\begin{array}{l} A : set \\ a : A \\ b : A \\ c : E(A, a, b) \end{array}}{a = b : A} \text{ (EE 1)} \qquad \frac{\begin{array}{l} A : set \\ a : A \\ b : A \\ c : E(A, a, b) \end{array}}{c = EI : E(A, a, b)} \text{ (EE 2)}$$

Naturally, since the EE-rules contain no eliminatory operator, there cannot be an E-equality rule.

C, n & o

(1) What is here referred to by 'E' (extensional identity set) is denoted by 'I' in versions (5) and (6) ('I' is used by Martin-Löf in version (8) for the intensional identity set). Nordström et al. use 'Eq' for 'E'. Moreover, Martin-Löf uses 'r' in versions (5) and (6) to denote EI and Nordström et al. use 'eq'.

(2)Nordström et al. mention that it is possible to derive structural induction rules for E corresponding to the rules (IE 1) and (IE 2); they will be called weak E-elimination axioms and rules in the sequel.[76]

The weak E-elimination axiom states that the selector or eliminatory operator for the set $E(A, a, b)$ is an object of the following type

$$\begin{aligned} EE : (X : set)(x : X)(Z : ((x : X)(y : X)E(X, x, y))set)((x : X)\\ (y : X)(z : E(X, x, y))Z(x, y, z))(z : E(X, x, y))Z(x, y, z) \end{aligned}$$

This E-elimination axiom can be expressed alternatively by the two rules

$$\frac{\begin{array}{l} A : set \\ a : A \\ b : A \\ C : ((x : A)(y : A)E(A, x, y))set \\ d : (w : A)C(w, w, EI) \\ c : E(A, a, b) \end{array}}{EE(A, a, b, C, d, c) : C(a, b, c)} \text{(wEE 1)}$$

$$\frac{\begin{array}{l} A : set \\ a = e : A \\ b = f : A \\ C : ((x : A)(y : A)E(A, x, y))set \\ d = g : (w : A)C(w, w, EI) \\ c = h : E(A, a, b) \end{array}}{EE(A, a, b, C, d, c) = EE(A, e, f, C, g, h) : C(a, b, c)} \text{(wEE 2)}$$

The E-equality axiom is

$$EE(A, a, b, C, d, EI) = d(a) : C(a, b, EI)$$

where $A : set$, $a : A$, $b : A$, $C : ((x : A)(y : A)E(A, x, y))set$ and $d : (w : A)C(w, w, EI)$.

This E-equality axiom can also be expressed by the rule

$$\frac{\begin{array}{l} A : set \\ a : A \\ b : A \\ C : ((x : A)(y : A)E(A, x, y))set \\ d : (w : A)C(w, w, EI) \end{array}}{EE(A, a, b, C, d, EI) = d(a) : C(a, b, EI)} \text{(wE-equ)}$$

The weak E-elimination rule (wEE 1) is derived from the strong rules (EE 1) and (EE 2) as follows:

$$\dfrac{\dfrac{\dfrac{\begin{array}{l} A : set \\ a : A \\ b : A \\ C : ((x : A)(y : A)E(A, x, y))set \end{array}}{C(a, b) : (E(A, a, b))set} \quad \dfrac{\begin{array}{l} A : set \\ a : A \\ b : A \\ c : E(A, a, b) \end{array}}{c = EI : E(A, a, b)}}{C(a, b, c) = C(a, b, EI) : set} \quad \dfrac{\dfrac{\begin{array}{l} A : set \\ a : A \\ b : A \\ c : E(A, a, b) \end{array}}{a = b : A} \quad d : (w : A)C(w, w, EI)}{d(a) : C(a, b, EI)}}{d(a) : C(a, b, c)}$$

(3) If the well-ordered set-operator W or the tree operator *Tree* are introduced into constructive type theory, then E-identity is required to get the (WE 1), (WE 2), (W-equ)- as well as the (TreeE 1), (TreeE 2)- and (Tree-equ)-rules. (cf. Sect. iv, Chapter 3)

(4) In C, n & o (5) to Π-equality, a second derived Π-equality rule (Π-equ 2) was mentioned, which is nothing else but η-conversion on a judgemental level; it was the rule

$$\frac{\begin{array}{l} A : set \\ B : (A)set \\ C : (\Pi(A, B))set \\ d : (y : (x : A)B(x))C(\Pi I(A, B, y)) \\ c : \Pi(A, B) \end{array}}{c = \Pi I(A, B, (x)\Pi E(A, B, C, d, c)) : \Pi(A, B)}$$

which can now be justified by means of E-identity.

Justification

By (Π-equ 1) one has $\Pi E(A, B, (z)B(a), (y)y(a), \Pi I(A, B, b)) = ((y)y(a))(b) = b(a) : B(a)$. Hence

$$\frac{\frac{\frac{x : A \vdash \Pi E(A, B, (z)B(x), (y)y(x), \Pi I(A, B, b)) = b(x) : B(x)}{(x)\Pi E(A, B, (z)B(x), (y)y(x), \Pi I(A, B, b)) = b : (x : A)B}}{\Pi I(A, B, (x)\Pi E(A, B, (z)B(x), (y)y(x), \Pi I(A, B, b))) = \Pi I(A, B, b) : \Pi(A, B)}}{EI : E(\Pi(A, B), \Pi I(A, B, (x)\Pi E(A, B, (z)B(x), (y)y(x), \Pi I(A, B, b))), \Pi I(A, B, b))}$$

$$\frac{\begin{array}{l} c : \Pi(A, B) \\ \hline \Pi E(A, B, E(\Pi(A, B), \Pi I(A, B, (x)\Pi E(A, B, (z)B(x), (y)y(x), \Pi I(A, B, b))), \Pi I(A, B, b)), \\ \quad EI, c) : E(\Pi(A, B), \Pi I(A, B, (x)\Pi E(A, B, (z)B(x), (y)y(x), c)), c) \end{array}}{\Pi I(A, B, (x)\Pi E(A, B, (z)B(x), (y)y(x), c)) = c : \Pi(A, B)}$$

□

Note that if only I-identity is available, one can only prove that there is a term t s.t.

$t : I(\Pi(A, B), \Pi I(A, B, (x)\Pi E(A, B, (z)B(x), (y)y(x), c)), c)$ provided $c : \Pi(A, B)$.[77]

By (Π-equ),

$$\Pi I(A, B, (x)\Pi E(A, B, (z)B(x), \Pi I(A, B, (y)y), x)) \\ = \Pi I(A, B, (y)y) : \Pi(A, B)$$

where $x : A$ and $x : A \vdash y(x) : B(x)$; then by (II)

$$II(\Pi(A, B), \Pi I(A, B, (x)\Pi E(A, B, (z)B(x), \Pi I(A, B, (y)y), x))) : \\ I(\Pi(A, B), \Pi I(A, B, (x)\Pi E(A, B, (z)B(x), \Pi I(A, B, (y)y), x)), \\ \Pi I(A, B, (y)y))$$

and by (ΠE)

$$\Pi E(A, B, I(\Pi(A, B), \Pi I(A, B, (x)\Pi E(A, B, (z)B(x), \Pi I(A, B, (y)y), x)), \\ II(\Pi(A, B), \Pi I(A, B, (x)\Pi E(A, B, (z)B(x), \Pi I(A, B, (y)y), x))), c) \\ : I(\Pi(A, B), \Pi I(A, B, (x)\Pi E(A, B, (z)B(x), (y)y(x), c)), c)$$

In C, n & o (3) to Σ-equality, a second derived Σ-equality rule (Σ-equ 2) was presented, namely

$$\frac{A : set \quad B : (A)set \quad c : \Sigma(A, B)}{c = \Sigma I(A, B, \Sigma E(A, B, A, (x)(y)x, c), \Sigma E(A, B, B, (x)(y)y, c)) : \Sigma(A, B)}$$

which can now be justified by means of E-identity as follows

Justification

$$\dfrac{\dfrac{\dfrac{\dfrac{A : set \quad B : (A)set \quad (x : A) \quad (y : B(x))}{p(A, B, \Sigma I(A, B, x, y)) = x : A} \qquad \dfrac{A : set \quad B : (A)set \quad (x : A) \quad (y : B(x))}{q(A, B, \Sigma I(A, B, x, y)) = y : B(x)}}{\Sigma I(A, B, p(A, B, \Sigma I(A, B, x, y)), q(A, B, \Sigma I(A, B, x, y))) = \Sigma I(A, B, x, y) : \Sigma(A, B)}}{EI : E(\Sigma(A, B), \Sigma I(A, B, p(A, B, \Sigma I(A, B, x, y)), q(A, B, \Sigma I(A, B, x, y))), \Sigma I(A, B, x, y)) \quad c : \Sigma(A, B)}}{\dfrac{\Sigma E(A, B, (x)(y)E(\Sigma(A, B), \Sigma I(A, B, p(A, B, \Sigma I(A, B, x, y)), q(A, B, \Sigma I(A, B, x, y))), \Sigma I(A, B, x, y)), (x)(y)EI, c) : E(\Sigma(A, B), \Sigma I(A, B, p(A, B, c), q(A, B, c)), c)}{\Sigma I(A, B, p(A, B, c), q(A, B, c)) = c : \Sigma(A, B)}}$$

□

(5) Martin-Löf remarks that whereas the existence of a fct. is given by some set elimination rule and the properties of this fct. are expressed by the equality rule that goes with the elimination rule, the uniqueness of a fct. is systematically demonstrated using the E-identity rules.

ix.9.1) What does a judgement of the form '$U : set$' mean? or What does it mean to be the set of small sets (or the first universe)?
To make the judgement '$U : set$' one has to know (i) what a canonical element of U is, and (ii) when two canonical elements of U are identical.

The formation axioms or rules for the composite set U are

$$U : set \text{ (UF 1)} \qquad U = U : set \text{ (UF 2)}$$

ix.9.2) What does a judgement of the form '$A : U$' mean? or What does it mean to be a small set (or an element of the first universe)?
To explain what a judgement of the form '$A : U$' means, one has to know that (i) A has the form of an object formed by one of the element-formation rules of U (A has the form of a canonical element of U) or (ii) A can be evaluated to one of those forms (A is a non-canonical element of U).

The element-formation or introduction axioms of U are

$UI1 : (U)U$	where $UI1$	can also be denoted by	'$\hat{S}$'
$UI2 : U$	where $UI2$	"	'$\hat{N}$'
$UI3 : (U)(U)U$	where $UI3$	"	'$\hat{+}$'
$UI4 : (X : U)((Set(X))U)U$	where $UI4$	"	'$\hat{\Pi}$'

$UI5 : (U)(U)U$	where $UI5$	can also be denoted by	'$\hat{\rightarrow}$'
$UI6 : (X : U)((Set(X))U)U$	where $UI6$	"	'$\hat{\Sigma}$'
$UI7 : (U)(U)U$	where $UI7$	"	'$\hat{\times}$'
$UI8 : (X : U)(Set(X))(Set(X))U$	where $UI8$	"	'$\hat{I}$'

The *Set* operator on U is (partially or conditionally) defined as follows:

typing axiom (rule) $\quad Set : (U)set$
definitional identity axioms

$Set(UI1(X)) = S(Set(X)) : set$	where $X : U$
$Set(UI2) = N : set$	
$Set(UI3(X, Y)) = Set(X) + Set(Y) : set$	where $X : U,\ Y : U$
$Set(UI4(X, Y)) = \Pi(Set(X), (v)Set(Y(v))) : set$	where $X : U, Y : (Set(X))U,\ v : Set(X)$
$Set(UI5(X, Y)) = Set(X) \rightarrow Set(Y) : set$	where $X : U,\ Y : U$
$Set(UI6(X, Y)) = \Sigma(Set(X), (v)Set(Y(v))) : set$	where $X : U, Y : (Set(X))U,\ v : Set(X)$
$Set(UI7(X, Y)) = Set(X) \times Set(Y) : set$	where $X : U,\ Y : U$
$Set(UI8(X, y, z)) = I(Set(X), y, z) : set$	where $X : U,\ y : Set(X), z : Set(X)$

Basically all these definitional identity axioms are given as rules in Nordström et al.[78] U and *Set* obviously have to be defined simultaneously.

The element-formation axioms of U can be expressed alternatively by the introduction rules

$$\frac{A : U}{UI1(A) : U}\ (\text{UI 1}) \qquad UI2 : U\ (\text{UI 2}) \qquad \frac{A : U \quad B : U}{UI3(A, B) : U}\ (\text{UI 3})$$

$$\frac{A : U \quad B : (Set(A))U}{UI4(A, B) : U}\ (\text{UI 4}) \qquad \frac{A : U \quad B : U}{UI5(A, B) : U}\ (\text{UI 5})$$

$$\frac{A : U \quad B : (Set(A))U}{UI6(A, B) : U}\ (\text{UI 6}) \qquad \frac{A : U \quad B : U}{UI7(A, B) : U}\ (\text{UI 7})$$

$$\frac{A : U \quad b : Set(A) \quad c : Set(A)}{UI8(A, b, c) : U}\ (\text{UI 8})$$

C, n & o

The idea behind all this is that each canonical element of U represents or codes a set formed by means of the set forming operators mentioned earlier, i.e. by S, N, $+$, Π, $\rightarrow$, Σ, $\times$ and I. Simultaneously with the def. of the canonical elements of U, an operator *Set* is defined which decodes the elements in the universe U to the sets they represent or code.[79] This is the so-called Tarski formulation of version (6).

ix.9.3) What does a judgement of the form '$A = B : U$' mean? or What does it mean for two small sets to be identical?
To explain what a judgement of the form '$A = B : U$' means, one has to know that (i) A and B have the form of two objects formed by one of the identity rules associated with the element-formation rules of U (A and B are identical canonical elements of U) or (ii) A and B can be evaluated to one of those forms (A or B are identical non-canonical elements of U).

The identity rules associated with the element-formation rules or introduction rules of U are

$$\frac{A = B : U}{UI1(A) = UI1(B) : U} \text{ (UI 9)} \qquad UI2 = UI2 : U \text{ (UI 10)}$$

$$\frac{A = C : U \quad B = D : U}{UI3(A, B) = UI3(C, D) : U} \text{ (UI 11)} \qquad \frac{A = C : U \quad B = D : (Set(A))U}{UI4(A, B) = UI4(C, D) : U} \text{ (UI 12)}$$

$$\frac{A = C : U \quad B = D : U}{UI5(A, B) = UI5(C, D) : U} \text{ (UI 13)} \qquad \frac{A = C : U \quad B = D : (Set(A))U}{UI6(A, B) = UI6(C, D) : U} \text{ (UI 14)}$$

$$\frac{A = C : U \quad B = D : U}{UI7(A, B) = UI7(C, D) : U} \text{ (UI 15)}$$

$$\frac{A = B : U \quad a = c : Set(A) \quad b = d : Set(A)}{UI8(A, a, b) = UI8(B, c, d) : U} \text{ (UI 16)}$$

C, n & o

The following type can be derived:

$$\begin{aligned}&(X : U)(Y : U)(Z : (U)set)((Z(X))Z(UI1(X)))(Z(UI2))((Z(X))(Z(Y))Z(UI3(X, Y)))\\ &((Z(X))((v : Set(X))Z(Y(v)))Z(UI4(X, Y)))((Z(X))(Z(Y))Z(UI5(X, Y)))((Z(X))\\ &((v : Set(X))Z(Y(v)))Z(UI6(X, Y)))((Z(X))(Z(Y))Z(UI7(X, Y)))((x : Set(X))\\ &(y : Set(X))(Z(X))Z(UI8(X, x, y)))(z : U)Z(z) : type\end{aligned}$$

Derivation: exercise

The U-elimination axiom states that the selector or eliminatory operator for the set U is just an object of this type, i.e.

$$\begin{aligned}UE : &(X : U)(Y : U)(Z : (U)set)((Z(X))Z(UI1(X)))(Z(UI2))((Z(X))(Z(Y))\\ &Z(UI3(X, Y)))((Z(X))((v : Set(X))Z(Y(v)))Z(UI4(X, Y)))((Z(X))(Z(Y))\\ &Z(UI5(X, Y)))((Z(X))((v : Set(X))Z(Y(v)))Z(UI6(X, Y)))((Z(X))(Z(Y))\\ &Z(UI7(X, Y)))((x : Set(X))(y : Set(X))(Z(X))Z(UI8(X, x, y)))(z : U)Z(z)\end{aligned}$$

The U-elimination axiom can be expressed alternatively by the two rules

$$\frac{\begin{array}{l}C : (U)set\\ d_1 : (X : U)(C(X))C(UI1(X))\\ d_2 : C(UI2)\\ d_3 : (X : U)(Y : U)(C(X))(C(Y))C(UI3(X, Y))\\ d_4 : (X : U)(Y : (Set(X))U)(C(X))((v : Set(X))C(Y(v)))C(UI4(X, Y))\\ d_5 : (X : U)(Y : U)(C(X))(C(Y))C(UI5(X, Y))\\ d_6 : (X : U)(Y : (Set(X))U)(C(X))((v : Set(X))C(Y(v)))C(UI6(X, Y))\\ d_7 : (X : U)(Y : U)(C(X))(C(Y))C(UI7(X, Y))\\ d_8 : (X : U)(y : Set(X))(z : Set(X))(C(X))C(UI8(X, y, z))\\ c : U\end{array}}{UE(C, d_1, d_2, d_3, d_4, d_5, d_6, d_7, d_8, c) : C(c)} \text{ (UE 1)}$$

$$\frac{\begin{array}{l} C : (U)set \\ d_1 = e_1 : (X : U)(C(X))C(UI1(X)) \\ d_2 = e_2 : C(UI2) \\ d_3 = e_3 : (X : U)(Y : U)(C(X))(C(Y))C(UI3(X, Y)) \\ d_4 = e_4 : (X : U)(Y : (Set(X))U)(C(X))((v : Set(X))C(Y(v)))C(UI4(X, Y)) \\ d_5 = e_5 : (X : U)(Y : U)(C(X))(C(Y))C(UI5(X, Y)) \\ d_6 = e_6 : (X : U)(Y : (Set(X))U)(C(X))((v : Set(X))C(Y(v)))C(UI6(X, Y)) \\ d_7 = e_7 : (X : U)(Y : U)(C(X))(C(Y))C(UI7(X, Y)) \\ d_8 = e_8 : (X : U)(y : Set(X))(z : Set(X))(C(X))C(UI8(X, y, z)) \\ c = f : U \end{array}}{UE(C, d_1, \ldots, d_8, c) = UE(C, e_1, \ldots, e_8, f) : C(c)} \quad \text{(UE 2)}$$

The meaning of the selector UE is explained as follows:

1. Evaluate c to canonical form.

2a. If $c \Longrightarrow UI1(A)$ then $UE(C, d_1, \ldots, d_8, c) \Longrightarrow d_1(A, UE(C, d_1, \ldots, d_8, A))$ and $UE(C, d_1, \ldots, d_8, c) = d_1(A, UE(C, d_1, \ldots, d_8, A))$.

2b. If $c \Longrightarrow UI2$ then $UE(C, d_1, \ldots, d_8, c) \Longrightarrow d_2$ and $UE(C, d_1, \ldots, d_8, c) = d_2$.

2c. If $c \Longrightarrow UI3(A, B)$ then $UE(C, d_1, \ldots, d_8, c) \Longrightarrow d_3(A, B, UE(C, d_1, \ldots, d_8, A), UE(C, d_1, \ldots, d_8, B))$ and $UE(C, d_1, \ldots, d_8, c) = d_3(A, B, UE(C, d_1, \ldots, d_8, A), UE(C, d_1, \ldots, d_8, B))$.

2d. If $c \Longrightarrow UI4(A, B)$ then $UE(C, d_1, \ldots, d_8, c) \Longrightarrow d_4(A, B, UE(C, d_1, \ldots, d_8, A), (v)UE(C, d_1, \ldots, d_8, B(v)))$ and $UE(C, d_1, \ldots, d_8, c) = d_4(A, B, UE(C, d_1, \ldots, d_8, A), (v)UE(C, d_1, \ldots, d_8, B(v)))$.

2e. If $c \Longrightarrow UI5(A, B)$ then $UE(C, d_1, \ldots, d_8, c) \Longrightarrow d_5(A, B, UE(C, d_1, \ldots, d_8, A), UE(C, d_1, \ldots, d_8, B))$ and $UE(C, d_1, \ldots, d_8, c) = d_5(A, B, UE(C, d_1, \ldots, d_8, A), UE(C, d_1, \ldots, d_8, B))$.

2f. If $c \Longrightarrow UI6(A, B)$ then $UE(C, d_1, \ldots, d_8, c) \Longrightarrow d_6(A, B, UE(C, d_1, \ldots, d_8, A), (v)UE(C, d_1, \ldots, d_8, B(v)))$ and $UE(C, d_1, \ldots, d_8, c) = d_6(A, B, UE(C, d_1, \ldots, d_8, A), (v)UE(C, d_1, \ldots, d_8, B(v)))$.

2g. If $c \Longrightarrow UI7(A, B)$ then $UE(C, d_1, \ldots, d_8, c) \Longrightarrow d_7(A, B, UE(C, d_1, \ldots, d_8, A), UE(C, d_1, \ldots, d_8, B))$ and $UE(C, d_1, \ldots, d_8, c) = d_7(A, B, UE(C, d_1, \ldots, d_8, A), UE(C, d_1, \ldots, d_8, B))$.

2h. If $c \Longrightarrow UI8(A, a, e)$ then $UE(C, d_1, \ldots, d_8, c) \Longrightarrow d_8(A, a, e, UE(C, d_1, \ldots, d_8, A))$ and $UE(C, d_1, \ldots, d_8, c) = d_8(A, a, e, UE(C, d_1, \ldots, d_8, A))$.

Justification
of (UE 1) As usual.
of (UE 2) Analogous. □

The U-equality axioms are

$$\begin{array}{l} UE(C, d_1, \ldots, d_8, UI1(A)) = d_1(A, UE(C, d_1, \ldots, d_8, A)) : C(UI1(A)) \\ UE(C, d_1, \ldots, d_8, UI2) = d_2 : C(UI2) \\ UE(C, d_1, \ldots, d_8, UI3(A, B)) = d_3(A, B, UE(C, d_1, \ldots, d_8, A), \end{array}$$

$$
\begin{aligned}
&UE(C, d_1, \ldots, d_8, B)) : C(UI3(A, B))\\
&UE(C, d_1, \ldots, d_8, UI4(A, D)) = d_4(A, D, UE(C, d_1, \ldots, d_8, A),\\
&\quad (v)UE(C, d_1, \ldots, d_8, D(v))) : C(UI4(A, D))\\
&UE(C, d_1, \ldots, d_8, UI5(A, B)) = d_5(A, B, UE(C, d_1, \ldots, d_8, A),\\
&\quad UE(C, d_1, \ldots, d_8, B)) : C(UI5(A, B))\\
&UE(C, d_1, \ldots, d_8, UI6(A, D)) = d_6(A, D, UE(C, d_1, \ldots, d_8, A),\\
&\quad (v)UE(C, d_1, \ldots, d_8, B(v))) : C(UI6(A, B))\\
&UE(C, d_1, \ldots, d_8, UI7(A, B)) = d_7(A, B, UE(C, d_1, \ldots, d_8, A),\\
&\quad UE(C, d_1, \ldots, d_8, B)) : C(UI7(A, B))\\
&UE(C, d_1, \ldots, d_8, UI8(A, a, e)) = d_8(A, a, e, UE(C, d_1, \ldots, d_8, A)) :\\
&\quad C(UI8(A, a, e))
\end{aligned}
$$

where $C : (U)set,\ A : U,\ B : U,\ D : (Set(A))U,$

$$
\begin{aligned}
&d_1 : (X : U)(C(X))C(UI1(X)),\ d_2 : C(UI2),\\
&d_3 : (X : U)(Y : U)(C(X))(C(Y))C(UI3(X, Y)),\\
&d_4 : (X : U)(Y : (Set(X))U)(C(X))((v : Set(X))C(Y(v)))C(UI4(X, Y)),\\
&d_5 : (X : U)(Y : U)(C(X))(C(Y))C(UI5(X, Y)),\\
&d_6 : (X : U)(Y : (Set(X))U)(C(X))((v : Set(X))C(Y(v)))C(UI6(X, Y)),\\
&d_7 : (X : U)(Y : U)(C(X))(C(Y))C(UI7(X, Y)),\\
&d_8 : (X : U)(y : Set(X))(z : Set(X))(C(X))C(UI8(X, y, z))
\end{aligned}
$$

The U-equality axioms can be expressed by the rules

$$
\frac{\begin{array}{l} C : (U)set\\ d_1 : (X : U)(C(X))C(UI1(X))\\ d_2 : C(UI2)\\ \vdots\\ d_8 : (X : U)(y : Set(X))(z : Set(X))(C(X))C(UI8(X, y, z))\\ A : U \end{array}}{UE(C, d_1, \ldots, d_8, UI1(A)) = d_1(A, UE(C, d_1, \ldots, d_8, A)) : C(UI1(A))}\ \text{(U-equ 1)}
$$

$$
\frac{\begin{array}{l} C : (U)set\\ d_1 : (X : U)(C(X))C(UI1(X))\\ d_2 : C(UI2)\\ \vdots\\ d_8 : (X : U)(y : Set(X))(z : Set(X))(C(X))C(UI8(X, y, z)) \end{array}}{UE(C, d_1, \ldots, d_8, UI2) = d_2 : C(UI2)}\ \text{(U-equ 2)}
$$

$$
\begin{array}{l} C : (U)set\\ d_1 : (X : U)(C(X))C(UI1(X))\\ d_2 : C(UI2)\\ \vdots\\ d_8 : (X : U)(y : Set(X))(z : Set(X))(C(X))C(UI8(X, y, z)) \end{array}
$$

$$\frac{\begin{array}{l} A : U \\ B : U \end{array}}{\begin{array}{l} UE(C, d_1, \ldots, d_8, UI3(A, B)) = d_3(A, B, UE(C, d_1, \ldots, d_8, A), \\ \quad UE(C, d_1, \ldots, d_8, B)) : C(UI3(A, B)) \end{array}} \quad \text{(U-equ 3)}$$

$$\frac{\begin{array}{l} C : (U)set \\ d_1 : (X : U)(C(X))C(UI1(X)) \\ d_2 : C(UI2) \\ \vdots \\ d_8 : (X : U)(y : Set(X))(z : Set(X))(C(X))C(UI8(X, y, z)) \\ A : U \\ B : (Set(A))U \end{array}}{\begin{array}{l} UE(C, d_1, \ldots, d_8, UI4(A, B)) = d_4(A, B, UE(C, d_1, \ldots, d_8, A), \\ \quad (v)UE(C, d_1, \ldots, d_8, B(v))) : C(UI4(A, B)) \end{array}} \quad \text{(U-equ 4)}$$

$$\frac{\begin{array}{l} C : (U)set \\ d_1 : (X : U)(C(X))C(UI1(X)) \\ d_2 : C(UI2) \\ \vdots \\ d_8 : (X : U)(y : Set(X))(z : Set(X))(C(X))C(UI8(X, y, z)) \\ A : U \\ B : U \end{array}}{\begin{array}{l} UE(C, d_1, \ldots, d_8, UI5(A, B)) = d_5(A, B, UE(C, d_1, \ldots, d_8, A), \\ \quad UE(C, d_1, \ldots, d_8, B)) : C(UI5(A, B)) \end{array}} \quad \text{(U-equ 5)}$$

$$\frac{\begin{array}{l} C : (U)set \\ d_1 : (X : U)(C(X))C(UI1(X)) \\ d_2 : C(UI2) \\ \vdots \\ d_8 : (X : U)(y : Set(X))(z : Set(X))(C(X))C(UI8(X, y, z)) \\ A : U \\ B : (Set(A))U \end{array}}{\begin{array}{l} UE(C, d_1, \ldots, d_8, UI6(A, B)) = d_6(A, B, UE(C, d_1, \ldots, d_8, A), \\ \quad (v)UE(C, d_1, \ldots, d_8, B(v))) : C(UI6(A, B)) \end{array}} \quad \text{(U-equ 6)}$$

$$\frac{\begin{array}{l} C : (U)set \\ d_1 : (X : U)(C(X))C(UI1(X)) \\ d_2 : C(UI2) \\ \vdots \\ d_8 : (X : U)(y : Set(X))(z : Set(X))(C(X))C(UI8(X, y, z)) \\ A : U \\ B : U \end{array}}{\begin{array}{l} UE(C, d_1, \ldots, d_8, UI7(A, B)) = d_7(A, B, UE(C, d_1, \ldots, d_8, A), \\ \quad UE(C, d_1, \ldots, d_8, B)) : C(UI7(A, B)) \end{array}} \quad \text{(U-equ 7)}$$

$$
\frac{\begin{array}{l} C : (U)set \\ d_1 : (X : U)(C(X))C(UI1(X)) \\ d_2 : C(UI2) \\ \vdots \\ d_8 : (X : U)(y : Set(X))(z : Set(X))(C(X))C(UI8(X, y, z)) \\ A : U \\ b : Set(A) \\ c : Set(A) \end{array}}{\begin{array}{l} UE(C, d_1, \ldots, d_8, UI8(A, b, c)) = d_8(A, b, c, UE(C, d_1, \ldots, d_8, A)) : \\ \quad C(UI8(A, b, c)) \end{array}} \quad \text{(U-equ 8)}
$$

Justification
of (U-equ 1) – (U-equ 8): straightforward by the meaning explanation of UE. □

C, n & o

(1) Instead of '$UI1$' Martin-Löf uses in version (6) infinitely many introductory operators 'n_0', 'n_1', 'n_2', …, corresponding to a set structure with infinitely many 0-ary set forming operations $N_0, N_1, N_2, \ldots$ This, however, prevents him from formulating a U-elimination rule of structural induction of such a kind as it was introduced for $S, N, +, \Pi, \rightarrow, \Sigma, \times$ and I. Instead of '$UI2$', '$UI3$', '$UI4$', '$UI6$' and '$UI8$' Martin-Löf uses in version (6) 'n', '$+$', 'π', 'σ' and 'i' resp. in their polymorphic form. The operator *Set* is denoted by 'T'. And, as has just been mentioned, there is no eliminatory operator UE for U.

Nordström et al. use '$\hat{0}$' and '$\hat{T}$' and a fct. constant 'N'' defined by '$\hat{0}$' and '$\hat{T}$' instead of '$UI1$', and '$\hat{N}$', '$\hat{+}$', '$\hat{\Pi}$', '$\hat{\Sigma}$', '$\hat{Id}$' instead of '$UI2$', '$UI3$', '$UI4$', '$UI6$' and '$UI8$' resp., again in polymorphic form. And instead of 'UE' they use '$urec$' in a polymorphic form.

(2) If one abstracts from universes, one only has a structure of finite types, since one can only iterate the given set forming operations N_i, $i \in \omega$, N, $+$, Π, $\rightarrow, \ldots, I$ a finite number of times. If universes are added to the language of constructive set theory, one can obtain transfinite types. The idea of a universe is obviously to define a set as the least set closed under certain specified set forming operations. The set of small sets obtained in this way reflects the set structure on the object level.

Sets of sets of this kind, i.e. universes, are needed in category theory, but they are also needed in programming for many specifications, when the most natural way of expressing a proposition is to use recursion or conditionals. Moreover, they are needed to prove inequalities – which are actually conditionals (a famous example is Peano's fourth axiom: $I(N, NI1, NI2(NI1)) \rightarrow N_0$) – and they are needed when defining abstract data types in constructive type theory.[80]

(3) The process of really defining the set of small sets can be iterated such as to obtain the set of large sets or the second universe U'.

ix.9.4) What does a judgement of the form 'U' : *set*' mean? or What does it mean to be the set of large sets (or the second universe)?

To make the judgement 'U' : *set*' one has to know (i) what a canonical element of U' is, and (ii) when two canonical elements of U' are identical.

The formation axioms or rules for the set U' are

$$U' : set \text{ (U'F 1)} \qquad\qquad U' = U' : set \text{ (U'F 2)}$$

ix.9.5) What does a judgement of the form '$a : U'$' mean? or What does it mean to be a large set (an element of the second universe)?
To explain what a judgement of the form '$a : U'$' means, one has to know that (i) a has the form of an object formed by one of the element-formation rules of U' (a has the form of a canonical element of U') or (ii) a can be evaluated to one of those forms (a is a non-canonical element of U').

The element-formation or introduction axioms of U' are

$$U'I1 : U' \qquad\qquad U'I2 : (U)U'$$

where $U'I1$ can also be denoted by '$\hat{U}$' and $U'I2$ by '$\hat{Set}$'.

These element-formation axioms of U' can be expressed alternatively by the introduction rules

$$U'I1 : U' \text{ (U'I 1)} \qquad\qquad \frac{A : U}{U'I2(A) : U'} \text{ (U'I 2)}$$

Simultaneously with the def. of the canonical elements of U', an operator *Set'* is defined which decodes the elements in the second universe U' to the sets that they represent or code:

typing axiom (rule) $\qquad Set' : (U')set$

definitional identity axioms (rules)

$$Set'(U'I1) = U : set$$
$$Set'(U'I2(X)) = Set(X) : set \qquad \text{where } X : U$$

ix.9.6) What does a judgement of the form '$a = b : U'$' mean? or What does it mean for two large sets to be identical?
To explain what a judgement of the form '$a = b : U'$' means, one has to know that (i) a and b have the form of two objects formed by one of the identity rules associated with the element-formation rules of U' (a and b are identical canonical elements of U') or (ii) a and b can be evaluated to one of those forms (a or b are identical non-canonical elements of U').

The identity rules associated with the element-formation rules or introduction rules of U' are

$$U'I1 = U'I1 : U' \text{ (U'I 3)} \qquad\qquad \frac{A = B : U}{U'I2(A) = U'I2(B) : U'} \text{ (U'I 4)}$$

C, n & o

The following judgement

$$(Z : (U')\mathit{set})(Z(U'I1))((X : U)Z(U'I2(X)))(z : U')Z(z) : \mathit{type}$$

can be derived as follows

$$\dfrac{\dfrac{U' : \mathit{set}}{U' : \mathit{type}} \quad \mathit{set} : \mathit{type}}{(U')\mathit{set} : \mathit{type}} \qquad \dfrac{\dfrac{\dfrac{(X : U)}{U'I2(X) : U'} \quad (Z : (U')\mathit{set})}{Z(U'I2(X)) : \mathit{set}}}{Z(U'I2(X)) : \mathit{type}} \qquad \dfrac{\dfrac{U' : \mathit{set}}{U' : \mathit{type}} \quad \dfrac{\dfrac{\begin{array}{l}(z : U')\\ (Z : (U')\mathit{set})\end{array}}{Z(z) : \mathit{set}}}{Z(z) : \mathit{type}}}{(z : U')Z(z) : \mathit{type}}$$

$$\dfrac{\dfrac{\dfrac{U : \mathit{set}}{U : \mathit{type}} \quad Z(U'I2(X)) : \mathit{type}}{(X : U)Z(U'I2(X)) : \mathit{type}} \quad (z : U')Z(z) : \mathit{type}}{((X : U)Z(U'I2(X)))(z : U')Z(z) : \mathit{type}}$$

$$\dfrac{(U')\mathit{set} : \mathit{type} \quad \dfrac{\dfrac{\dfrac{\begin{array}{l}U'I1 : U'\\ (Z : (U')\mathit{set})\end{array}}{Z(U'I1) : \mathit{set}}}{Z(U'I1) : \mathit{type}} \quad ((X : U)Z(U'I2(X)))(z : U')Z(z) : \mathit{type}}{(Z(U'I1))((X : U)Z(U'I2(X)))(z : U')Z(z) : \mathit{type}}}{(Z : (U')\mathit{set})(Z(U'I1))((X : U)Z(U'I2(X)))(z : U')Z(z) : \mathit{type}}$$

The U'-elimination axiom states that the selector or eliminatory operator for the set U' is an object of this type, i.e.

$$U'E : (Z : (U')\mathit{set})(Z(U'I1))((X : U)Z(U'I2(X)))(z : U')Z(z)$$

The U'-elimination axiom can be expressed alternatively by the two rules

$$\dfrac{\begin{array}{l} C : (U')\mathit{set} \\ d : C(U'I1) \\ e : (X : U)C(U'I2(X)) \\ c : U' \end{array}}{U'E(C, d, e, c) : C(c)} \text{ (U'E 1)}$$

$$\dfrac{\begin{array}{l} C : (U')\mathit{set} \\ d = f : C(U'I1) \\ e = g : (X : U)C(U'I2(X)) \\ c = h : U' \end{array}}{U'E(C, d, e, c) = U'E(C, f, g, h) : C(c)} \text{ (U'E 2)}$$

The meaning of the selector $U'E$ is explained as follows:

1. Evaluate c to canonical form.
2a. If $c \Longrightarrow U'I1$ then $U'E(C, d, e, c) \Longrightarrow d$ and $U'E(C, d, e, c) = d$.
2b. If $c \Longrightarrow U'I2(A)$ for some $A : U$ then $U'E(C, d, e, c) \Longrightarrow e(A)$ and $U'E(C, d, e, c) = e(A)$.

The U'-equality axioms are

$$U'E(C, d, e, U'I1) = d : C(U'I1)$$
$$U'E(C, d, e, U'I2(A)) = e(A) : C(U'I2(A))$$

where $C : (U')set$, $d : C(U'I1)$ and $e : (X : U)C(U'I2(X))$, $A : U$.

These U'-equality axioms can also be expressed by the rules

$$\frac{\begin{array}{l} C : (U')set \\ d : C(U'I1) \\ e : (X : U)C(U'I2(X)) \end{array}}{U'E(C, d, e, U'I1) = d : C(U'I1)} \text{ (U'-equ 1)}$$

$$\frac{\begin{array}{l} C : (U')set \\ d : C(U'I1) \\ e : (X : U)C(U'I2(X)) \\ A : U \end{array}}{U'E(C, d, e, U'I2(A)) = e(A) : C(U'I2(A))} \text{ (U'-equ 2)}$$

Justification

of (U'E 1), (U'E 2), (U'-equ 1) and (U'-equ 2) as usual. □

From the set of large sets or from the second universe, one can generate in an analogous way the set of extralarge sets or third universe U'', etc. etc.

(4) It is possible to slightly generalize the approach to universes in another direction: instead of introducing after the small the large sets, then the extralarge and the extraextralarge sets, etc., one can extend the concept of smallness from sets to families of sets:

A set A is said to be small or a U-set if there is an element

$$D : U \text{ s.t. } Set(D) = A : set.$$

A family of sets $x_1 : A_1, \ldots, x_n : A_n(x_1, \ldots, x_{n-1}) \vdash A(x_1, \ldots, x_n) : set$ is said to be small or a U-family of sets if there exists some indexing fct. $D(x_1, \ldots, x_n)$ onto U, i.e. $x_1 : A_1, \ldots, x_n : A_n(x_1, \ldots, x_{n-1}) \vdash D(x_1, \ldots, x_n) : U$ s.t. $x_1 : A_1, \ldots, x_n : A_n(x_1, \ldots, x_{n-1}) \vdash Set(D(x_1, \ldots, x_n)) = A(x_1, \ldots, x_n) : set$ where $x_1 : A_1, \ldots, x_n : A_n(x_1, \ldots, x_{n-1}) \vdash Set(D(x_1, \ldots, x_n)) : set$ is defined to be $Set(D(a_1, \ldots, a_n)) : set$ provided $a_1 : A_1, \ldots, a_n : A_n(a_1, \ldots, a_{n-1})$.

(5) A particularly interesting example of a large set is the set V. V has been used by Aczel to give meaning to a constructive version of ZF set theory via constructive type theory.[81] (cf. C, n & o (1) iii) ad 2) to vi.1).)

Whereas V was taken by Aczel to be a ground set, Martin-Löf showed that it is just a large set, i.e. it can be defined by means of U as follows

$$\frac{\begin{array}{l} X : U \vdash Set(X) : set \\ U : set \end{array}}{\begin{cases} V : set \\ V = (WX : U)Set(X) : set \end{cases}}$$

where W is the well-ordering operator (cf. Chapter 3, Sect. iv).

Another interesting example using the first universe is Nordström et al.'s example:

$$A : U, B : U, x : N_0 \vdash A = B : U$$

which essentially presupposes extensional identity E. This example is interesting, because it allows us to exhibit a canonical element of a $\rightarrow$-*set*, i.e. a lambda expression, whose body does not terminate.[82]

(6) Jacob has carried out a perspicuous analysis of the consequences of higher order extensions of constructive type theory.[83]

If higher order quantification is wanted, one must at least be able to quantify over all propositions, i.e. over the type *prop* ; hence the following has to be an axiom

$$prop : set \quad \text{(HO)}$$

for Higher Order. Since it is not known what the *prop*-introduction rules are (and thus the *prop*-elimination and -equality rules neither), there is no genuine meaning explanation of (HO). Note that by means of (HO) one can introduce all higher order propositions and propositional fcts. But if (HO) is added to constructive type theory, impredicativity sneeks into the resulting theory. So, let the theory obtained from adding (HO) to constructive type theory be called higher order type theory.

One can then prove that higher order type theory is inconsistent. One can show that a version of Mirimanoff's paradox can be derived in higher order type theory, and Girard was the first to discover this paradox. Jacobs has very neatly explained the state of affairs as follows:

(a) the operations Σ and $\exists$ of constructive type theory constitute some form of comprehension.
(b) In version (1) of (constructive) type theory 'type' was understood in the sense of *set*, and the fundamental axiom of version (1) is *set* : *set* or, since the type of sets was then denoted by 'U', $U : U$.

In constructive type theory every proposition is interpreted or identified with the set of its proof-objects, hence $prop \subseteq set$. Since in higher order type theory one also has (HO) *prop* : *set*, it is possible to quantify over all sets which are propositions. This, together with (a) yields comprehension over that part of the universe *set* of sets which consists of propositions.

(c) Define the proposition T as follows:

$$T = (\Sigma A : prop)(\Sigma <_A : A \to (A \to prop))WF(A, <_A) : prop$$

where $WF(A, <_A)$ is an abbreviation of

$$(\forall r : A \to prop)((\exists x : A)rx \,\&\, ((\forall x : A)(rx \supset (\exists y : A)(ry \,\&\, y < x))) \supset \bot)$$

Then T can be informally interpreted as the collection of propositions A with the well-founded orderings $<_A$ on the sets of their proof-objects. A proposition becomes a set equipped with a well-founded binary relation. Note that the Σ-quantifications in the def. of T, as well as the first $\forall$-quantification in the def. of $WF(A, <_A)$ clearly presuppose (HO); and the def. of T uses comprehension as mentioned in (a) above.

Finally, an ordering $<_T$ on the set of proof-objects of T can be defined s.t. a proof-object $(A, (<_A, r_A)) : T$ stands in the relation $<_T$ to proof-object $(B, (<_B, r_B)) : T$ if there is a fct. $f : A \to B$ which respects the ordering and stays below a certain point in B. Let p and q be the left and right projections defined by Σ-elimination; if $u : T$ then $p(u) : prop$ and $p(q(u)) : p(u) \to (p(u) \to prop)$ is an ordering on $p(u)$. Then

$$\begin{aligned} <_T \;=\; & \lambda u \lambda v(\exists f : p(u) \to p(v))(\exists z : p(v))(\forall x : p(u)) \\ & (\forall y : p(u))x(p(q(u)))y \supset fx(p(q(v)))fy \,\& \\ & (\forall x : p(u))fx(p(q(v)))z : T \to (T \to prop) \end{aligned}$$

Now, a proof-object $r_T : WF(T, <_T)$ can be found, whence $WF(T, <_T)$ is true. Write $t \equiv (T, (<_T, r_T)) : T$. But also a proof-object $r^* : t <_T t$ can be found. But the latter gives rise to $\cdots <_T t <_T t <_T t$, i.e. an infinitely descending $<_T$-chain of proof-objects of T, which contradicts the true proposition $WF(T, <_T)$.

The paradox essentially consists of constructing an $x : T$ with $x <_T x$ (since $<_T$ is a transitive and well-founded $\in$-relation, this means that $x \in x$). E.g. if the Russell set R is defined by $R = (\exists x : T)x <_T x \supset \bot : prop$ and $<_R = \lambda u \lambda v(p(u) <_T p(v)) : T \to (T \to prop)$, then one can show in a similar way that there is a proof-object $r_R : WF(R, <_R)$. Write $r^\circ \equiv (R, (<_R, r_R)) : T$. The fact that one can find a proof-object of $r^\circ <_T r^\circ : prop$ leads to a contradiction.

The conclusion is that higher order type theory is inconsistent, because one has a form of naive set theory in type theory with full comprehension.

It is also possible to extend (constructive) type theory so that it contains second order logic only. Let this extension be called second order type theory. Second order type theory consists of all the axioms and rules of constructive type theory to which are added two new proposition-forming operations (second order quantifiers) $\forall^2$ and $\exists^2$ which are objects of the following types:

$$\begin{aligned} \forall^2, \exists^2 \;\; & : \;\; (r : prop)((r)prop)prop \\ & : \;\; (r : A \to prop)((r)prop)prop \\ & : \;\; (r : A \to (A \to prop))((r)prop)prop \\ & \;\;\;\;\; etc. \end{aligned}$$

as well as the following rules of inference concerning these new operations; e.g.

$\forall^2$-formation

$$\frac{\begin{array}{l} A \to prop : set \\ B : (A \to prop)prop \end{array}}{(\forall^2 r : A \to prop)B(r) : prop} \;(\forall^2\text{F})$$

$\forall^2$-introduction

$$\frac{\begin{array}{l} A \to prop = C \to prop : set \\ B = D : (A \to prop)prop \\ b : (r : A \to prop)B(r) \end{array}}{\forall^2 I(A \to prop, B, b) : (\forall^2 r : C \to prop)D(r)} \;(\forall^2\text{I})$$

$\forall^2$-elimination

$$\frac{\begin{array}{l} A \to prop : set \\ B : (A \to prop)prop \\ f : (\forall^2 r : A \to prop)B(r) \\ a : A \to prop \end{array}}{app^2(A \to prop, B, f, a) : B(a)} \;(\forall^2\text{E})$$

$\forall^2$-equality

$$\frac{\begin{array}{l} A \to prop : set \\ B : (A \to prop)prop \\ f : (\forall^2 r : A \to prop)B(r) \\ a : A \to prop \end{array}}{app^2(A \to prop, B, \forall^2 I(A \to prop, B, f), a) = f(a) : B(a)} \;(\forall^2\text{-equ})$$

$\exists^2$-formation

$$\frac{\begin{array}{l} A \to prop : set \\ B : (A \to prop)prop \end{array}}{(\exists^2 r : A \to prop)B(r) : prop} \;(\exists^2\text{F})$$

$\exists^2$-introduction

$$\frac{\begin{array}{l} A \to prop = C \to prop : set \\ B = D : (A \to prop)prop \\ a : A \to prop \\ b : B(a) \end{array}}{\exists^2 I(A \to prop, B, a, b) : (\exists^2 r : C \to prop)D(r)} \;(\exists^2\text{I})$$

$\exists^2$-elimination

$$\frac{\begin{array}{l} A \to prop : set \\ B : (A \to prop)prop \\ C : ((\exists^2 r : A \to prop)B(r))prop \\ g : (r : A \to prop)(s : B(r))C(\exists^2 I(A \to prop, B, r, s)) \\ c : (\exists^2 r : A \to prop)B(r) \end{array}}{\exists^2 E(A \to prop, B, C, g, c) : C(c)} \;(\exists^2\text{E})$$

$\exists^2$-equality

$$\frac{\begin{array}{l} A \rightarrow prop \\ B : (A \rightarrow prop)prop \\ C : ((\exists^2 r : A \rightarrow prop)B(r))prop \\ g : (r : A \rightarrow prop)(s : B(r))C(\exists^2 I(A \rightarrow prop, B, r, s)) \\ a : A \rightarrow prop \\ b : B(a) \end{array}}{\begin{array}{l} \exists^2 E(A \rightarrow prop, B, C, g, \exists^2 I(A \rightarrow prop, B, a, b)) = g(a, b) : \\ \quad C(\exists^2 I(A \rightarrow prop, B, a, b)) \end{array}} \quad (\exists^2\text{-equ})$$

Note: 1) A : *set*, and 2) all the identity rules associated with ($\forall^2$F), ($\forall^2$I), ($\forall^2$E) as well as with ($\exists^2$F), ($\exists^2$I) and ($\exists^2$E) have been omitted.

A quick inspection of the def. of T reveals that Mirimanoff's or Russell's paradoxes and other paradoxes can still be derived in second order type theory (cf. the typing of $\forall^2$ and $\exists^2$, and the higher order quantifications in the def. of T). Thus, second order type theory is inconsistent.

Now, constructive type theory contains a strong notion of existential quantification (cf. Sect. x, C, n & o (2) to x.6.3)) or of disjoint union of a family of sets which allows us to define the left and right projections p and q resp. (cf. C, n & o (2) to ix.6.3)) Standard second order logic contains only a weak notion of existential quantification (which does not allow us to define these projections) and can be shown to be consistent.

The conclusion is that higher order quantification and a strong notion of $\exists$ or Σ are incompatible. Jacobs remarks that one has to choose. For a constructivist things are clear: she or he wants a constructive or strong notion of Σ or $\exists$ and can hardly make sense of higher order quantification. For a computer scientist, however, there may be reasons to prefer higher order quantification to a strong notion of Σ or $\exists$.

x CONSTRUCTIVE LOGIC

In Sect. vii. it was only really defined what it means to be a proposition, but nothing was said about what propositions there are. In the sequel, two ground propositions, namely the propositional constants falsehood and truth will be introduced and then the propositional operations of disjunction, universal quantification, implication, existential quantification, conjunction and intensional identity (in this very order) will be listed which allow us to generate a finite propositional structure. (*Note*: since quantified propositions are obviously propositions, what is here called a 'propositional structure' is not the structure of propositional logic, but rather the structure of constructive quantificational or first order logic.)

As in the case of the sets, four kinds of axioms or rules will be given with each of the following propositions or propositional operations: (i) the formation axioms or rules which say how a proposition can be formed or constructed from other propositions, (ii) the introduction axioms or rules which provide the meaning of the proposition by indicating which are the canonical and identical canonical proof-objects of the proposition, (iii) the elimination axioms or rules which introduce non-canonical proof-objects and an eliminatory operator or selector playing a role analogous to the one played by the selector for sets, and (iv) the equality rules which determine the meaning of the selectors and show how these selectors operate on proof-objects in canonical form of the major premiss.

x.1.1) What does a judgement of the form '$\bot : prop$' mean? or What does it mean to be a falsehood (absurdity)?
To explain what it means to be a falsehood $\bot$, one has to know (i) what a canonical proof-object of $\bot$ is, and (ii) when two canonical proof-objects of $\bot$ are identical.

The formation rule for the ground proposition $\bot$ is the axiom or rule with 0 premisses

$$\bot : prop \qquad (\bot\text{F})$$

To be an entity at all under the given requirements, $\bot$ has to have an identity which is expressed by the judgement $\bot = \bot : prop$ and which is nothing but an instance of the prop type identity $prop = prop : type$.

It is pointless to ask what the meaning of '$c : \bot$' or of '$c = d : \bot$' is, i.e. to ask what a proof-object of $\bot$ is and when two proof-objects of $\bot$ are identical, since falsehood is on no condition true, which means that it has under no circumstances a proof-object, be it canonical or non-canonical. Hence, there are of course no $\bot$-introduction rules.

The following is a type, $(Z : (\bot)prop)(x : \bot)Z(x) : type$, as can be seen from the derivation

$$\frac{\begin{matrix}(x : \bot)\\ (Z : (\bot)prop)\end{matrix}}{Z(x) : prop} \qquad \dfrac{\dfrac{\dfrac{\bot : prop}{\bot : type} \quad prop : type}{(\bot)prop : type} \quad \dfrac{\dfrac{\bot : prop}{\bot : type} \quad \dfrac{\begin{matrix}|\\ Z(x) : prop\end{matrix}}{Z(x) : type}}{(x : \bot)Z(x) : type}}{(Z : (\bot)prop)(x : \bot)Z(x) : type}$$

The selector or eliminatory operator $\bot E$ of the proposition $\bot$ is an object of this type

$$\bot E : (Z : (\bot)prop)(x : \bot)Z(x)$$

and this is the $\bot$-elimination axiom which can alternatively be expressed by the following rules

$$\frac{c : \bot \quad C : (\bot)prop}{\bot E(C, c) : C(c)} \ (\bot\text{E 1})$$

$$\frac{c = c' : \bot \quad C : (\bot)prop}{\bot E(C, c) = \bot E(C, c') : C(c)} \ (\bot\text{E 2})$$

Justification

of ($\bot$E 1) As was explained above, there is no proof-object of falsehood $\bot$. Thus, vacuously, if something is taken to be a proof-object of $\bot$, it can be taken as or turned into a proof-object of any propositional fct. C whatsoever.

of ($\bot$E 2) Analogously to ($\bot$E 1) □

Finally, there are no $\bot$-equality axioms or rules since there are no $\bot$-introduction rules.

x.2.1) What does a judgement of the form '$\top$: *prop*' mean? or What does it mean to be a truth?
To explain what it means to be a truth $\top$, one has to know (i) what a canonical proof-object of $\top$ is, and (ii) when two canonical proof-objects of $\top$ are identical.

The formation rule for the ground proposition $\top$ is the axiom or rule with 0 premisses

$$\top : prop \quad (\text{TF 1})$$

By the rule (prop type id) one also has

$$\top = \top : prop \quad (\text{TF 2})$$

x.2.2) What does a judgement of the form '$c : \top$' mean? or What does it mean to be a proof-object of truth?
To explain what a judgement of the form '$c : \top$' means, one has to know that (i) c has the form of an object formed by the proof-object formation rule of $\top$ (i.e. c has the form of a canonical proof-object of $\top$), or (ii) c can be evaluated to that form (c is a non-canonical proof-object of $\top$).

The proof-object formation rule or introduction axiom/rule of $\top$ is

$$\top I : \top \quad (\text{TI 1})$$

x.2.3) What does a judgement of the form '$c = d : \top$' mean? or What does it mean for two proof-objects of truth to be identical?
To explain what a judgement of the form '$c = d : \top$' means, one has to know that (i) c and d have the form of two objects formed by the identity rule associated with the proof-object formation rule of $\top$ (c and d are identical canonical proof-objects of $\top$) or (ii) c and d can be evaluated to that form (c or d are identical non-canonical proof-objects of $\top$).

The identity rule associated with the proof-object formation rule or introduction rule of $\top$ is

$$\top I = \top I : \top \quad (\text{TI 2})$$

The type $(Z : (\top)prop)(Z(\top I))(x : \top)Z(x)$ can be derived as follows

$$\dfrac{\dfrac{\begin{array}{c}\top I : \top \\ (Z : (\top)prop)\end{array}}{Z(\top I) : prop}}{Z(\top I) : type} \qquad\qquad \dfrac{\dfrac{\top : prop}{\top : type} \quad \dfrac{\dfrac{\begin{array}{c}(x : \top) \\ (Z : (\top)prop)\end{array}}{Z(x) : prop}}{Z(x) : type}}{(x : \top)Z(x) : type}$$

$$\dfrac{\dfrac{\dfrac{\top : prop}{\top : type} \quad prop : type}{(\top)prop : type} \quad \dfrac{\begin{array}{c}|\\ Z(\top I) : type\end{array} \quad \begin{array}{c}|\\ (x : \top)Z(x) : type\end{array}}{(Z(\top I))(x : \top)Z(x) : type}}{(Z : (\top)prop)(Z(\top I))(x : \top)Z(x) : type}$$

The selector or eliminatory operator $\top E$ of the proposition $\top$ is an object of this type and the $\top$-elimination axiom/rule states just this fact

$$\top E : (Z : (\top)prop)(Z(\top I))(x : \top)Z(x)$$

The $\top$-elimination axiom can be expressed alternatively by the two rules

$$\frac{\begin{array}{l} C : (\top)prop \\ d : C(\top I) \\ c : \top \end{array}}{\top E(C, d, c) : C(c)} \text{ (TE 1)}$$

$$\frac{\begin{array}{l} C : (\top)prop \\ d = f : C(\top I) \\ c = e : \top \end{array}}{\top E(C, d, c) = \top E(C, f, e) : C(c)} \text{ (TE 2)}$$

The meaning of the selector is explained by the following algorithm:

1. Evaluate c to canonical form.
2. If $c \Longrightarrow \top I$ then $\top E(C, d, c) \Longrightarrow d$ and the value of $\top E(C, d, c)$ is identical with the value of d, i.e. $\top E(C, d, c) = d$.

Justification
of (TE 1) Assuming the premisses one has to show that the value of $\top E(C, d, c)$ is a canonical proof-object of $C(c)$.
If $c \Longrightarrow \top I$, then $\top E(C, d, c) = d$, by the meaning explanation of $\top E$, and the value of $\top E(C, d, c)$ is identical with the value of d which, by the second premiss, is a canonical proof-object of $C(\top I)$. From the first premiss and the fact that $c = \top I : \top$ follows, that $C(\top I) = C(c) : prop$ and hence that the value of $\top E(C, d, c)$ is a canonical proof-object of $C(c)$.
of (TE 2) Analogous to (TE 1). □

The $\top$-equality axiom is

$$\top E(C, d, \top I) = d : C(\top I)$$

where $C : (\top)prop$, $d : C(\top I)$.

This axiom can be expressed by the following rule

$$\frac{C : (\top)prop \quad d : C(\top I)}{\top E(C, d, \top I) = d : C(\top I)} \text{ (T-equ)}$$

Justification
of ($\top$-equ) is straightforward by the meaning explanation of $\top E$. □

x.3.1) What does a judgement of the form '$A \vee B : prop$' mean? or What does it mean to be the disjunction of two propositions?
To explain what it means to be the disjunction $A \vee B$ of two propositions A, B, one has to know (i) what a canonical proof-object of $A \vee B$ is, and (ii) when two canonical proof-objects of $A \vee B$ are identical.

The formation axiom for the composite proposition $A \vee B$ is

$$\vee : (prop)(prop)prop$$

where clearly $(prop)(prop)prop : type$.

Equivalently, the formation axiom can be given by the two rules

$$\frac{A : prop \quad B : prop}{A \vee B : prop}\ (\vee\text{F } 1) \qquad \frac{A = C : prop \quad B = D : prop}{A \vee B = C \vee D : prop}\ (\vee\text{F } 2)$$

x.3.2) What does a judgement of the form '$c : A \vee B$' mean? or What does it mean to be a proof-object of the proposition $A \vee B$?
To explain what a judgement of the form '$c : A \vee B$' means, one has to know that (i) c has the form of an object formed by one of the proof-object-formation rules of $A \vee B$ (c has the form of a canonical proof-object of $A \vee B$) or (ii) c can be evaluated to one of those forms (c is a non-canonical proof-object of $A \vee B$).

The proof-object formation or introduction axioms of $A \vee B$ are

$$\vee I1 : (X : prop)(Y : prop)(X)X \vee Y$$
$$\vee I2 : (X : prop)(Y : prop)(Y)X \vee Y$$

which can be expressed alternatively by the introduction rules

$$\frac{A = C : prop \quad B = D : prop \quad a : A}{\vee I1(A, B, a) : C \vee D}\ (\vee\text{I } 1)$$

$$\frac{A = C : prop \quad B = D : prop \quad b : B}{\vee I2(A, B, b) : C \vee D}\ (\vee\text{I } 2)$$

x.3.3) What does a judgement of the form '$c = d : A \vee B$' mean? or What does it mean for two proof-objects of $A \vee B$ to be identical?
To explain what a judgement of the form '$c : A \vee B$' means, one has to know that (i) c and d have the form of two objects formed by one of the identity rules associated with the proof-object formation rules of $A \vee B$ (c and d are identical canonical proof-objects of $A \vee B$) or (ii) c and d can be evaluated to one of those forms (c or d are identical non-canonical proof-objects of $A \vee B$).

The identity rules associated with the proof-object formation rules of $A \vee B$, i.e. the introduction rules of $A \vee B$ are

$$\frac{A = C : prop \quad B = D : prop \quad a = c : A}{\vee I1(A, B, a) = \vee I1(C, D, c) : A \vee B}\ (\vee\text{ I } 3)$$

$$\frac{A = C : prop \quad B = D : prop \quad b = d : B}{\vee I2(A, B, b) = \vee I2(C, D, d) : A \vee B}\ (\vee\text{ I } 4)$$

The following judgement

$$(X : prop)(Y : prop)(Z : (X \vee Y)prop)((x : X)Z(\vee I1(X, Y, x)))$$
$$((y : Y)Z(\vee I2(X, Y, y)))(z : X \vee Y)Z(z) : type$$

is readily verified (derived).

The $\vee$-elimination axiom states that the selector or eliminatory operator $\vee E$ for the proposition $A \vee B$ is an object of the type just mentioned, i.e.

$$\vee E : (X : prop)(Y : prop)(Z : (X \vee Y)prop)((x : X)Z(\vee I1(X, Y, x)))$$
$$((y : Y)Z(\vee I2(X, Y, y)))(z : X \vee Y)Z(z)$$

The $\vee$-elimination axiom can be expressed alternatively by the two rules

$$\frac{\begin{array}{l} A : prop \\ B : prop \\ C : (A \vee B)prop \\ d : (x : A)C(\vee I1(A, B, x)) \\ e : (y : B)C(\vee I2(A, B, y)) \\ c : A \vee B \end{array}}{\vee E(A, B, C, d, e, c) : C(c)} \; (\vee \text{E } 1)$$

$$\frac{\begin{array}{l} A : prop \\ B : prop \\ C : (A \vee B)prop \\ d = g : (x : A)C(\vee I1(A, B, x)) \\ e = h : (y : B)C(\vee I2(A, B, y)) \\ c = f : A \vee B \end{array}}{\vee E(A, B, C, d, e, c) = \vee E(A, B, C, g, h, f) : C(c)} \; (\vee \text{E } 2)$$

The meaning of the selector $\vee E$ is explained as follows

1. Evaluate c to canonical form.

2a. If $c \Longrightarrow \vee I1(A, B, a)$ then $\vee E(A, B, C, d, e, c) \Longrightarrow d(a)$ and $\vee E(A, B, C, d, e, c) = d(a)$.

2b. If $c \Longrightarrow \vee I2(A, B, b)$ then $\vee E(A, B, C, d, e, c) \Longrightarrow e(b)$ and $\vee E(A, B, C, d, e, c) = e(b)$.

Justification

of ($\vee$E 1) If $c \Longrightarrow \vee I1(A, B, a)$ then, by the meaning explanation of $\vee E$, the value of $\vee E(A, B, C, d, e, c)$ is identical with the value of $d(a)$ which, by the fourth premiss, is a canonical proof-object of $C(\vee I1(A, B, a))$. Moreover, by the third premiss and the fact that $c = \vee I1(A, B, a) : A \vee B$ the value of $\vee E(A, B, C, d, e, c)$ is a canonical proof-object of $C(c)$.

And if $c \Longrightarrow \vee I2(A, B, b)$ proceed similarly.

of ($\vee$E 2) Analogous to ($\vee$E 1). □

The $\vee$-equality axioms are

$$\vee E(A, B, C, d, e, \vee I1(A, B, a)) = d(a) : C(\vee I1(A, B, a))$$
$$\vee E(A, B, C, d, e, \vee I2(A, B, b)) = e(b) : C(\vee I2(A, B, b))$$

where $A : prop$, $B : prop$, $C : (A \vee B)prop$, $d : (x : A)C(\vee I1(A, B, x))$, $e : (y : B)C(\vee I2(A, B, y))$, $a : A$, $b : B$.

These axioms can be expressed by the following two rules

$$\frac{\begin{array}{l} A : prop \\ B : prop \\ C : (A \vee B)prop \\ d : (x : A)C(\vee I1(A, B, x)) \\ e : (y : B)C(\vee I2(A, B, y)) \\ a : A \end{array}}{\vee E(A, B, C, d, e, \vee I1(A, B, a)) = d(a) : C(\vee I1(A, B, a))} \; (\vee\text{-equ 1})$$

$$\frac{\begin{array}{l} A : prop \\ B : prop \\ C : (A \vee B)prop \\ d : (x : A)C(\vee I1(A, B, x)) \\ e : (y : B)C(\vee I2(A, B, y)) \\ b : B \end{array}}{\vee E(A, B, C, d, e, \vee I2(A, B, b)) = e(b) : C(\vee I2(A, B, b))} \; (\vee\text{-equ 2})$$

Justification
of ($\vee$-equ 1) and ($\vee$-equ 2): straightforward by the meaning explanation of $\vee E$. □

C, n & o

(1) The above rules ($\vee$I 1), ($\vee$I 2), ($\vee$E 1), ($\vee$-equ 1) and ($\vee$-equ 2) are a novel and original combination of ideas of Gentzen, Heyting and Prawitz (cf. Subsect. ii.1, Chapter 4): Gentzen provided the first system of deduction for the logical operators, i.e. his system of natural deduction; his rules for $\vee$-introduction and $\vee$-elimination are (in the more formal notation of constructive type theory)

$\vee$-introduction $\qquad \dfrac{A \; true}{A \vee B \; true} \qquad \dfrac{B \; true}{A \vee B \; true}$

$\vee$-elimination $\qquad \dfrac{A \vee B \; true \quad \begin{array}{c}(A \; true)\\ C \; true\end{array} \quad \begin{array}{c}(B \; true)\\ C \; true\end{array}}{C \; true}$

(for the more original notation, cf. Subsect. ii.1 of Chapter 4)

Heyting was the first logician to informally introduce proof-objects in the semantics of the logical operators. His semantical explanation of disjunction is

A proof of $A \vee B$ consists of a proof of A or a proof of B together with an indication as to which of the disjuncts is proven.

put into more mathematical terms

If $a : A$ (a is a proof-object of A) then $i(a) : A \vee B$ (the left injection from A to $A \vee B$ applied to proof-object a is a proof-object of $A \vee B$) or if $b : B$ (b is a proof-object of B) then $j(b) : A \vee B$ (the right injection from B to $A \vee B$ applied to proof-object b is a proof-object of $A \vee B$)

Finally, Prawitz discovered the deeper connection between the introduction and the elimination rules for the logical operators, namely that a proof of the conclusion of an elimination rule, is so to speak, already contained in the proofs of the premisses whenever the major premiss of this elimination rule is inferred by the corresponding introduction rule (Prawitz calls this sort of correspondence between the introduction and elimination rules for a logical operator the inversion principle) and he exploits this fact in the formulation of his reduction rules. Note that the logical constant $\perp$ constitutes an exception to the general scheme. (cf. Subsect. ii.1 of Chapter 4.)

The reduction rules for disjunction are

$$\dfrac{\dfrac{\begin{matrix}\Pi\\ A\end{matrix}}{A \vee B} \quad \begin{matrix}(A)\\ \Sigma_1\\ C\end{matrix} \quad \begin{matrix}(B)\\ \Sigma_2\\ C\end{matrix}}{C} \quad = \quad \begin{matrix}\Pi\\ A\\ \Sigma_1\\ C\end{matrix}$$

$$\dfrac{\dfrac{\begin{matrix}\Pi\\ B\end{matrix}}{A \vee B} \quad \begin{matrix}(A)\\ \Sigma_1\\ C\end{matrix} \quad \begin{matrix}(B)\\ \Sigma_2\\ C\end{matrix}}{C} \quad = \quad \begin{matrix}\Pi\\ B\\ \Sigma_2\\ C\end{matrix}$$

the linearized versions of which are the $\vee$-equality axioms.

(2) The introduction of proof-objects does not automatically lead to, but it allows for the strengthening of the elimination rules (not the introduction rules) for the logical operators. This can be seen as follows for the case of disjunction:

If Gentzen's $\vee$-elimination rule is translated into constructive logic, one does not get the $\vee$-elimination axiom or the corresponding rules ($\vee$E 1) and ($\vee$E 2), one only gets the (Gentzen) $\vee$-elimination axiom

$$\vee E_G : (X : prop)(Y : prop)(Z : prop)((x : X)Z)((y : Y)Z)Z$$

or, alternatively, the two rules

$$\dfrac{\begin{matrix}A : prop\\ B : prop\\ C : prop\\ d : (A)C\\ e : (B)C\\ c : A \vee B\end{matrix}}{\vee E_G(A, B, C, d, e, c) : C}$$

$$\frac{\begin{array}{l} A : prop \\ B : prop \\ C : prop \\ d = g : (A)C \\ e = h : (B)C \\ c = f : A \vee B \end{array}}{\vee E_G(A, B, C, d, e, c) = \vee E_G(A, B, C, g, h, f) : C}$$

and the (Gentzen) $\vee$-equality axioms

$$\vee E_G(A, B, C, d, e, \vee I1(A, B, a)) = d(a) : C$$
$$\vee E_G(A, B, C, d, e, \vee I2(A, B, b)) = e(b) : C$$

where $A : prop$, $B : prop$, $C : prop$, $d : (x : A)C$, $e : (y : B)C$, $a : A$ and $b : B$.

Whereas C here is any proposition, C in ($\vee$E 1), ($\vee$E 2), ($\vee$-equ 1) and ($\vee$-equ 2) is a propositional fct. over proofs of disjunction propositions. It has become common to call ($\vee$E 1), ($\vee$E 2) the strong $\vee$-elimination rules and the corresponding (Gentzen) rules the weak $\vee$-elimination rules.

x.4.1) What does a judgement of the form '$(\forall x : A)B : prop$' (or '$\forall(A, (x)B) : prop$') mean? or What does it mean to be the universal quantification of a propositional fct.?
To explain what it means to be the universal quantification $(\forall x : A)B$ (or $\forall(A, (x)B)$) of a propositional fct. B, one has to know (i) what a canonical proof-object of $(\forall x : A)B$ is, and (ii) when two canonical proof-objects of $(\forall x : A)B$ are identical.

The formation axiom for the composite proposition $(\forall x : A)B$ is

$$\forall : (X : set)(Y : (X)prop)prop$$

The formation axiom is tantamount to the following two rules

$$\frac{A : set \quad x : A \vdash B : prop}{(\forall x : A)B : prop} \ (\forall\text{F } 1)$$

$$\frac{A = C : set \quad x : A \vdash B = D : prop}{(\forall x : A)B = (\forall x : C)D : prop} \ (\forall\text{F } 2)$$

or equivalently to the following two rules

$$\frac{A : set \quad B : (A)prop}{\forall(A, B) : prop} \qquad \frac{A = C : set \quad B = D : (A)prop}{\forall(A, B) = \forall(C, D) : prop}$$

x.4.2) What does a judgement of the form '$c : (\forall x : A)B$' (or '$c : \forall(A, (x)B)$') mean? or What does it mean to be a proof-object of the universal quantification of a propositional fct.?

To explain what a judgement of the form '$c : (\forall x : A)B$' (or '$c : \forall(A, (x)B)$') means, one has to know that (i) c has the form of an object formed by the proof-object formation rule of $(\forall x : A)B$ (c has the form of a canonical proof-object of $(\forall x : A)B$) or (ii) c can be evaluated to that form (c is a non-canonical proof-object of $(\forall x : A)B$).

The proof-object formation or introduction axiom of $(\forall x : A)B$ is

$$\forall I : (X : set)(Y : (X)prop)((x : X)Y(x))\forall(X, Y)$$

which can be alternatively expressed by the introduction rule

$$\frac{\begin{array}{l} A = C : set \\ x : A \vdash B = D : prop \\ x : A \vdash b : B \end{array}}{\forall I(A, (x)B, (x)b) : (\forall x : C)D} \ (\forall I\ 1)$$

or reformulated in a higher level language

$$\frac{\begin{array}{l} A = C : set \\ B = D : (A)prop \\ b : (x : A)B(x) \end{array}}{\forall I(A, B, b) : \forall(C, D)}$$

C, n & o
The first two premisses are explained the same way as the resp. premisses of rule (ΠI 1) in C, n & o attached to ix.4.2).

x.4.3) What does a judgement of the form '$c = d : (\forall x : A)B$' (or '$c = d : \forall(A, (x)B)$') mean? or What does it mean for two proof-objects of $(\forall x : A)B$ to be identical?
To explain what a judgement of the form '$c = d : (\forall x : A)B$' ('$c = d : \forall(A, (x)B)$') means, one has to know that (i) c and d have the form of two proof-objects formed by the identity rule associated with the proof-object formation rule of $(\forall x : A)B$ (c and d are identical canonical proof-objects of $(\forall x : A)B$) or (ii) c and d can be evaluated to that form (c or d are identical non-canonical proof-objects of $(\forall x : A)B$).

The identity rule associated with the proof-object formation rule of $(\forall x : A)B$, i.e. the introduction rule of $\forall(A, (x)B)$ is

$$\frac{\begin{array}{l} A = C : set \\ x : A \vdash B = D : prop \\ x : A \vdash b = d : B \end{array}}{\forall I(A, (x)B, (x)b) = \forall I(C, (x)D, (x)d) : (\forall x : A)B} \ (\forall I\ 2)$$

or reformulated

$$\frac{\begin{array}{l} A = C : set \\ B = D : (A)prop \\ b = d : (x : A)B(x) \end{array}}{\forall I(A, B, b) = \forall I(C, D, d) : \forall(A, B)}$$

The following judgement can fairly easily be verified (derived)

$$(X : set)(Y : (X)prop)(Z : (\forall(X, Y))prop)((y : (x : X)Y(x))$$
$$Z(\forall I(X, Y, y)))(z : \forall(X, Y))Z(z) : type$$

The $\forall$-elimination axiom states that the selector or eliminatory operator $\forall E$ for the proposition $(\forall x : A)B$ (or $\forall(A, (x)B)$) is an object of the type just indicated, i.e.

$$\forall E : (X : set)(Y : (X)prop)(Z : (\forall(X, Y))prop)((y : (x : X)Y(x))$$
$$Z(\forall I(X, Y, y)))(z : \forall(X, Y))Z(z)$$

and this axiom can be formulated alternatively in terms of two rules

$$\frac{\begin{array}{l} A : set \\ B : (A)prop \\ C : (\forall(A, B))prop \\ d : (y : (x : A)B(x))C(\forall I(A, B, y)) \\ c : \forall(A, B) \end{array}}{\forall E(A, B, C, d, c) : C(c)} \ (\forall E\ 1)$$

$$\frac{\begin{array}{l} A : set \\ B : (A)prop \\ C : (\forall(A, B))prop \\ d = f : (y : (x : A)B(x))C(\forall I(A, B, y)) \\ c = e : \forall(A, B) \end{array}}{\forall E(A, B, C, d, c) = \forall E(A, B, C, f, e) : C(c)} \ (\forall E\ 2)$$

The meaning of the selector $\forall E$ is explained as follows:

1. Evaluate c to canonical form.
2. If $c \Longrightarrow \forall I(A, B, b)$ then $\forall E(A, B, C, d, c) \Longrightarrow d(b)$ and $\forall E(A, B, C, d, c) = d(b)$.

Justification

of (∀E 1) If $c \Longrightarrow \forall I(A, B, b)$, then, by the meaning explanation of $\forall E$, the value of $\forall E(A, B, C, d, c)$ is identical with the value of $d(b)$ which is a canonical proof-object of $C(\forall I(A, B, b))$ by premiss 4. By the fact that $\forall I(A, B, b) = c : \forall(A, B)$ and by premiss 3, the value of $\forall E(A, B, C, d, c)$ is a canonical proof-object of $C(c)$.

of (∀E 2) Analogous to (∀E 1). □

The ∀-equality axiom is

$$\forall E(A, B, C, d, \forall I(A, B, b)) = d(b) : C(\forall I(A, B, b))$$

where A : *set*, $B : (A)prop$, $C : (\forall(A, B))prop$, $d : (y : (x : A)B(x))C(\forall I(A, B, y))$ and $b : (x : A)B(x)$.

This ∀-equality axiom can also be expressed by the following rule

$$\frac{\begin{array}{l} A : set \\ B : (A)prop \\ C : (\forall(A, B))prop \\ d : (y : (x : A)B(x))C(\forall I(A, B, y)) \\ b : (x : A)B(x) \end{array}}{\forall E(A, B, C, d, \forall I(A, B, b)) = d(b) : C(\forall I(A, B, b))} \ (\forall\text{-equ})$$

Justification
of (∀-equ): straightforward by the meaning explanation of $\forall E$. □

C, n & o
(1) The above rules (∀I) and (∀E) are only partially based on ideas of Gentzen's and Heyting's (cf. C, n & o (2) to x.3.3)). Gentzen's ∀-elimination rule differs considerably in form from his ∨-elimination rule. Schroeder-Heister uniformized the form of the elimination rules of all the logical operators by adapting their form to the one of ∨-elimination (or ∃-elimination).[84]

The Gentzen ∀-introduction and the Schroeder-Heister ∀-elimination rule are (in the more formal notation of constructive type theory)

∀-introduction
$$\frac{\begin{array}{l}(x : A) \\ B\ true\end{array}}{(\forall x : A)B\ true}$$

∀-elimination
$$\frac{(\forall x : A)B\ true \qquad \begin{array}{c}\begin{pmatrix}(x : A) \\ B\ true\end{pmatrix} \\ C\ true\end{array}}{C\ true}$$

Heyting's semantical explanation of universal quantification is

A proof of $(\forall x : A)B$ consists of a fct. which takes a variable $x : A$ into a proof of B

put into more mathematical terms

If $x : A \vdash b : B$ (b is a proof-object of B provided x is an element of A) then $\lambda((x)b) : (\forall x : A)B$ (the lambda abstraction of b is a proof-object of $(\forall x : A)B$).

Prawitz's reduction rule for universal quantification is based on Gentzen's ∀-introduction and ∀-elimination rule (cf. note (3) below).

(2) If Schoeder-Heister's ∀-elimination rule is translated into constructive logic, one does not get the ∀-elimination axiom or the corresponding rules (∀E 1) and (∀E 2); one only obtains the weaker (Schroeder-Heister) ∀-elimination axiom

$$\forall E_{SH} : (X : set)(Y : (X)prop)(Z : prop)(((x : X)Y(x))Z)(\forall(X, Y))Z$$

or, alternatively, the two rules

$$\frac{A : set \quad B : (A)prop \quad C : prop \quad d : ((x : A)B(x))C \quad e : \forall(A, B)}{\forall E_{SH}(A, B, C, d, e) : C}$$

$$\frac{A : set \quad B : (A)prop \quad C : prop \quad d = f : ((x : A)B(x))C \quad e = g : \forall(A, B)}{\forall E_{SH}(A, B, C, d, e) = \forall E_{SH}(A, B, C, f, g) : C}$$

and the (Schroeder-Heister) ∀-equality axiom

$$\forall E_{SH}(A, B, C, d, \forall I(A, B, b)) = d(b) : C$$

where $A : set,\ B : (A)prop,\ C : prop,\ d : ((x : A)B(x))C$ and $b : (x : A)B(x)$.

Whereas C is here any proposition, C in (∀E 1), (∀E 2) and (∀-equ) is a propositional fct. over proofs of universal propositions, and (∀E 1) and (∀E 2) are clearly a strengthening of the (Schroeder-Heister) ∀-elimination rules which have been called the weak ∀-elimination rules, as opposed to the strong ∀-elimination rules (∀E 1) and (∀E 2).

(3) Gentzen's ∀-elimination rule is (in the formal notation of constructive type theory)

$$\frac{(\forall x : A)B\ true \quad a : A}{B(a/x)true}$$

and Prawitz's reduction rule for universal quantification is

$$\cfrac{\cfrac{\begin{matrix}\Pi \\ B\end{matrix}}{(\forall x : A)B}}{B(a/x)} \quad = \quad \frac{\Pi(a/x)}{B(a/x)}$$

These rules can of course be mimicked in constructive type theory as follows: If one writes $\forall E_G$ for the Gentzen selector or eliminatory operator for $(\forall x : A)B$, the Gentzen ∀-elimination axiom states

$$\forall E_G : (X : set)(Y : (X)prop)(\forall(X, Y))(x : X)Y(x)$$

which can be formulated in terms of two rules

$$\frac{A : set \quad x : A \vdash B : prop \quad b : (\forall x : A)B \quad a : A}{\forall E_G(A, (x)B, b, a) : B(a/x)}$$

$$\frac{\begin{array}{l} A : set \\ x : A \vdash B : prop \\ b = d : (\forall x : A)B \\ a = c : A \end{array}}{\forall E_G(A, (x)B, b, a) = \forall E_G(A, (x)B, d, c) : B(a/x)}$$

The meaning of the selector $\forall E_G$ is explained as follows:

1. Evaluate b to canonical form.
2. If $b \Longrightarrow \forall I(A, B, b)$ then $\forall E_G(A, (x)B, b, a) \Longrightarrow b(a)$ and $\forall E_G(A, (x)B, b, a) = b(a)$.

This meaning explanation of $\forall E_G$ justifies the two (Gentzen) ∀-elimination rules just given, as well as the following (Prawitz) ∀-equality axiom/rule

$$\forall E_G(A, (x)B, \forall I(A, (x)B, b), a) = b(a) : B(a/x)$$

where $A : set$, $x : A \vdash B : prop$, $b : (x : A)B$ and $a : A$.

This (Prawitz) ∀-equality axiom is equivalently expressed by the rule

$$\frac{\begin{array}{l} A : set \\ x : A \vdash B : prop \\ b : (\forall x : A)B \\ a : A \end{array}}{\forall E_G(A, (x)B, \forall I(A, (x)B, b), a) = b(a) : B(a/x)}$$

In other terms, the (Prawitz) ∀-equality axiom is the linearized version and the (Prawitz) ∀-equality rule is the tree-like version of the Prawitz reduction rule for ∀.

(4) Note that the Schroeder-Heister and the Gentzen ∀-elimination rules are equivalent.

G-∀-elimination is $$\frac{(\forall x : A)B \quad a : A}{B(a/x)}$$

SH-∀-elimination is $$\frac{(\forall x : A)B \qquad \begin{array}{c} \begin{pmatrix} (x : A) \\ B \end{pmatrix} \\ C \end{array}}{C}$$

For a proof of this equivalence, cf. Schroeder-Heister (1984), p. 417.

x.5.1) What does a judgement of the form '$A \supset B : prop$' mean? or What does it mean to be the (material) implication of two propositions?
To explain what it means to be the (material) implication $A \supset B$ of two propositions A and B, one has to know (i) what a canonical proof-object of $A \supset B$ is, and (ii) when two canonical proof-objects of $A \supset B$ are identical.

The formation axiom for the composite proposition $A \supset B$ is

$$\supset: (prop)(prop)prop$$

where obviously $(prop)(prop)prop : type$.

Equivalently, the formation axiom can be given by the two rules

$$\frac{A : prop \quad B : prop}{A \supset B : prop} (\supset F\ 1) \qquad \frac{A = C : prop \quad B = D : prop}{A \supset B = C \supset D : prop} (\supset F\ 2)$$

x.5.2) What does a judgement of the form '$c : A \supset B$' mean? or What does it mean to be a proof-object of the proposition $A \supset B$?
To explain what a judgement of the form '$c : A \supset B$' means, one has to know that (i) c has the form of an object formed by the proof-object formation rule of $A \supset B$ (c has the form of a canonical proof-object of $A \supset B$), or (ii) c can be evaluated to that form (c is a non-canonical proof-object of $A \supset B$).

The proof-object formation or introduction axiom of $A \supset B$ is

$$\supset I : (X : prop)(Y : prop)((X)Y)X \supset Y$$

which can be expressed alternatively by the introduction rule

$$\frac{\begin{array}{l} A = C : prop \\ B = D : prop \\ b : (A)B \end{array}}{\supset I(A, B, b) : C \supset D} (\supset I\ 1)$$

x.5.3) What does a judgement of the form '$c = d : A \supset B$' mean? or What does it mean for two proof-objects of $A \supset B$ to be identical?
To explain what a judgement of the form '$c = d : A \supset B$' means, one has to know that (i) c and d have the form of two objects formed by the identity rule associated with the proof-object formation rule of $A \supset B$ (c and d are identical canonical proof-objects of $A \supset B$) or (ii) c and d can be evaluated to that form (c or d are identical non-canonical proof-objects of $A \supset B$).

The identity rule associated with the proof-object formation rule of $A \supset B$, i.e. the second introduction rule of $A \supset B$ is

$$\frac{\begin{array}{l} A = C : prop \\ B = D : prop \\ b = d : (A)B \end{array}}{\supset I(A, B, b) = \supset I(C, D, d) : A \supset B} (\supset I\ 2)$$

The $\supset$-elimination axiom states that the selector or eliminatory operator $\supset E$ for the set $A \supset B$ is an object of the following type

$$\supset E : (X : prop)(Y : prop)(Z : (X \supset Y)prop)((v : (X)Y)$$
$$Z(\supset I(X, Y, v)))(z : X \supset Y)Z(z)$$

This ⊃-elimination axiom can be expressed alternatively by the two rules

$$\frac{\begin{array}{l} A : prop \\ B : prop \\ C : (A \supset B)prop \\ d : (v : (A)B)C(\supset I(A, B, v)) \\ c : A \supset B \end{array}}{\supset E(A, B, C, d, c) : C(c)} \ (\supset\text{E } 1)$$

$$\frac{\begin{array}{l} A : prop \\ B : prop \\ C : (A \supset B)prop \\ d = f : (v : (A)B)C(\supset I(A, B, v)) \\ c = e : A \supset B \end{array}}{\supset E(A, B, C, d, c) = \supset E(A, B, C, f, e) : C(c)} \ (\supset\text{E } 2)$$

The meaning of the selector $\supset E$ is explained as follows:

1. Evaluate c to canonical form.
2. If $c \Longrightarrow \supset I(A, B, b)$ then $\supset E(A, B, C, d, c) \Longrightarrow d(b)$ and $\supset E(A, B, C, d, c) = d(b)$.

Justification
of (⊃E 1) and (⊃E 2) as usual. □

The ⊃-equality axiom is

$$\supset E(A, B, C, d, \supset I(A, B, b)) = d(b) : C(\supset I(A, B, b))$$

where $A : prop$, $B : prop$, $C : (A \supset B)prop$, $d : (v : (A)B)C(\supset I(A, B, v))$ and $b : (A)B$.

This ⊃-equality axiom can also be expressed by the rule

$$\frac{\begin{array}{l} A : prop \\ B : prop \\ C : (A \supset B)prop \\ d : (v : (A)B)C(\supset I(A, B, v)) \\ b : (A)B \end{array}}{\supset E(A, B, C, d, \supset I(A, B, b)) = d(b) : C(\supset I(A, B, b))} \ (\supset\text{-equ})$$

Justification
of (⊃-equ): obvious by the meaning explanation of $\supset E$. □

C, n & o

(1) Similarly to note (1) of C, n & o to x.4.3), the above rules are inspired by ideas of Gentzen, Schroeder-Heister, Heyting and Prawitz.

The Gentzen ⊃-introduction rule and the Schroeder-Heister ⊃-elimination rule are (in the notation of constructive type theory)

$\supset$-introduction
$$\frac{\begin{array}{c}(A\ true)\\ B\ true\end{array}}{A \supset B\ true}$$

$\supset$-elimination
$$\frac{A \supset B\ true \qquad \begin{array}{c}\left(\begin{array}{c}(A\ true)\\ B\ true\end{array}\right)\\ C\ true\end{array}}{C\ true}$$

Heyting's semantical explanation of (material) implication is

A proof of $A \supset B$ consists of a fct. which takes a proof of A into a proof of B

put into more mathematical terms

If $x : A \vdash b : B$ (b is a proof-object of B provided x is a proof-object of A) then $\lambda((x)b) : A \supset B$ (the λ-abstraction of b is a proof-object of $A \supset B$)

Prawitz's reduction rule for (material) implication is based on Gentzen's $\supset$-introduction and $\supset$-elimination rule and hence only indirectly inspired the present $\supset$-equality axiom.

(2) If Schroeder-Heister's $\supset$-elimination rule is translated into constructive logic, one does not get the $\supset$-elimination axiom or the corresponding rules ($\supset$E 1) and ($\supset$E 2), but only the (Schroeder-Heister) $\supset$-elimination axiom

$$\supset E_{SH} : (X : prop)(Y : prop)(Z : prop)(((X)Y)Z)(X \supset Y)Z$$

or, alternatively, the two rules

$$\frac{\begin{array}{l}A : prop\\ B : prop\\ C : prop\\ d : ((A)B)C\\ e : A \supset B\end{array}}{\supset E_{SH}(A, B, C, d, e) : C}$$

$$\frac{\begin{array}{l}A : prop\\ B : prop\\ C : prop\\ d = f : ((A)B)C\\ e = g : A \supset B\end{array}}{\supset E_{SH}(A, B, C, d, e) = \supset E_{SH}(A, B, C, f, g) : C}$$

and the (Schroeder-Heister) $\supset$-equality axiom

$$\supset E_{SH}(A, B, C, d, \supset I(A, B, b)) = d(b) : C$$

where $A : prop$, $B : prop$, $C : prop$, $d : ((A)B)C$, and $b : (A)B$.

Again, the so-called strong rules ($\supset$E 1) and ($\supset$E 2) are a strengthening of the so-called weak (Schroeder-Heister) $\supset$-elimination rules.

(3) Gentzen's $\supset$-elimination rule is the well-known MP

$$\frac{A \supset B\ \mathit{true} \quad A\ \mathit{true}}{B\ \mathit{true}}$$

and Prawitz's reduction rule for (material) implication is

$$\frac{\dfrac{\begin{array}{c} A \\ \Pi \\ B \end{array}}{A \supset B} \quad \begin{array}{c} \Sigma \\ A \end{array}}{B} \quad = \quad \begin{array}{c} \Sigma \\ A \\ \Pi \\ B \end{array}$$

Again, it is possible to mimick these rules in constructive logic:

If one writes $\supset E_G$ for the Gentzen selector or eliminatory operator for $A \supset B$, the Gentzen $\supset$-elimination axiom states

$$\supset E_G : (X : prop)(Y : prop)(X \supset Y)(X)Y$$

which can be formulated in terms of two rules

$$\frac{\begin{array}{l} A : prop \\ B : prop \\ b : A \supset B \\ a : A \end{array}}{\supset E_G(A, B, b, a) : B}$$

$$\frac{\begin{array}{l} A : prop \\ B : prop \\ b = d : A \supset B \\ a = c : A \end{array}}{\supset E_G(A, B, b, a) = \supset E_G(A, B, d, c) : B}$$

The meaning of the selector $\supset E_G$ is explained as follows:

1. Evaluate b to canonical form.
2. If $b \Longrightarrow \supset I(A, B, b)$ then $\supset E_G(A, B, b, a) \Longrightarrow b(a)$ and $\supset E_G(A, B, b, a) = b(a)$.

The meaning explanation of $\supset E_G$ justifies at once the two Gentzen $\supset$-elimination rules above and the following (Prawitz) $\supset$-equality axiom/rule

$$\supset E_G(A, B, \supset I(A, B, b), a) = b(a) : B$$

where $A : prop$, $B : prop$, $b : (A)B$, and $a : A$.

The corresponding (Prawitz) $\supset$-equality rule is

$$\frac{\begin{array}{l} A : prop \\ B : prop \\ b : (A)B \\ a : A \end{array}}{\supset E_G(A, B, \supset I(A, B, b), a) = b(a) : B}$$

The (Prawitz) $\supset$-equality axiom and $\supset$-equality rule are nothing but the linearized and tree-like versions resp. of Prawitz's reduction rule for $\supset$.

(4) Note that the Schroeder-Heister and the Gentzen $\supset$-elimination rules are equivalent. The same applies to their resp. $\supset$-equality rules. For a proof of these equivalences, cf. Schroeder-Heister (1984A), pp. 1294/1295.

x.6.1) What does a judgement of the form '$(\exists x : A)B : prop$' (or '$\exists(A, (x)B) : prop$') mean? or What does it mean to be an existential quantification of a propositional fct.?
To explain what it means to be the existential quantification $(\exists x : A)B$ (or $\exists(A, (x)B)$) of a propositional fct. B, one has to know (i) what a canonical proof-object of $(\exists x : A)B$ is, and (ii) when two canonical proof-objects of $(\exists x : A)B$ are identical.

The formation axiom for the composite proposition $(\exists x : A)B$ is

$$\exists : (X : set)(Y : (X)prop)prop$$

This formation axiom can equally be expressed by the following two rules

$$\frac{A : set \quad x : A \vdash B : prop}{(\exists x : A)B : prop} \ (\exists\text{F } 1)$$

$$\frac{A = C : set \quad x : A \vdash B = D : prop}{(\exists x : A)B = (\exists x : C)D : prop} \ (\exists\text{F } 2)$$

or reformulated in a higher level language

$$\frac{A : set \quad B : (A)prop}{\exists(A, B) : prop} \qquad \frac{A = C : set \quad B = D : (A)prop}{\exists(A, B) = \exists(C, D) : prop}$$

x.6.2) What does a judgement of the form '$c : (\exists x : A)B$' (or '$c : \exists(A, (x)B)$') mean? or What does it mean to be a proof-object of an existential quantification of a propositional fct.?
To explain what a judgement of the form '$c : (\exists x : A)B$' (or '$c : \exists(A, (x)B)$') means, one has to know that (i) c has the form of an object formed by the proof-object formation rule of $(\exists x : A)B$ (c has the form of a canonical proof-object of $(\exists x : A)B$) or (ii) c can be evaluated to that form (c is a non-canonical proof-object of $(\exists x : A)B$).

The proof-object formation or introduction axiom of $(\exists x : A)B$ is

$$\exists I : (X : set)(Y : (X)prop)(x : X)(Y(x))\exists(X, Y)$$

which can be alternatively expressed by the introduction rule

$$\frac{\begin{array}{l} A = C : set \\ x : A \vdash B = D : prop \\ a : A \\ b : B(a/x) \end{array}}{\exists I(A, (x)B, a, b) : (\exists x : C)D} \ (\exists\text{I } 1)$$

or reformulated

$$\frac{\begin{array}{l} A = C : set \\ B = D : (A)prop \\ a : A \\ b : B(a) \end{array}}{\exists I(A, B, a, b) : \exists(C, D)}$$

x.6.3) What does a judgement of the form '$c = d : (\exists x : A)B$'. (or '$c = d : \exists(A, (x)B)$') mean? or What does it mean for two proof-objects of $(\exists x : A)B$ to be identical?
To explain what a judgement of the form '$c = d : (\exists x : A)B$' (or '$c = d : \exists(A, (x)B)$') means, one has to know that (i) c and d have the form of two objects formed by the identity rule associated with the proof-object formation rule of $(\exists x : A)B$ (c and d are identical canonical proof-objects of $(\exists x : A)B$) or (ii) c and d can be evaluated to that form (c or d are identical non-canonical proof-objects of $(\exists x : A)B$).

The identity rule associated with the proof-object formation rule of $(\exists x : A)B$, i.e. the second introduction rule of $(\exists x : A)B$ is

$$\frac{\begin{array}{l} A = C : set \\ x : A \vdash B = D : prop \\ a = c : A \\ b = d : B(a/x) \end{array}}{\exists I(A, (x)B, a, b) = \exists I(C, (x)D, c, d) : (\exists x : A)B} \; (\exists\text{I } 2)$$

or reformulated

$$\frac{\begin{array}{l} A = C : set \\ B = D : (A)prop \\ a = c : A \\ b = d : B(a) \end{array}}{\exists I(A, B, a, b) = \exists I(C, D, c, d) : \exists(A, B)}$$

The following judgement can relatively easily be derived

$$(X : set)(Y : (X)prop)(Z : (\exists(X, Y))prop)((x : X)(y : Y(x))$$
$$Z(\exists I(X, Y, x, y)))(z : \exists(X, Y))Z(z) : type$$

The $\exists$-elimination axiom states that the selector or eliminatory operator $\exists E$ for the proposition $(\exists x : A)B$ (or $\exists(A, (x)B)$) is an object of the type just mentioned, i.e.

$$\exists E : (X : set)(Y : (X)prop)(Z : (\exists(X, Y))prop)((x : X)(y : Y(x))$$
$$Z(\exists I(X, Y, x, y)))(z : \exists(X, Y))Z(z)$$

and this axiom can be formulated alternatively in terms of two rules

$$\frac{\begin{array}{l} A : set \\ B : (A)prop \\ C : (\exists(A, B))prop \\ d : (x : A)(y : B(x))C(\exists I(A, B, x, y,)) \\ c : \exists(A, B) \end{array}}{\exists E(A, B, C, d, c) : C(c)} \; (\exists \text{E } 1)$$

$$\frac{\begin{array}{l} A : set \\ B : (A)prop \\ C : (\exists(A, B))prop \\ d = f : (x : A)(y : B(x))C(\exists I(A, B, x, y)) \\ c = e : \exists(A, B) \end{array}}{\exists E(A, B, C, d, c) = \exists E(A, B, C, f, e) : C(c)} \; (\exists \text{E } 2)$$

The meaning of the selector $\exists E$ is explained as follows:

1. Evaluate c to canonical form.
2. If $c \Longrightarrow \exists I(A, B, a, b)$ then $\exists E(A, B, C, d, c) \Longrightarrow d(a, b)$ and $\exists E(A, B, C, d, c) = d(a, b)$.

Justification

of (∃E 1) If $c \Longrightarrow \exists I(A, B, a, b)$ then, by the meaning explanation of $\exists E$, the value of $\exists E(A, B, C, d, c)$ is identical with the value of $d(a, b)$ which, by premiss 4, is a canonical proof-object of $C(\exists I(A, B, a, b))$. By the fact that $\exists I(A, B, a, b) = c : \exists(A, B)$ and premiss 3, the value of $\exists E(A, B, C, d, c)$ is a canonical proof-object of $C(c)$.

of (∃E 2) Analogous to (∃E 1). □

The ∃-equality axiom is

$$\exists E(A, B, C, d, \exists I(A, B, a, b)) = d(a, b) : C(\exists I(A, B, a, b))$$

where $A : set$, $B : (A)prop$, $C : (\exists(A, B))prop$, $d : (x : A)(y : B(x))C(\exists I(A, B, x, y))$, $a : A$ and $b : B(a)$.

This ∃-equality axiom can also be expressed by the rule

$$\frac{\begin{array}{l} A : set \\ B : (A)prop \\ C : (\exists(A, B))prop \\ d : (x : A)(y : B(x))C(\exists I(A, B, x, y)) \\ a : A \\ b : B(a) \end{array}}{\exists E(A, B, C, d, \exists I(A, B, a, b)) = d(a, b) : C(\exists I(A, B, a, b))} \; (\exists\text{-equ})$$

Justification

of (∃-equ): by the meaning explanation of $\exists E$. □

C, n & o

(1) The axioms and rules above are, as before, motivated by ideas of Gentzen, Heyting and Prawitz. Gentzen rules for ∃-introduction and ∃-elimination are (in the notation of constructive type theory)

$$\exists\text{-introduction} \qquad \frac{a : A \quad B(a/x)\ \mathit{true}}{(\exists x : A)B\ \mathit{true}}$$

$$\exists\text{-elimination} \qquad \frac{(\exists x : A)B\ \mathit{true} \qquad \begin{array}{c}(x : A,\ B\ \mathit{true})\\ C\ \mathit{true}\end{array}}{C\ \mathit{true}}$$

Heyting's semantical explanation of existential quantification is:

A proof of $(\exists x : A)B$ consists of an element a of a set A and a proof of $B(a/x)$

put into more mathematical terms

If $a : A$ (a is an element of A) and $b : B(a/x)$ (b is a proof-object of the proposition $B(a/x)$) then $(a, b) : (\exists x : A)B$ (the ordered pair (a, b) is a proof-object of the proposition $(\exists x : A)B$).

Prawitz's reduction rule for existential quantification is

$$\frac{\dfrac{\begin{array}{c}\Pi\\ B(a/x)\end{array}}{(\exists x : A)B} \qquad \begin{array}{c}(B)\\ \Sigma\\ C\end{array}}{C} \quad = \quad \begin{array}{c}\Pi\\ B(a/x)\\ \Sigma(a/x)\\ C\end{array}$$

of which (∃-equ) can be considered a linearized version.

(2) If Gentzen's ∃-elimination rule is translated into the language of constructive logic, the result is not the ∃-elimination axiom or the associated rules (∃E 1) and (∃E 2), but rather the following (Gentzen) ∃-elimination axiom

$$\exists E_G : (X : \mathit{set})(Y : (X)\mathit{prop})(Z : \mathit{prop})((x : X)(Y(x))Z)(\exists(X, Y))Z$$

or alternatively, the two rules

$$\frac{\begin{array}{l}A : \mathit{set}\\ B : (A)\mathit{prop}\\ C : \mathit{prop}\\ d : (x : A)(B(x))C\\ c : \exists(A, B)\end{array}}{\exists E_G(A, B, C, d, c) : C}$$

$$\frac{\begin{array}{l}A : \mathit{set}\\ B : (A)\mathit{prop}\\ C : \mathit{prop}\\ d = f : (x : A)(B(x))C\\ c = e : \exists(A, B)\end{array}}{\exists E_G(A, B, C, d, c) = \exists E_G(A, B, C, f, e) : C}$$

and the (Gentzen) ∃-equality axiom

$$\exists E_G(A, B, C, d, \exists I(A, B, a, b)) = d(a, b) : C$$

where $A : set$, $B : (A)prop$, $C : prop$, $d : (x : A)(B(x))C$, $b : B(a)$ and $a : A$.

Whereas C is here any proposition, C in (∃E 1) and (∃E 2) and (∃-equ) is a propositional fct. of proofs of existential propositions, and the strong rules (∃E 1) and (∃E 2) are a strengthening of the weak (Gentzen) ∃-elimination rules.

(3) The strengthening of the (Gentzen) ∃-elimination rules mentioned in the preceding note (2) allows us to prove the axiom of choice AC

$$(\forall x : A)(\exists y : B(x))C(x, y) \supset (\exists f : (\Pi x : A)B(x))(\forall x : A)$$
$$C(x, app(A, (\Pi x : A)B(x), f, x)) \; true$$

or written in the language of constructive set theory according to the Curry-Howard isomorphism (cf. Sect. xi)

$$(\Pi x : A)(\Sigma y : B(x))C(x, y) \rightarrow (\Sigma f : (\Pi x : A)B(x))(\Pi x : A)$$
$$C(x, app(A, (\Pi x : A)B(x), f, x)) \; true$$

Note that AC generalizes AC-NF where $A \equiv N$ and $B(x) \equiv N(x)$, and it naturally also generalizes AC-N where $A \equiv N$ and $B(x) \equiv D$, and of course AC-NN where $A \equiv N$ and $B(x) \equiv N$.[85]

As for the proof, cf. version (6) of constructive type theory.

As can be seen from the proof of AC, it relies essentially on two applications of (ΣE) which in the language of constructive logic are the strengthened rules of ∃-elimination (∃E).

In 1904 Zermelo introduced AC. With this new axiom, new fcts. could at once be defined which were not computable. In 1908 Brouwer realized that it was the principle of excluded middle PEM together with AC which gave rise to uncomputable fcts., and he subsequently discovered that PEM alone gave rise to such fcts. Once classical logic is replaced by intuitionistic logic, AC is valid, which can easily be justified intuitionistically:

To validate AC assume that there is a proof of the antecedent $(\forall x : A)(\exists y : B(x))C(x, y)$; by the intuitionistic explanation of ∀ (cf. note (1) of C, n & o to x.4.3)), this proof consists of a fct. which when applied to $x : A$ yields a proof of $(\exists y : B(x))C(x, y)$; by the intuitionistic explanation of ∃ (cf. note (1) of C, n & o to x.6.3)), the proof of $(\exists y : B(x))C(x, y)$ consists of an ordered pair which consists of an element $y : B(x)$ and a proof of $C(x, y)$. Let f denote the fct. which to an arbitrary element $x : A$ assigns the resp. $y : B(x)$. Then $C(x, f(x))$ holds for an arbitrary $x : A$ and so does the consequent of AC, $(\exists f : (\Pi x : A)B(x))(\forall x : A)C(x, app(A, (\Pi x : A)B(x), f, x))$.

In (1967) Bishop also confirmed that AC is valid under a constructive interpretation of the logical operators. He remarks: 'A choice function exists in constructive mathematics, because a choice is implied by the very meaning of existence.'[86] Moreover, AC is necessary to develop intuitionistic mathematics. One needs AC e.g. in order to prove the completeness of the real number system or in order to find a partial inverse of a surjective fct. etc. Finally, it may be observed that there is no proof of AC in ZF set theory and that is why it has to be added to the theory as an axiom, for which on the other hand it is anything but easy to claim self-evidence. And as far as Heyting Arithmetic of finite type N-HA$^{\omega}$ is concerned, AC is expressible but not provable in the theory, and that is why it also has to be added to the theory in order to develop intuitionistic mathematics with it.

Finally, Swaen has made some interesting observations on AC for a version of constructive type theory, which however contains extensional E-identity instead of intensional I-identity. He starts off by stating that as a consequence of the strong Σ-rules of constructive type theory, its logic, i.e. constructive logic, is generally stronger than the logic of traditional formal systems. He then observes that there are two possibilities to reconcile the constructive set-approach and the traditional proposition-approach to logic: either constructive set theory is weakened s.t. fewer sets are inhabited, or traditional logic is strengthened by adding choice-like schemes s.t. more propositions can be proved. – Apart from this weakening and strengthening of the logic, there are two ways of internally characterizing strong $\exists$- or Σ-rules: 1) strong $\exists$-elimination ($\exists$E) can be characterized by a so-called $\exists$-exhaustion principle, saying that any inhabitant of a $\exists$-set (or a Σ-set) is a canonical inhabitant. (cf. Sect. vii., Chapter 3). 2)Let AC^{+} be the rule

$$\frac{x_1 : A_1, \ldots, x_n : A_n \vdash t : (\forall x : A)(\exists y : B(x))C(x, y)}{x_1 : A_1, \ldots, x_n : A_n \vdash a(t) : (\exists f : (\Pi x : A)B(x))(\forall x : A)C(x, app(f, x))}$$

then it is proved that if D is a Σ-free set, then D is inhabited in the theory with strong Σ-rules iff D is inhabited in the theory with weak Σ-rules plus AC^{+} as an extra rule. (Remember that the results 1) and 2) presuppose that constructive type theory contains E-identity.)[87]

x.7.1) What does a judgement of the form '$A \,\&\, B\,:\,prop$' mean? or What does it mean to be the conjunction of two propositions?
To explain what it means to be the conjunction $A \,\&\, B$ of two propositions A and B one has to know (i) what a canonical proof-object of $A \,\&\, B$ is, and (ii) when two canonical proof-objects of $A \,\&\, B$ are identical.

The formation axiom for the composite proposition $A\&B$ is

$$\& : (prop)(prop)prop$$

where $(prop)(prop)prop : type$.

Equivalently, the formation axiom can be given by the two rules

$$\frac{A : prop \quad B : prop}{A \,\&\, B : prop}\ (\&F\ 1) \qquad \frac{A = C : prop \quad B = D : prop}{A \,\&\, B = C\&D : prop}\ (\&F\ 2)$$

x.7.2) What does a judgement of the form '$c : A \,\&\, B$' mean? or What does it mean to be a proof-object of a proposition $A \,\&\, B$?
To explain what a judgement of the form '$c : A \,\&\, B$' means, one has to know that (i) c has the form of an object formed by the proof-object formation rule of $A \,\&\, B$ (c has the form of a canonical proof-object of $A \,\&\, B$), or (ii) c can be evaluated to that form (c is a non-canonical proof-object of $A \,\&\, B$).

The proof-object formation or introduction axiom of $A\&B$ is

$$\&I : (X : prop)(Y : prop)(X)(Y)X\&Y$$

which can be expressed alternatively by the introduction rule

$$\frac{\begin{array}{l} A = C : prop \\ B = D : prop \\ a : A \\ b : B \end{array}}{\&I(A, B, a, b) : C \,\&\, D} \ (\&I\ 1)$$

x.7.3) What does a judgement of the form '$c = d : A \,\&\, B$' mean? or What does it mean for two proof-objects of $A \,\&\, B$ to be identical?
To explain what a judgement of the form '$c = d : A \,\&\, B$' means, one has to know that (i) c and d have the form of two objects formed by the identity rule associated with the proof-object-formation rule of $A \,\&\, B$ (c and d are identical canonical proof-objects of $A \,\&\, B$) or (ii) c and d can be evaluated to that form (c or d are identical non-canonical proof-objects of $A \,\&\, B$).

The identity rule associated with the proof-object-formation rule of $A \,\&\, B$, i.e. the second introduction rule of $A \,\&\, B$ is

$$\frac{\begin{array}{l} A = C : prop \\ B = D : prop \\ a = c : A \\ b = d : B \end{array}}{\&I(A, B, a, b) = \&I(C, D, c, d) : A \,\&\, B} \ (\&I\ 2)$$

The &-elimination axiom states that the selector or eliminatory operator $\&E$ for the proposition $A\&B$ is an object of the following type

$$\&E : (X : prop)(Y : prop)(Z : (X \,\&\, Y)prop)((x : X)(y : Y)$$
$$Z(\&I(X, Y, x, y)))(z : X \,\&\, Y)Z(z)$$

This &-elimination axiom can be expressed alternatively by the two rules

$$\frac{\begin{array}{l} A : prop \\ B : prop \\ C : (A \,\&\, B)prop \\ d : (x : A)(y : B)C(\&I(A, B, x, y)) \\ c : A \,\&\, B \end{array}}{\&E(A, B, C, d, c) : C(c)} \ (\&E\ 1)$$

$$\frac{\begin{array}{l} A : prop \\ B : prop \\ C : (A \,\&\, B)prop \\ d = f : (x : A)(y : B)C(\&I(A, B, x, y)) \\ c = e : A \,\&\, B \end{array}}{\&E(A, B, C, d, c) = \&E(A, B, C, f, e) : C(c)} \ (\&E\ 2)$$

The meaning of the selector $\&E$ is explained as follows:

1. Evaluate c to canonical form.
2. If $c \Longrightarrow \&I(A, B, a, b)$ then $\&E(A, B, C, d, c) \Longrightarrow d(a, b)$ and $\&E(A, B, C, d, c) = d(a, b)$.

Justification
of (&E 1) and (&E 2) as usual. □

The &-equality axiom is

$$\&E(A, B, C, d, \&I(A, B, a, b)) = d(a, b) : C(\&I(A, B, a, b))$$

where $A : prop$, $B : prop$, $C : (A \,\&\, B)prop$, $d : (x : A)(y : B)C(\&I(A, B, x, y))$, $a : A$ and $b : B$.

This &-equality axiom can also be expressed by the rule

$$\frac{\begin{array}{l} A : prop \\ B : prop \\ C : (A \,\&\, B)prop \\ d : (x : A)(y : B)C(\&I(A, B, x, y)) \\ a : A \\ b : B \end{array}}{\&E(A, B, C, d, \&I(A, B, a, b)) = d(a, b) : C(\&I(A, B, a, b))}\ (\&\text{-equ})$$

Justification
of (&-equ): straightforward by the meaning explanation of $\&E$. □

C, n & o

(1) The axioms and rules above are once more inspired by ideas of Gentzen, Schroeder-Heister, Heyting and Prawitz. Gentzen rule for &-introduction and Schroeder-Heister's rule for &-elimination are (in the notation of constructive type theory)

&-introduction $\dfrac{A\ true \quad B\ true}{A \,\&\, B\ true}$

&-elimination $\dfrac{A \,\&\, B\ true \qquad \begin{array}{c}(A\ true,\ B\ true) \\ C\ true\end{array}}{C\ true}$

Heyting's semantical explanation of conjunction is

A proof of $A\&B$ consists of a proof of A and a proof of B

or in more mathematical terms

If $a : A$ and $b : B$ (a is a proof-object of A and b is a proof-object of B) then $(a, b) : A\&B$ (the ordered pair (a, b) is a proof-object of $A \,\&\, B$).

Prawitz's reduction rule for conjunction is based on Gentzen's &-introduction and &-elimination rule and hence enters here only indirectly.

(2) If Schroeder-Heister's &-elimination rule is translated into the language of constructive logic, one as usual does not get the &-elimination axiom or the corresponding rules (&E 1) and (&E 2), but only the (Schroeder-Heister) &-elimination axiom

$$\&E_{SH} : (X : prop)(Y : prop)(Z : prop)((X)(Y)Z)(X \,\&\, Y)Z$$

or, alternatively, the two rules

$$\frac{\begin{array}{l} A : prop \\ B : prop \\ C : prop \\ d : (A)(B)C \\ e : A \,\&\, B \end{array}}{\&E_{SH}(A, B, C, d, e) : C}$$

$$\frac{\begin{array}{l} A : prop \\ B : prop \\ C : prop \\ d = f : (A)(B)C \\ e = g : A \,\&\, B \end{array}}{\&E_{SH}(A, B, C, d, e) = \&E_{SH}(A, B, C, f, g) : C}$$

and the (Schroeder-Heister) &-equality axiom

$$\&E_{SH}(A, B, C, d, \&I(A, B, a, b)) = d(a, b) : C$$

where $A : prop,\ B : prop,\ C : prop,\ d : (A)(B)C,\ a : A$ and $b : B$.

Again, the strong rules (&E 1) and (&E 2) are a strengthening of the weak (Schroeder-Heister) &-elimination rules.

(3) Gentzen's &-elimination rules are (in the notation of constructive type theory)

$$\frac{A \,\&\, B\ true}{A\ true} \qquad \frac{A \,\&\, B\ true}{B\ true}$$

and Prawitz's reduction rules for conjunction are

$$\cfrac{\cfrac{\begin{array}{cc} \Pi & \Sigma \\ A & B \end{array}}{A\&B}}{A} \quad = \quad \begin{array}{c} \Pi \\ A \end{array} \qquad\qquad \cfrac{\cfrac{\begin{array}{cc} \Pi & \Sigma \\ A & B \end{array}}{A\&B}}{B} \quad = \quad \begin{array}{c} \Sigma \\ B \end{array}$$

It is of course possible to mimic these rules in constructive logic: If one writes $\&E_G$ for the Gentzen selector or eliminatory operator for $A \,\&\, B$, the (Gentzen) &-elimination axioms state

$$\&E1_G : (X : prop)(Y : prop)(X \,\&\, Y)X$$
$$\&E2_G : (X : prop)(Y : prop)(X \,\&\, Y)Y$$

which can be formulated in terms of four rules

$$\frac{A : prop \quad B : prop \quad c : A \,\&\, B}{\&E1_G(A, B, c) : A} \qquad \frac{A : prop \quad B : prop \quad c = f : A \,\&\, B}{\&E1_G(A, B, c) = \&E1_G(A, B, f) : A}$$

$$\frac{A : prop \quad B : prop \quad c : A \,\&\, B}{\&E2_G(A, B, c) : B} \qquad \frac{A : prop \quad B : prop \quad c = f : A \,\&\, B}{\&E2_G(A, B, c) = \&E2_G(A, B, f) : B}$$

The meaning of the selectors $\&E1_G$ and $\&E2_G$ is explained as follows:

1. Evaluate c to canonical form.
2. If $c \Longrightarrow \&I(A, B, a, b)$ then $\&E1(A, B, c) \Longrightarrow a$ and $\&E1_G(A, B, c) = a$; and then $\&E2_G(A, B, c) \Longrightarrow b$ and $\&E2_G(A, B, c) = b$.

This meaning explanation of $\&E1_G$ and $\&E2_G$ justifies at once the two (Gentzen) &-elimination rules above, as well as the following (Prawitz) &-equality axioms/rules

$$\&E1_G(A, B, \&I(A, B, a, b)) = a : A$$
$$\&E2_G(A, B, \&I(A, B, a, b)) = b : B$$

where $A : prop$, $B : prop$, $a : A$ and $b : B$.

The corresponding (Prawitz) equality rules are

$$\frac{A : prop \quad B : prop \quad a : A \quad b : B}{\&E1_G(A, B, \&I(A, B, a, b)) = a : A}$$

$$\frac{A : prop \quad B : prop \quad a : A \quad b : B}{\&E2_G(A, B, \&I(A, B, a, b)) = b : B}$$

The (Prawitz) &-equality axioms and &-equality rules are nothing but the linearized and tree-like versions resp. of Prawitz's reduction rules for &.

(4) The Schroeder-Heister and Gentzen &-elimination rules are equivalent. The same holds for their resp. &-equality rules. For a proof of these equivalences, cf. Schroeder-Heister (1984A), pp. 1293, 1294.

x.8.1) What does a judgement of the form '$I(A, a, b) : prop$' mean? or What does it mean to be the propositional identity between two elements of a given set?

To explain what it means to be the propositional identity $I(A, a, b)$ between two elements a and b of a set A, one has to know (i) what a canonical proof-object of $I(A, a, b)$ is, and (ii) when two canonical proof-objects of $I(A, a, b)$ are identical.

The formation axiom for the composite proposition $I(A, a, b)$ is

$$I : (X : set)(X)(X)prop$$

where $(X : set)(X)(X)prop : type$

Equivalently, the formation axiom can be given by the two rules

$$\frac{A : set \quad a : A \quad b : A}{I(A, a, b) : prop} \text{(IF 1)} \qquad \frac{A = C : set \quad a = c : A \quad b = d : A}{I(A, a, b) = I(C, c, d) : prop} \text{(IF 2)}$$

x.8.2) What does a judgement of the form '$c : I(A, a, b)$' mean? or What does it mean to be a proof-object of the proposition $I(A, a, b)$?
To explain what a judgement of the form '$c : I(A, a, b)$' means, one has to know that (i) c has the form of an object formed by the proof-object formation rule of $I(A, a, b)$ (c has the form of a canonical proof-object of $I(A, a, b)$) or (ii) c can be evaluated to that form (c is a non-canonical proof-object of $I(A, a, b)$).

The proof-object formation or introduction axiom of $I(A, a, b)$ is

$$II : (X : set)(x : X)I(X, x, x)$$

which can be expressed alternatively by the introduction rule

$$\frac{A = B : set \quad a : A}{II(A, a) : I(B, a, a)} \text{(II 1)}$$

C, n & o

(1) The proof-object formation axiom (rule) of $I(A, a, b)$ is motivated by the introduction rule for identity in predicate logic, called the law of identity, which is simply $I(A, a, a)$.

(2) Propositional identity I can readily be seen to mimic judgemental identity inside constructive logic. (Cf. note (1) of C, n & o to ix.8.2).)

x.8.3) What does a judgement of the form '$c = d : I(A, a, b)$' mean? or What does it mean for two proof-objects of $I(A, a, b)$ to be identical?
To explain what a judgement of the form '$c = d : I(A, a, b)$' means, one has to know that (i) c and d have the form of two objects formed by the identity rule associated with the proof-object formation rule of $I(A, a, b)$ (c and d are identical canonical proof-objects of $I(A, a, b)$) or (ii) c and d can be evaluated to that form (c or d are identical non-canonical proof-objects of $I(A, a, b)$).

The identity rule associated with the proof-object formation rule of $I(A, a, b)$, i.e. the second introduction rule of $I(A, a, b)$ is

$$\frac{A = B : set \quad a = b : A}{II(A, a) = II(B, b) : I(A, a, b)} \text{ (II 2)}$$

The I-elimination axiom states that the selector or eliminatory operator IE for the proposition $I(A, a, b)$ is an object of the following type

$$IE : (X : set)(x : X)(y : X)(Z : (x : X)(y : X)(I(X, x, y))prop)((w : X) \\ Z(w, w, II(X, w)))(z : I(X, x, y))Z(x, y, z)$$

The I-elimination axiom can be expressed alternatively by the two rules

$$\frac{\begin{array}{l} A : set \\ a : A \\ b : A \\ C : (x : A)(y : A)(I(A, x, y))prop \\ d : (w : A)C(w, w, II(A, w)) \\ c : I(A, a, b) \end{array}}{IE(A, a, b, C, d, c) : C(a, b, c)} \text{ (IE 1)}$$

$$\frac{\begin{array}{l} A : set \\ a = e : A \\ b = f : A \\ C : (x : A)(y : A)(I(A, x, y))prop \\ d = g : (w : A)C(w, w, II(A, w)) \\ c = h : I(A, a, b) \end{array}}{IE(A, a, b, C, d, c) = IE(A, e, f, C, g, h) : C(a, b, c)} \text{ (IE 2)}$$

The meaning of the selector IE is explained as follows:

1. Evaluate c to canonical form.
2. If $c \Longrightarrow II(A, a)$ then $IE(A, a, b, C, d, c) \Longrightarrow d(a)$ and $IE(A, a, b, C, d, c) = d(a)$.

Justification
of (IE 1) and (IE 2): cf. the justification of the homonymous rules in the explanations following ix.8.3). □

The I-equality axiom is

$$IE(A, a, b, C, d, II(A, a)) = d(a) : C(a, b, II(A, a))$$

where $A : set$, $a : A$, $b : A$, $C : (x : A)(y : A)(I(A, x, y))prop$ and $d : (w : A)C(w, w, II(A, w))$.

This I-equality axiom can also be expressed by the rule

$$\frac{\begin{array}{l} A : set \\ a : A \\ b : A \\ C : (x : A)(y : A)(I(A, x, y))prop \\ d : (w : A)C(w, w, II(A, w)) \end{array}}{IE(A, a, b, C, d, II(A, a)) = d(a) : C(a, b, II(A, a))} \text{(I-equ)}$$

Justification
of (I-equ): obvious by the meaning explanation of IE. □

C, n & o
(1) The elimination axiom for $I(A, a, b)$ is motivated by the elimination rule for identity in predicate logic, called Leibniz's principle of the indiscernibility of identicals, i.e.

$$\frac{I(A, a, b) \quad C(a)}{C(b)}$$

If one adds the proof-objects to the premisses of this natural deduction style rule of inference

$$\frac{\left.\begin{array}{c} \vdots \\ I(A, a, b) \end{array}\right\} c \qquad \left.\begin{array}{c} \vdots \\ C(a) \end{array}\right\} d(a)}{C(b)}$$

one obtains $IE(A, a, b, C, d, c)$ as a proof-object for $C(b)$. It is then easy to see that the rule (IE 1) constitutes a strengthening of the rule just mentioned, along the lines sketched out in x.3 – x.7, namely in the sense that the premiss $C : (A)prop$ in this latter rule is replaced by the premiss $C : (x : A)(y : A)(I(A, x, y))prop$. This strengthened or generalized I-elimination rule is due to Paulin-Mohring.

(2) Frege's axioms in his *Grundgesetze der Arithmetik* (1893) are (in slightly modernized notation):[88]

- (I) and (II): logical laws concerning propositions and universal quantification
- (III) $g(x = y) \supset g((\forall f)(f(y) \supset f(x)))$ which is Frege's way of stating the identity law
- (IV) $\sim(A \supset\subset \sim B) \supset (A \supset\subset B)$
- (V) $\breve{\varepsilon} f(\varepsilon) = \breve{\alpha} g(\alpha) \supset\subset (\forall x)(f(x) = g(x))$ i.e. two course-of-values are identical iff their corresponding fcts. are extensionally identical.
- (VI) $a = \backslash\breve{\varepsilon}(a = \varepsilon)$

Note that whereas Frege's axioms (I)–(III) are valid in constructive type theory, axioms (IV) and (V) are not. Axiom (IV) is obviously intuitionistically not valid. Axiom (V) is rewritten in the notation of constructive type theory by means of the Curry–Howard isomorphism (cf. Sect. xi.) as follows:

$$I(\Pi(A,B),\Pi I(A,B,f),\Pi I(A,B,g)) \supset\subset (\forall x:A)I(B(x),f(x),g(x))$$

Observe that $f(x)$ is definitionally identical with $app(A,B,\Pi I(A,B,f),x)$ and that $g(x)$ is def. identical with $app(A,B,\Pi I(A,B,g),x)$.

Let $c: I(\Pi(A,B),\Pi I(A,B,f),\Pi I(A,B,g))$. Then

$$\frac{\begin{array}{l}\Pi(A,B):set\\ \Pi I(A,B,f):\Pi(A,B)\\ \Pi I(A,B,g):\Pi(A,B)\\ f:\Pi(A,B), g:\Pi(A,B)\vdash \Pi(A,(x)I(B(x),app(A,B,f,x),app(A,B,g,x))):prop\\ h:\Pi(A,B)\vdash k(h):\Pi(A,(x)I(B(x),app(A,B,h,x),app(A,B,h,x)))\\ c:I(\Pi(A,B),\Pi I(A,B,f),\Pi I(A,B,g))\end{array}}{\begin{array}{l}IE(\Pi(A,B),\Pi I(A,B,f),\Pi I(A,B,g),(f)(g)(x)I(B(x),app(A,B,f,x),app(A,B,g,x)),k,c):\\ \quad\Pi(A,(x)I(B(x),app(A,B,\Pi I(A,B,f),x),app(A,B,\Pi I(A,B,g),x)))\end{array}}$$

hence

$$\frac{\begin{array}{l}I(\Pi(A,B),\Pi I(A,B,f),\Pi I(A,B,g)):prop\\ \Pi(A,(x)I(B(x),app(A,B,\Pi I(A,B,f),x),app(A,B,\Pi I(A,B,g),x))):prop\\ e:(I(\Pi(A,B),\Pi I(A,B,f),\Pi I(A,B,g)))\Pi(A,(x)I(B(x),app(A,B,\Pi I(A,B,f),x),\\ \quad app(A,B,\Pi I(A,B,g),x)))\end{array}}{\begin{array}{l}\supset I(I(\Pi(A,B),\Pi I(A,B,f),\Pi I(A,B,g))),\Pi(A,(x)I(B(x),app(A,B,\Pi I(A,B,f),x),\\ \quad app(A,B,\Pi I(A,B,g),x))),e):\\ \qquad I(\Pi(A,B),\Pi I(A,B,f),\Pi I(A,B,g))\supset\Pi(A,(x)I(B(x),app(A,B,\Pi I(A,B,f),x),\\ \qquad\quad app(A,B,\Pi I(A,B,g),x)))\end{array}}$$

which is nothing else than

$$I(\Pi(A,B),\Pi I(A,B,f),\Pi I(A,B,g)) \supset (\forall x:A)I(B(x),f(x),g(x))\ true.$$

But the converse, namely

$$\begin{array}{l}(\forall x:A)I(B(x),app(A,B,\Pi I(A,B,f),x),app(A,B,\Pi I(A,B,g),x))\supset\\ \quad I(\Pi(A,B),\Pi I(A,B,f),\Pi I(A,B,g))\end{array}$$

does not hold in constructive type theory which can be seen as follows:

Since $I(\Pi(A,B),\Pi I(A,B,f),\Pi I(A,B,g))$ does not depend on any assumptions, this proposition is true iff f and g are convertible in the sense of combinatory logic (this can actually be shown by a metamathematical argument). Hence it is decidable whether $I(\Pi(A,B),\Pi I(A,B,f),\Pi I(A,B,g))$ is true or not. On the other hand, one cannot even expect $(\forall x:N)I(N,app(N,N,\Pi I(N,N,f),x),app(N,N,\Pi I(N,N,g),x))$ to be decidable, never mind $(\forall x:A)I(B(x),app(A,B,\Pi I(A,B,f),x),app(A,B,\Pi I(A,B,g),x))$, for arbitrary $A: set$, $B:(A)set$.[89] Thus, there is no justification for Frege's axiom (V) if identity is interpreted as extensional identity.

If however one considers definitional or intensional identity, which Frege did not have, then Frege's axiom holds:

$$\frac{f=g:(x:A)B(x)}{\overline{\Pi I(A,B,f)=\Pi I(A,B,g):\Pi(A,B)}}$$

are valid inferences because of the def. of identity between canonical elements of $\Pi(A, B)$. (Cf. also C, n & o (3) to Π-equality.)

General observations on notation in constructive logic
(1) Gentzen's introduction rules for the logical operators are, as has already been mentioned, meaning explanations of these operators. Gentzen's and Heyting's meaning explanations of the logical operators are essentially the same, they only differ w.r.t. their notation: whereas Gentzen uses a tree-like notation, Heyting uses a linear notation, but there is an isomorphism between these two notations.

Gentzen's tree-like notation of the meaning explanation of the logical operators	Heyting's linear (monomorphic) notation of the meaning explanation of the logical operators
$\dfrac{\left.\begin{array}{c}\vdots\\ A\end{array}\right\}a \qquad \left.\begin{array}{c}\vdots\\ B\end{array}\right\}b}{A \,\&\, B}$	$\&I(A, B, a, b) : A \,\&\, B$
$\dfrac{\left.\begin{array}{c}\vdots\\ A\end{array}\right\}a}{A \vee B} \qquad \dfrac{\left.\begin{array}{c}\vdots\\ B\end{array}\right\}b}{A \vee B}$	$\vee I1(A, B, a) : A \vee B$ $\vee I2(A, B, b) : A \vee B$
$\dfrac{\left.\begin{array}{c}(A)\\ \vdots\\ B\end{array}\right\}b}{A \supset B}$	$\supset I(A, B, (x)b) : A \supset B$
$\dfrac{a : A \qquad \left.\begin{array}{c}(x : A)\\ \vdots\\ B\end{array}\right\}b}{(\exists x : A)B}$	$\exists I(A, (x)B, a, (x)b) : (\exists x : A)B$
$\dfrac{\left.\begin{array}{c}(x : A)\\ \vdots\\ B\end{array}\right\}b}{(\forall x : A)B}$	$\forall I(A, (x)B, (x)b) : (\forall x : A)B$
$\dfrac{a : A}{I(A, a, a)}$	$II(A, a) : I(A, a, a)$

Note that in Gentzen's original notation the proof-objects have no name.

(2) Several different notations are being used in the literature on constructive type theory in order to denote the canonical proof-objects of the various kinds of propositions. Here is an overview of some of these notations.

The kind of proposition	Mathematical notation for the canonical proof-object	Notation inspired by the calculus of natural deduction/ polymorphic notation	Monomorphic notation
$\bot$	—	—	—
$\top$	?	$\top I$	$\top I$
$A \vee B$	$i(a)$ $j(b)$	$\vee I1(a)$ $\vee I2(b)$	$\vee I1(A, B, a)$ $\vee I2(A, B, b)$
$(\forall x : A)B$ $\forall(A, B)$	$\lambda((x)b)$ $\lambda(b)$	$\forall I((x)b)$ $\forall I(b)$	$\forall I(A, (x)B, (x)b)$ $\forall I(A, B, b)$
$A \supset B$	$\lambda((x)b)$	$\supset I((x)b)$	$\supset I(A, B, (x)b)$
$(\exists x : A)B$ $\exists(A, B)$	(a, b) (a, b)	$\exists I(a, b)$ $\exists I(a, b)$	$\exists I(A, (x)B, a, b)$ $\exists I(A, B, a, b)$
$A\&B$	(a, b)	$\&I(a, b)$	$\&I(A, B, a, b)$
$I(A, a, b)$	?	$II(a)$	$II(A, a)$

(more about polymorphism/monomorphism, cf. Subsubsect. ii.3.3, Chapter 2).

xi THE CURRY–HOWARD ISOMORPHISM OR THE PROPOSITIONS-AS-SETS INTERPRETATION

Sets and propositions look like they behave in a very similar way:

(i) both are types, i.e. $prop : type$ and $set : type$, with objects A, i.e. $A : prop$ and $A : set$.

(ii) both give rise to dependent types, namely
$$X : prop \vdash X : type$$
$$X : set \vdash X : type$$

(iii) the objects A of each type *prop* and *set* are types as well with objects $a : A$, i.e. a is a proof-object of A where $A : prop$, and a is an element of A where $A : set$.

(iv) the theories of each of these types, i.e. logic and set theory, have similar ground and composite entities:

Constructive logic	Constructive set theory
$\bot$	$\{\}$ or $\emptyset$ or N_0
$\top$	N_1
$A \vee B$	$A + B$
$(\forall x : A)B(x)$	$\Pi_{x\in A} B_x$ or $(\Pi x : A)B(x)$
$A \supset B$	B^A or $A \to B$
$(\exists x : A)B(x)$	$\Sigma_{x\in A} B_x$ or $(\Sigma x : A)B(x)$
$A \,\&\, B$	$A \times B$
$I(A, a, b)$	$I(A, a, b)$

(v) A comparison of the axioms and rules for the logical operations on the one hand and the set-theoretical operations on the other, shows that there exists a precise correspondence (an isomorphism) between them.

Put $prop = set : type$ which means that $A : prop$ is interpreted as $A : set$ and $A = B : prop$ is interpreted as $A = B : set$. Then one has

Constructive logic	Constructive set theory
($\bot$F 1), ($\bot$F 2)	(N_0F 1), (N_0F 2) resp.
($\bot$E 1), ($\bot$E 2)	(N_0E 1), (N_0E 2)
($\top$F 1), ($\top$F 2)	(N_1F 1), (N_1F 2)
($\top$I 1) ($\top$I 2)	(N_1I 1) (N_1I 2)
($\top$E 1), ($\top$E 2)	(N_1E 1), (N_1E 2)
($\top$-equ)	(N_1-equ)
($\vee$F 1), ($\vee$F 2)	(+F 1), (+F 2)
($\vee$I 1), ($\vee$I 2), ($\vee$I 3), ($\vee$I 4)	(+I 1), (+I 2), (+I 3), (+I 4)
($\vee$E 1), ($\vee$E 2)	(+E 1), (+E 2)
($\vee$-equ 1), ($\vee$-equ 2)	(+-equ 1), (+-equ 2)
($\forall$F 1), ($\forall$F 2)	(ΠF 1), (ΠF 2)
($\forall$I 1), ($\forall$I 2)	(ΠI 1), (ΠI 2)
($\forall$E 1), ($\forall$E 2)	(ΠE 1), (ΠE 2)
($\forall$-equ)	(Π-equ)
($\supset$F 1), ($\supset$F 2)	($\to$F 1), ($\to$F 2)
($\supset$I 1), ($\supset$I 2)	($\to$I 1), ($\to$I 2)
($\supset$E 1), ($\supset$E 2)	($\to$E 1), ($\to$E 2)
($\supset$-equ)	($\to$-equ)
($\exists$F 1), ($\exists$F 2)	(ΣF 1), (ΣF 2)
($\exists$I 1), ($\exists$I 2)	(ΣI 1), (ΣI 2)
($\exists$E 1), ($\exists$E 2)	(ΣE 1), (ΣE 2)

(∃-equ)	(Σ-equ)
(&F 1), (&F 2)	(×F 1), (×F 2)
(&I 1), (&I 2)	(×I 1), (×I 2)
(&E 1), (&E 2)	(×E 1), (×E 2)
(&-equ)	(×-equ)
(IF 1), (IF 2)	(IF 1), (IF 2)
(II 1), (II 2)	(II 1), (II 2)
(IE 1), (IE 2)	(IE 1), (IE 2)
(I-equ)	(I-equ)

Of course, the question can be raised: What about the interpretation of the sets N_n, $n > 1$, N and U in constructive logic? This question can be answered as follows:

1. Strictly speaking the Curry–Howard isomorphism holds between constructive logic and a fragment of constructive set theory, namely the fragment built up from N_0, N_1, $+$, Π, $\rightarrow$, Σ, $\times$ and I.
2. It is possible to give a logical interpretation of N_n, $n > 1$, N and U, but these logical interpretations would be very artificial.

If one takes the idea seriously that a set is really defined by the formation rules of its canonical elements and its identical canonical elements, and that a proposition is really defined by the formation rules of its canonical proof-objects and its identical canonical proof-objects, then it seems clear that any separation of the notions of set and proposition only leads to unnecessary reduplication.

At the end of note (3) of C, n & o to x.6.3) there are a few comments on Swaen's investigation of weak and strong sum elimination in constructive type theory. There it was noted that as a consequence of the strong Σ- and $+$-rules of constructive set theory, its logic is generally stronger than the logic of traditional formal systems. This can be regarded as a conflict between sets and propositions, which runs counter to the identification of propositions with sets. Swaen then considers throughout his thesis (1989) various possibilities to reconcile the constructive set-approach and the traditional proposition-approach, or simply to reconcile set theory (in the sense of constructive type theory) and logic (in the traditional sense). This is what makes him write that one of the main questions of his thesis is, to what extent the correspondence between sets and propositions is actually an isomorphism.[90] Since traditional logic is here a topic for the C, n & o's only, and since constructive set theory is here based on intensional I-sets rather than on extensional E-sets (as is the case with Swaen's version of constructive type theory), the same cannot be said about the present work.

Reductionism or complementarity ?

Does this mean a reduction of constructive logic to constructive set theory, or conversely, a reduction of constructive set theory to constructive logic? If it meant a reduction of constructive set theory to constructive logic, this would be a revival of logicism. But this is quite impossible, since sets are necessary in (quantificational) logic anyway. (Cf. also note (2) in C, n & o to vi.1).) The propositions-as-sets interpretation shows that Heyting's constructive interpretation of propositional and predicate logic is set-theoretically valid, which means that one has rather a set-theoretical reduction of all the laws of logic.

But Martin-Löf prefers to conceive of this interpretation in terms of an isomorphism between constructive logic and constructive set theory rather than in terms of a reduction. The idea is that proposition and set (in a constructive sense) are basically one and the same notion. And evidence is provided by the way in which the questions: What is a proposition? and What is a set? and related questions are answered. (Frege developed (modern) logic at about the same time as Cantor developed set theory (basically logic free), and about a century later these two independent conceptions are fused on a deeper level into one conception. For a more recent historical side of this issue, cf. Subsect. ii.2, Chapter 4)

xii AN EXTENSION OF THE CURRY–HOWARD ISOMORPHISM TO SEQUENTS OR THE SEQUENTS-AS-TYPES INTERPRETATION

At first the sequent calculus will be presented as an extension of the calculus of propositions. The formulation of the sequent calculus requires three new forms of categorical judgement, namely

i. $\alpha : seq$, i.e. α is a sequent
ii. $\alpha = \beta : seq$, i.e. α and β are identical sequents, and
iii. α *true*, provided α is a sequent.

Moreover, three new kinds of hypothetical judgement are needed which in their most general form are

i. $x_1 : \alpha_1, \ldots, x_m : \alpha_m \vdash \alpha : seq$
ii. $x_1 : \alpha_1, \ldots, x_m : \alpha_m \vdash \alpha = \beta : seq$
iii. $x_1 : \alpha_1, \ldots, x_m : \alpha_m, \beta_1\ true, \ldots, \beta_n\ true \vdash \alpha\ true$

The sequent calculus

The sequent calculus is built up of two kinds of rules, the sequent formation and the sequent proof rules.

The sequent formation rules are answers to the question: What does the form of judgement '$\alpha : seq$' mean?, and they are the following judgements and rules:

$$\frac{A : prop}{A : seq} \qquad\qquad \frac{A = B : prop}{A = B : seq}$$

Alternatively, these formation rules for sequents can be given by the rule with 0 premisses

$$X : prop \vdash X : seq \quad (\text{seq})$$

The formation rule for 'dependent fct. sequents' is the following rule

$$\frac{x:\alpha\vdash\beta:seq\qquad \alpha:type}{|_{x:\alpha}\beta:seq}\ \text{(dep fct seq)}$$

and the associated identity rules are the following two rules

$$\frac{x:\alpha\vdash\beta:seq\quad \alpha=\gamma:type}{|_{x:\alpha}\beta=|_{x:\gamma}\beta:seq}\ \text{(dep fct. seq id. 1)}$$

$$\frac{x:\alpha\vdash\beta=\delta:seq\quad \alpha:type}{|_{x:\alpha}\beta=|_{x:\alpha}\delta:seq}\ \text{(dep fct. seq id. 2)}$$

The formation rules for the 'non-dependent fct. sequents' are

$$\frac{\alpha:seq\quad \beta:seq}{\alpha|\beta:seq}\ \text{(nondep fct. seq 1)}$$

$$\frac{\alpha=\gamma:seq\quad \beta=\delta:seq}{\alpha|\beta=\gamma|\delta:seq}\ \text{(nondep fct. seq 2)}$$

The following judgement $\beta_1,\ldots,\beta_n|_{x_1:\alpha_1,\ldots,x_m:\alpha_m}\gamma : seq$, where $\alpha_1 : type,\ldots,\alpha_m :$ *type* and $\beta_1 : seq,\ldots,\beta_n : seq$, can be nominally defined in the calculus of sequents as follows

$$\frac{\begin{array}{l} x_1:\alpha_1,\ldots,x_m:\alpha_m\vdash\beta_1:seq\\ \vdots\\ x_1:\alpha_1,\ldots,x_m:\alpha_m\vdash\beta_n:seq\\ x_1:\alpha_1,\ldots,x_m:\alpha_m\vdash\gamma:seq\end{array}}{\begin{cases}\beta_1,\ldots,\beta_n|_{x_1:\alpha_1,\ldots,x_m:\alpha_m}\gamma:seq\\ \beta_1,\ldots,\beta_n|_{x_1:\alpha_1,\ldots,x_m:\alpha_m}\gamma=|_{x_1:\alpha_1}\cdots|_{x_m:\alpha_m}(\beta_1|\ldots(\beta_n|\gamma)\ldots):seq\end{cases}}$$

C, n & o

(1) Gentzen's notation for $\beta_1,\ldots,\beta_n|\gamma$ is $\beta_1,\ldots,\beta_n\to\gamma$, and Schroeder-Heister's notation for the same expression is $\beta_1,\ldots,\beta_n\Rightarrow\gamma$. The term 'sequent' for these expressions stems from Gentzen who also called $\beta_1,\ldots,\beta_n$ the antecedents, and γ the succedent of the sequent.

(2) Applying the rule (seq) $n+1$ times and the rule (nondep fct. seq 1) n times yields the ordinary Gentzen sequents $A_1,\ldots,A_n\to B : seq$ for $A_1 : prop,\ldots,A_n : prop$ and $B : prop$ (in the present notation $A_1|(A_2|(\ldots(A_n|B)\ldots)))$.

The sequent proof rules are answers to the question: What does the form of judgement 'α *true*', for $\alpha : seq$, mean?, and they are the following judgements and rules:

the assumption rule $$\frac{\alpha:seq}{\alpha\ true}\ \text{(seq ass)}$$

or in sequent form

$$x_1:\alpha_1,\ldots,x_m:\alpha_m,\beta_1\ true,\ldots,\beta_n\ true,\gamma\ true\vdash\gamma\ true\qquad\text{(gen seq ass)}$$

the rules which will be called abstraction rules

$$\frac{x : \alpha \vdash \beta\ true}{|_{x:\alpha}\beta\ true}\ (\text{seq abs 1}) \qquad \frac{\alpha\ true \vdash \beta\ true}{\alpha|\beta\ true}\ (\text{seq abs 2})$$

the rules which will be called application rules

$$\frac{a : \alpha \quad |_{x:\alpha}\beta\ true}{\beta(a/x)\ true}\ (\text{seq app 1}) \qquad \frac{\alpha\ true \quad \alpha|\beta\ true}{\beta\ true}\ (\text{seq app 2})$$

and finally the identity of sequents rule

$$\frac{\alpha\ true \quad \alpha = \beta : seq}{\beta\ true}\ (\text{id. of seq})$$

C, n & o

(1) The fact that a formula of first order logic is logically valid in Tarski's sense, i.e. true in all models, can be formally expressed in the sequent calculus by means of a sequent.[91]

If $A_1, \ldots, A_n$ and B are formulas of first order logic, then the fact that B is a logical consequence of $A_1, \ldots, A_n$ in Tarski's sense, i.e. B is true in all models in which all of $A_1, \ldots, A_n$ are true, can be formally expressed in the sequent calculus by means of a fct. sequent, namely the sequent $A_1, \ldots, A_n|_{x_1:\alpha_1,\ldots,x_m:\alpha_m} B$ formally expressing the logically valid formula of first order logic $A_1 \supset (\ldots(A_n \supset B)\ldots)$ in the sequent calculus.

(2) If one drops the constants $\bot E_G : (Z : prop)(\bot)Z$, $\top I$, $\top E_G : (Z : prop)(Z)(\top)Z$, $\vee I1$, $\vee I2$, $\vee E_G$, $\forall I$, $\forall E_{SH}$, $\supset I$, $\supset E_{SH}$, $\exists I$, $\exists E_G$, $\& I$, $\& E_{SH}$ from their resp. typings (cf. x.2.2), and the C, n & o's to x.3.3), x.4.3), x.5.3), etc.) and one adapts the notation to the one of the sequent calculus, one obtains the Gentzen or Schroeder-Heister inference rules of constructive first order logic turned into (logically valid) closed axioms of the sequent calculus.

(3) If the procedure of (2) is applied to the constants $N_n E$ and NE of Sect. ix., one obtains the corresponding inference rules of constructive set theory transformed into closed axioms of the sequent calculus.

The preceding notes (2) and (3) suggest or actually presuppose that sequents are interpreted as types. This leads to

The sequents-as-types interpretation

Just as propositions can syntactically and semantically be interpreted as sets, namely as the sets of their resp. proofs (proof-objects), so can sequents be syntactically and semantically explained as types.

Put

- $seq = type$ which means that $\alpha : seq$ is interpreted as $\alpha : type$ and $\alpha = \beta : seq$ is interpreted as $\alpha = \beta : type$.
- $|_{x:\alpha}\beta = (x : \alpha)\beta : type$
- $\alpha|\beta = (\alpha)\beta : type$

- $\beta_1, \ldots, \beta_n|_{x_1:\alpha_1,\ldots,x_m:\alpha_m}\gamma = (x_1 : \alpha_1)\ldots(x_m : \alpha_m)(\beta_1)\ldots(\beta_n)\gamma$: *type*

Sequent calculus	Constructive type theory
(seq)	(dep prop type)
(dep fct seq)	(fct type)
(dep fct seq id. 1)	(fct type id. 1)
(dep fct seq id. 2)	(fct type id. 2)
(nondep fct seq)	(nondep fct type)
(seq ass)	(ass)
(gen seq ass)	(gen ass)
(seq abs 1)	(abs 1)
(seq abs 2)	$\dfrac{x : \alpha \vdash b : \beta}{(x)b : (\alpha)\beta = (x : \alpha)\beta}$
(seq app 1)	(app 1)
(seq app 2)	$\dfrac{b : (\alpha)\beta \quad a : \alpha}{b(a) : \beta}$
(id. of seq)	(id. of types 1)

The notion of truth is extended from propositions (cf. C, n & o to vii.3)) to sequents in the following way

- α *true* is interpreted as a shorthand form of $a : \alpha$, for some a, i.e. there exists a proof-object a of the sequent α
- $x_1 : \alpha_1, \ldots, x_m : \alpha_m, \beta_1$ *true*$, \ldots, \beta_n$ *true* $\vdash \gamma$ *true* is interpreted as a shorthand form of $x_1 : \alpha_1, \ldots, x_m : \alpha_m, y_1 : \beta_1, \ldots, y_n : \beta_n \vdash c : \gamma$, for some c, i.e. there exists a proof-object c of the sequent γ depending on the variables $x_1 : \alpha_1, \ldots, x_m : \alpha_m, y_1 : \beta_1, \ldots, y_n : \beta_n$.

NOTES

1 Dubislav (1931), pp. 2–7.

2 Curry (1977), Chapter 3, Sect. C, or Curry and Feys (1958), Chapt. 2, Sect. E.

3 To see how Curry inspired the conception of def. identity in constructive type theory, cf. Sect. viii. of this chapter.

4 Solomon (1970), pp. 398–401.

5 Scholz (1930), pp. 38–40.

6 In his (1983A), p. 235, he advocates this order (for the first time) for all the concepts or notions to be really defined in that paper. In his (1990), p. 141, he appears to generalize this statement to include all concepts whatsoever.

7 Cf. Dummett (1981).

8 Patzig (1973), pp. 61,62.

9 p. 76; in the same direction cf. Frede (1987). The bulk of Frede's and Patzig's papers is critical in the sense of a discussion and critique of earlier interpretations of Artistotle's categories, and the positive attempt at an interpretation of their own is tentative and fragmentary.

10 Cf. Körner (1955), Chapter 3.

11 Cf. Bar-Hillel (1957) and Thompson (1967).

[12] Quine (1969), p. 18.
[13] p. 3.
[14] p. 9.
[15] p. 18.
[16] p. 18.
[17] p. 19.
[18] p. 23.
[19] Gottlieb (1979), p. 81.
[20] p. 84.
[21] p. 92.
[22] Geach (1967, 1973).
[23] Griffin (1977), p. 9.
[24] pp. 10–12.
[25] pp. 13–17.
[26] p. 21.
[27] Cf. also Stegmüller (1978).
[28] Cf. Def. 5.1 in Thompson (1991), p. 131.
[29] Apart from Leibniz, the following quotes are taken from Rüthing (1984).
[30] Pfeiffer and Dahan-Dalmedico (1986), pp. 217–218, and also Chapt. 6, §§7–15.
[31] Rüthing (1984), p. 75.
[32] Cf. also Martin-Löf (1986).
[33] Cf. Hatcher (1982), Chapter 4, Sect. 5.
[34] Cf. de Bruijn (1970, 1980).
[35] Curry and Feys (1958), cf. Introduction.
[36] This whole section is based on Tasistro (1993).
[37] Tasistro (1993), p. 18.
[38] For the details, cf. pp. 17–18.
[39] p. 18.
[40] p. 19.
[41] p. 19.
[42] Cf. pp. 14–16. For the details of these justifications, cf. pp. 20, 21.
[43] p. 22.
[44] p. 22.
[45] p. 23.
[46] pp. 25, 26.
[47] pp. 25, 26.
[48] p. 23.
[49] For details, cf. p. 24.
[50] p. 26/7.
[51] p. 26.
[52] p. 27.
[53] p. 27.
[54] Bocheński (1970), §40.
[55] The following quotes are taken from Hallett (1984), pp. 33, 34, 45.
[56] Parsons (1967), p. 195.
[57] Cf. van Dalen et al. (1978), Chapter 2, Sect. 4. For an informal account, cf. Boolos (1971).
[58] Van Dalen et al. (1978), Chapter 1, Sect. 20.
[59] Aczel (1978), p. 61.
[60] pp. 61, 62.
[61] Cf. Hatcher (1982), Chapter 4, Sect. 2.
[62] Hatcher (1982), p. 109.
[63] Cf. also Jacobs (1989), pp. 402, 403.
[64] p. 404.
[65] This note (1) makes use of Sundholm (1993).
[66] Cf. e.g. Quine (1969).
[67] Berg (1962), p. 51.

68 Cf. Martin-Löf's examples in version (7), pp. 25, 26.
69 Dummett (1975).
70 For these notes, cf. Thompson (1991), pp. 68, 74.
71 Cf. Nordström et al. (1990), pp. 103, 104.
72 For details, cf. version (6), pp. 74ff.
73 Cf. Nordström et al. (1990), pp. 72,73.
74 For an example, cf. Nordström et al. (1990), p. 51.
75 Cf. Nordström et al. (1990), p. 63.
76 p. 66.
77 p. 67.
78 pp. 93–95.
79 Cf. p. 93.
80 p. 93.
81 Aczel (1978), p. 61.
82 Cf. Nordström et al. (1990), pp. 100, 101.
83 Cf. Jacobs (1989).
84 Cf. Schroeder-Heister (1984, 1984A).
85 Cf. for these distinctions Troelstra and van Dalen (1988), pp. 189, 190.
86 Bishop (1967), p. 9.
87 Swaen (1989), esp. pp. 10–12.
88 Frege (1893), p. 61.
89 Cf. Nordström et al. (1990), pp. 84, 66.
90 Swaen (1989), pp. 65, 99.
91 For an example cf. version (7), p. 94.

CHAPTER 2
METATHEORETICAL CONSIDERATIONS

There is a wealth of metatheoretical or philosophical considerations behind the technicalities of constructive type theory. The purpose of this chapter is to cast at least some light on at least a few of those considerations and to draw attention to the philosophical underpinnings and intricacies of type theory.

These issues have been selected for consideration: the philosophical evergreen called identity (Sect. i), one of the few main pillars of any logic, namely the notion of judgement, which is also elucidated by means of the theory of knowledge (Sect. ii), the great tradition of metalogic with its principal targets: consistency, completeness and decidability (Sect. iii), and last but not least the positioning of constructive type theory within the philosophy or foundations of mathematics (Sect. iv). Needless to write, the real bulk of metatheoretical or philosophical consideration remains still to be carried out – the following is but an initiation.

i THE IDENTITY ISSUE: THE MANY FACES OF IDENTITY IN CONSTRUCTIVE TYPE THEORY

Identity in constructive type theory actually consists of a whole family of (identity) relations. The first subsection concerned with this issue attempts to bring some order into this family by distinguishing various levels (syntactical–semantical, formal–informal), sorts of entities (objects and types) and notions of identity (syntactically induced, semantical or definitional and set-theoretical/propositional). The second subsection casts some light onto the fundamental relationship between application or typing and semantical or intensional identity; it also draws attention to the fact that (the transitive closure of) semantical or definitional identity constitutes the bridge between entities in non-canonical and entities in canonical form. In the third subsection it is shown how the relation of semantical or intensional identity gives rise to the notions of intensional transparency and intensional opacity which are compared to those of referential (or extensional) transparency/opacity and syntactical transparency/opacity. After a short historical digression on Frege and Quine concerning identity (Subsect. 4), the fifth subsection deals with the question as to which of the identity relations mentioned in Subsect. 1 are decidable and which are not. All these explanations are by no means intended to dissolve the considerable complexity of the identity issue of constructive type theory into 'pure simplicity', but their purpose is to at least reduce the degree of this complexity.

i.1 *Various Kinds of Identity*

In digression Sect. viii., Chapt. 1 the natural distinction between the semantical level, i.e. the level of types and their resp. objects, and the syntactical level, i.e. the level of the expressions for these entities, was explained. Subsequently, two kinds of semantical identity, namely definitional or semantical identity $=$ and syntactically induced identity $\equiv$, and two kinds of syntactical identity were introduced. Finally, one could distinguish each of definitional and syntactically induced identity according to whether they constitute a relation between objects or between types.

In an earlier paper on the notion of definitional equality, Martin-Löf defines the relation of definitional or semantical identity as follows:[1] Definitional identity is the equivalence relation between entities generated by the two and only the two principles that (1) a definiens (entity) is always definitionally identical with a definiendum (entity), and (2) the principle of substituting identicals for identicals. (Actually, note that in this paper Martin-Löf defines the syntactical counterpart of the relation just defined, i.e. the semantically induced relation of identity or the relation of synonymy.)

So far the following forms of judgement express identity on the semantical level (the notation is here taken over from Sect. viii., Chapt. 1): $a = b : \alpha$, $a \equiv b : \alpha$, $a = b : type$ and $a \equiv b : type$; all these kinds of identity can be called kinds of judgemental identity. But since judgemental identity and in particular definitional identity between objects cannot be used to construct sets or propositions, it is necessary to have an elementary set/proposition expressing the definitional identity between two elements of a set/two proof-objects of a proposition. In Sect. ix, Chapt. 1, point ix.8 two kinds of set-theoretical/propositional identity have been introduced in order to mimic definitional judgemental identity between elements/proof-objects of a given set/proposition inside constructive set theory/constructive logic. $a = b : A$ where $A : set/prop$ can be expressed in constructive set theory/constructive logic by $I(A, a, b)$ or by $E(A, a, b)$. The sets/propositions $I(A, a, b)$ and $E(A, a, b)$ differ essentially in their elimination rules, where $E(A, a, b)$ has much stronger elimination rules than $I(A, a, b)$. Since the derivability of $c : I(A, a, b)$ (provided the derivation is closed) can be understood as convertibility of a and b, the set-theoretical/propositional identity I is called intensional. Since it is possible to show that extensionally identical fcts. are identical in the sense of E, the set-theoretical/propositional identity E is called extensional. (More on convertibility, intension and extension, cf. below.)

There is an order among these different notions of identity on the semantical level. Concerning two objects of a given type, one has for $A : set$, hence $A : type$, $a : A$ and $b : A$

(i) syntactically induced identity $a \equiv b : A$
(ii) definitional or semantical identity $a = b : A$
(iii) set-theoretical/propositional identity $c : I(A, a, b)$, for some object c

which are arranged in order of decreasing strength, i.e. one can always pass from the stronger to the weaker identity, but not conversely.

Example The three identity notions applied to fcts. in the modern sense.
Let $A : set$, $B : (A)\ set$, $f : (x : A)B(x)$, and $g : (x : A)B(x)$

(i) syntactically induced identity $f \equiv g : (x : A)B(x)$
(ii) definitional or semantical identity $f = g : (x : A)B(x)$
(iii) set-theoretical/propositional identity $c : I((\forall x : A)B(x), f, g)$, for some object c

As far as the extensional set-theoretical/propositional identity relation between objects is concerned, it fits in between definitional or semantical identity and set-theoretical/ propositional identity I, whereby definitional or semantical identity and extensional set-theoretical/propositional identity are of equal strength.

Example Above
extensional set-theoretical/propositional identity $d : (\forall x : A)E(B(x), f(x), g(x))$, for some object d
which can be proved to be equivalent to $EI : E(\Pi(A, B), \Pi I(f), \Pi I(g))$.[2]
Concerning two types one has

(i) syntactically induced identity $A \equiv B : type$
(ii) definitional or intensional or semantical identity $A = B : type$
(iii) extensional identity $A \supset\subset B : type$

which are again arranged in order of decreasing strength.

Examples

1. The three identity notions applied to propositions
   Obvious; use $A : prop$, $B : prop$ instead of $A : type$, $B : type$; extensional identity then yields the concept of material equivalence between two propositions.
2. The three identity notions applied to sets
   Analogous to 1.; use $A : set$, $B : set$ instead of $A : type$, $B : type$, and extensional identity (iii) can be notated by $A \overset{\rightarrow}{\leftarrow} B : set$ which yields the concept of equipotency of two sets.
3. The three identity notions applied to classes
   Let $A : set$, $B : class(A)$ and $C : class(A)$.

   (i) syntactically induced identity $B \equiv C : class(A)$
   (ii) definitional or intensional or semantical identity $B = C : class(A)$
   (iii) extensional identity $c : (\forall x : A)(B(x) \supset\subset C(x))$, for some object c

4. The three identity notions applied to relations
   Let $A : set$, $B : set$, $C : rel(A, B)$, $D : rel(A, B)$

   (i) syntactically induced identity $C \equiv D : rel(A, B)$
   (ii) definitional or intensional or semantical identity $C = D : rel(A, B)$
   (iii) extensional identity $c : (\forall x : A)(\forall y : B)(C(x, y) \supset\subset D(x, y))$, for some object c

All these cases of identity $=$, I and E satisfy the criteria of definitional or semantical identity laid down above. All three are or give rise to equivalence relations, all three satisfy principle (1) and all three satisfy principle (2) of the substitutivity of identicals by identicals; cf. (id. of types 1) and (id. of types 2) in the case of = ; note (1) on Leibniz principle in C, n & o to x.8.3), Chapt. 1 in the case of I; and the derivation of Leibniz principle in version (6) in the case of E.[3] This ends the discussion of the forms of identity on the semantical level.

As for the syntactical level two kinds of identity are being distinguished, as has already been mentioned at the beginning of this section (and in Sect. viii., Chapt. 1), namely the syntactical identity $\equiv$ and the semantically induced identity or synonymy relation =. Naturally, one could also on this level distinguish each of these kinds of identity according to whether they constitute a relation between object expressions or between type expressions. But the more interesting identity relation on this level is clearly the relation of semantically induced identity or identity of meaning or synonymy, and the following discussion will focus on it.

In his (1972A) paper on the notion of definitional equality, Martin-Löf also distinguishes between informal and formal concepts which occur in the context of constructive type theory. There he presents a dictionary which is slightly modified here in order to avoid certain ambiguities (concerning the syntactic and semantic levels) and to adapt it to the present terminology:

Informal level	Formal level
propositional expression	formula
?	derivation, proof figure
type expression	type symbol
expression for mathematical object	term
defining equation	rule of conversion
definiendum (expression)	redex
definiens (expression)	contractum
definitional equality or semantically induced identity	convertibility

The formal counterpart of the semantically (or definitionally) induced identity, as it has been called so far, is thus the relation of convertibility studied in combinatory logic, proof theory and programming language theory. The convertibility relation *conv* is defined to be the equivalence relation between terms generated by the two and only the two conversion rules (principles) that (1) redex *conv* contractum, and (2) $\dfrac{'a' \; conv \; 'c'}{'b[a]' \; conv \; 'b[c]'}$ where $'a'$ and $'c'$ must have the same type symbol as the variable $'x'$ in $'b[x]'$. This def. is in perfect agreement with the informal syntactical counterpart of the def. of semantical or definitional identity.

It is possible to prove on the informal level that

$$\text{if } 'a' =' b' \; : \; 'A' \text{ then } 'I(A, a, b)' \; \textit{true}$$

which on the formal level means that[4]

$$\text{if } 'a' \; conv \; 'b' \text{ then } 'c' \; : \; 'I(A, a, b)' \text{ for some term } 'c'$$

But Martin-Löf proves likewise a theorem according to which[5]

$$\text{if } 'c' : 'I(A, a, b)', \text{ for some closed (!) term } 'c' \text{ then } 'a' \textit{ conv } 'b'$$

Of course, as is made explicit by Nordström et al., this theorem holds only for a version of constructive type theory in which no E-sets/propositions are involved.

These results then provide the following order between the different notions of identity on the syntactical level. Concerning two terms, both of which have a given type symbol, one has for $'A' : \textit{type symbol}$, $'a' : 'A'$ and $'b' : 'A'$

(i) syntactical identity $'a' \equiv 'b' : 'A'$
(ii) semantically or definitionally induced identity $'a' =' b' : 'A'$
(iii) convertibility $'a' \textit{ conv } 'b'$
(iv) syntactic intensional set-theoretical/propositional identity $'c' : 'I(A, a, b)'$, for some term $'c'$

which are arranged in order of decreasing strength except for the identity relations (ii) and (iii) which are of equal strength.

Note that, as Martin-Löf points out in his paper,[6] Curry's rule of conversion

$$'\lambda((x)b(x))' \textit{ conv } 'b' \quad (\eta\textit{-rule})$$

provided x does not occur free in the term $'b'$, is not part of a correct formalization of the notion of semantically or definitionally induced identity. It is possible to prove the hypothetical judgement

$$'f' : '\Pi(A, B)' \vdash 'c' : 'I(\Pi(A, B), \lambda((x)app(f, x)), f)' \text{ for some open term } 'c'$$

but there is no way one can derive the judgement

$$'f' : '\Pi(A, B)' \vdash '\lambda((x)app(f, x))' = 'f' : '\Pi(A, B)'$$

in constructive type theory unless one uses E-sets/propositions, i.e. extensional set-theoretical/propositional identity (on the syntactical level).[7]

i.2 *Digging Deeper into Identity*

According to the doctrine of types endorsed by constructive type theory, no entity is without type: either something is an object of a given type, or a type is an object of a universe and thus becomes itself an object of a type, or a universe is an object of a higher universe and thus becomes again an object of a type. Each entity is thus an entity of a certain type. Moreover, another device is endorsed by constructive type theory, namely no type without semantical or intensional identity. This device expresses the necessity of endowing or equipping each entity with a definite semantical or intensional identity. The doctrine of types and this device together actually entail Quine's slogan: No entity without identity – be this entity an object or a type. The fact that each (non-dependent, dependent or fct.) type is given with a semantical or intensional identity, provides the criterion for determining the semantical or intensional identity between any two (non-dependent, dependent or fct.) objects of that type. This, plus the criterion of application of the resp. type, provide the criterion for determining the extensional identity between any two types.

Quine's reasons in favor of his slogan could not be considered very convincing (cf. C, n & o to i.1), Chapt. 1). However, if, as in Martin-Löf's case, the slogan is put forward for logic and mathematics, the need and to a certain extent the necessity of precise notions and clearly defined entities becomes much more obvious. There is a great risk of confusion, imprecision and misconception, if notions such as fct., class, proposition, proof-object, etc. are not exactly defined. And it is understandable that Quine should reject the adoption of some of those entities, given that he couldn't determine them and they had never been determined with utmost precision. Furthermore, according to Martin-Löf's def. of synonymy, it is necessary to have such a semantical or intensional identity between any two objects of a given type, as there is no synonymy without semantical or intensional identity.

Not only is there no type without semantical or intensional identity, but there is also no semantical or intensional identity without type; i.e. semantical or intensional identity is always typed. The reason is that in constructive type theory every (well-formed) expression should express something, every (well-formed) expression of constructive type theory should be meaningful. Now, when Curry started developing combinatory logic, he gave priority to semantical or intensional identity, or rather its formal and syntactical counterpart i.e. conversion, over application or typing. This led unavoidably to the formation of meaningless expressions, i.e. to expressions (of combinatory logic) which do not express anything – which of course is somewhat nonsense. The lesson to be learnt from this development of combinatory logic was that semantical or intensional identity does not come before application/typing, and since application/typing does not come before semantical or intensional identity, this means that the two notions are interconnected and hence have to come simultaneously.

Each non-dependent and non-fct. object, as well as each type which is not a dependent or fct. type, is not merely equipped with a simple semantical (or intensional) or definitional identity, according to what has been explained above, but it is equipped with a definitional chain, i.e. a whole chain of semantical or definitional identities supplying the resp. object or type with a (semantic) value. This means that each such object or type is either in a specific normal or canonical form, and it is then merely evaluated to itself, or else it is in a non-normal or non-canonical form and can be evaluated via its definitional chain to its normal or canonical form. If a is such a non-normal object (or A is such a non-normal type), then there is a chain $a \equiv a_0 = \cdots = a_n \equiv b$ (a chain $A \equiv A_0 = \cdots = A_n \equiv B$ resp.) s.t. b is the normal form of a (B is the normal form of A resp.) and a is related to b via n semantical or definitional identities, with $n > 1$. Hence, since $=$ is a transitive relation, a non-normal object a (type A) can be said to be semantically or definitionally identical to its normal form counterpart b (or B resp.).

Finally, each type is not only equipped with a semantical or intensional identity (as well as a syntactically induced identity), but also with a set-theoretical/propositional (intensional) identity which is weaker than the semantical or intensional one. Note that the same does not hold for every object, in the sense that there is no such distinction for objects which are neither dependent nor fct. objects.

i.3 *Intensionality and Extensionality/Transparency and Opacity*

In (1960) Quine calls a mode of containment Φ referentially transparent (in the terminology of Russell/Whitehead's PM) if, whenever an occurrence of a singular term t is purely referential in a term or sentence $\psi(t)$, it is purely referential also in the containing term or sentence $\Phi(\psi(t))$. O.w. a mode of containment Φ is called referentially opaque.

To understand this definition it ought to be explained what it means for an occurrence of a singular term t to be purely referential in a term or sentence $\psi(t)$.

An occurrence of a singular term t is called purely referential in a term or sentence $\psi(t)$ if $\psi(t')$ is codesignative or materially equivalent resp. with $\psi(t)$ whenever t' is codesignative w.r.t. t (i.e. t' is referring to the same object as t). In other words, an occurrence of a singular term t is called purely referential in a term or sentence $\psi(t)$ if the following two rules hold resp.:

$$\frac{\psi(t) : B \quad codesig(t, t')}{codesig(\psi(t), \psi(t'))}$$

$$\frac{\psi(t) : sentence \quad codesig(t, t')}{\psi(t) \supset\subset \psi(t')\ true}$$

This can be expressed in constructive type theory as follows:

An occurrence of an object expression $'a'$ is called purely referential in an (object or proposition) expression ψ if the following two rules hold for $'A' : 'set'$, $'B' : 'set'$:

$$\frac{\psi : '(A)B' \quad 'a' \simeq 'b' : 'A'}{\psi('a') \simeq \psi('b') : 'B'}$$

$$\frac{\psi : '(A)prop' \quad 'a' \simeq' b' : 'A'}{'c' : \psi('a') \supset\subset \psi('b')}$$

where $\simeq$ denotes codesignation or extensional identity between two objects of a given type.

It can quite easily be verified that the $'c'$ in question is just the object expression $\&I(\psi('a') \supset \psi('b'), \psi('b') \supset \psi('a'), \supset I(\psi('a'), \psi('b'), (x)x), \supset I(\psi('b'), \psi('a'), (x)x))$. Note that the second of these two rules is often referred to as substitutivity of identity salva veritate (also by Quine). The former case could be called analogously the substitutivity of identity salva designatione (or salva referentia).

Now, it was said earlier on that every object of some type and every type in constructive type theory is equipped with a semantical or intensional identity. This allows us to introduce into this theory in addition to the referential transparency the notion of intensional transparency.

Let a containment expression Φ be called intensionally transparent if whenever an occurrence of an object expression $'a'$ is purely intensional in an (object or proposition) expression $\psi('a')$, it is purely intensional also in the containing (object or proposition) expression $\Phi(\psi('a'))$. O.w. the containment expression Φ is called intensionally opaque.

And, an occurrence of an object expression $'a'$ is called purely intensional in an (object or proposition) expression $\psi('a')$ if $\psi('b')$ is cointensional or semantically induced identical with $\psi('a')$ whenever $'a'$ and $'b'$ are semantically induced identical. In other terms, an occurrence of an object expression $'a'$ is called purely intensional in an (object or proposition) expression $\psi('a')$ if the following two rules hold resp.:

$$\frac{\psi : '(A)B' \quad 'a' = 'b' : 'A'}{\psi('a') = \psi('b') : 'B'}$$

$$\frac{\psi : '(A)prop' \quad 'a' = 'b' : 'A'}{\psi('a') = \psi('b') : 'prop'}$$

This principle could be called in analogy with the previous two principles, the principle of substitutivity of identity salva identitate (intensionale).[8]

Examples of containment expressions and their transparency/opacity

1. Quotation: is referentially as well as intensionally opaque
2. Truth functions: are referentially as well as intensionally transparent
3. General terms predicatively used: are referentially and intensionally transparent
4. Propositional attitudes: are referentially and intensionally opaque
5. Modal operators or functions: are referentially opaque and intensionally transparent

Lastly, attention should be drawn to the fact that one could even define the notions of syntactical transparency and syntactical opacity in a way similar to the definitions given for referential and intensional transparency/opacity. But this will not be done here and moreover it is rather trivial since every mode of containment Φ is syntactically transparent.

As mentioned above, the set-theoretical/propositional identity which yields the extensional identity of fcts. in the sense of Church has therefore been called extensional (cf. $(\forall x : A)E(B(x), app(A, B, f, x), app(A, B, g, x))$ iff $E(\Pi(A, B), f, g)$), whereas the identity which is the informal set-theoretical/propositional equivalent of convertibility (and which mimics the informal semantical or intensional identity) has been called intensional.

i.4 *Digression on Frege and Quine Concerning Identity or Intension*

i.4.1 *Frege*

In Frege's *Begriffsschrift* §8 appears the term 'Inhaltsgleichheit' which is denoted by $\equiv$ and related to the one of 'Urteilsinhalt'. 'Urteilsinhalt' corresponds in constructive type theory to 'propositional expression' and Frege's 'Inhaltsgleichheit' corresponds on the formal level to convertibility and on the informal level to syntactic intensional identity.

But this interpretation does not work any longer as soon as the following axiom of identity is introduced in §21: $a \equiv b \supset (A(a) \supset A(b))$. Since, as Martin-Löf pointed out,[9] if $\equiv$ is regarded as a relation between two expressions, and in particular $a \equiv b$ as a proposition, it is certainly not on a par with propositions such as $A(a)$ or $A(b)$ which are proved in the formal theory by means of possibly logically complicated proofs. Thus, it cannot be combined with these latter propositions into compound propositions by means of the logical operations.

Hence in his *Grundgesetze der Arithmetik* (1893–1903) (and already before) Frege changes his identity tune. He abandons the equality of content $\equiv$ and replaces it by the binary relation of identity $=$ on the objects of his universal universe of objects. He defines two fcts. to be intensionally or definitionally identical if they are extensionally identical (§9), i.e. two course-of-values are definitionally or semantically identical iff their corresponding fcts. are extensionally identical. Since intensional and extensional identity coincide w.r.t. objects, Frege's overall conception of identity on the semantical level can be regarded as consistently extensional.

i.4.2 *Quine*

In (1960) and in numerous other publications Quine tries to show that most of the interesting proposals to determine sentence synonymy, analyticity of sentences, the identity of propositions, and the concept of class have failed (cf. §14, §40, §42, and §55).

Constructive type theory challenges all of Quine's attempts to dismantle the just mentioned concepts and to dismiss them as undetermined (or underdetermined) and unclear.

To start with, it ought to be observed that the concept of meaning of an object expression or type expression in constructive type theory is an immediate and intuitive one: a given object expression signifies or stands for or expresses its meaning which is an object, and a type expression signifies or stands for, etc., its meaning which is a type. The reference of an expression whose meaning is a non-dependent and non-fct. object or type is then given in terms of the meaning of the expression: it is the result of the evaluation or computation of the meaning of the resp. expression. (Note that dependent or fct. objects or types have no reference or value.)

The synonymy or identity of meaning of two expressions $'a'$ and $'b'$ is determined in terms of the meaning a and b of the resp. expressions and the definitional or semantical identity $a = b$ between these meanings. This definitional or semantical identity between the two meanings is given simultaneously with the meanings themselves (remember: no entity without identity). As was said earlier on, synonymy is nothing but semantically or definitionally induced identity between expressions.

Whereas 'synonymy' applies e.g. to propositional expressions, 'analytic' applies to judgements. A can be judged to be a type, i.e. A : *type* (A : *prop*), if one can indicate what the (canonical proof) objects and the identical (canonical proof) objects of A are. Thus, the judgements $a : A$, $a = b : A$ and $A = B$: *type* ($A = B$: *prop*) are all analytic, because they all presuppose the meaning explanation of the judgement A : *type* (A : *prop*), and 'A : *type*' means exactly to know all these things.

These propositions are defined very precisely: 'A is a proposition' is explained by explaining what the canonical proof-objects of A are. The identity $A = B : prop$ between two propositions A and B is then readily explained in terms of $A : prop$ and $B : prop$. Note that there is not only an extensional identity between two propositions, but also an intensional one.

Finally, the classes are equally well-defined, once the propositions are defined; a class over a set A is nothing but a fct. type, namely $class(A) = (A)prop : type$. And, as in the case of propositions, one has immediately an extensional as well as an intensional identity between two classes.

i.5 *Decidability/undecidability of the Various Identity Relations*

As far as syntactical identity $\equiv$ (and hence also syntactically induced identity) is concerned, it is plainly decidable. Given two expressions, one only has to check whether or not they have identical syntactical form as well as syntactically identical parts.

Whether or not definitional or semantical identity $=$ (and hence definitionally or semantically induced identity) is decidable, depends on the kind of identity sets/propositions which are assumed in constructive set theory/constructive logic. If constructive set theory or logic merely contains I-sets/propositions as identity sets/propositions, then definitional or semantical identity is decidable. However, if constructive set theory or logic contains E-sets/propositions, then definitional or semantical identity is undecidable. The reason is, that it is then possible to prove identity judgements by reasoning using sets or propositions.[10]

The decidability of the definitional or semantical identity in constructive set theory or logic based only on I-sets/propositions hinges on the decidability of the convertibility relation *conv*. The decidability of *conv* is proved by normalization. The equal strength of definitional or semantical identity within the framework of I-sets and convertibility conveys decidability onto the former identity relation. (cf. Subsect. i.1)

In order to preserve decidability of definitional or semantical identity as much as possible it is necessary to avoid extensional set-theoretical/propositional identity, i.e. E-sets/propositions as much as possible.

Appendix

The notation used in this presentation is essentially the one of version (8). What is called in Nordström et al.

- (syntactical) definitional or intensional equality[11] corresponds here to definitionally or semantically induced identity, denoted by $=$;
- judgemental equality,[12] denoted by $=$, corresponds here to definitional or semantical identity, denoted by $=$;
- intensional equality,[13] denoted by Id, corresponds here to intensional set-theoretical/propositional identity, denoted by I;
- extensional equality,[14] denoted by Eq, corresponds here to extensional set-theoretical/propositional identity, denoted by E.

In version (6)

- (syntactical) definitional or intensional equality or equality of meaning (synonymy), denoted by $\equiv$ or $=_{def}$, corresponds here to definitionally or semantically induced identity, denoted by $=$;
- equality between objects and equality between types, denoted by =, corresponds here to definitional or semantical identity, denoted by $=$;
- propositional equality, denoted by I, corresponds here to extensional set-theoretical/propositional identity, denoted by E.

In Thompson

- (syntactical) definitional equality,[15] denoted by $\equiv$, corresponds here to definitionally or semantically induced identity, denoted by $=$;

  The definitional equality $\equiv$ is the reflexive and transitive closure of the (1-step) (syntactical) definitional equality,[16] denoted by $\equiv_{def}$;
- the relation of convertibility,[17] denoted by $\leftrightarrow$, is the smallest equivalence relation extending the relation of reduction, denoted by $\twoheadrightarrow$, which is the reflexive and transitive closure of the relation of computation (or evaluation),[18] denoted by $\rightarrow$; all these relations express definitionally or semantically induced identity and are here denoted by $=$;
- propositional equality,[19] denoted by $=_A$ for $A : prop/set$ or by I, corresponds here to intensional set-theoretical/propositional identity, denoted likewise by I;
- extensional equality,[20] denoted by $\simeq_A$ where $A : prop/set$, corresponds here to extensional set-theoretical/propositional identity, denoted by E.

ii THE JUDGEMENT ISSUE

Judgements are one of the few most basic ingredients of logic in general and of constructive type theory in particular. It is therefore their due to be taken into appropriate historical and philosophical consideration.

The first subsection points out the different forms of judgement which can be found throughout the history of logic, from Aristotle's logic to Gentzen's logic. The second subsection harps on the fundamental logical distinction between judgements and propositions (which has disappeared from modern mainstream logic). The third subsection consists of an extensive discussion of two kinds of judgement, namely analytic and synthetic ones: it contains a brief presentation of the analytic and synthetic forms of judgement (in constructive type theory) as well as of the logic of analytic judgements and the logic of synthetic judgements, and it also contains a new explanation of undecidability and incompleteness phenomena which are typical for the logic of synthetic judgements; moreover, suggestions can be found concerning the distinction of various degrees of analyticity of analytic judgements through considerations regarding monomorphism and polymorphism; and finally, existence, truth and falsity are exposed as different kinds of synthetic judgement. The fourth subsection deals with two senses of judgement: with judgement in the sense of object or result of an act of judgement, and judgement in the sense of act of judging, i.e. judgement as demonstration; with both these senses can be associated a concept of 'truth' which is called correctness for the judgements in the first sense and validity for judgements in the second sense. Subsect. 5 endeavors to clarify the notion of judgement by the equally fundamental notions of knowledge and knowability. The last Subsect. 6 summarizes various readings of the forms of judgement of constructive type theory: it extends the propositions-as-sets interpretation to intentions, problems, specifications and games.

ii.1 *Judgements as One of the Basic Ingredients of Logic*

The subject matter of logic is constituted by the forms of judgement and the forms of inference. The main historical examples are perhaps the following:

(i) *Aristotle's logic*
Putting aside the indefinite judgement, Aristotle recognizes three forms of judgement which affirm or deny a predicate of a subject, namely a singular and two general, i.e. the universal and the particular. The affirmation '*S* is *P*' consists of subject, copula, predicate, where in a singular judgement *S* is a singular term, and in a general judgement the subject is actually 'Every *S*' (universal) or 'Some *S*' (particular) and *S* is a general term.

The most important ones are no doubt the four forms of general judgement:

- Every *S* is *P* (universal affirmative) (A)
- No *S* is *P* (universal negative) (E)
  (or Every *S* is not *P*)
- Some *S* is *P* (particular affirmative) (I)
- Some *S* is not *P* (particular negative) (O)

where *S* and *P* are terms for which there should be given formation rules if the system were to be completely formal.

The forms of inference are the syllogisms.[21]

(ii) *Stoic logic*
The stoic forms of judgement are

- *A* and *B* (conjunction)
- *A* or *B* but not both (exclusion)
  (later on also *A* or *B* (disjunction))
- Exactly either *A* or *B* (contravalence)
- If *A* then *B* (material implication)
- Not *A* (negation was not defined to be a judgement, but at the same time the stoics 'realized that a negative proposition could not properly be treated as co-ordinate with the proposition of which it was a negative'.[22])

where *A* and *B* are propositions for which the stoics would have had to give formation rules if their system were to be completely formal.

The forms of inference are the five indemonstrable moods (modus ponendo ponens etc.).[23]

(iii) *Medieval logic*
The medieval forms of judgement are those of Aristotle's logic as well as those of the stoic logic.

The forms of inference are accordingly the syllogisms and the stoic moods.[24]

(In the late scholastic development of logic artificial quantifiers a, b, c and d were introduced which allowed to form judgements of the form aS is every P (or: aS is not P), No S is bP, or, Some S is not bP, Of some S every P is cQ, Every S of every P is dQ etc.[25]).

(iv) *Fregean logic*
Frege's only form of judgement is $\vdash A$, where A is a proposition.

His forms of inference in his *Begriffsschrift* are explicitly modus ponens MP and implicitly a principle of substitution and two quantifier rules.[26]

(v) *Gentzen's logic*
Under Gentzen's formalistic interpretation the forms of judgement become formulas A or sequents $A_1, \ldots, A_n \rightarrow B_1, \ldots, B_m$ where the A_i's, $1 \leq i \leq n$, and B_j's, $1 \leq j \leq m$, are formulas for which formation rules are given.

The forms of inference in the first case (i.e. formulas) are the rules of the calculus of natural deduction NJ or NK, and in the second case (i.e. sequents) the rules of the sequent calculus LJ or LK resp.

(vi) *Constructive type theory*
The forms of judgement of constructive type theory are of course

- $x_1 : \alpha_1, \ldots, x_n : \alpha_n \vdash a : \alpha$
- $x_1 : \alpha_1, \ldots, x_n : \alpha_n \vdash a = b : \alpha$

where 'a' or 'b' denote particular types and 'α' stands for '*type*', or 'a' or 'b' denote objects of some type denoted by 'α', and in either case each of 'α_1', …,'α_n' denotes some type.

Frege's break-up with the traditional form of judgements
The basic forms of judgement of Aristotle's logic and subsequently in traditional logic (i.e. medieval logic, Leibniz, even Boolean logic) are: (1) S is P (affirmation), and (2) S is not P (denial) (cf. Subsubsect. (i)).

Frege changes this traditional analysis of the forms of judgement radically. He replaces it with an analysis in terms of truth and propositional content and an analysis of the propositional content in terms of 'function' and 'argument' and arrives at a new account of general judgements (as it is well-known formalistically from first order logic). Moreover, he abandons the symmetrical treatment of affirmation and denial and makes denial depend on affirmation (by treating denial of A as affirmation of $\sim A$). Frege's new forms of judgement have since become standard in what has been called modern logic. (The reasons for Frege's change will hardly concern us here; they basically are the failure to handle multiply general judgements by traditional logic, and the design of a formal language of pure thought, independent of the grammar of historically grown languages and subjective, psychological features. Nor will the reasons for arriving at the Aristotelian analysis concern us; they appear to be misguidance by the structure of Indo-European languages, and misguidance by psychological considerations concerning subjective attitudes and expectations of speaker and hearer.)[27]

ii.2 *Judgements vs. Propositions*

In constructive type theory it is of crucial importance to distinguish between propositions and judgements (or assertions).

What the logical operators operate on and what is held to be true are propositions. When a proposition is held to be true, then a judgement is made.

$$\overbrace{A \text{ is true}}^{\text{judgement}}$$

$$\nearrow$$

$$\text{proposition}$$

What the rules of inference operate on are judgements. That is, the premisses and conclusion of a logical inference are judgements: E.g. they express that a proposition is true (conclusion) provided some other propositions are true (premisses).

The objective of constructive type theory is to formalize informal mathematical reasoning as well as possible (cf. formalization, Subsect. viii.1, Chapt. 1). In the usual natural deduction style the rules of inference are not entirely formal, as e.g. the rule

$$\frac{A \quad B}{A\&B}\ \text{\&-introduction}$$

takes it for granted that A and B are well-formed formulas and only then does it say that one can infer $A\&B$ is true provided A is true and B is true. A completely formal way of writing this rule of inference thus looks as follows:

$$\frac{A : prop \quad B : prop \quad A\ true \quad B\ true}{A\&B\ true}$$

or, if written by means of the assertion sign

$$\frac{A : prop \quad B : prop \quad \vdash A \quad \vdash B}{\vdash A\&B}$$

(Martin-Löf takes the judgement stroke $\vdash$ and the phrase 'is true' to express the same thing, whereas according to Dummett's *Frege* they operate in quite a different way.[28])

Historically, this distinction between proposition and judgement was made from Frege to Russell/Whitehead. (Bolzano already made a similar distinction between a thought and a judgement: a thought is a concrete instance of a proposition produced and thought by some person at some time, and a judgement is a concrete instance of a proposition produced and accepted by some person at some time.[29])

With the rise of metamathematics these notions were replaced by the formalistic notions of formula (or well-formed formula) and theorem (in a formal system) resp. But whereas the formulas form a set (in the sense of constructive type theory) the propositions obviously do not – they only form a type.[30]

ii.3 *Two Kinds of Judgement: Analytic vs. Synthetic*

The distinction between analytic and synthetic judgements was originally made and discussed by Leibniz, Locke and Hume, but it became a major philosophical issue only through its treatment by Kant.

Kant's classical explanation of analytic judgement according to which its predicate is contained in its subject gave rise to the following definition:

A judgement is analytic if it is evident in virtue of the meanings of the terms that occur in it.

And the explanation of synthetic judgement is:

A judgement is synthetic if it is not evident solely in virtue of the meanings of the terms involved, but it is evident only if one goes beyond what is contained entirely within the judgement.

Kantian *examples* of synthetic judgements are:

- empirical judgements: what one goes to, beyond what is contained entirely within the judgement, is experience
- mathematical judgements: what one goes to, beyond what is contained entirely within the judgement, is a mathematical construction, or, in Kant's terms, a construction of concepts.

What one goes to, beyond what is contained entirely within the judgement in order to make a synthetic judgement evident, will be shown below to be a piece of evidence or a proof-object.

Discussing empirical judgements Martin-Löf proposes to extend the notion of judgement such as to include complexes consisting of two parts, namely a linguistic part, i.e. a proposition, and a non-linguistic part which is a certain experience.

Example The empirical judgement: It is raining.

Consider the complex consisting of a non-linguistic part, the experience of falling rain, and a linguistic part, the proposition that it is raining, as a judgement.

Then it is natural to say that a complex consisting of a linguistic part, i.e. a proposition A, and a non-linguistic part which is an experience fitting A, and considered as a judgement, is an analytic judgement, since it is evident in virtue of its constituents. *Note*: This treatment of the linguistic and non-linguistic on a par in the conception of judgements is something which would have been alien to Kant.

If synthetic judgements are generally treated the way Martin-Löf proposes to treat the empirical judgements, then every synthetic judgement is grounded on an analytic judgement. In that sense the analytic judgements are more basic than the synthetic ones. Furthermore, the analytic judgement on which the synthetic judgement is grounded is obtained from the synthetic judgement by fitting in the piece of evidence (the proof-object) which has been suppressed. In other terms, it is the mechanism of evidence or proof suppression which gives rise to synthetic judgements. And the piece of evidence or proof for a synthetic judgement cannot be obtained by conceptual analysis alone, but it has to be constructed.

ii.3.1 *Analytic and synthetic forms of judgement in constructive type theory*

The two fundamental forms of categorical judgement in constructive type theory, i.e. $a : \alpha$ and $a = b : \alpha$, where α : *type*, are both analytic. That is, if a judgement of one of these two forms is evident at all, it is evident solely by virtue of the meanings of the terms that occur in it. The same holds of course for the two more general forms of hypothetical judgement,

$$x_1 : \alpha_1, \ldots, x_n : \alpha_n \vdash a : \alpha$$
$$x_1 : \alpha_1, \ldots, x_n : \alpha_n \vdash a = b : \alpha$$

But a moment's reflection shows that the following forms of hypothetical judgement are likewise analytic

$$x_1 : \alpha_1, \ldots, x_n : \alpha_n \vdash \alpha : type$$
$$x_1 : \alpha_1, \ldots, x_n : \alpha_n \vdash \alpha = \beta : type$$
$$x_1 : \alpha_1, \ldots, x_n : \alpha_n \vdash \alpha : seq$$
$$x_1 : \alpha_1, \ldots, x_n : \alpha_n \vdash \alpha = \beta : seq$$

E.g. Take the judgement $\alpha : type$ (for simplicity's sake the corresponding categorical instead of the hypothetical form is chosen). If one knows what it means to be a type and what it means to be an α, and α is a type at all, then α : *type* is evident (solely by virtue of the meanings of the terms that occur in this judgement). Analogously, if one knows what it means for two types to be identical and what it means to be an α and what it means to be a β, and if α and β are identical types at all, then $\alpha = \beta$: *type* is evident (solely by virtue of the meanings of the terms that occur in this judgement). Etc.

If one does not care what exactly this object a of type α is, but is merely interested in the existence of an object of type α, one can introduce a new form of judgement, namely α *exists*, i.e. the existential form of judgement. (Equivalent readings of α *exists* are: The concept α has existence, or, There exists an α, or, α is, which can be found in Bolzano and Brentano.)

The semantical explanation of this new form of judgement answers the question

What does a judgement of the form 'α *exists*' mean?
To explain what it means that a type α exists one has to know what an object a of type α is.

Alternatively, 'α *exists*' is really defined by the following rule

$$\frac{a : \alpha}{\alpha\ exists}$$

The existential form of judgement α *exists* includes the form of judgement A *true*, where A : *prop*, as a special case. This can be seen as follows:

One of the basic rules of constructive type theory is $\dfrac{A : prop}{proof(A) : type}$ also written as $\dfrac{A : prop}{A : type}$ (cf. vii.5), Chapt. 1). Now, if A : *prop* and $a : A$ then $\dfrac{\dfrac{a : A}{a : proof(A)}}{proof(A)\ exists}$

where '*proof*(A) *exists*' means that there exists a proof (object) or construction of the proposition A. But according to the intuitionistic explanation of truth (cf. C, n & o to vii.3), Chapt. 1)

$$A \; true = \text{ there exists a proof (object) of } A$$

whence, '*proof*(A) *exists*' means A *true*. Thus, the form of judgement A *true* is indeed a special case of the existential form of judgement α *exists*.

The existential form of judgement α *exists* also includes the form of judgement A *false*, where $A : prop$, as a special case.

Let f be a proof (object) of $\sim A$, where $A : prop$, i.e.

$$\frac{\dfrac{f : A \supset \bot}{f : proof(A \supset \bot)}}{proof(A \supset \bot)\; exists} \qquad \text{or} \qquad \frac{\dfrac{f : A \supset \bot}{f : (A)\bot}}{(A)\bot \; exists}$$

where '*proof*$(A \supset \bot)$ *exists*' means that there exists a proof (object) or construction of the proposition $A \supset \bot$. But according to the intuitionistic explanation of falsehood

$$\begin{aligned} A\; false \;&=\; \text{there exists a disproof (object) of } A \\ &=\; \text{there exists a proof (object) of } A \supset \bot \text{ or of } (A)\bot \text{ resp.} \end{aligned}$$

whence, '*proof*$(A \supset \bot)$ *exists*' means A *false*, and the form of judgement A *false* is a special case of the existential form of judgement as well.

The synthetic form of judgement is exactly the existential form of judgement just introduced. To see this one asks what this judgement's evidence is based on. The evidence of a judgement of the form α *exists* clearly rests on the construction of a proof object a of α which thus implies that one has to go beyond what is contained in the judgement itself. *Note*: Kant considered existential judgements to be synthetic (but not necessarily conversely).

Another *example* of synthetic judgements of Martin-Löf's

Logical laws in their usual formulation: they all say that an arbitrary proposition of a certain form is true; e.g. $A \supset (B \supset A\&B)$ *true*, for $A : prop$, $B : prop$.

By the intuitionistic analysis of the notion of truth this means that $proof(A \supset (B \supset A\&B))$ *exists*. And one convinces oneself of the correctness of this latter judgement by making a little construction, e.g. a derivation of the proposition $A \supset (B \supset A\&B)$ in natural deduction style; i.e. one has to construct an object c which is a proof of the proposition $A \supset (B \supset A\&B)$, i.e. $c : proof(A \supset (B \supset A\&B))$.

Observe that standard propositional and predicate logic consist not only of logical laws but also of rules of term formation and rules of formula formation. The following four forms of judgement occur in propositional and predicate logic:

$$\begin{array}{ll} t : elem(A) & \text{where } A : \; set \\ A : prop & \\ A \; true & \text{where } A : prop \\ A \; false & \text{where } A : prop \end{array}$$

whereby the first two forms are special cases of the form of judgement $a : \alpha$ which means that they are analytic; and only the latter two are forms of synthetic judgement.

Note: Kant (according to Körner's interpretation) did not consider all mathematical judgements to be synthetic, but only the interesting ones, and the interesting ones are exactly the existential ones. The difficulty of a mathematical proof is to find or construct an object which falls under the concept to which the existential judgement ascribes existence.

Lastly, under the sequents-as-types interpretation $seq = type$ (cf. Sect. xii., Chapt. 1) the notion of truth is extended from propositions to sequents and thus becomes a concept coextensional with existence.

What does a judgement of the form 'α *true*' (where $\alpha : seq$) mean?
To explain what it means that a sequent α is true, one has to know how to construct a proof-object a of α.

Alternatively, 'α *true*' is really explained by the following rule

$$\frac{a : \alpha}{\alpha \; true}$$

Hence the form of synthetic judgement α *true* is a logical interpretation of the type-theoretical form of synthetic judgement α *exists*.

The most general form of synthetic judgement in constructive type theory is

$$x_1 : \alpha_1, \ldots, x_m : \alpha_m, \beta_1 \; exists, \ldots, \beta_n \; exists \vdash \beta \; exists$$

of which

$$x_1 : A_1, \ldots, x_m : A_m, B_1 \; true, \ldots, B_n \; true \vdash B \; true$$

is merely a special case.

And the most general form of synthetic judgement in constructive type theory can alternatively be expressed by

$$x_1 : \alpha_1, \ldots, x_m : \alpha_m, \beta_1 \; true, \ldots, \beta_n \; true \vdash \beta \; true$$

ii.3.2 *The logic of analytic judgements and the logic of synthetic judgements*

The logic of analytic judgements is the logic for deriving judgements of the six analytical forms (whereby the first two forms are the most important ones).

The logic of synthetic judgements is the logic for deriving judgements of synthetical as well as analytical form.

The logic of analytic judgements is decidable in the sense that for any judgement of one of the two forms $a : \alpha$ and $a = b : \alpha$ (or any of the other analytical forms) it can be effectively decided whether or not that judgement is derivable by means of the formal rules of this logic.

The logic of analytic judgements is complete in the sense that given any judgement of one of the two forms $a : \alpha$ and $a = b : \alpha$ (or any of the other analytical forms) which contains certain constants expressing certain concepts, no other laws or rules are needed to derive that judgement than the laws/rules which concern precisely those concepts, i.e. those laws/rules are complete for deriving the judgement.

It is now possible to gain a new understanding of the incompleteness and undecidability phenomena which turned up in traditional logical calculi: These traditional logical calculi are formal systems within the logic of synthetic judgements. As has been explained above, in the synthetic or existential form of judgement α *exists* the object a of α which is claimed to exist, has been suppressed (cf. the mechanism of generating synthetic forms of judgement described above).

In the passage from the analytic $a : \alpha$ to the synthetic α *exists*, decidability is lost: for to prove that α *exists* one has to search for an object a of α and even if one limits oneself to a particular formal system, that search procedure will not necessarily terminate, let alone then, if no limitation is imposed on the axioms which can be invoked in the course of that procedure. In particular, if the type is of the form *proof*(A), for A : *prop*, it is the fact that the proof-object a of *proof*(A) has been suppressed in the synthetic form of judgement *proof*(A) *exists* or, in other words, A is provable, or what amounts to the same, A *true* (cf. the explanation above of A *true* in terms of the intuitionistic conception of truth), which is the source of the undecidability.

Example Church's undecidability theorem

- Church's original formal system is, say, a consistent extension T of first order arithmetic
- Church constructed a propositional fct. representing 'is a proof in T of X', X : *prop*, (i.e. '$proof_T(X)$') in T. This propositional fct. can be shown to be decidable. From this propositional fct. he constructed a judgement representing '$proof_T(A)$ *exists*' in T.
- By means of the judgement representing '$proof_T(A)$ *exists*' in T, for A : *prop*, Church proved (using a diagonal argument) this judgement and T to be undecidable.

Similarly with incompleteness: consider again the simplest and most important case of a synthetic judgement of the form A *true* where the proposition A has been formulated in the language of a certain formal system. Then A may be unprovable in that system simply because there is no proof-object or construction a : *proof*(A) that can be expressed in it. However, it may well be that if one goes from the original system to some extension of it by introducing new concepts, then one will be able to construct a proof-object a : *proof*(A) in the extended system, although that was impossible in the original system in which the proposition A was expressed.

Example Gödel's incompleteness theorem

- Gödel's original formal system is, say, first order arithmetic.
- Gödel constructed a particular first order arithmetical proposition A which is s.t. there is no proof of A in first order arithmetic; but at the same time he showed informally that there exists nevertheless a proof of A which means that A is true. Hence first order arithmetic is incomplete.
- If first order arithmetic is extended by adding e.g. a reflection principle to it, a proof-object a of *proof*(A) can actually be constructed in this extended formal system.

In short, the suppression of the proof-objects is the source of undecidability and incompleteness of the logic of synthetic judgements. Thus, it is wrong to interpret Gödel's incompleteness result as saying that formal reasoning is always incomplete w.r.t. informal reasoning. A formal language is incomplete because it can always be extended by the introduction of new concepts as well as axioms or rules governing them. But no sooner have the latter been formulated with full rigor, will the extended language be just as formal as the original one.

ii.3.3 *Monomorphism and polymorphism: a further refinement of analytic and non-analytic judgements in constructive type theory*

A (constructive) type theory is said to be monomorphic if every object (term) is of one and only one type (type expression). If the theory has objects (terms) which have different types (type expressions), then it is called polymorphic.

In a paper 'Polymorphism and Monomorphism in Martin-Löf's Type Theory', A. Salvesen mentions two advantages of using a polymorphic type theory in connection with programming and theorem proving: (a) to have type consistency, reusability and flexibility w.r.t. terms, and (b) to simplify the task of writing down programs. However, a condition to be met by any simplification of monomorphic type theory is that it should be safe, i.e. it shouldn't loose any of its essential properties.[31]

Martin-Löf stresses in version (8) that the limit of a simplification of monomorphic type theory is given by the preservation of its essential property of decidability of type checking and of definitional identity checking. (cf. next Sect. iii.). It ought to be investigated how much type information can be thrown out of the element of a set or out of the proof-object of a proposition without losing the property just mentioned.

Now, Salvesen proves in her paper

1i.) in monomorphic type theory the judgements $A : \mathit{set}$, $A = B : \mathit{set}$, $a : A$ and $a = b : A$ are decidable;[32]

1ii.) in monomorphic type theory it is decidable for any a of the form $c(t_1, \ldots, t_n)$, where c is an introductory or eliminatory operator, whether or not a has a type and it is possible to compute this type unique modulo type identity;[33]

2) in polymorphic type theory the judgements $A : \mathit{set}$, $A = B : \mathit{set}$, $a : A$ and $a = b : A$ are in general not decidable.[34]

In the light of these results she answers more specifically the question whether polymorphic type theory is just a simplification (or a feasible simplification according to Martin-Löf) of monomorphic type theory as follows: Let a fct. which to expressions of monomorphic type theory assigns expressions of polymorphic type theory by stripping off type information be called a stripping fct. Then: If polymorphic type theory is a simplification of monomorphic type theory by applying the stripping fct. to all of the latter's expressions, then this simplification is not safe (not feasible according to Martin-Löf, since the essential property of decidability of unstripped judgements is lost with/does not hold for stripped judgements).[35]

Salvesen proceeds to showing that polymorphic type theory (in this stripped sense) and monomorphic type theory are actually different type theories. She proves this by proving

3) there are judgements of the form $a = b : A$ which are derivable in polymorphic type theory, but which cannot be obtained from derivable judgements in monomorphic type theory by stripping (Thm. 4);[36]

4) there are judgements of the form $a : A$ derivable in polymorphic type theory which do not come from judgements derivable in monomorphic type theory by stripping (Thm. 5);[37]
5) there are judgements of the form $A = B : set$ and $A : set$ derivable in polymorphic type theory which do not come from judgements derivable in monomorphic type theory by stripping (Cor. 6 to Thm. 4 and Thm. 5);[38]

The original aim of simplifying monomorphic type theory in order to simplify programming and theorem proving remains. It was shown that polymorphic type theory (in the sense of a stripped monomorphic type theory) is not only not a safe simplification of monomorphic type theory, but as a matter of fact is not a simplification at all. But with the original aim the question remains as to how much type information can be removed from the objects (terms) of monomorphic type theory without losing essential information. Salvesen makes a suggestion which is essentially a modification of the stripping fct. mentioned above: the new stripping fct. behaves like an identity fct. on all the eliminatory operators, and like the original stripping fct., on all the other expressions.[39]

The operators	The stripping fct. $*$ applied $\ldots * =$	The new stripping fct. $\star$ applied $\ldots \star =$
N_n	$= N_n$	$= N_n$
$N_n I0$	$= N_n I0$	$= N_n I0$
$\vdots$	$\vdots$	$\vdots$
$N_n In-1$	$= N_n In-1$	$= N_n In-1$
$N_n E(C, c_0, \ldots, c_{n-1}, c)$	$= N_n E(c_0^*, \ldots, c_{n-1}^*, c^*)$	$= N_n E(C^\star, c_0^\star, \ldots, c_{n-1}^\star, c^\star)$
N	$= N$	$= N$
$NI1$	$= NI1$	$= NI1$
$NI2(a)$	$= NI2(a^*)$	$= NI2(a^\star)$
$NE(C, d, e, a)$	$= NE(d^*, e^*, a^*)$	$= NE(C^\star, d^\star, e^\star, a^\star)$
$A + B$	$= A^* + B^*$	$= A^\star + B^\star$
$+I1(A, B, a)$	$= +I1(a^*)$	$= +I1(a^\star)$
$+I2(A, B, b)$	$= +I2(b^*)$	$= +I2(b^\star)$
$+E(A, B, C, d, e, c)$	$= +E(d^*, e^*, c^*)$	$= +E(A^\star, B^\star, C^\star, d^\star, e^\star, c^\star)$
$\Pi(A, B)$	$= \Pi(A^*, B^*)$	$= \Pi(A^\star, B^\star)$
$\Pi I(A, B, b)$	$= \Pi I(b^*)$	$= \Pi I(b^\star)$
$\Pi E(A, B, C, d, c)$	$= \Pi E(d^*, c^*)$	$= \Pi E(A^\star, B^\star, C^\star, d^\star, c^\star)$
$(A \to B)$	$= A^* \to B^*$	$= A^\star \to B^\star$
$\to I(A, B, b)$	$= \to I(b^*)$	$= \to I(b^\star)$
$\to E(A, B, C, d, c)$	$= \to E(d^*, c^*)$	$= \to E(A^\star, B^\star, C^\star, d^\star, c^\star)$
$\Sigma(A, B)$	$= \Sigma(A^*, B^*)$	$= \Sigma(A^\star, B^\star)$
$\Sigma I(A, B, a, b)$	$= \Sigma I(a^*, b^*)$	$= \Sigma I(a^\star, b^\star)$
$\Sigma E(A, B, C, d, c)$	$= \Sigma E(d^*, c^*)$	$= \Sigma E(A^\star, B^\star, C^\star, d^\star, c^\star)$
$A \times B$	$= A^* \times B^*$	$= A^\star \times B^\star$
$\times I(A, B, a, b)$	$= \times I(a^*, b^*)$	$= \times I(a^\star, b^\star)$
$\times E(A, B, C, d, c)$	$= \times E(d^*, c^*)$	$= \times E(A^\star, B^\star, C^\star, d^\star, c^\star)$
$I(A, a, b)$	$= I(A^*, a^*, b^*)$	$= I(A^\star, a^\star, b^\star)$

$II(A, a)$	$= II(a^*)$	$= II(a^\star)$
$IE(A, a, b, C, d, c)$	$= IE(a^*, b^*, d^*, c^*)$	$= IE(A^\star, a^\star, b^\star, C^\star, d^\star, c^\star)$
	moreover,	
	$A^* = A^*$ if $A : type$	
	provided $A : set$	
	$x^* = x$	
	$((x_1) \ldots (x_n)b)^*$	
	$= (x_1) \ldots (x_n)b^*$	
	$((m)n)^* = (m^*)n^*$	

Salvesen then proves that the resulting type theory (the theory resulting from an application of the new stripping fct. $\star$ to monomorphic type theory) is a safe rule-to-rule simplification of monomorphic type theory modulo set identity in decorations, and she also proves that claim 1ii.) (above) does not hold for this resulting theory. Finally, she conjectures that this suggestion of hers is probably not minimal w.r.t. type information in the objects (terms) to be preserved.[40]

Now, constructive type theory is the logic of analytic judgements as explained above (this is especially emphasized in version (7)). This is not only a descriptive, but also a normative statement. An essential property of the logic of analytic judgements is its decidability (cf. above). If the analyticity of judgements is to imply their decidability, then any simplification of monomorphism has to preserve analyticity. These observations suggest the following distinctions:

- Judgements of the form of analytic judgements in monomorphic notation are clearly analytic.
- Judgements of the form of analytic judgements in polymorphic notation are clearly not analytic anymore.

The idea is that there are judgements of the form of analytic judgements which are analytic without being in strictly monomorphic notation. In other terms, the range of analytic judgements (of the form of analytic judgements) is greater than that of the judgements of analytic form in monomorphic notation. Monomorphism implies analyticity but not vice versa.

E.g. one could consider newly stripped judgements (i.e. monomorphic judgements stripped by Salvesen's new stripping fct. $\star$) to still be analytic. As already mentioned, proposition 1ii.) does not hold for the corresponding type theory. Salvesen suggests that this fact may be of little importance w.r.t. practical matters. But it may be of greater importance w.r.t. philosophical matters and it may indeed disqualify newly stripped judgements as candidates for analytic judgements of the form of analytic judgements which are not in monomorphic notation.

Finally it should be pointed out that the analytic judgements under discussion here are excellent candidates for Sundholm's alternative 'Sätze der Logik', the fourth and fifth features of which reflect the problem which is here being discussed from a speculative point of view.[41]

ii.3.4 *More on the intuitionistic meaning of existence, truth and falsity*

Existence

In the previous subsection the existential form of judgement α *exists* was introduced. Its meaning in intuitionism is that of an incomplete or abbreviated judgement, i.e.

α *exists* $=$ there exists an object of type α

C, n & o

(1) It has been mentioned earlier on (cf. C, n & o (1) to vii.3), Chapt. 1) that the concept of existence used here is not the one expressed by the existential quantifier. It is rather the traditional philosophical notion of existence of a concept (or existence of an essence).

(2) Note on an unprecise way of rendering the intuitionistic meaning of existence

Many intuitionists write somewhat misleadingly that

'α *exists*' says/expresses/means that an object of type α exists

But Martin-Löf corrects this explanation in the sense that 'α *exists*' does not say/express/mean anything like that

α *exists* shows that an object of type α exists

This distinction is based on Wittgenstein's Tractarian distinction between saying and showing.

(3) (Historical) The intuitionistic meaning explanation of 'α *exists*' can already be found in Hilbert's talk of incomplete communication, or in Weyl's talk of partial judgement or judgement abstract. Kleene made the incomplete judgement/communication issue famous with his realizability interpretation (but this realizability interpretation is a non-standard, or unintended interpretation of intuitionistic logic and thus actually no example for the meaning intended here).[42]

Martin-Löf differentiates between three notions of existence

(1) existence simpliciter or bare existence

α *exists* $=$ there exists an object of type α

(2) actual existence

α *exists actually* $=$ an object of type α is known

$=$ an object of type α has been constructed

(3) potential existence

α *exists potentially* $=$ an object of type α can be known

$=$ an object of type α can be constructed

$=$ an object of type α can be known actually/can be constructed actually

The third rendering of α *exists potentially* shows that actual existence is conceptually prior to potential existence, and that generally actuality precedes potentiality in the conceptual order of priority (in accordance with Aristotle, Thomas, Albertus and others). That α *exists potentially* means that any time α can (be made to) exist actually. And simplicity is conceptually prior to actuality, i.e. that α *exists actually* presupposes that α *exists*.

Truth (as a special case of existence)
In the previous subsection truth has already been shown to be a special case of existence. One can define

A true = *A exists*, for *A* : *prop*
= there exists a proof (object) of proposition *A*

The second rendering of *A true* is called by Martin-Löf the verification principle of truth (cf. C, n & o to vii.3), Chapt. 1 concerning the use of 'verification' in this context). This verification principle of truth has to be clearly distinguished from what he calls the verification principle of meaning which says that a proposition is defined by its method (means) of verification (in the sense of proof conditions).

C, n & o

(1) It has been pointed out earlier on (cf. C, n & o to vii.3), Chapt. 1) that the word 'proof' in the intuitionistic meaning explanation of '*A true*' is used differently from when it is applied to a mathematical theorem. It is here referring to a mathematical object which can also be called a 'verification' of the resp. proposition.

(2) Note on an unprecise way of rendering the intuitionistic meaning of truth

The comment made on an unprecise way of rendering the intuitionistic meaning of existence carries over mutatis mutandis to truth.

According to the three notions of existence, one can distinguish three notions of truth

(1) bare truth or truth simpliciter
A true = there exists a proof (object) of proposition *A*

(2) actual truth
A actually true = proposition *A* is known to be true
= a proof (verification) of proposition *A* has been constructed

(3) potential truth
A potentially true = proposition *A* can be known to be true
= a proof (verification) of proposition *A* can be constructed
= a proof of proposition *A* can actually be constructed

Actual truth precedes potential truth in the conceptual order of priority for the very same reasons for which actual existence is conceptually prior to potential existence. And the same holds for bare truth conceptually preceding actual truth.

(2) and (3) can be compared as follows:

ad (2) Actual truth is knowledge dependent and tensed: proposition *A* is known to be true (by the first rendering), and a proof-object of proposition *A* has been constructed at a certain moment in the course of time (by the second rendering).

Because of this heavy reliance on knowledge, the notion of actual truth could be called a more idealist notion of truth.

ad (3) Potential truth is knowledge independent and tenseless: proposition *A* can be known to be true (by the first rendering) implies that it is free of actually being known to be true, and that a proof-object of proposition *A* can be constructed, means that a proof-object of *A* can be constructed at any moment in time and is thus independent of any particular time (by the second rendering). Observe though that potential truth is knowledge independent in a weak sense, i.e. it is not totally independent of knowledge but only independent of actual knowledge. In particular, it is conceptually dependent on the notion of knowledge. Observe also that potential truth is tenseless in a strong sense. Prawitz has shown that *A potentially true*, or equivalently, proposition *A* can be proved, is not equivalent with proposition *A* has been, is being or will be proved at some time in the course of history; since, whereas proposition *A* has been, is being or will be proved at some time in the course of history entails proposition *A* can be proved, the converse is clearly not the case.[43]

Because of this relative independence of knowledge, the notion of potential truth could be called a more realist notion of truth.

C, n & o

(1) (Historical) The notion of potential truth has been introduced and discussed in the late 70s by Dummett,[44] by Prawitz,[45] and by Putnam.[46]

(2) Note on a frequent objection to the intuitionistic notion of truth:

The objection is that it is counter-intuitive to say that a proposition *A* becomes true when *A* is proved.

The intuitionist replies that this objection only concerns the notion of actual truth and not the notion of potential truth (potential truth was characterized as being tenseless).

(3) Note on true and unprovable propositions:

On the analysis of potential truth presented here, it is obvious that there are no propositions which are (potentially) true but cannot be proved (because potential truth of a proposition has been analyzed in terms of potential existence of a proof of the resp. proposition).

This note can also be understood as saying that there are no propositions which are true without them being knowable to be true. Hence the intuitionistic notion of (actual or potential) truth is knowledge dependent (in a strong or a weak sense resp.). This is the result of the intuitionistic analysis of truth, according to which the concept of knowing a proof of a proposition is conceptually prior to the concept of knowing that a proposition is true. This makes intuitionism a philosophy of mathematics with a strong idealist flavor.

(4) (Historical) A similar position was upheld by the early as well as the late Husserl. He maintained that the notion of proof or evidence is conceptually prior to the notion of truth. Whereas the early Brentano started off with the opposite view according to which the notion of truth is conceptually prior to the notion of evidence, his analyses forced him to change his stance and the late Brentano's position is similar to that of Husserl's. However, the philosopher who influenced Brouwer (and intuitionism) was

neither Husserl nor Brentano, but Kant, whose position is that the notions of proof or natural experience are conceptually prior to the notion of truth of a mathematical or an empirical proposition resp.

A truth-maker theory of truth simpliciter can be conceived as an abstract version of the correspondence theory of truth. According to this truth-maker theory, a judgement '*A true*' is defined as follows:

A true = there exists a truth-maker for *A*

and the theory involves three elements:

(i) the category of truth-bearers *A*
(ii) for each object *A* of the category of truth-bearers, a category TM_A of truth-makers for *A*
(iii) the relevant notion of existence for the category TM_A

The intuitionistic conception of truth is obviously an instance of the truth-maker theory of truth, where '*A true*' is defined as

A true = there exists a proof (object) of the proposition *A*

and where

(i) the category of truth-bearers *A* is the category or type *prop* of propositions
(ii) for each $A : prop$, the category TM_A for *A* is *A* (remember that if $A : prop$, then $A : type$)
(iii) the relevant notion of existence is the traditional philosophical notion of existence simpliciter of a concept (or existence simpliciter of an essence)[47]

Since truth in intuitionism is but a special case of existence, one may consider construing a general existence-maker theory of existence. According to such a theory, a judgement 'α *exists*' is defined as follows:

α *exists* = there is an object of α (or simply, there is an α)

and this theory involves as before three ingredients:

(i) the existence-bearers α
(ii) for each existence-bearer α a collection EM_α of existence-makers for α
(iii) the relevant notion of existence for the collection EM_α

Now, constructive type theory can easily be seen to be an instance of such a general existence-maker theory of existence. Observe that the meaning of the judgement 'α *exists*' is indeed defined by

α *exists* = there is an object *a* of type α

where

(i) the existence-bearers α are the types
(ii) for each $\alpha : type$, the collection EM_α for α is α itself
(iii) the relevant notion of existence is once again the traditional philosophical notion of existence of a concept (or existence of an essence)

Falsity (as a special case of existence)

In the previous subsection falsity also has been shown to be a special case of existence.

One can define

$$\begin{aligned} A\,\textit{false} &= (A)\bot\ \textit{exists}, \text{ for } A : \textit{prop} \\ &= \text{there exists a disproof (object) of proposition } A \\ &= \text{there exists a proof (object) of proposition } A \supset \bot \end{aligned}$$

According to the three notions of existence, three notions of falsity can be distinguished:

(1) bare falsity or falsity simpliciter

$$A\,\textit{false} = \text{there exists a disproof (object) of proposition } A$$

(2) actual falsity

$$\begin{aligned} A\ \textit{actually false} &= \text{proposition } A \text{ is known to be false} \\ &= \text{a disproof (falsification) of proposition } A \text{ has been constructed} \end{aligned}$$

(3) potential falsity

$$\begin{aligned} A\ \textit{potentially false} &= \text{proposition } A \text{ can be known to be false} \\ &= \text{a disproof (falsification) of proposition } A \text{ can be constructed} \end{aligned}$$

Bare falsity precedes actual falsity which precedes potential falsity in the conceptual order of priority, again for the same reasons for which existence simpliciter is conceptually prior to actual existence which is conceptually prior to potential existence.

ii.4 *Correctness, Evidence and Validity of Judgements*

In C, n & o to vii.3), Chapt. 1 the following distinction was made w.r.t. judgements: 'judgement' can be understood as an act of judgement or judging (cf. meaning (a) there), and 'judgement' can be understood as the object or the result of an act of judgement (cf. meaning (b) there). The first part of the following explanations will deal with the second meaning of 'judgement' and the second part with the first meaning.

ii.4.1 *Judgement (in the sense of object or product of an act of judgement)*

The 'proof' of a judgement will in the sequel be called a demonstration (in accordance with C, n & o to vii.3), Chapt. 1) and the 'truth' of a judgement will be referred to as its correctness. Just as in the preceding subsect. 3 notions of truth of a proposition were distinguished, 3 notions of correctness of a judgement can be distinguished as follows:

(1) bare correctness or correctness simpliciter
  A judgement is correct = there is a demonstration of a judgement
(2) actual correctness
  A judgement is actually correct = a judgement has been demonstrated
(3) potential correctness
  A judgement is potentially correct = a judgement can be demonstrated/is demonstrable

(There is no further parallelism between correctness and truth; cf. end of this subsection.)

Martin-Löf now introduces the following terminology:

A judgement is evident = a judgement has been demonstrated

which means that an actual demonstration is what makes a judgement evident or constitutes the evidence for a judgement. He then presents the following chain of equalities:

evident = actually correct = known = demonstrated/justified/evidenced

A judgement is correct = a judgement can be demonstrated

which gives rise to the chain of equalities:

correct = potentially correct = knowable = demonstrable/justifiable/evidenceable

He thus maintains the principle of identity of bare and potential correctness:

$$\left.\begin{array}{c}\text{Existence of a demonstration}\\ \text{of a judgement}\end{array}\right\} = \left\{\begin{array}{c}\text{possibility of a demonstration}\\ \text{of a judgement}\end{array}\right.$$

An immediate consequence of this principle is:

The evidence of a judgement entails its correctness.

Justification

If a judgement has been demonstrated, it obviously can be demonstrated (which is just an instance of the medieval principle: *Ab esse ad posse valet consequentia*). □

C, n & o

(Terminological) Dummett and Prawitz call subjectively correct what is here called an evident judgement, and they call objectively correct what is here simply termed a correct judgement.[48]

From what has been said above it follows that

A judgement is incorrect	=	a judgement is not correct
	=	no demonstration exists for a judgement
	=	a judgement cannot be demonstrated = known
A judgement is blind	=	a judgement is correct but not evident to the person who makes it

C, n & o

(1) (Terminological) The term 'blind judgement' was used in pretty much the same sense by Brentano.

(2) Note on judgements evident in themselves or known in themselves

It is out of the question that there are any judgements which are evident in themselves, i.e. independently of anybody's cognitive activity. To speak of judgements evident in themselves is as absurd as to speak of judgements not known to anybody but known in themselves. 'To be evident' is as relative an expression as is the expression 'to be known', and they ought to be read as to be evident to s.o. or to be known by s.o. resp. That is, there is no evidence outside s.o.'s experience of it.

That is a basic tenet of intuitionism which Brouwer expressed using 'truth' for 'evident judgement' as follows: There are no unexperienced truths (which, interpreted into the present terminology, reads as follows: All evident judgements of the form A *true* (where $A : prop$) are experienced).

(3) It would be a mistake or non-sense to transfer the comments of the preceding note (2) to the level of the truth of propositions: there can well be propositions which are true but haven't been demonstrated to be true yet (namely the potentially true propositions).

(4) Martin-Löf's views on correctness and evidence of judgements are inspired by Husserl's views on these topics:

> Thus, Husserl believes that what makes a correct judgement a knowing one is evidence. He claims that what distinguishes the evident and the blind judgement is a claim involving the certainty on the part of the epistemic subject that the correctness-claim common to both is justified, which is made in the former but not in the latter case. Husserl's conception of evidence has to be understood in the context of a theory of justification (or, as it has been called here, a theory of demonstration).

Husserl also makes a couple of distinctions which do not appear in, but may be of interest to constructive type theory: 1) On the one hand Husserl states that evidence is the actual experience of correctness of a judgement. As has been seen above: hence the evidence of a judgement entails its correctness. If Pietersma writes that for Husserl the evident is (though not infallible) the correct, then he or Husserl pays no attention to the distinction between the actual and the potential w.r.t. the demonstration of a judgement. On the other hand Husserl does distinguish between the evident and the correct judgement, namely there where he insists that there is a correctness in itself. A single experience of evidence cannot be taken to be an indefeasible grasp of correctness in itself. Every single experience of evidence is always contextual in at least two ways: (a) it is a function within the total context of consciousness, and (b) it includes reference to what Husserl calls a synthesis of recognition, i.e. the cognitive subject experiencing the evident judgement presumes he can return and be in a position to find it still evident once more. – Here Husserl does make a distinction between the actual and the potential, but in a different way from the one explained above: The 'correctness in itself' may not include reference to any particular actual experience of evidence, but it does include reference to potentialities of repetition (of such experiences) grounded in the life of the transcendental ego. Correctness is then demonstrability in the sense of performability of the act of demonstration on the part of transcendental subjectivity.[49] 2) Husserl also assumes a whole cognitive progression from a so-called 'empty' intention (which corresponds to the absence of any justification for a judgement) to a 'fulfilled' intention (which corresponds to the actual presence of a demonstration). Pietersma interprets Husserl's cognitive progression as follows: epistemic situations are regarded as occupying different places on a scale which is supposed to measure something called 'epistemic distance'. In an epistemic situation in which the cognitive subject has an 'empty' intention, the epistemic distance is maximal, whereas in an epistemic situation in which the cognitive subject has a 'fulfilled' intention, the epistemic distance is minimal. Moreover, it can be said that the smaller the epistemic distance of an epistemic situation of a cognitive subject passing a judgement, the greater is the epistemic worth of the judgement. There is a whole range of cognitive moves (implicit or suggested by a judgement passed or to be passed) that have to be taken by a cognitive subject in order to move up the cognitive progression towards the optimal epistemic situation (i.e. the epistemic situation with 'fulfilled' intention).

The more possible cognitive moves are taken by a cognitive subject, the shorter is the epistemic distance. If all possible cognitive moves have been taken, the cognitive subject has attained the optimal epistemic situation. This optimal epistemic situation is also characterized by Husserl as 'seeing the object itself' or 'primordially presentative seeing'. Now, to attain an optimal epistemic situation is an achievement. Pietersma suggests that a cognitive subject cannot have a sense of achievement if he has had no sense of failure or if he does not conceive of failure as a possibility. This implies that a claim to be seeing the object itself or to be passing an evident judgement, can be subject to critical evaluation. What the cognitive subject takes to be the optimal epistemic situation for passing a correct judgement may not in fact be optimal. 3) Lastly, Husserl distinguishes two kinds of evidence: (a) adequate evidence (of a judgement) if nothing whatever is ascribed to the object in the judgement beyond what is actually seen (adequate evidence seems to coincide with evidence as the actual demonstration of a judgement); (b) apodictic evidence (of, or insight into a judgement) if the person judging is unable to judge otherwise and this inability is due to an 'objective-ideal necessity'. This kind of evidence characterizes especially the optimal epistemic situation in all inquiry directed to essences. Note that according to 2) an apodictic or an adequate evidence may be revealed as error.

(Føllesdal disagrees with Pietersma in that he interprets Husserl as saying that there is no such thing as adequate evidence. Adequate evidence or the optimal epistemic situation is only a regulative idea guiding the research (the cognitive progression) of some cognitive subject, which can never be attained by him.[50]

According to Patzig, Husserl tries in the *Logische Untersuchungen* (1900–01) to reduce evidence to correctness, but in view of the difficulties of such a reduction, he could have found it more promising to reverse the reduction and instead reduce correctness to a hierarchy of evidences; and this is what he does in the *Ideen* (1913) and especially in the *Cartesianische Meditationen* (1950).[51])

(*Note*: the terminology of these explanations has been adapted to the terminology of constructive type theory. In particular, 'truth' and 'true' have all over been replaced by 'correctness' and 'correct' resp.)

ii.4.2 *Demonstration (in the sense of act of judging)*

Another characterization of evident judgements which Martin-Löf uses quite often is the following:

A judgement is evident = a judgement has been understood/grasped/ comprehended/ known (in the sense of gotten to know)

This leads to the identification

demonstration of a judgement = act of understanding/grasping/ comprehending/ getting to know a judgement

This identification stresses the fact that a demonstration is not an object, but an act. And Brouwer was referring to this fact when he wrote that a proof (here: demonstration) is a mental construction (act).

The concept of correctness applies to 'judgement' in both senses, i.e. to judgements as well as to demonstrations. The 'correctness' of a demonstration will in the sequel be called its validity.

The very basic fact behind many of the preceding distinctions is the fact that human beings (and that includes the mathematicians) make mistakes, they err. This basic fact necessitates the following distinction:

A demonstration is valid	=	a demonstration is free of errors or mistakes
A demonstration is invalid	=	a demonstration contains at least one error or mistake

Now, what has been referred to in the first part of this subsection as a demonstration is of course a valid demonstration.

The concept of validity of a demonstration is conceptually prior to the concept of correctness of a judgement.

Justification
When one asks whether a mathematical theorem (the object or result of an act of mathematical judgement) is correct, one naturally turns to the demonstration of the theorem and one has to check whether this demonstration is valid. For a theorem cannot be incorrect except as the result of a mistake in its demonstration. □

Hence the notion of validity of a judgement (demonstration) is one of the most fundamental notions of all: it is more basic than the notions of correctness or evidence of a judgement which are more basic than the notion of truth of a proposition.

More terminology:

If the act of demonstrating a judgement (the inference) is atomic or indivisible, then the demonstration is called immediate (intuitive).

If the demonstration of a judgement consists of a whole sequence or chain of atomic acts, it is called mediate (discursive). A mediate demonstration is thus a chain of immediate demonstrations.

If the actual demonstration of a judgement is immediate, the judgement is said to be immediately evident. An immediately evident judgement is what is called an axiom.

If the actual demonstration of a judgement is mediate, the judgement is said to be mediately evident. A mediately evident judgement is what is called a theorem.

C, n & o
(1) Another source of inspiration for Martin-Löf's views on correctness, evidence and validity of judgements have been the late Brentano's views on evidence and truth.

According to the late Brentano, every evident judgement is either immediately or mediately evident. The evidence of mediately evident judgements is conferred, ultimately, by judgements that are immediately evident or have 'inner correctness'. Now, the immediately evident judgements are of two sorts: (a) judgements of 'inner perception', and (b) judgements of reason or insight (which greatly resemble analytical judgements in the sense of being evident by virtue of the meanings of the terms occurring in the judgement). The late Brentano also claims that every evident judgement is correct, but not every correct judgement is evident, and he calls a judgement which is correct but not evident, a blind judgement. Examples of blind judgements are: a) judgements of 'outer perception', and b) judgements based on memory.

Finally, correctness is characterized by reference to evidence: Correctness pertains to the judgement of a person who judges validly . . . hence it pertains to the judgement of one who asserts what the person who judges on the base of an actual (valid) demonstration, i.e. with evidence, would assert. Hence, the evidence of a judgement entails its correctness.[52]

(*Note*: the terminology of these explanations has been adapted to the terminology of constructive type theory. In particular, Brentano's terms 'direct' and 'indirect' have been replaced by 'immediate' and 'mediate' resp., 'truth' and 'true' have been replaced by 'correctness' and 'correct' resp., and 'correct' has been replaced by 'valid'.)

(2) Note on the notion of formal demonstration or formal proof (derivation) in modern textbooks of mathematical logic

The notion of formal proof (derivation) in modern textbooks of mathematical logic is arrived at by

i) formalistically interpreting what is meant by an immediate demonstration or inference
ii) forgetting about the difference between judgement and proposition
iii) interpreting the notion of proposition formalistically as (well-formed) formula

From the point of view of constructive type theory, the truth-maker theory is a theory of truth of propositions (Martin-Löf writes in this context about the correspondence theory of truth, which does not seem entirely appropriate). (cf. the previous Subsubsect. ii.3.4, part *Truth*) But w.r.t. judgements, something like a correctness-maker theory of judgements does not work. What is suitable in this case is rather a coherence notion of correctness, where, by 'incoherence' is meant the existence of some kind of inconsistency.

ii.5 *Judgement as Knowledge, and Knowability*

ii.5.1 *Judgement as Knowledge*

Martin-Löf once remarked that when dealing with such basic notions such as judgement or demonstration, there is no way of reducing them to any other even more basic notions; one has to content oneself with the fact that these notions are the same as certain other notions and that there can only be a way of clarifying the former ones by means of the latter ones or of revealing the structure into which these notions fit. This is what the following explanations are all about.

As was mentioned at the beginning of the previous Subsect. ii.4, 'judgement' can be understood as an act of judging or as that which is judged, i.e. the object or result of an act of judging. Now, if to know is considered to be the same notion as to have judged, written, to have judged = to know, then the distinction concerning the two meanings of 'judgement' is transferred onto 'knowledge':

the act of judging	=	the act of knowing, or more precisely, of getting to know
that which is judged (the object or result of judging)	=	that which is known (the piece or object of knowledge)

'Knowledge' is of course a prim. concept which, as has been explained above, can be clarified but certainly not reduced to any other concepts. Moreover, 'to know' is a success word whereas 'to have judged' is not; if 'to know' is taken in a Husserlian sense, then it is a relative success word and 'to have judged' can be regarded as being one as well.

C, n & o
This identification of the two notions of judgement and knowledge explains the frequent use which was made of 'to know', 'knowable' or 'known' in the preceding two subsections.

Parallel to the verification principle of meaning according to which a proposition is defined by its truth conditions, intuitionistically elaborated into proof (verification) conditions, one could speak of a validation principle of meaning in the case of judgements, according to which a judgement is defined by its assertion conditions:

Validation principle of meaning

A judgement is defined by laying down what it means to make the judgement (or alternatively, by laying down what one has to know in order to have the right to make the judgement).

C, n & o
Since this def. is independent of the actual object of the judgement and merely concerns the form of the judgement, the principle above actually says:

A judgement as an instance of a form of judgement is defined by what it means to make a judgement of that form.

And parallel to the verification principle of truth according to which a proposition is true if there exists a proof of it, one could speak of a validation principle of correctness in the case of judgements:

Validation principle of correctness

A judgement is correct	if	there exists a (valid) demonstration of it, i.e.
	if	it can be (validly) demonstrated

The notion of judgement defined by the validation principle of meaning is just one notion of judgement, namely the one of bare judgement or judgement simpliciter. But as a matter of fact, 3 notions of judgement can be distinguished:

(1) bare judgement or judgement simpliciter
    A judgement is defined by its assertion-conditions.
(2) actual judgement
    an actual judgement = a judgement which has been made
(3) potential judgement
    a potential judgement = a judgement that can be made

The conceptual order of priority is as in so many of the earlier analogous cases: simplicity precedes actuality precedes potentiality in the same order.

It is certainly the case that one cannot know something except as the result of having gotten to know it. Hence, if demonstration is the very act or process of getting to know something, one cannot know something except as the result of having demonstrated it. This means that knowledge is by its very nature demonstrative knowledge. These explanations and observations concerning judgement, knowledge and demonstration provide moreover the framework for answers to traditional epistemological questions:

1) What do you know? A judgement.
2) How do you know it? Through a demonstration.

C, n & o
(Historical) The notion of judgement as knowledge is at least implicitly due to Kant. Explicitly, the notions of judgement and knowledge were connected by Bolzano who simply defined knowledge as (objectively) correct judgement.

ii.5.2 *Knowability*

According to the chain of equalities

$$\text{correct} = \text{potentially correct} = \text{knowable} = \text{demonstrable}/\ldots$$

presented in subsubsect. ii.4.1, correctness of a judgement is knowability. This can be expressed by a

Principle of knowability

A judgement is correct iff it can be known (it is knowable/ it is possible to know it)

If this principle is applied to particular sorts of judgement, it yields

i) If a judgement of the form $A : set$ is correct then A can be known to be a set;
ii) If a judgement of the form A *true* is correct then A can be known to be true (where A : *prop*);

etc.

This principle of knowability allows us to justify some laws of knowability, namely:
1. Law of knowability: Absolute consistency

The judgement '$\perp$ *true*' is unknowable (cannot be known to be true).

Justification
To know that A is true, for A : *prop*, one has to know a proof-object of A. Now, $\perp$: *prop* was really defined s.t. it has no proof-objects. Thus, the judgement $\perp$ *true* is incorrect and hence cannot be known. □

Corollary: Law of contradiction

There is no pair of judgements 'A *true*' and 'A *false*' s.t. 'A *true*' is knowable and 'A *false*' is knowable.

Justification
Straightforward. □

C, n & o
Note the difference between this 'metaregulative' law of contradiction and the logical laws in the language of constructive type theory also called laws of contradiction, i.e. $A\&{\sim}A$ *false*, or, ${\sim}(A\&{\sim}A)$ *true*.

2. Law of knowability: Unknowability of truth entails knowability of falsity

> That the judgement 'A *true*' is unknowable entails that the judgement 'A *false*' is knowable.

Justification
When the judgement 'A *true*' is unknowable, then it is incorrect, i.e. the proposition A has no proof-object. Then the judgement '$r : (A)\bot$' is vacuously correct whereby r serves as a disproof object of A. Whence the judgement 'A *false*' is correct and 'A *false*' is knowable. □

Corollary: Law of excluded middle

> There is no proposition A s.t. neither is the judgement 'A *true*' knowable nor is the judgement 'A *false*' knowable.
> In other words: There are no absolutely undecidable propositions.

Justification
E.g. if the judgement 'A *true*' is unknowable, then the judgement 'A *false*' is knowable.

This corollary can alternatively be justified in a more roundabout way, following Brouwer's justification of the judgement ${\sim}{\sim}(A \vee {\sim}A)$ *true*:

A is an absolutely undecidable proposition	iff	'$A \vee {\sim}A$ *true*' is not knowable
	iff	'$A \vee {\sim}A$ *false*' is knowable/ correct

If '$A \vee {\sim}A$ *false*' is correct then '${\sim}A$ *true*' is correct (since if 'A *true*' is correct '$A \vee {\sim}A$ *true*' is correct), and if '$A \vee {\sim}A$ *false*' is correct, then '${\sim}A$ *false*' is correct (since if 'A *false*' is correct then '$A \vee {\sim}A$ *true*' is correct), hence '${\sim}A$ *true*' and '${\sim}A$ *false*' are both correct judgements which contradicts the law of contradiction above.[53] □

C, n & o
(1) For Brentano, logic is the 'theory of correct judgement'. It is based on general 'apodictic' principles about the correctness of judgements. Now, Brentano divides the judgements into two exclusive classes, namely the affirmative judgements and the negative judgements. Some of the basic principles of the theory of correct judgements are the following:

(i) It is impossible for there to be anything which is s.t. it may be correctly affirmed and also not correctly affirmed.

(ii) It is impossible for there to be anything which is s.t. it may be correctly denied and also not correctly denied.
(iii) It is impossible for there to be anything which is s.t. it may be neither correctly affirmed nor not correctly affirmed.
(iv) It is impossible for there to be anything which is s.t. it may be neither correctly denied nor not correctly denied.

According to Chisholm, principles (i) and (ii) constitute the ground of the law of contradiction, and principles (iii) and (iv) constitute the ground of the law of the excluded middle. If one abstracts from Brentano's traditional division of the judgements into affirmative and negative judgements and if one invokes the principle of knowability introduced above, then the corollaries of the law of contradiction and of the law of the excluded middle above are nothing but Brentano's principles (i)/(ii) and (iii)/(iv) resp.[54]

(2) Boolos defines by means of provability logic extremely undecidable propositions as follows:[55] Let

$\phi, \psi, \chi, \ldots$	be variables ranging over fcts. from the propositional variables $p_0, p_1, \ldots, p_n, \ldots$ of propositional modal logic to propositions (or rather closed wffs) of PA
$A, B, C, \ldots$ (in this note)	be variables ranging over propositions (or wffs) of propositional modal logic.

A proposition (closed wff) S of PA is undecidable in PA if for any ϕ, $\nvdash_{PA} \phi(p_0)$ and $\nvdash_{PA} \sim\phi(p_0)$ where $S = \phi(p_0)$.

A proposition (closed wff) S of PA is extremely undecidable, if for all A containing no propositional variables other than p_0 if for some ψ, $\nvdash_{PA} \psi(A)$, then for any ϕ, $\nvdash_{PA} \phi(A)$, where $S = \phi(p_0)$.

Roughly speaking, a proposition of PA is said to be extremely undecidable, if it can be proved to have only those modal-logically characterizable properties that every undecidable proposition of PA can be proved to have.

Boolos' main result is that there exist infinitely many inequivalent extremely undecidable propositions of PA. To dwell on the obvious, these extremely undecidable propositions of PA are extremely undecidable in PA, but not absolutely undecidable altogether. As a matter of fact, they can be decided in a simple extension of PA such as PA+Reflection. Thus, there is no conflict between what has been said before and Boolos' result.

ii.6 *Various Readings of the Forms of Judgement*

The simplest forms of categorical judgement are $\alpha : type$, $\alpha = \beta : type$, $a : \alpha$ and $a = b : \alpha$. In the sequel the various readings of only these forms of judgement will be given, and as a matter of fact only of the first and third forms among them, as those are readily transferred onto the second and fourth forms. Moreover, the readings of the categorical forms of judgement can easily also be extended to the hypothetical forms of judgement.

Each of the following forms of judgement admits of different readings in constructive type theory:

$A : \alpha$	$a : A$	A *exists*
(i) A is a set	a is an element of set A	A is non-empty or inhabited
(ii) A is a proposition	a is a proof (verification) of proposition A	A is true
(iii) A is an intention (expectation)	a is a method of fulfilling (realizing) the intention (expectation) A	A is fulfillable(realizable)
(iv) A is a problem (task)	a is a method of solving the problem (doing the task) A	A is solvable
(v) A is a specification	a is a program that satisfies the specification A	A is programmable
(vi) A is a game (in the sense of Game-Theoretical-Semantics GTS)	a is an effective winning strategy of Myself in A	A is effectively winnable

and the reading in standard presentations of first order predicate calculus is

A is a formula	a is an individual term if A is chosen to be the type of individuals	A is true

The equivalence of the readings (i) and (ii) is implicit in the Brouwer–Heyting interpretation of the logical operators and was made explicit in the Curry–Howard isomorphism. (cf. Sect. xi. Chapt. 1).

The equivalence of the readings (ii) and (iii) was shown by Heyting in his (1931).

The equivalence of the readings (ii) and (iv) is due to Kolmogorov and was demonstrated by him in his (1932).

The equivalence of the readings (i), (ii) and (v) is mentioned for the first time by Martin-Löf in his version (6) of constructive type theory.

The equivalence of the readings (i), (ii) and (vi) is due to Ranta and explained by him in his (1988).

Note that there is no equivalence between any of (i) – (vi) and the reading in standard presentations of mathematical logic.

iii TRADITIONAL METAMATHEMATICAL ISSUES: CONSISTENCY, COMPLETENESS, DECIDABILITY, IN A NEW FRAMEWORK

This section confronts the metamathematical treatment of three 'classical' metalogical issues: consistency, completeness and decidability, with a new, so-called direct treatment of these issues. In each of these cases it is shown that the metamathematical treatment ultimately presupposes a direct one.

iii.1 *Consistency*

iii.1.1 *Metamathematical consistency*

To prove mathematically the consistency of a theory T, one considers another theory T′ which contains

i) codes for the propositions of T
ii) a predicate $Der(x)$ s.t. $Der(\ulcorner A \urcorner)$ expresses the fact that the proposition A of T with code $\ulcorner A \urcorner$ in T′ is derivable in T

Then consistency of T is defined as follows:

$$Cons =_{\text{def}} \sim Der(\ulcorner \bot \urcorner) =_{\text{def}} Der(\ulcorner \bot \urcorner) \supset \bot$$

and it is proved by showing that the proposition $Cons$ is true in T′.

C, n & o
This way of proceeding seems to be necessary or unavoidable when there is no hope or belief that a semantical justification of the axioms or rules of inference of a theory T is possible.

iii.1.2 *Direct consistency*

Direct consistency of a theory T means that the judgement $\bot$ *true* is not demonstrable in T, or that it will never be possible to (correctly) judge $\bot$ *true* by means of the axioms or rules of inference of T.

Explanation
If the judgement $\bot$ *true* can be derived in T, then there is a proof-object c s.t. $c : \bot$; and then c can be evaluated to a canonical proof-object $d : \bot$. But this is impossible, since the proposition $\bot$ has by its real def. no canonical proof-objects (no introduction rules). Thus, it is not possible to judge $\bot$ *true* by means of the axioms or rules of inference of T if these axioms or rules of inference are correct or valid resp. Hence, if one hit upon a demonstration of $\bot$ *true* in T, one would know that either (i) there is a formal error or mistake in the demonstration, or (ii) in the case of a formally correct demonstration of $\bot$ *true* in T, at least one of the axioms or rules of inference of T is not correct resp. valid.

Example Reflecting on the meaning of each of the rules of constructive type theory one convinces oneself that they are valid (cf. the justification of the various rules). Therefore it will never be possible to find a formally correct demonstration of $\bot$ *true* in constructive type theory, which means that constructive type theory is directly consistent.

C, n & o
This means that the direct consistency problem for a theory T is really the problem of the correctness and validity of the axioms and rules of inference resp. of T.

iii.1.3 *On the relationship between metamathematical consistency and direct consistency*

In order to obtain the direct consistency of a theory T from its metamathematical consistency, one has to rely on the direct consistency of the theory T′. Once $c : Cons$ is demonstrated in T′, for some proof-object c, one must argue as follows: if T were not directly consistent, one would have a demonstration of $\perp$ *true* in T, i.e. a judgement $a : \perp$, for some proof-object a. Codifying this judgement in T′ yields $\ulcorner a \urcorner : Der(\ulcorner \perp \urcorner)$. Then, $\supset E_G(Der(\ulcorner \perp \urcorner), \perp, c, \ulcorner a \urcorner) : \perp$ and $\perp$ *true* would be correct judgements of T′. To conclude that $\perp$ *true* is not demonstrable in T, one has to be convinced that $\perp$ *true* is not demonstrable in T′, i.e. that T′ is directly consistent.

The same link that was observed between direct consistency and the validity of the rules of inference exists also between metamathematical consistency and validity: Suppose one proves in a theory T′ the consistency of a theory T in a metamathematical way. That consistency proof relies on the validity of the axioms and rules of inference of T′ applied in the proof. Whence the proof of metamathematical consistency also relies on the validity of certain rules of inference.

iii.2 *Validity and Completeness*

iii.2.1 *Metamathematical validity and completeness (model theory)*

A first order language $\mathcal{L}$ has the following symbols

a) the variables $x, y, z, x', y', z', x'', y'', z'', \ldots$
b) for each $n \in \omega$, n-ary fct. symbols $f, \ldots$
c) for each $n \in \omega$, n-ary predicate symbols $P, \ldots$
d) the symbols $\sim, \vee, \exists$

A structure $\mathcal{A}$ for $\mathcal{L}$ consists of the following items

i) a non-empty set $|\mathcal{A}|$ called the domain or universe of $\mathcal{A}$. The elements of $|\mathcal{A}|$ are called the individuals of $\mathcal{A}$.
ii) for each n-ary fct. symbol f of $\mathcal{L}$, an n-ary fct. $f_{\mathcal{A}} : |\mathcal{A}|^n \to |\mathcal{A}|$ (in particular for each constant e of $\mathcal{L}$, $e_{\mathcal{A}}$ is an individual of $\mathcal{A}$)
iii) for each n-ary predicate symbol P of $\mathcal{L}$ other than $=$, an n-ary predicate $P_{\mathcal{A}} \subseteq |\mathcal{A}|^n$.

For each individual a of $\mathcal{A}$ choose a new constant, called the name of a, whereby it is understood that different individuals have different names. The first order language obtained from $\mathcal{L}$ by adding all the names of individuals of $\mathcal{A}$ is designated by $\mathcal{L}(\mathcal{A})$. Let i, j be variables for those names.

Suppose the formulas of $\mathcal{L}$ are inductively defined as usual. It will then be defined what it means for a formula of $\mathcal{L}$ to be true:

For every term t of $\mathcal{L}$, let $\mathcal{A}(t) := [\![t]\!]$ be defined as follows:

$$\begin{aligned}
&[\![e]\!] = e_{\mathcal{A}} \\
&[\![f(t_1,\ldots,t_n)]\!] = [\![f]\!]([\![t_1]\!],\ldots,[\![t_n]\!]) \\
&\qquad = f_{\mathcal{A}}([\![t_1]\!],\ldots [\![t_n]\!])
\end{aligned}$$

Now, lets define the truth $\mathcal{A}(A) = [\![A]\!]$ of A, for every formula A of $\mathcal{L}$ (Tarski's truth def.):

$$\begin{aligned}
[\![P(t_1,\ldots,t_n)]\!] &= [\![P]\!]([\![t_1]\!],\ldots,[\![t_n]\!]) \\
&= P_{\mathcal{A}}([\![t_1]\!],\ldots,[\![t_n]\!]) \\
[\![t_1 = t_2]\!] &= [\![t_1]\!] \text{ equals } [\![t_2]\!] \\
[\![A \vee B]\!] &= [\![A]\!] \text{ or } [\![B]\!] \\
[\![\sim A]\!] &= \text{not } [\![A]\!] \\
[\![(\exists x)A]\!] &= \text{for some } i,\ [\![A(i/x)]\!]
\end{aligned}$$

Tarski read $[\![A]\!]$ as A is true, i.e. $[\![A]\!] = \mathit{True}(A)$

If A is a formula of $\mathcal{L}$, an $\mathcal{A}$-instance of A is a closed formula of the form $A(i_1,\ldots,i_n)$ of $\mathcal{L}(\mathcal{A})$.

- A formula A of $\mathcal{L}$ is said to be valid in $\mathcal{A}$ if $[\![A']\!]$, for every $\mathcal{A}$-instance A' of A.
- A rule of inference is called valid in $\mathcal{A}$ if its conclusion is valid in $\mathcal{A}$ provided its premisses are valid in $\mathcal{A}$.

A first order theory, or simply a theory, is a formal system T s.t.

(i) the language of T is a first order language $\mathcal{L}$
(ii) the rules of inference are the logical rules of inference $\mathcal{L}$(T) and certain further rules of inference, the non-logical rules of inference

Obviously, a theory T here is always a natural deduction style formalization of what is commonly called a theory. The logical rules of inference are readily seen to be valid in every structure for $\mathcal{L}$.

Let a structure $\mathcal{A}$ for $\mathcal{L}$(T), in which all the non logical rules of inference of T are valid, be called a model of T.

A formula A of $\mathcal{L}$ is said to be valid in T if it is valid in every model $\mathcal{A}$ of T.

The Validity theorem

If T is a theory, then every formula of $\mathcal{L}$ derivable by means of the rules of inference of T is valid in T.

The Completeness theorem

If T is a theory, then every formula of $\mathcal{L}$ valid in T is derivable by means of the rules of inference of T.

iii.2.2 *Direct validity and direct completeness (direct semantics)*

The language L of constructive set theory has the following symbols

a) the variables (of some type) $x, y, z, w, \ldots\ f, g, h, \ldots, X, Y, Z, \ldots$
b) the set constants: N_n, N, +, Π, $\rightarrow$, Σ, $\times$, I, U,...

c) the element constants: $N_nIm, 0 \leq m < n, NI1, NI2, +I1, +I2, \Pi I, \rightarrow I, \Sigma I, \times I, II, UI1$–$UI7, \ldots, N_nE, NE, +E, \Pi E, \rightarrow E, \Sigma E, \times E, IE, UE, \ldots$

The meanings of all these symbols are directly explained in the meaning explanations (or real definitions) of all the judgements of the form $A : set$, $A = B : set$, $a : A$ and $a = b : A$ (provided $A : set$), where A and a are constructed from set constants and from element constants resp. in such a way as is indicated by the formation and associated definitional identity rules for all these constants resp. (the same holds more generally for the hypothetical forms of judgement of L).

Observe that as a new constant is introduced into constructive set theory, its meaning is directly explained (i.e. there is no division of formation rules and interpretation rules as is the case with model theory).

A (formation or def. identity) rule of constructive set theory is said to be valid iff its conclusion judgement is analytical w.r.t. its premiss judgements, i.e. if the conclusion is evident on the assumption that the premisses are evident.

Since constructive type theory as it is considered here is what has been called the logic of analytic judgements, truth does not enter and hence no rules for truth are needed.

All the formation and def. identity rules associated with the constants N_n, N_nIm, $0 \leq m < n$, N_nE, N, $NI1$, $NI2$, NE, $+$, $+I1, \ldots, U, UI1, \ldots, UI7, UE, \ldots$ are readily seen to be valid in the sense just explained.

Martin-Löf's validity thesis

Every judgement of L which is derived by means of the formation and def. identity rules associated with the constants $N_n, N_nIm, N_nE, \ldots, U, UI1, \ldots, UI7, UE, \ldots$ is evident and hence correct, i.e. constructive set theory is valid.

Martin-Löf's completeness thesis

Every correct judgement of L is derivable by means of the formation and def. identity rules associated with the constants $N_n, N_nIm, N_nE, \ldots, U, UI1, \ldots, UI7, UE, \ldots$, i.e. constructive set theory is complete.

C, n & o

(1) Martin-Löf's two theses cannot be proved mathematically, because (as is the case with Church's thesis) they identify a precise formal notion with an informal one, namely derivability in constructive set theory with the correctness of judgements of L. The theses are warranted by a lot of reasons and examples. But if anything, they can only be refuted by means of counterexamples. The 'real' thesis is of course the completeness thesis, whereas the validity thesis is substantiated by the meaning theory.

(2) An extension of constructive set theory by the addition of new constant and def. identity rules is conservative since, by Martin-Löf's completeness thesis, constructive set theory is already complete.

(3) In a rather trivial sense constructive set theory can be said to be incomplete: it can always be extended by new concepts (constants) and formation and def. identity rules governing them. But if this extension is carried out accurately, the extended

constructive set theory can again be called complete (in the sense of Martin-Löf's completeness thesis).

(4) After some notational and terminological changes, the above explanations apply of course equally well to constructive logic by the Curry–Howard isomorphism.

iii.2.3 *On the relationship between metamathematical semantics (model theory) and direct semantics*

The modelling of a first order language $\mathcal{L}$ described in Subsubsect. iii.2.1 is a way of giving meaning to (interpreting) the symbols of $\mathcal{L}$. Modelling not only applies to first order, but to a wide range of formal languages.

In model theory the structures $\mathcal{A}$ are studied from a linguistic point of view, i.e. the language in which $\mathcal{A}$ is expressed is the object of inquiry, whereas in mainstream mathematics (e.g. algebra, geometry, number theory) the structures $\mathcal{A}$ are studied from an object-oriented or 'natural' point of view. (cf. Sect. viii., Chapt. 1)

Now, whereas for the model theorist as well as for the mainstream mathematician a structure $\mathcal{A}$ is the main focus of attention (although approached from a different perspective), a proof theorist focuses mainly on the language which expresses a structure $\mathcal{A}$ (in other terms, a model theorist is more interested in the semantical side of a formal language expressing a structure, whereas a proof theorist is more interested in the syntactical side of such a language). But then, what is modelling in model theory actually becomes a translation from one language into another language in proof theory, and modelling and translation are one and the same thing, depending on whether one adopts a more semantical or a more syntactical attitude towards interpretation.

Although modelling or translating is a suitable way to proceed in many situations, it is obvious that one cannot always give meaning to a formal language in this way, i.e. one cannot indefinitely translate from a language $\mathcal{L}$ to another language $\mathcal{L}'$ and again to another language $\mathcal{L}''$, etc. in order to give meaning to language $\mathcal{L}$. Eventually, one needs a language which is not given meaning by translating it into yet another language, but which has to be given meaning in some other, namely in a direct way. This means that the symbols of that language have to be directly semantically explained. And this is exactly the procedure of constructive set theory. Thus, the metamathematical semantics is ultimately based on a direct semantics.[56]

iii.3 *Decidability*

iii.3.1 *Metamathematical decidability*

Let T be a theory with language $\mathcal{L}$, and let $Der_T(x)$ be the predicate true of Gödel numbers (codes) of formulas of $\mathcal{L}$ derivable in T. It is readily shown that T has a decision method iff $Der_T(x)$ is calculable.

A theory T is said to be decidable if its predicate $Der_T(x)$ is rec. O.w. T is said to be undecidable.

Then, by Church's thesis, T is decidable iff T has a decision method.

iii.3.2 *Direct decidability*

Constructive set theory is said to be decidable if for any analytic judgement of L it can be checked or decided whether or not it is derivable by means of the (set or element) formation rules of L.

The method for checking or deciding the derivability or well-formedness of a judgement of L of one of the two forms

$$x_1 : A_1, \ldots, x_n : A_n \vdash A : set$$
$$x_1 : A_1, \ldots, x_n : A_n \vdash a : A$$

is sometimes called (by computer scientists) the method of type (or set) checking.

The method for checking or deciding the derivability or well-formedness of a judgement of L of one of the two forms

$$x_1 : A_1, \ldots, x_n : A_n \vdash A = B : set$$
$$x_1 : A_1, \ldots, x_n : A_n \vdash a = b : A$$

can analogously be called the method of definitional identity checking.

C, n & o

(1) The method of type (or set) checking and the method of def. identity checking together form the formal or logical heart of computer proof systems.

(2) Because of the rule $\dfrac{a : A \quad A = B : set}{a : B}$ of constructive set theory (i.e. the rule (id. of types 1) applied to sets), type (or set) checking is inextricably linked with def. identity checking.

In version (7) of constructive type theory, Martin-Löf sketched such a method of type (or set) checking and a method of def. identity checking for terms in $\beta\eta$-normal form. They will here be presented together in a slightly more formal way (which borrows terminology and notation from Magnusson's Ph.D. thesis (1994). *Note*: There's no question of beginning the method by reducing an arbitrary element-expression to $\beta\eta$-normal form, since that reduction would not be guaranteed to terminate.

Let a term (i.e. an element-expression) be in $\beta\eta$-normal form, i.e. in the form

$$(x_1) \ldots (x_m)c(b_1) \ldots (b_n)$$

where

either c is a variable which is assigned a type (set) in the given context
or c is a constant whose type (set) is known outright from the axiom associated with it.

i) Check whether or not the judgement

$$(x_1) \ldots (x_m)c(b_1) \ldots (b_n) : (x_1 : A_1) \ldots (x_p : A_p)\genfrac{}{}{0pt}{}{set}{A} \quad \text{is well-formed}$$

where $(x_1 : A_1) \ldots (x_p : A_p)\genfrac{}{}{0pt}{}{set}{A}$ stands for $(x_1 : A_1) \ldots (x_p : A_p)set$ or $(x_1 : A_1) \ldots (x_p : A_p)A$ where $x_1 : A_1, \ldots, x_p : A_p \vdash A : set$

$$\frac{IC(m, p) \Longrightarrow \xi_1 \quad TC(c(b_1) \ldots (b_n), \alpha, [x_1 : A_1, \ldots, x_p : A_p]) \Longrightarrow \xi_2}{TC((x_1) \ldots (x_m)c(b_1) \ldots (b_n), (x_1 : A_1) \ldots (x_p : A_p)\alpha, [\,]) \Longrightarrow \xi_1 \times \xi_2}$$

where $IC(m, p)$ means check whether or not $m = p$, $\xi_i \in \{Succ, Fail\}$, $i = 1, 2$, $\alpha \in \{set, A\}$, and $\xi_1 \times \xi_2$ is here defined as follows:

$$\xi_1 \times \xi_2 = \begin{cases} Fail & \text{if} \quad \xi_i = Fail, \text{ for } i = 1, 2 \\ Succ & \text{o.w.} \end{cases}$$

ii) Check whether or not the judgement

$$c(b_1)\dots(b_n) : \overset{set}{\underset{A}{}} \text{ is well-formed}$$

Let $c : (y_1 : B_1)\dots(y_q : B_q)^{set}_{B}$ where $y_1 : B_1, \dots, y_q : B_q \vdash B : set$.

$$\frac{\begin{array}{l} TC(c, (y_1 : B_1)\dots(y_q : B_q)set, [\,]) \Longrightarrow \xi_1 \\ IC(n, q) \Longrightarrow \xi_2 \\ TC(b_1, B_1, [\,]) \Longrightarrow \xi_3 \\ \vdots \\ TC(b_n, B_q(b_1, \dots, b_{n-1}/y_1, \dots, y_{n-1}), [\,]) \Longrightarrow \xi_{q+2} \end{array}}{TC(c(b_1)\dots(b_n), set, [x_1 : A_1, \dots, x_p : A_p]) \Longrightarrow \xi_1 \times \dots \times \xi_{q+2}}$$

and

$$\frac{\begin{array}{l} TC(c, (y_1 : B_1)\dots(y_q : B_q)B, [\,]) \Longrightarrow \xi_1 \\ IC(n, q) \Longrightarrow \xi_2 \\ TC(b_1, B_1, [\,]) \Longrightarrow \xi_3 \\ \vdots \\ TC(b_n, B_q(b_1, \dots, b_{n-1}/y_1, \dots, y_{n-1}), [\,]) \Longrightarrow \xi_{q+2} \\ TConv(A, B(b_1, \dots, b_n/y_1, \dots, y_n), [\,]) \Longrightarrow \xi_{q+3} \end{array}}{TC(c(b_1)\dots(b_n), A, [x_1 : A_1, \dots, x_p : A_p]) \Longrightarrow \xi_1 \times \dots \times \xi_{q+3}}$$

iii) That the judgement

$$c : (y_1 : B_1)\dots(y_q : B_q)\overset{set}{\underset{B}{}} \text{ is well-formed}$$

is easily checked since

either c is a variable, in which case its type can be looked up in the context in which one finds oneself at the moment

or c is a constant in which case its type is known outright

(Note: ii) is also the point where conversion comes in.)

iv) To check def. identity judgements (type (or set) or term (or element) identity) reduce the two members of the identity to normal form.

v) Check whether or not the normal forms of the two members of this identity are α-convertible. Proceed in a lazy manner.

iii.3.3 *On the relationship between metamathematical decidability and direct decidability*

Essentially, there seems to be no difference between metamathematical decidability and direct decidability: both approaches define a (or the) theory to be decidable if there exists a decision method (or decision methods) for the resp. theory.

But whereas decision methods are directly presented for constructive set theory, the metamathematical proof of the decidability /undecidability of a theory T proceeds via an arithmetization of the syntax of $\mathcal{L}$ of T. Of course, ultimately metamathematical decidability is, by Church's thesis, also based on a more intuitive or direct notion of decidability.

iv CONSTRUCTIVE TYPE THEORY AND THE TRADITIONAL POINTS OF VIEW ON THE FOUNDATIONS OF MATHEMATICS

The following section purports to present a few main theses of three main movements in the philosophy of mathematics, namely logicism, formalism and intuitionism. All of these are subjected to a critical evaluation from the point of view of constructive type theory.

iv.1 *Logicism (Frege, Russell/Whitehead, Church, Quine)*

i) Pure mathematics can be reduced to logic – and in particular arithmetic can be reduced to logic.

To verify this claim (a) it has to be shown that pure mathematics only operates with concepts which can be explicitly defined by means of a small number of fundamental logical concepts, and (b) it has to be shown that all sentences of pure mathematics can be logically deduced from a small number of logical principles. The mathematical sentences are thus analytic sentences (in Frege's sense).

Critical evaluation

1) Most logicists admit that a deductively hereditary property of 'being logical' or 'having a logical character' does not exist.

Moreover, since the discovery of the logical antinomies every logicist system had to assume at least one axiom which is not a logically evident principle, but could only be justified pragmatically. Hence no existing logicist system is based on a set of axioms, all of which can immediately be seen to be of a logical character.

2) Arithmetic cannot be reduced to logic unless the principles of set theory involved count as logical principles. But if logic consists of the necessary principles of all coherent reasoning, it is evident that the stronger principles of set theory are not logical and it seems even doubtful, that the weaker principles can be considered to be logical. Hence to choose the logical notions as the basic ones does not work; if anything is basic, it is rather the set-theoretical notions (cf. C, n & o (2) to vi.1), Chapt. 1).

3) To draw a sharp dividing line between logical and non-logical notions seems rather artificial. Moreover, the Curry–Howard isomorphism (i.e. the correspondence

between logical and set-theoretical notions cf. Sect. xi. Chapt. 1), shows that this kind of reductionist thinking of logicism is obsolete, has lost its interest.

ii) Transfinite concepts are admitted and hence have an 'ontological' status (transfinitism).

Critical evaluation

Frege and Russell have in an uncritical and naive way almost completely integrated Cantor's transfinite mathematics into their logicist systems. Frege's system was soon discovered to lead to logical antinomies, and so of course was Cantor's set theory. Logicism lacks any real philosophy of infinity. (Carnap seems to contradict this criticism with his claim that the 'constructivist' method is part of logicism.[57])

iii) Logic is conceived of in an antipsychological, objectivist way.

Frege makes a clear separation between

psychological	logical
subjective	objective
act	object

and the logicist is to treat logic on the right side, avoiding and excluding any reference to anything on the left side. Frege and Husserl showed that the intrusion of empirical psychology into logic was an error.

Critical evaluation

On the one hand it is not possible to ban the subjective or the act entirely from logic. If one considers the notion of inference as one of the most basic notions of logic (cf. Sect. ii.1 of this Chapter), then inference contains a reference to someone demonstrating a judgement or to someone judging, etc. (cf. ii.4.2 above).

The subjective manifests itself in logic also in various kinds of evidence: systematic evidence, i.e. evidence concerning the choice of axioms and rules of inference; structural evidence, i.e. evidence concerning the equivalence of definiendum and definiens in a definition; symbolic evidence, i.e. evidence concerning the correctness of a formal deduction; metatheoretical evidence, i.e. evidence concerning certain ways of intuitive reasoning used in the metatheory [58] (cf. Subsect. ii.5).

On the other hand Frege and Husserl were right insofar as logic indeed does not rely on empirical psychology. Hence acts of inference, of demonstration, of judgement, etc. have to be explained in a non-empirically psychological way. This could perhaps take place in a logical or pure psychology such as Husserl conceived of it in his *Die Krisis der europäischen Wissenschaften und die transzendentale Phänomenologie* (1954).

iv) The logicist goal is a comprehensive formal system with a natural interpretation (interpreted formal system) (cf. Subsect. i.1, Chapt. 4).

iv.2 *Formalism (Hilbert, Curry, Gentzen, Gödel, Kleene, von Neumann, et al.)*

i) According to Hilbert one has to distinguish between formal mathematics, which consists of clearly delimited concrete objects, i.e. formalisms (= procedures for the formal or mechanical derivation of formulas) and metamathematics, which consists of an intuitive and finitary mathematical theory of formal mathematics. Now it is understood that, even though under a certain interpretation the formulas of formal mathematics become sentences of classical mathematics, formal mathematics is totally independent of this interpretation.

The relationship between classical, formal and metamathematics can be explained as follows: if part of the 'contentful' (inhaltlich) classical mathematics (called first order mathematics by Brouwer) is formalized, the result of this formalization procedure is formal mathematics; this formal mathematics can be investigated – total abstraction being made from this formalization procedure – by a 'contentful', intuitive, finitary mathematical theory (called second order mathematics by Brouwer). Now, second order mathematics is obviously part of first order mathematics, and their sentences in common are called real sentences by Hilbert. Those sentences, concepts or inferences, however, which are part of first but not of second order mathematics are called ideal concepts, ideal sentences or ideal inferences by Hilbert.

The finitary mathematical or metamathematical sentences are thus conceived as empirical sentences.

Note that if the primary content of mathematics is 'contentful' mathematical sentences, and that formulas and formalisms are but secondary means by means of which the preexisting 'contentful' mathematical theories are investigated, then formulas are formalized sentences and formalisms are formalized theories (moderate formalism).

i′) According to Curry mathematics is the science of formalisms, i.e. formalisms or formal mathematics is a substitute for classical 'contentful' mathematics and there is nothing next to or behind the formalisms that these formalisms could possibly formalize. Moreover, a proof of consistency of a formalism is neither necessary nor sufficient for the usefulness of the formalism (strict formalism).

Critical evaluation

Frege criticized that metamathematical concepts and operations are not empirical – and thus nor are metamathematical sentences – but rather idealizations of empirical concepts and operations. That is, the separation of sensible perception and mathematics does not start with the passage from elementary to classical arithmetic (or from second order to first order mathematics) only, but it starts at the very beginning of elementary arithmetic (or second order mathematics resp.).

ii) The first aim of a (moderate) formalist is to completely reconstruct 'contentful' classical mathematical theories as formalisms. The second aim of a (moderate) formalist is to provide a ('contentful') justification of the 'contentful' classical mathematical theories (1) by proving their consistency by means of a proof of the consistency of their corresponding formal mathematical theories (their formalizations); or (2) by reducing their corresponding formal mathematical theories to formalized finitary mathematical theories.

Critical evaluation

1) The formalist's aims or expectations were disappointed by Gödel's incompleteness thms: (1′) it is merely possible to work with formal systems embodying (and not even completely formalizing) only part of classical mathematics, and (2′) the proofs of consistency of these formalisms have to rely on more abstract but still constructive notions.

2) Brouwer maintained that the justification of classical 'contentful' mathematics by means of a metamathematical proof of its consistency contained a vicious circle, since such a justification makes use of ideal elements which are exactly in need of justification (namely the ideal sentence or principle of the excluded middle PEM).

iii) Transfinite concepts are admitted, but have no "ontological" status; they are simply useful for the purpose of simplifying and systematizing mathematical theories. And they can be admitted only insofar as the theories using them can be proved by constructive/finitary methods to be consistent (methodological transfinitism).

Critical evaluation

The way out of the troubles of formalism caused by Gödel's incompleteness theorems consisted in allowing transfinitary methods of inference to become part of metamathematics. But this obviously comes pretty close to abandoning methodological transfinitism.

iv) Logic is conceived as a minimal system of empirical sentences describing perceptable features of metamathematical constructions.

Critical evaluation

A formalist cannot only rely on empirical sentences (if they were empirical at all, cf. *critical evaluation* to i) above) of the kind 'a given construction of perceptable objects with features of perception k have ipso facto also features l', but she/he also presupposes that the construction is correct. But any judgement concerning the correctness of a metamathematical construction is based on logical principles. Thus, a formalist not only relies on empirical sentences of the kind just mentioned, but also on certain logical principles.

The only way to dispense with the logical principles consists of the assumption that the medium in which the metamathematical constructions take place is of such a special kind that they can immediately be described by general and necessary sentences without thereby raising the question of whether or not a particular construction is correct (e.g. the pure intuition of the intuitionists is such a medium).

v) Further contributions to formalism/Hilbert's program are:

- Gentzen's analysis of first order proofs and his improvement of the logical symbolism: the systems of natural deduction (cf. Subsect. ii.1., Chapt. 4)
- Gödel's attempt to replace Heyting's abstract notion of constructive proof by something more definite, less abstract (more finitist), i.e. by the notion of computable functional of finite type, thereby showing that formal proofs and functionals are in many ways similar (the *Dialectica*- interpretation)

- Prawitz's proof-theoretical analyses of Gentzen's systems of natural deduction (normalization) and his extension of these analyses to more comprehensive systems (cf. Subsect. ii.1, Chapt. 4)
- Howard's establishment of the Curry–Howard isomorphism (cf. Subsect. ii.2, Chapt. 4)

iv.3 *Intuitionism (Brouwer, Heyting, Kolmogorov)*

i) Mathematics is essentially a languageless mental activity whose origin lies in the perception of temporal change. Thus, two kinds of activities have to be distinguished sharply: on the one hand the activity of mathematical constructions, and on the other hand the linguistic activity of reporting about these constructions (i.e. sentences about the results of such constructions and the application of logical rules of inference to such sentences). The mathematical sentences are thus synthetic and *a priori* (non-empirical) sentences.

Brouwer makes a clear separation between

real	mental
external	internal
formal	contentual

and the intuitionist is to treat mathematics on the right side, avoiding and excluding any reference to anything on the left side.

Critical evaluation
On the one hand Brouwer makes the mistake of considering the thought or the mental to be the content of a linguistic expression instead of considering the objects as such (such as e.g. Frege did) (cf. Subsect. viii.1, Chapt. 1). On the other hand Brouwer is right in that mathematical objects are not just formulas, formal derivations or formalisms which are manipulated, but rather the contents or meanings of mathematical expressions.

ii) The objective of an intuitionist is to practice intuitionistic mathematics, i.e. to perceive and construct mathematical objects in pure intuition. Mathematical existence means effective constructability, i.e. only constructed objects have mathematical existence.

iii) Transfinite concepts are not admitted; only concepts of a potential infinite are admitted and have an 'ontological' status (moderate finitism).

iv) Logic is a (post factum) registration of the principles of inference applied in mathematical constructions. Logic is thus the logic of mathematics (cf. Heyting's and Kolmogorov's formalization of intuitionistic logic).

v) The logical operators have a particular semantical explanation, namely the Heyting–Kolmogorov explanation using proof-objects (also called Brouwer–Heyting–Kolmogorov or BHK-explanation by Troelstra/van Dalen) (cf. Sect. x., Chapt. 1).

vi) Every true mathematical proposition can be justified by a construction which is (a) an evident experience, and (b) no external or sensible perception. Mathematical truth is interpreted as verifiability or provability, i.e. a mathematical proposition is true if there is a proof of it. Furthermore, mathematical constructions are regarded by the intuitionists as intersubjective experiences and their evidence as being essentially part of these experiences. A lack of evidence manifests itself in differing reports on one and the same experience either concerning their content or concerning their evidence (cf. Sect. vii., Chapt. 1).

Critical evaluation

An argument based on contradictory accounts on apparently evident mathematical constructions has a very critical or disturbing effect on the safety of intuitionistic mathematics (an effect which has been compared to the one of the logical antinomies on naive set theory and classical mathematics).

Since there are different intuitionistic accounts on the experience of negation, it appears that no construction involving negation can claim to be evident. Likewise, since there are different accounts by the intuitionist and the strict finitist on the experience of potentially infinite sequences (concerning their evidence), again it appears that no construction involving such sequences can claim to be evident.[59]

NOTES

1 Martin-Löf (1972A).

2 Cf. the theorem on extensional equality of fcts., Nordström et al. (1990), pp. 84,85, stating that extensionally identical fcts. are identical in the sense of E.

3 Version (6), p. 62.

4 Martin-Löf (1972A), p. 102.

5 pp. 102,103.

6 p. 100.

7 Nordström et al. (1990), p. 67.

8 What has been called an expression or mode of containment Φ is just an abbreviatory formulation of the phrase 'a mode of containment of singular terms or sentences in singular terms or sentences'. Cf. Quine (1960), p. 144.

9 Martin-Löf (1972A), pp. 104,105.

10 Cf. the example in Nordström et al. (1990), p. 66.

11 Nordström et al. (1990), p. 19.

12 E.g. p. 61.

13 p. 61.

14 p. 61.

15 Thompson (1991), e.g. p. 166.

16 E.g. pp. 133, 166.

17 p. 119.

18 E.g. p. 118.

19 p. 110.

20 p. 174.

21 Kneale and Kneale (1978), p. 55.

22 p. 160.

23 pp. 160–163.

24 Broadie (1993), chapts. 8 and 7, resp.

25 Broadie (1993), pp. 100–109.

[26] Kneale and Kneale (1978), p. 489.
[27] Sluga (1980), pp. 83, 84.
[28] Dummett (1981), p. 316.
[29] Berg (1962), p. 64.
[30] For the historical evolution of the terms 'proposition', 'judgement', 'assertion', cf. generally Martin-Löf (1983A), pp. 205–211.
[31] Salvesen (1989), p. 62.
[32] p. 67.
[33] p. 68.
[34] p. 69.
[35] p. 72.
[36] p. 79.
[37] pp. 79, 80.
[38] p. 82.
[39] p. 71 and pp. 83, 84.
[40] pp. 84, 85.
[41] Sundholm (1990), pp. 60, 61, esp. p. 61.
[42] For more history on this point, cf. Note 30 in Sundholm (1994).
[43] Cf. Martin-Löf's short discussion of Lovejoy's principle of plenitude and the medieval Ab posse ad esse non valet consequentia, in Martin-Löf (1990), p. 143.
[44] E.g. in Dummett (1977).
[45] Prawitz (1980).
[46] Putnam (1981).
[47] All these explanations can be found almost literally in Sundholm (1994), p. 117 and pp. 121–123; note, however, that what Sundholm calls a 'general truth-maker analysis of truth' is here called a truth-maker theory of truth.
[48] Dummett (1976) and Prawitz (1987).
[49] About the need for further transcendental exploration of evidence, cf. Pietersma (1977), pp. 47–49.
[50] Føllesdal (1991), pp. 27, 28.
[51] Patzig (1973A).
[52] These explanations are based on Chisholm (1967), p. 367, and Chisholm (1982).
[53] This alternative justification was suggested to me by A.S. Troelstra.
[54] Cf. Chisholm (1982).
[55] Cf. Boolos (1982).
[56] Cf. Martin-Löf (1985).
[57] Cf. Carnap (1931).
[58] Stegmüller (1969), pp. 265–270.
[59] The main sources for Sect. iv. are Körner (1960) and Martin-Löf's version (8) of constructive type theory. Other auxiliary sources are Heyting (1934) and (1971), von Neumann (1931), Parsons (1967) and Prawitz (1971).

CHAPTER 3

HISTORY OF CONSTRUCTIVE TYPE THEORY (1970–95)

The history of constructive type theory is written along eight parameters; that is, it is written by following the historical development of eight of the most important features of this theory. These parameters are either different sorts of rules or else various more general features of type theory.

One parameter consists of the form of symbolic expressions or judgements of the theory (Sect. i), another one of the specific conception of type (Sect. ii), a third one of the kind of semantics which goes along with the theory (Sect. iii), a fourth one of the question as to whether the theory is monomorphic or polymorphic and whether type checking is decidable or undecidable (Sect. viii). All the other parameters regard different sorts of rules: a fifth parameter consists of the basic stock rules for Π, Σ, $+$, N_n, and N (Sect. vii), a sixth one of the rules for the universe(s) (Sect. vi), a seventh one of the rules for supplementary gadgets of the theory, i.e. for set-theoretic operations not belonging to the basic stock (Sect. iv). The eighth and last parameter concerns a certain sort of rule as well as a general feature of type theory, namely equality or identity relations (Sect. v).

i FORMS OF SYMBOLIC EXPRESSIONS/SENTENCES/JUDGEMENTS

The following little story concerning the change of terminology as well as the change of the actual forms of judgement used in the various versions of constructive type theory, means to convey an idea of the process of trial and error which eventually led up to the present version of type theory.

In versions (1) and (2) a formal system is set up whose derivations are certain tree-like arrangements of symbolic expressions of the form $a \in A$. When $a \in A$ has been derived, it may be read as 'a denotes an object whose type is denoted by A', or for short, 'a is a term with type symbol A'. $a \in A$ thus expresses the relation that a is an object of type A, or if one thinks of A as denoting a proposition rather than a type, it expresses the relation that a is a proof of the proposition A. Martin-Löf comments on this form of symbolic expression as follows: 'In intuitionistic mathematics the relation $a \in A$ is more fundamental than the assertion relation $\vdash A$ which merely expresses that A holds or is true, that is, that there is a proof a of the proposition A, because half, and, usually, the more valuable half, of the information contained in $a \in A$ is lost when passing to $\vdash A$. I think that $a \in A$ may be regarded as the general form of the theorem of intuitionistic mathematics.'[1] Only on the metatheoretical level does another form of expression, namely $a = b$, appear. This relation between two expressions a and b called definitional or intensional equality, expresses that a and b denote the same object.

The formal system in version (3) is based on rules of inference for the derivation of two forms of symbolic expressions, namely $a \in A$ and a *conv* b. Whereas the first form reads as in versions (1) and (2), the second one reads as a converts into b and its intended interpretation is that the expressions a and b are definitionally equal in the sense of versions (1) and (2). Thus, contrary to versions (1) and (2), in version (3) definitional or intensional equality has become part of the theory in the form of conversion. Now, the rules of inference whose conclusion is of the form a *conv* b (the rules of conversion) are such that it is straightforward that parallel to a derivation of a *conv* b there are derivations of $a \in A$ and $b \in B$ for some A and B. Hence, whereas the explicit second form of symbolic expression is a *conv* b a symbolic expression of the form $A \ni a$ *conv* $b \in B$ is always implicitly given with it.

In version (4) we have a multiplying of forms of symbolic expressions, now called forms of sentences, of version (3), namely the following ones: $x \in X$, X is a type, $\bar{a} \in \bar{A}$, $\bar{A}$ is a type, $\bar{A} \ni \bar{a} = \bar{b} \in \bar{B}$ and $\bar{A} = \bar{B}$. The first form of symbolic expression of (3) is now split in two forms of sentences $x \in X$ and $\bar{a} \in \bar{A}$ which read as follows: x is an object of type X, or, x is an element of the basic set X, or, x is a (canonical) proof of proposition X, and, a is an $\bar{A}$-valued function, or, the value of a is an object (proof) of the type (proposition) which is the value of A, resp. There are not only the objects anymore denoted by the terms $a, b, \ldots$ in versions (1)–(3), but they are now distinguished from (object-valued) functions which yield objects when they are evaluated. The second form of symbolic expression in (3) takes on the form in version (4) $\bar{A} \ni \bar{a} = \bar{b} \in \bar{B}$, which reads: the $\bar{A}$-valued function a and the $\bar{B}$-valued function b are definitionally equal. It does not express the relation between two terms or object expressions denoting the same object anymore, but the relation between two object-valued functions which yield the same object when they are evaluated. In addition, version (4) has three forms of sentences which (run parallel to the ones for terms and which) read: X is a basic set (category), or X is a proposition, and, A is a type-valued function, i.e. a function whose values are types, or, A is a propositional function, i.e. a function whose values are propositions, and, the type-valued functions A and B are definitionally equal. Version (4) thus explicitly introduces types into the theory, and it also introduces the distinction between objects and object-valued functions as well as the one between types and type-valued functions. Finally, it introduces a switch from the linguistic to the object-oriented or natural attitude. (cf. Sect. viii., Chapt. 1)

In version (5) the rules of inference of the theory allow us to derive forms of judgement – as they are called in this new version – of the following four forms: $a \in A$, A is a type, $a = b \in A$ and $A = B$. This means that the forms of judgement $x \in X$ and $\bar{a} \in \bar{A}$ as well as the two forms of judgement X is a type and $\bar{A}$ is a type of version (4) are now contracted, each into one form of judgement, i.e. $a \in A$ and A *type* resp. This economy in the forms of judgement is made possible by the following measure: in version (4) to each object expression corresponds an object-valued functional expression which, when it is evaluated, denotes the same object as the object expression; and to each type expression corresponds a type-valued functional expression which, when it is evaluated, denotes the same type as the type expression. In version (5), however, every expression eventually denoting an object is either an object expression or an object-valued functional expression, and this is so according to the form of the resp. expression; the same applies to expressions eventually denoting a type. Hence, if in $a \in A$, a has the form of an object expression, then $a \in A$ clearly is the same as $x \in X$, and if a has the form of an object-valued functional expression, $a \in A$ is the same as $\bar{a} \in \bar{A}$. The same holds *mutatis mutandis* for the form of judgement A *type*. But a judgement of any of the four forms of version (5) is in general hypothetical, i.e. made under assumptions or in a context $x_1 \in A_1, \ldots, x_n \in A_n$. In such a context it is always the case that A_1 is a type, A_2 is a type under the assumption $x_1 \in A_1, \ldots, A_n$ is a type under the preceding assumptions that $x_1 \in A_1, \ldots, x_{n-1} \in A_{n-1}$. In addition to the four previously mentioned categorical forms of judgement (i.e. without assumptions), there are now four new hypothetical forms of judgement (i.e. with assumptions or in a context): $a \in A\,(x_1 \in A_1, \ldots, x_n \in A_n)$, A *type* $(x_1 \in A_1, \ldots, x_n \in A_n)$, $a = b \in A\,(x_1 \in A_1, \ldots, x_n \in A_n)$, $A = B\,(x_1 \in A_1, \ldots, x_n \in A_n)$. Version (6) is essentially the same as version (5). Instead of using the form of judgement A *type* of version (5), the form of judgement A *set* is adopted. This change in notation is due to the fact that in version (6) a distinction is made between sets and types, and what was indiscriminately called type or set in version (5) and earlier versions is now specified to be a set (cf. Sect. ii). Another minor difference consists in the following: whereas version (5) presents immediately the most general form of judgements made in a context, version (6) first introduces the forms of judgement made under one assumption, i.e. $b(x) \in B(x)\,(x \in A)$, $B(x)$ *set* $(x \in A)$, $b(x) = d(x) \in B(x)\,(x \in A)$, and $B(x) = D(x)\,(x \in A)$, and then indicates how to inductively build up the forms of judgement made under n assumptions or in a general context.

Version (7) has all the forms of judgement that version (6) has, with the notational difference that a colon : is used instead of the epsilon $\in$. But since version (7) contains a higher level type structure, these forms of judgement are but particular forms of the following general forms of judgement without assumptions $a : \alpha, \alpha : type, a = b : \alpha$, $\alpha = \beta : type$, and the following general forms of judgement with assumptions $a : \alpha(x_1 : \alpha_1, \ldots, x_n : \alpha_n)$, $\alpha : type\ (x_1 : \alpha_1, \ldots, x_n : \alpha_n)$, $a = b : \alpha(x_1 : \alpha_1, \ldots, x_n : \alpha_n)$, $\alpha = \beta : type\ (x_1 : \alpha_1, \ldots, x_n : \alpha_n)$. Hence the distinction between sets and types introduced in version (6) is developed further. But version (7) also has specific particular forms of judgement, namely $c : (x : \alpha)\beta$, $(x : \alpha)\beta : type$, $c = d : (x : \alpha)\beta$, $(x : \alpha)\beta = (x : \gamma)\beta : type$ and $(x : \alpha)\beta = (x : \alpha)\delta : type$ involving fct. types, as well as $set : type$ and $X : type(X : set)$. All these new particular forms of judgement allow us to build up constructive set theory the same way as in the previous versions – although in version (7), within the framework of a higher level type structure. Moreover, three more general forms of judgement without assumptions $\alpha : seq, \alpha = \beta : seq, \alpha$ *true* and three more general forms of judgement with assumptions $\alpha : seq(x_1 : \alpha_1, \ldots, x_n : \alpha_n)$, $\alpha = \beta : seq(x_1 : \alpha_1, \ldots, x_n : \alpha_n)$, $\alpha : seq(x_1 : \alpha_1, \ldots, x_n : \alpha_n, \beta_1\ true, \ldots, \beta_m\ true)$ are discussed in version (7). They are based on an extension of propositions to sequents (of propositions) in Gentzen's sense. Finally, all these forms of judgement are classified into analytic and synthetic ones, whereby the last mentioned categorical as well as hypothetical forms of judgement concerning sequents are the only synthetic ones. Version (8) is essentially the same as version (7).

Note: In version (6) one also finds, somewhat by the way or unsystematically, the forms of judgement A *prop* and A *true*, provided A is a proposition. In versions (7) and (8) these forms of judgement are presented in a completely systematic way. The first of these forms of judgement is based on assuming implicitly the axiom $prop = set : type$. One obtains subsequently the particular forms of judgement $a : A$ (where A is a proposition), $A : prop$, $a = b : A$ (where A is a proposition), $A = B : prop$ (where A, B are propositions), $a : A\ (x_1 : A_1, \ldots, x_n : A_n)$, $A : prop\ (x_1 : A_1, \ldots, x_n : A_n)$, $a = b : A(x_1 : A_1, \ldots, x_n : A_n)$, $A = B : prop$ $(x_1 : A_1, \cdots, x_n : A_n)$. And they are based explicitly on the two forms of judgement $prop : type$ and $X : type(X : prop)$ of versions (7) and (8).

Actually version (8) has a notational difference to version (7): Families of types are written $x_1 : \alpha_1, \cdots, x_n : \alpha_n \vdash a : \alpha$, $x_1 : \alpha_1, \cdots, x_n : \alpha_n \vdash \alpha : type$, $x_1 : \alpha_1, \cdots, x_n : \alpha_n \vdash a = b : \alpha$, $x_1 : \alpha_1, \cdots, x_n : \alpha_n \vdash \alpha = \beta : type$, and similarly for families of sets and families of propositions.

ii THE NOTION OF TYPE AND THE SPLITTING OF TYPE (SET/TYPE)

For anybody somewhat familiar with constructive type theory of today, the difference between sets and types, the identification of propositions with sets, the distinction between canonical and non-canonical sets (propositions), etc. are commonplace. The following representations are meant to tell the story of the gradual establishment of this commonplace.

In the versions (1)–(3) the concept of type is always introduced by stating that every mathematical object is of a certain kind or type. A type is then said to be defined by prescribing what one has to do in order to construct an object of that type. This is essentially the notion of set given by Bishop in (1967).

In version (1) (but not in versions (2) and (3)) a type is moreover considered to be a mathematical object, and thus, itself to be of a certain type. That is, a type is the same as an object of the type of types. This conception, however, was abandoned for reasons explained elsewhere (cf. Sect. vi)

In versions (2) and (3) an alternative characterization of types crops up: A type is said to be well-defined if one understands or grasps, in Kreisel's (1970) terms, what it means to be an (arbitrary) object of that type. (This second definition of type is given for the first time in Martin-Löf's (1971B).) These two characterizations of types are taken as two ways of expressing the same thing.

Version (4) contains a new approach to the question of what a type is. A type is there conceived as the meaning of a type expression, or in other words, a type is the sort of thing a type expression stands for. And something is a type expression either by virtue of its having one of the forms stipulated for a type expression and object expressions as its constituents, or by virtue of being the value of a type-valued functional expression under an assignment of object expressions to its variables. (*Note:* Since the basic attempt of version (4) at finding a direct semantics for constructive type theory did not work out (cf. Sect. iii), the def. of type in this version remains incomplete.)

Version (5) goes back to the first definition of type which is now extended in the following way: A canonical type A is defined by prescribing (i) how a canonical object of type A is constructed, and (ii) how two identical canonical objects of type A are constructed. – Version (5) differs from versions (2) and (3) on the one hand by introducing a criterion of identity for types, and on the other hand by skipping their second definition of type. Moreover, A is defined to be a non-canonical type if A has a canonical type as value (or in other words, if A denotes a canonical type). Thus, version (5) takes over the distinction between types and type-valued functions in the form of canonical and non-canonical types.

In version (6) a (canonical or non-canonical) set is defined in exactly the same way as a (canonical or non-canonical) type in version (5). There is the more terminological difference that a type in version (5) is now called a set and an object in version (5) is now called an element. However, this difference is quite crucial in the sense that version (6) splits the term 'type' into two terms: the term 'set' and the term 'type' or 'category' (in a philosophical or logical sense). In version (6) a type is defined by understanding or explaining

(i) what an (arbitrary) object of that type is, and

(ii) when two objects of that type are identical.

The definition of a type consists not only of a criterion of application, but also of a criterion of identity, and thus extends the second definition of type presented in the versions (2) and (3). But since a type (category) clearly need not be a set, since one can grasp what it means to be an (arbitrary) object of a given type (category) without providing exhaustive rules for constructing its objects, the first and second definition of type in versions (2) and (3) turn out not to be equivalent. But the definitions of set and type make it obvious that each set determines a type (category), namely the type of elements of the set, but not conversely (the standard examples being the type of propositions and the type of sets). (In Martin-Löf's paper (1971B) it is observed that if

(a) only sets (then called types) constitute domains of quantification,

(b) the sets and propositions are identified, and

(c) the propositions are assumed to form a set,

then one ends up with a set of sets or a type of types. Hence, after giving up the assumption (c) the propositions could only be explicitly and formally reintroduced into constructive type theory after the introduction of the distinction between sets and types.)

Versions (7) and (8) take over the definitions of set and type from version (6) and add the following two comments: First, a set is always an inductively defined or generated set. Second, the identity criterion of a type has to be defined in such a way that it is reflexive, symmetric and transitive. Furthermore, a type structure is built up by means of ground types and function types (composite or higher types). The function types are again divided into dependent and independent function types: in the second case one has a functional composition of two independent types, whereas in the first case one has a functional composition of a type with another type which depends on the first one. The dependent function types were introduced into type theory for the first time simultaneously by Martin-Löf and de Bruijn. (Note that there is already an allusion to function or higher types in version (6a), but no systematic account of them is given.) Versions (7) and (8) finally make a distinction between syntactical types (in version (7) called syntactical categories) and semantical types (in version (7) called semantical categories): an expression is an object of a syntactical type, whereas its meaning, i.e. the thing it stands for, is an object of a semantical type. According to Martin-Löf the distinction is highlighted by the fact that a syntactical type is always relative to the formation rules of a given language or formal system, whereas the semantic types are absolute. Thus, in versions (7) and (8) a distinction which is common to versions (1)–(4), but which disappeared in versions (5) and (6), surfaces again, namely the distinction between type expression (syntactical type) and type (semantical type); however, it now appears not on the level of the sets (in versions (1)–(4) called types), but on the level of the types proper.

To summarize: in versions (7) and (8) one has on the one hand sets, i.e. canonical and non-canonical sets, and on the other hand types, i.e. syntactical and semantical types, whereby among the syntactical as well as the semantical types one can distinguish ground types and function types (and of course things are even a bit more complicated considering that each set determines a type).

iii FROM METAMATHEMATICAL SEMANTICS TO DIRECT SEMANTICS AND OTHER SEMANTICAL MATTERS

The beginning of constructive type theory took place in a metamathematical setting. Martin-Löf used therefore normalization, that is the proof that reduction or computation of a (well-typed) term always terminates, to obtain a metamathematical intuitionistic model, or more precisely, an intuitionistic term model for his type theory as well as other metamathematical results (Church–Rosser property, decidability etc.)

The normalization theorem of version (2) is of interest for exactly these reasons. A flashback to version (1) shows that its normalization theorem has a couple of advantages over that of version (2) which, however, are paid with a reflection principle rendering version (1) inconsistent. But with the normalization proof of version (2) two other problems surfaced: the one is connected with the conversion on raw terms approach, the other one with eager evaluation. Martin-Löf solved these problems by proving normalization for a new version (version (3)) of his type theory with a conversion as judgement approach and with a combinator formulation instead of the λ-abstraction formulation of versions (1) and (2).

But this solution did not convince him entirely either, as he started to question the value of normalization proofs and implicitly given intuitionistic term models altogether: he realized that modelling a formal theory boils down to translating the formal language of that theory into another formal language (the point of view of the proof theorist) and that this procedure will not work anymore as soon as one is supposed to give meaning to the syntactical entities of a formal language containing the most primitive notions of a theory. One must then proceed to a semantical procedure of a radically new sort, which Martin-Löf has termed direct (as opposed to the indirect metamathematical one). In version (4) he tried to directly explain the meaning of the type-theoretic operations not by their introduction, but by their elimination rules. This attempt had to fail, and the same new direct semantical procedure is applied in version (5), rightly so, to the introduction rules. Since version (5) and all later versions of constructive type theory abandon the combinator formulation in order to return to the old λ-abstraction formulation, the old problem connected with eager evaluation resurfaced. Martin-Löf solves this problem anew by adopting a so called normal form or lazy evaluation. In version (5) he develops the new direct semantics which is essentially taken over in the versions (6) to (8).

(1) In version (2) Martin-Löf proves the *Normalization theorem*:

Every term is reducible (computable) to a normal term.

His proof of normalization uses an extension of the method of computability introduced in Tait (1967) in order to prove normalization of Gödel's theory T. The proof consists basically of the definition of a computability predicate Φ_A which expresses the computability (reducibility) of a term of type A, and of the proof that Φ_A is defined and that $\Phi_A(a)$ (i.e. a is a computable (reducible) term of type A) if $a \in A$ is derivable.

The notions of reduction and conversion are introduced as follows:

Rules of contraction. 'An expression a is contracted to an expression b' is written $a\ contr\ b$.

$$
\begin{array}{rll}
(\lambda x)b[x](a) & contr & b[a] \\
E((a,b),(\lambda x)(\lambda y)d[x,y]) & contr & d[a,b] \\
D(i(a),(\lambda x)d[x],(\lambda y)e[y]) & contr & d[a] \\
D(j(b),(\lambda x)d[x],(\lambda y)e[y]) & contr & e[b] \\
R_n(1,c_1,\cdots,c_n) & contr & c_1 \\
\vdots & & \\
R_n(n,c_1,\ldots,c_n) & contr & c_n \\
R(0,d,(\lambda x)(\lambda y)e[x,y]) & contr & d \\
R(s(a),d,(\lambda x)(\lambda y)e[x,y]) & contr & e[a,R(a,d,(\lambda x)(\lambda y)e[x,y])]
\end{array}
$$

An expression a is said to 1-step reduce to an expression b, written $a\ red'\ b$ iff

- for closed expressions a and b, if $a\ contr\ b$ then $a\ red'\ b$;
- for c an expression with n free variables, if $a_i\ red'\ b_i$ or if $a_i \equiv b_i$, $1 \leq i \leq n$ and for at least one i, $a_i\ red'\ b_i$, then $c(a_1,\cdots,a_n)\ red'\ c(b_1,\cdots,b_n)$;
- for a, b expressions with n free variables if $a(x_1,\cdots,x_n)\ red'\ b(x_1,\cdots,x_n)$ then $a\ red'\ b$.

An expression is said to n-step reduce to an expression b, written a *red* b iff

- if a *red′* b then a *red* b;
- a *red* a;
- if a *red* b and b *red* c then a *red* c.

An expression a is said to convert into an expression b, written a *conv* b, if there exists an expression c s.t. a *red* c and b *red* c.

An expression a is said to be normal or irreducible if there exists no expression b s.t. a *red′* b. And an expression is said to be normalizable if a can be n-step reduced to a normal expression. Sometimes it will be said that 'a computes to b' instead of 'a reduces to b'.

In order to prove that a *conv* b is an equivalence relation, and in particular to prove that *conv* is transitive, one has to prove the Church–Rosser property: If a *red* b and a *red* c then there is an expression d s.t. b *red* d and c *red* d.[2]

Since version (2) contains constants of every closed type, it is not necessary to consider in the proof of the normalization theorem the case of the open terms – their case can be "reduced" to the one of closed terms.

Martin-Löf calls a term on introduction or on elimination form if it has been formed by one of the introduction or by one of the elimination rules resp., and he observes that, unless a closed term is a constant, it is necessarily either on introduction or on elimination form.

1. **Definition of the partial computability function/predicate** Φ defined only for those syntactic expressions which can be proved to be closed types A.

   1.1 $\Phi_{P(a_1,\ldots,a_n)}$ is the species of normalizable closed terms of type $P(a_1, \ldots, a_n)$.

   1.2 If Φ_A has been defined and $\Phi_{B[a]}$ has been defined for all closed terms $a \in A$ s.t. $\Phi_A(a)$, then $\Phi_{(\Pi x \in A)B[x]}$ is defined as the least predicate s.t.

      1. if $(\lambda x)b[x] \in (\Pi x \in A)B[x]$ is a closed term and $\Phi_{B[a]}(b[a])$ for all closed terms $a \in A$ s.t. $\Phi_A(a)$, then $\Phi_{(\Pi x \in A)B[x]}((\lambda x)b[x])$.
      2. $c \in (\Pi x \in A)B[x]$ is a closed normal term s.t. c is not on λ-form, then $\Phi_{(\Pi x \in A)B[x]}(c)$.
      3. if $b \in (\Pi x \in A)B[x]$ is a closed term on elimination form and a is a term s.t. $\Phi_{(\Pi x \in A)B[x]}(a)$ and b *red* a, then $\Phi_{(\Pi x \in A)B[x]}(b)$.

   1.3 If Φ_A has been defined and $\Phi_{B[a]}$ has been defined for all closed terms $a \in A$ s.t. $\Phi_A(a)$, then $\Phi_{(\Sigma x \in A)B[x]}$ is defined to be the least predicate s.t.

      1. if $a \in A$ and $b \in B[a]$ are closed terms s.t. $\Phi_A(a)$ and $\Phi_{B[a]}(b)$, then $\Phi_{(\Sigma x \in A)B[x]}((a, b))$.
      2. if $c \in (\Sigma x \in A)B[x]$ is a closed normal term which is not on (,)-form, then $\Phi_{(\Sigma x \in A)B[x]}(c)$.
      3. if $b \in (\Sigma x \in A)B[x]$ is a closed term on elimination form and a is a term s.t. $\Phi_{(\Sigma x \in A)B[x]}(a)$ and b *red* a, then $\Phi_{(\Sigma x \in A)B[x]}(b)$.

   1.4 If Φ_A and Φ_B have been defined, then Φ_{A+B} is defined as the least predicate s.t.

1. if $a \in A$ is a closed term s.t. $\Phi_A(a)$, then $\Phi_{A+B}(i(a))$, if $b \in B$ is a closed term s.t. $\Phi_B(b)$, then $\Phi_{A+B}(j(b))$.
2. if $c \in A + B$ is a closed normal term which is not on either i- or j-form, then $\Phi_{A+B}(c)$.
3. if $b \in A + B$ is a closed term on elimination form and a is a term s.t. $\Phi_{A+B}(a)$ and b *red* a, then $\Phi_{A+B}(b)$.

1.5 Φ_V is defined as the least predicate s.t.

1. if $C \in V$ is a closed normal term which is not on Π-, Σ- or $+$-form, then $\Phi_V(C)$ and Φ_C is defined to be the species of normalizable closed terms of type C.
2. if $\Phi_V(A)$ and $\Phi_V(B[a])$ for all closed terms $a \in A$ s.t. $\Phi_A(a)$, then $\Phi_V((\Pi x \in A)B[x])$ and $\Phi_{(\Pi x \in A)B[x]}$ is defined as in 1.2.
3. if $\Phi_V(A)$ and $\Phi_V(B[a])$ for all closed terms $a \in A$. s.t. $\Phi_A(a)$, then $\Phi_V((\Sigma x \in A)B[x])$ and $\Phi_{(\Sigma x \in A)B[x]}$ is defined as in 1.3.
4. if $\Phi_V(A)$ and $\Phi_V(B)$, then $\Phi_V(A + B)$ and Φ_{A+B} is defined as in 1.4.
5. if $B \in V$ is a closed term on elimination form and A is a term s.t. $\Phi_V(A)$ and B *red* A, then $\Phi_V(B)$ and Φ_B is a set equal to Φ_A (and thus defined).

Note that the meaning of '$\Phi_V(A)$' and the predicate Φ_A are defined simultaneously by transfinite induction, and thus Φ_A does not only depend on A but also on the proof of $\Phi_V(A)$, but the notation doesn't show this explicitly.

1.6 If $A \in V$ is a term s.t. $\Phi_V(A)$, then Φ_A is the predicate associated in 1.5.

The proof of the second part of the normalization proof requires two lemmata:

2. **Lemma** When defined, the computability predicate Φ_A has three properties:

2.1 If $a \in A$ is a closed normal term which is not on introduction form, then $\Phi_A(a)$.
2.2 If $b \in A$ is a closed term on elimination form and a is a term s.t. $\Phi_A(a)$ and b *red* a, then $\Phi_A(b)$.
2.3 If $\Phi_A(a)$, then a is normalizable.

The proofs of 2.1 and 2.2 are straightforward by Def. 1 and the proof of 2.3 is fairly straightforward too.

3. **Lemma** If Φ_A and Φ_B are both defined and A *conv* B, then $\Phi_A(a)$ iff $\Phi_B(a)$, i.e. Φ_A and Φ_B are extensionally identical computability predicates.
4. **Proof that if $a \in A$ is a closed term, then Φ_A is defined and $\Phi_A(a)$**

4.1 (Ass. rule) If $x \in A$ is introduced, by IH, Φ_A is defined. One has to show that Φ_A is defined and that if $a \in A$ is a closed term s.t. $\Phi_A(a)$, then $\Phi_A(a)$. Obvious.
4.2 (Constant rule) If $c \in A$ is introduced, by IH, Φ_A is defined. One has to show that Φ_A is defined and that $\Phi_A(c)$ which follows from Lemma 2.1.
4.3 (ΠI) By IH, Φ_A is defined, $\Phi_{B[a]}$ is defined for all closed terms $a \in A$ s.t. $\Phi_A(a)$, and $\Phi_{B[a]}(b[a])$ for all such terms a. Hence, by Def. 1.2, $\Phi_{(\Pi x \in A)B[x]}$ is defined and by Def. 1.2.1, $\Phi_{(\Pi x \in A)B[x]}((\lambda x)b[x])$.

4.4 (ΠE) By IH, Φ_A and $\Phi_{(\Pi x \in A)B[x]}$ are defined and $\Phi_A(a)$ and $\Phi_{(\Pi x \in A)B[x]}(b)$. Corresponding to the defining clauses of $\Phi_{(\Pi x \in A)B[x]}$ one has to distinguish:

1. if b is on form $(\lambda x)d[x]$ and for all terms a s.t. $\Phi_A(a)$, $\Phi_{B[a]}(d[a])$, then $b(a)$ is on elimination form and $b(a)$ *red* $d[a]$ so that by Lemma 2.2 $\Phi_{B[a]}(b(a))$.
2. if b is closed normal and not on λ-form, and if c is the normal form of a (which exists by IH $\Phi_A(a)$ and Lemma 2.3), then $b(a)$ *red* $b(c)$ which is normal and on elimination form. By Lemma 2.1 $\Phi_{B[c]}(b(c))$ and by Lemmata 2.2 and 3, $\Phi_{B[a]}(b(a))$.
3. if b is on elimination form and there is a term d s.t. $\Phi_{(\Pi x \in A)B[x]}(d)$ and b *red* d, then $b(a)$ *red* $d(a)$. It has already been shown that $\Phi_{B[a]}(d(a))$, hence by Lemma 2.2, $\Phi_{B[a]}(b(a))$.

4.5 (ΣI) $\Phi_{(\Sigma x \in A)B[x]}$ is defined by IH and Def. 1.3, and $\Phi_{(\Sigma x \in A)B[x]}((a, b))$ by Def. 1.3.1.

4.6 (ΣE) By IH, $\Phi_{(\Sigma x \in A)B[x]}$ is defined and $\Phi_{(\Sigma x \in A)B[x]}(c)$, and $\Phi_{C[e]}$ is defined for all terms e s.t. $\Phi_{(\Sigma x \in A)B[x]}(e)$, and for all a and b s.t. $\Phi_A(a)$ and $\Phi_{B[a]}(b)$, $\Phi_{C[(a,b)]}(d[a, b])$. Corresponding to the three defining clauses of $\Phi_{(\Sigma x \in A)B[x]}$ one has to distinguish

1. if c is on form (a, b) where $\Phi_A(a)$ and $\Phi_{B[a]}(b)$, then $E(c, (\lambda x)(\lambda y) d[x, y])$ is on elimination form and $E(c, (\lambda x)(\lambda y)d[x, y])$ *red* (a, b). But by IH $\Phi_{C[(a,b)]}(d[a, b])$ and hence by Lemma 2.2 $\Phi_{C[c]}(E(c, (\lambda x)(\lambda y) d[x, y]))$.
2. if c is normal and not a (,)-form, then let $a \in A$ and $b \in B[a]$ be constants. By Lemma 2.1, $\Phi_A(a)$ and $\Phi_{B[a]}(b)$. Hence, by IH, $\Phi_{C[(a,b)]}(d[a, b])$ so that by Lemma 2.3 there is a normal term $g[a, b]$ s.t. $d[a, b]$ *red* $g[a, b]$. But then $E(c, (\lambda x)(\lambda y)d[x, y])$ *red* $E(c, (\lambda x)(\lambda y)g[x, y])$ which is normal and on elimination form, so that by Lemma 2.1 $\Phi_{C[c]}(E(c, (\lambda x)(\lambda y) g[x, y]))$ and by Lemma 2.2 $\Phi_{C[c]}(E(c, (\lambda x)(\lambda y)d[x, y]))$.
3. if c is on elimination form and there is a term f s.t. $\Phi_{(\Sigma x \in A)B[x]}(f)$ and c *red* f, then $E(c, (\lambda x)(\lambda y)d[x, y])$ *red* $E(f, (\lambda x)(\lambda y)d[x, y])$ for which it has already been shown $\Phi_{C[f]}(E(f, (\lambda x)(\lambda y)d[x, y]))$. Hence by Lemmata 2.2 and 3, $\Phi_{C[c]}(E(c, (\lambda x)(\lambda y)d[x, y]))$.

4.7 $(+I)$ Φ_{A+B} is defined by IH and Def. 1.4 and $\Phi_{A+B}(i(a))$ and $\Phi_{A+B}(j(b))$ by Def. 1.4.1.

4.8 $(+E)$ By IH, Φ_{A+B} is defined and $\Phi_{A+B}(c)$, and for all terms a and b s.t. $\Phi_A(a)$ and $\Phi_B(b)$, $\Phi_{C[i(a)]}(d[a])$ and $\Phi_{C[j(b)]}(e[b])$, and for all c s.t. $\Phi_{A+B}(c)$, $\Phi_{C[c]}$ is defined. According to the three defining clauses of Φ_{A+B} the following cases have to be distinguished:

1. if c is on form $i(a)$ and $\Phi_A(a)$, then $D(c, (\lambda x)d[x], (\lambda y)e[y])$ is on elimination form and $D(c, (\lambda x)d[x], (\lambda y)e[y])$ *red* $d[a]$ where $\Phi_{C[i(a)]}(d[a])$ by IH. By Lemma 2.2, $\Phi_{C[c]}(D(c, (\lambda x)d[x], (\lambda y)e[y]))$. Similarly, if c is on form $j(b)$ and $\Phi_B(b)$.

2. if c is normal and not on i- or j-form, then let $a \in A$ and $b \in B$ be constants. By Lemma 2.1 $\Phi_A(a)$ and $\Phi_B(b)$. Hence $\Phi_{C[i(a)]}(d[a])$ and $\Phi_{C[j(b)]}(e[b])$ by IH, so that by Lemma 2.3 there are normal terms $g[a]$ and $h[b]$ s.t. $d[a]$ *red* $g[a]$ and $e[b]$ *red* $h[b]$. But then $D(c, (\lambda x)d[x], (\lambda y)e[y])$ *red* $D(c, (\lambda x)g[x], (\lambda y)h[y])$ which is normal and on elimination form so that $\Phi_{C[c]}(D(c, (\lambda x)g[x], (\lambda y)h[y]))$ by Lemma 2.1 and $\Phi_{C[c]}(D(c, (\lambda x)d[x], (\lambda y)e[y]))$ by Lemma 2.2.
3. if c is on elimination form and there is a term f s.t. $\Phi_{A+B}(f)$ and c *red* f, then $D(c, (\lambda x)d[x], (\lambda y)e[y])$ *red* $D(f, (\lambda x)d[x], (\lambda y)e[y])$ for which it has already been shown that $\Phi_{C[f]}(D(f, (\lambda x)d[x], (\lambda y)e[y]))$. Hence by the Lemmata 2.2 and 3, $\Phi_{C[c]}(D(c, (\lambda x)d[x], (\lambda y]e[y]))$.

4.9 $(N_n I)$ By Def. 1.5.1 Φ_{N_n} is defined to be the species of normalizable closed terms of type N_n. Hence $\Phi_{N_n}(1), \ldots, \Phi_{N_n}(n)$.

4.10 $(N_n E)$ By IH, for all terms c s.t. $\Phi_{N_n}(c)$, $\Phi_{C[c]}$ is defined. Moreover, $\Phi_{N_n}(c)$, $\Phi_{C[1]}(c_1), \ldots, \Phi_{C[n]}(c_n)$. Distinguish three cases according to the form of c.

1. if c is on form m, $m \in N_n$, then $R_n(c, c_1, \ldots, c_n)$ is on elimination form and $R_n(c, c_1, \ldots, c_n)$ *red* c_m where by IH $\Phi_{C[m]}(c_m)$. Hence by Lemma 2.2 $\Phi_{C[c]}(R_n(c, c_1, \ldots, c_n))$.
2. if c is normal and not one of $1, \ldots, n$; then by IH and Lemma 2.3, there are normal terms $f_1, \ldots, f_n$ s.t. c_1 *red* $f_1, \ldots, c_n$ *red* f_n, and hence $R_n(c, c_1, \ldots, c_n)$ *red* $R_n(c, f_1, \ldots, f_n)$ for which it has been shown that $\Phi_{C[c]}(R_n(c, f_1, \ldots, f_n))$. By Lemmata 2.2 and 3, $\Phi_{C[c]}(R_n(c, c_1, \ldots, c_n))$.
3. if c is not normal and there is a normal term f s.t. c *red* f, then $R_n(c, c_1, \ldots, c_n)$ *red* $R_n(f, c_1, \ldots, c_n)$ for which it has been shown that $\Phi_{C[f]}(R_n(f, c_1, \ldots, c_n))$. By Lemmata 2.2 and 3, $\Phi_{C[c]}(R_n(c, c_1, \ldots, c_n))$.

4.11 (NI) By Def. 1.5.1 Φ_N is defined to be the species of normalizable closed terms of type N. Hence $\Phi_N(0)$ and if $\Phi_N(a)$ then $\Phi_N(s(a))$.

4.12 (NE) By IH for all terms a s.t. $\Phi_N(a)$, $\Phi_{C[a]}$ is defined and $\Phi_N(c)$ and $\Phi_{C[0]}(d)$ and for a and b s.t. $\Phi_N(a)$ and $\Phi_{C[a]}(b)$, $\Phi_{C[s(a)]}(e[a, b])$. According to how $\Phi_N(c)$ was derived, distinguish four cases

1. if c is on form 0, then $R(c, d, (\lambda x)(\lambda y)e[x, y])$ *red* d, where by IH $\Phi_{C[0]}(d)$. By Lemma 2.2 $\Phi_{C[c]}(R(c, d, (\lambda x)(\lambda y)e[x, y]))$.
2. if c is on form $s(a)$, then by IH $\Phi_N(a)$ and $\Phi_{C[a]}(R(a, d, (\lambda x)(\lambda y) e[x, y]))$ so that again by IH $\Phi_{C[s(a)]}(e[a, R(a, d, (\lambda x)(\lambda y)e[x, y])])$. But $R(c, d, (\lambda x)(\lambda y)e[x, y])$ *red* $e[a, R(a, d, (\lambda x)(\lambda y)e[x, y])]$ so that by Lemma 2.2 $\Phi_{C[c]}(R(c, d, (\lambda x)(\lambda y)e[x, y]))$.
3. if c is normal and on elimination form, then from $\Phi_{C[0]}(d)$ (IH) follows by Lemma 2.3 that there is a normal term g s.t. d *red* g. And if $a \in N$ and $b \in C[a]$ are constants, then $\Phi_N(a)$ and $\Phi_{C[a]}(b)$ by Lemma 2.1. Hence by IH $\Phi_{C[s(a)]}(e[a, b])$ so that, again by Lemma 2.3 there is a normal term $h[a, b]$ s.t. $e[a, b]$ *red* $h[a, b]$. But then $R(c, d, (\lambda x)(\lambda y)e[x, y])$ *red* $R(c, g, (\lambda x)(\lambda y)h[x, y])$ which is normal and on elimination form, and

thus by Lemma 2.1 $\Phi_{C[c]}(R(c, g, (\lambda x)(\lambda y)h[x, y]))$ and by Lemma 2.2 $\Phi_{C[c]}(R(c, d, (\lambda x)(\lambda y)e[x, y]))$.

4. if c is on elimination form and there is a term f s.t. $\Phi_N(f)$ and c *red* f, then $R(c, d, (\lambda x)(\lambda y)e[x, y])$ *red* $R(f, d, (\lambda x)(\lambda y)e[x, y])$ for which it has already been shown that $\Phi_{C[f]}(R(f, d, (\lambda x)(\lambda y)e[x, y]))$. Hence by the Lemmata 2.2 and 3, $\Phi_{C[c]}(R(c, d, (\lambda x)(\lambda y)e[x, y]))$.

4.13 (VI) By Def. 1.5.1 $\Phi_V(N_0)$, $\Phi_V(N_1,), \ldots, \Phi_V(N)$. If $\Phi_V(A)$ and for all a s.t. $\Phi_A(a)$, $\Phi_V(B[a])$, then by Def. 1.5.2 $\Phi_V((\Pi x \in A)B[x])$ and by Def. 1.5.3 $\Phi_V((\Sigma x \in A)B[x])$. If $\Phi_V(A)$ and $\Phi_V(B)$, then by Def. 1.5.4 $\Phi_V(A + B)$.

4.14 (Type conversion) By IH Φ_A and Φ_B are defined and $\Phi_A(a)$. By lemma 3, $\Phi_B(a)$.

This ends the proof of the normalization theorem.

Now, the following problem arises in version (2). The relation of convertibility is intended as the formal counterpart of the informal definitional or intensional identity relation. But according to Lemma 3 above: if A *conv* B then Φ_A and Φ_B are extensionally identical, which shows that the term model provided by the normalization proof is not a model intended by the theory. The problem lies in the definition of the convertibility relation, which is too liberal in allowing unrestricted conversions under the λ-sign and other variable binding operations. Martin-Löf solved the problem by abandoning the conversion on raw terms approach and by adopting a conversion defined only on well-typed terms (conversion as judgement) approach to type theory. (cf. Subsect. (3) below)

There is another semantical problem arising in version (2) connected with eager evaluation of terms or what has been called computation from within (cf. Subsect. (5) below). In Tarskian semantics, for D an individual domain, $d \in D$, c_d a term and *val* an evaluation function, the following two facts hold:

1) $val(f(t_1, \ldots, t_n)) = val(f)(val(t_1), \ldots, val(t_n))$

and if $t(x_1, \ldots, x_n)$ is an open term

2) $val(t(t_1, \ldots, t_n)) = val(t(c_{d_1}, \ldots, c_{d_n}))$ where $val(t_1) = d_1, \ldots, val(t_n) = d_n$

where 1) expresses eager or applicative order evaluation.

For e.g. λ-terms one has the corresponding

1) $val((a_1/x_1, \ldots, a_n/x_n)(\lambda x)b) = val((val(a_1)/x_1, \ldots, val(a_n)/x_n)(\lambda x)b)$

2) $val((a_1/x_1, \ldots, a_n/x_n)(\lambda x)b) = val((\lambda x)(a_1/x_1, \ldots, a_n/x_n)b)$

3) $val((\lambda x)b) = (\lambda x)b$

But then the problem arises that

$$\begin{aligned} val((a_1/x_1, \ldots, a_n/x_n)(\lambda x)b) &= val((\lambda x)(a_1/x_1, \ldots, a_n/x_n)b) && \text{by 2)} \\ &= (\lambda x)(a_1/x_1, \ldots, a_n/x_n)b && \text{by 3)} \end{aligned}$$

and

$$\begin{aligned} val((a_1/x_1,\dots,a_n/x_n)(\lambda x)b) &= val((val(a_1)/x_1,\dots,val(a_n)/x_n)(\lambda x)b) && \text{by 1)} \\ &= val((\lambda x)(val(a_1)/x_1,\dots,val(a_n)/x_n)b) && \text{by 2)} \\ &= (\lambda x)(val(a_1)/x_1,\dots,val(a_n)/x_n)b && \text{by 3)} \end{aligned}$$

hence $(\lambda x)(a_1/x_1,\dots,a_n/x_n)b\ conv\ (\lambda x)(val(a_1)/x_1,\dots,val(a_n)/x_n)b$ which is clearly not the case. The same problem turns up with other variable binding operations.

Martin-Löf solved this problem by abolishing bound variables altogether and by adopting a combinator formulation (i.e. explicit definitions and parameters) of type theory. Then

$$(\lambda x)b = \lambda((x)b) = f(x_1,\dots,x_n)$$

$$(a_1/x_1,\dots,a_n/x_n)(\lambda x)b = f(a_1,\dots,a_n) \text{ and}$$

$$val(f(a_1,\dots,a_n)) = f(val(a_1),\dots,val(a_n)).$$

(2) A flashback to version (1).
In version (1) Martin-Löf proves the normalization theorem or *Hauptsatz*:
Every term reduces to a normal term.

The proof of this theorem uses also an extension of the method of computability introduced by Tait, and it consists essentially 1) of the definition of an abstract object α_a of an abstract type τ_A for every term $a \in A$, where α_a contains a computability predicate ϑ_a, and 2) of the proof that for every $a \in A$, α_A is a well-defined object of τ_V, and α_a is a well-defined object of τ_A, as well as that $\vartheta_A(c, \alpha_a)$ for some term c to which a reduces (i.e. the property $\vartheta_A(c, \alpha_a)$) is to imply that c is computable or normalizable (and thus so is a).

The notions of contraction, reduction and definitional equality (conversion) are introduced as follows: “An expression a is contracted to an expression b” is written $a \geq^1 b$. The only rule of contraction is

$$(\lambda x \in A)b(x)a \geq^1 b(a).$$

An expression a is said to reduce to an expression b, written $a \geq b$ iff

- if $a \geq^1 b$ then $a \geq b$
- $a \geq a$
- if $a \geq b$ and $b \geq c$ then $a \geq c$

Two expressions are said to be definitionally equal (convert), written $a = b$ iff there exists an expression c s.t. $a \geq c$ and $b \geq c$. An expression a is said to be normal or irreducible if there exists no expression b s.t. $a \geq^1 b$. And an expression is said to be normalizable if it can be reduced to a normal expression.

As in version (2), in order to prove that = is an equivalence relation, one has to prove the Church–Rosser property. (*Note:* these definitions of contraction and reduction are very slight adaptions of those of version (1) to the presentation of the analogous notions of version (2).)

1. **Definition of the abstract object α_a for each term** a (by induction on the construction of a) **and thus of the computability predicate ϑ_A for each type symbol** A

1.1 $\alpha_V \equiv (\tau_V, \omega_V, \vartheta_V)$ where

τ_V is the type of triples $(\tau, \omega, \vartheta)$ s.t. τ is a type, ω is an object of τ, ϑ is a 2-ary predicate with first argument an expression and second argument an object of type τ.

$\omega_V \equiv (\tau, \omega, \vartheta)$ where τ is some fixed non-empty type, ω is some fixed object of τ, $\vartheta(a, \xi)$ is the proposition that a is normalizable

ϑ_V is the 2-ary propositional function whose value for an expression A and an object $(\tau, \omega, \vartheta)$ of τ_V is the conjunction of the following three propositions

1.1.1. A is normal

1.1.2. $\vartheta(a, \xi)$ implies that a is normalizable

1.1.3. $\vartheta(a, \omega)$ provided a is a normal expression of the form $ya_1, \ldots, a_n$ where y is a variable

1.2 $\alpha_x(\xi) \equiv \xi$ (where ξ is an arbitrary abstract object of some abstract type τ).

1.3 If α_A is of type τ_V and for ξ of type τ_A, $\alpha_{B(x)}(\xi)$ is of type τ_V, then

$$\alpha_{(\Pi x \in A)B(x)} \equiv \left(\tau_{(\Pi x \in A)B(x)}, \omega_{(\Pi x \in A)B(x)}, \vartheta_{(\Pi x \in A)B(x)}\right)$$

where $\tau_{(\Pi x \in A)B(x)}$ is the type of functions η which to an object ξ of τ_A assigns an object $\eta(\xi)$ of $\tau_{B(x)}(\xi)$.

$\omega_{(\Pi x \in A)B(x)}$ is the function which to an object ξ of τ_A assigns $\omega_{B(x)}(\xi)$.

$\vartheta_{(\Pi x \in A)B(x)}(b, \eta)$ is the proposition that, for all expressions $a \in A$ and objects ξ of τ_A, if $\vartheta_A(a, \xi)$ then there is an expression d s.t. $ba \geq d$ and $\vartheta_{B(x)}(\xi)(d, y(\xi))$, and if $b \geq (\lambda x \in C)d(x)$, then C is normalizable.

1.4 If α_A is a well-defined object of τ_V and for ξ of τ_A, $\alpha_{b(x)}(\xi)$ is well-defined, then $\alpha_{(\lambda x \in A)b(x)}$ is the function which to an object ξ of τ_A assigns $\alpha_{b(x)}(\xi)$.

1.5 If α_b is a function and α_a is an object whose type is the domain of that function, then $\alpha_{ba} \equiv \alpha_b(\alpha_a)$ i.e. the result of applying α_b to α_a.

The second half of the proof of the normalization theorem requires two lemmata:

2. **Substitution property** $\alpha_{b(x)}(\alpha_a) = \alpha_{b(a)}$ in the sense that if one side is well-defined, then so is the other and the two sides are definitionally equal.
3. **Compatibility with equality** If α_a and α_b are both well-defined and $a = b$, then $\alpha_a = \alpha_b$.

4. **Proof that if, $a \in A$, then α_a and α_A are well-defined objects and there exists a c s.t. $a \geq c$ and $\vartheta_A(c, \alpha_a)$**
More precisely, Martin-Löf proves that

$$\text{if} \quad \alpha_{A_1} \text{ is a well-defined object of } \tau_V$$

$$\vdots$$

$$\alpha_{A_n(x_1,\ldots,x_{n-1})}(\xi_1, \ldots, \xi_{n-1}) \text{ is a well-defined object of } \tau_V$$
$$\text{with } \xi_1 \text{ of } \tau_{A_1}, \ldots, \xi_{n-1} \text{ of } \tau_{A_{n-1}(x_1,\ldots,x_{n-2})}(\xi_1, \ldots, \xi_{n-2})$$

then $\alpha_{A(x_1,\ldots,x_n)}(\xi_1, \ldots, \xi_n)$ is a well-defined object of τ_V and $\alpha_{a(x_1,\ldots,x_n)}(\xi_1, \ldots, \xi_n)$ is a well-defined object of $\tau_{A(x_1,\ldots,x_n)}(\xi_1, \ldots, \xi_n)$ s.t. if $\vartheta_{A_1}(a_1, \xi_1), \ldots, \vartheta_{A_n(x_1,\ldots,x_{n-1})}(\xi_1, \ldots, \xi_{n-1})(a_n, \xi_n)$ then there is an expression c s.t. $a(a_1, \ldots, a_n) \geq c$ and $\vartheta_{A(x_1,\ldots,x_n)}(\xi_1, \ldots, \xi_n)$ $(c, \alpha_{a(x_1,\ldots,x_n)}(\xi_1, \ldots, \xi_n))$ by induction on the length of a derivation

$$\begin{array}{ccc} x_1 \in A_1 & \ldots\ldots & x_n \in A_n(x_1, \ldots, x_{n-1}) \\ \ddots & & \cdot^{\cdot^{\cdot}} \\ & a(x_1, \ldots, x_n) \in A(x_1, \ldots, x_n) & \end{array}$$

4.1 (Basic axiom) $V \in V$
That $\alpha_V \equiv (\tau_V, \omega_V, \vartheta_V)$ is an object of τ_V is straightforward.
$\vartheta_V(V, \alpha_V)$ is the conjunction of three propositions:

4.1.1 V is normal. Clear
4.1.2 $\vartheta_V(A, \alpha)$ implies that A is normalizable. Clear, since by Def. 1.1, $\vartheta_V(A, \alpha)$ is the conjunction of three propositions, the first of which says that A is normal.
4.1.3 $\vartheta_V(A, \omega_V)$ provided A is a normal expression of the form $ya_1, \ldots, a_n$. Clear, since in that case A is normal and by Def. 1.1. of $\omega_V \equiv (\tau, \omega, \vartheta)$, it has the property that $\vartheta(a, \xi)$ implies that a is normalizable and by 1.1.3 it also has the property that $\vartheta(a, \omega)$ provided a is a normal expression of the form $ya_1, \ldots, a_n$.

4.2 (Ass. rule) $x \in A$
By IH, α_A is a well-defined object of τ_V. Trivially, α_A is a well-defined object of τ_V. And for ξ of type τ_A, $\alpha_x(\xi)$ is by Def. 1.2 easily seen to be a well-defined object of τ_A.

If $\vartheta_A(a, \xi)$ then there exists an expression c s.t. $a \geq c$ and $\vartheta_A(c, \alpha_x(\xi))$, which is readily seen if one takes c to be a itself.

4.3 (ΠF)

$$\frac{\begin{array}{cc} & [x \in A] \\ \vdots & \vdots \\ A \in V & B(x) \in V \end{array}}{(\Pi x \in A)B(x) \in V}$$

By IH, α_A is a well-defined object of τ_V and there is an expression C s.t. $A \geq C$ and $\vartheta_V(C, \alpha_A)$; also for ξ of τ_A, $\alpha_{B(x)}(\xi)$ is a well-defined object of τ_V s.t. $\vartheta_A(a,\xi)$ implies that there is an expression D s.t. $B(a) \geq D$ and $\vartheta_V(D, \alpha_{B(x)}(\xi))$.

By Defs. 1.3 and 1.1, $\alpha_{(\Pi x \in A)B(x)}$ is a well-defined object of τ_V.

It has to be verified that $\vartheta_V(E, \alpha_{(\Pi x \in A)B(x)})$ for some expression E s.t. $(\Pi x \in A)B(x) \geq E$. This means to verify that

4.3.1 E is normal. By IH, $\vartheta_V(C, \alpha_A)$ so that by Def. 1.1.1 C is normal and $\vartheta_A(x, \omega_A)$. Also, $\vartheta_A(x, \omega_A)$ implies, by IH, that $B(x) \geq D(x)$ s.t. $\vartheta_V(D(x), \alpha_{B(x)}(\omega_A))$, so that by 1.1.1 $D(x)$ is normal as well. If one defines $E \equiv (\Pi x \in C)D(x)$, then clearly $(\Pi x \in A)B(x) \geq E$ and E is normal.

4.3.2 $\vartheta_{(\Pi x \in A)B(x)}(b, \eta)$ implies that b is normalizable. By Def. 1.3, $\vartheta_{(\Pi x \in A)B(x)}(b, \eta)$ is the proposition: for all $a \in A$ and ξ of τ_A, if $\vartheta_A(a, \xi)$ then there is a d s.t. $ba \geq d$ and $\vartheta_{B(x)}(\xi)(d, \eta(\xi))$, and if $b \geq (\lambda x \in C)d(x)$ then C is normalizable. ω_A is an object of τ_A s.t. $\vartheta_A(x, \omega_A)$. Hence, by the meaning of $\vartheta_{(\Pi x \in A)B(x)}(b, \eta)$, $bx \geq d$ and $\vartheta_{B(x)}(\omega_A)(d, \eta(\omega_A))$. Since by IH, $B(x) \geq D(x)$ and $\vartheta_V(D(x), \alpha_{B(x)}(\omega_A))$, by Def. 1.1.2, d is normalizable. By IH and Def. 1.1.1, C is normalizable. Hence $(\lambda x \in C)d(x)$ is normalizable, and since $b \geq (\lambda x \in C)d(x)$, b is normalizable.

4.3.3 $\vartheta_{(\Pi x \in A)B(x)}(b, \omega_{(\Pi x \in A)B(x)})$ provided b is a normal expression of the form $ya_1 \ldots a_n$. Now, by Def. 1.3 $\omega_{(\Pi x \in A)B(x)}(\xi)$ is defined to be $\omega_{B(x)}(\xi)$ for ξ of τ_A. Therefore one has to verify that if $\vartheta_A(a, \xi)$ then there is a d s.t. $ba \geq d$ and $\vartheta_{B(x)}(\xi)(d, \omega_{B(x)}(\xi))$. Since by IH, $A \geq C$ s.t. $\vartheta_V(C, \alpha_A)$, by Def. 1.1.2, $\vartheta_A(a, \xi)$ implies that a is normalizable. Let d be the normal expression $ya_1 \ldots a_n a_{n+1}$ where a_{n+1} is the normal form of a. Because by IH, $B(a) \geq D$ s.t. $\vartheta_V(D, \alpha_{B(x)}(\xi))$, $\vartheta_{B(x)}(\xi)(d, \omega_{B(x)}(\xi))$ and by Def. 1.3 $\vartheta_{(\Pi x \in A)B(x)}(b, \omega_{(\Pi x \in A)B(x)})$.

4.4 (ΠI)

$$\frac{\begin{matrix} & [x \in A] \\ \vdots & \vdots \\ A \in V & b(x) \in B(x) \end{matrix}}{(\lambda x \in A)b(x) \in (\Pi x \in A)B(x)}$$

By IH, α_A is a well-defined object of τ_V and there is an expression C s.t. $A \geq C$ and $\vartheta_V(C, \alpha_A)$; also, for ξ of τ_A, $\alpha_{B(x)}(\xi)$ is a well-defined object of τ_V and $\alpha_{b(x)}(\xi)$ is a well-defined object of type $\tau_{B(x)}(\xi)$ s.t. if $\vartheta_A(a, \xi)$, then there is an expression d s.t. $b(a) \geq d$ and $\vartheta_{B(x)}(\xi)(d, \alpha_{b(x)}(\xi))$. Now, by Defs. 1.3 and 1.4 and IH it is clear that $\alpha_{(\Pi x \in A)B(x)}$ is a well-defined object of τ_V and $\alpha_{(\lambda x \in A)b(x)}$ is a well-defined object of $\tau_{(\Pi x \in A)B(x)}$.

To show that $\vartheta_{(\Pi x \in A)B(x)}((\lambda x \in A)b(x), \alpha_{(\lambda x \in A)b(x)})$ it has to be shown according to Def. 1.3, that for all $a \in A$ and ξ of τ_A, if $\vartheta_A(a, \xi)$ then there is some d s.t. $(\lambda x \in A)b(x)a \geq d$ and $\vartheta_{B(x)}(\xi)(d, \alpha_{b(x)}(\xi))$, and if $(\lambda x \in A)b(x) \geq (\lambda x \in C)d(x)$ then C is normalizable. By the contraction rule and IH, $(\lambda x \in A)b(x)a \geq b(a) \geq d$, and by Def. 1.4 $\alpha_{(\lambda x \in A)b(x)}(\xi)$ is defined to be $\alpha_{b(x)}(\xi)$ for ξ of τ_A. Moreover, by IH, A is normalizable and $A \geq C$; hence, by the Church–Rosser property, C is normalizable. Thus, by Def. 1.3, $\vartheta_{(\Pi x \in A)B(x)}((\lambda x \in A)b(x), \alpha_{(\lambda x \in A)b(x)})$.

4.5 (ΠE)

$$\frac{\begin{matrix} \vdots & & \vdots \\ b \in (\Pi x \in A)B(x) & & a \in A \end{matrix}}{ba \in B(a)}$$

By IH, α_A and $\alpha_{(\Pi x \in A)B(x)}$ are well-defined objects of τ_V; also α_a is a well-defined object of τ_A and α_b is a well-defined object of $\tau_{(\Pi x \in A)B(x)}$; and there are expressions c and d s.t. $a \geq c$ and $b \geq d$ and $\vartheta_A(c, \alpha_a)$ and $\vartheta_{(\Pi x \in A)B(x)}(d, \alpha_b)$. $\alpha_{B(a)} = \alpha_{B(x)}(\alpha_a)$ is a well-defined object of τ_V by the substitution property 2. To verify $\vartheta_{B(a)}(f, \alpha_{ba})$ for some f s.t. $ba \geq f$, observe that by Def. 1.3 of $\vartheta_{(\Pi x \in A)B(x)}$, there is an expression f s.t. $dc \geq f$ and $\vartheta_{B(x)}(\alpha_a)(f, \alpha_b(\alpha_a))$. Since by Def. 1.5 $\alpha_{ba} \equiv \alpha_b(\alpha_a)$ and by the substitution property 2, $\alpha_{B(x)}(\alpha_a) = \alpha_{B(a)}$, conclude $\vartheta_{B(a)}(f, \alpha_{ba})$ and $ba \geq dc \geq f$.

4.6 (Conversion rule)

$$\frac{\begin{matrix} \vdots & & \vdots \\ a \in A & & B \in V \end{matrix}}{a \in B} \qquad \text{where} \quad A = B$$

By IH, α_A is a well-defined object of τ_V and α_a is a well-defined object of τ_A; and there is an expression c s.t. $a \geq c$ and $\vartheta_A(c, \alpha_a)$; also α_B is a well-defined object of τ_V. And by the compatibility of equality 3. and $A = B$, $\alpha_A = \alpha_B$.

Trivially, α_B is a well-defined object of τ_V. By IH, $\alpha_A = \alpha_B$, hence $\tau_A = \tau_B$ and α_a is a well-defined object of τ_B by IH; hence also $\vartheta_A = \vartheta_B$, and by IH, $\vartheta_B(c, \alpha_a)$, for $a \geq c$.

Suppose there to be given an arbitrary derivation

$$\begin{matrix} x_1 \in A_1 & \cdots\cdots & x_n \in A_n(x_1, \ldots, x_{n-1}) \\ \ddots & & \iddots \\ & a(x_1, \ldots, x_n) \in A(x_1, \ldots, x_n) & \end{matrix}$$

with accompanying derivations

$$\begin{matrix} & & & x_1 \in A_1 & \cdots\cdots & x_{n-1} \in A_{n-1}(x_1, \ldots, x_{n-2}) \\ & x_1 \in A_1 & & \ddots & & \iddots \\ \vdots & \vdots & \cdots & & & \\ A_1 \in V & A_2(x_1) \in V & & & A_n(x_1, \ldots, x_{n-1}) \in V & \end{matrix}$$

Then, by IH,

α_{A_1} is a well-defined object of τ_V and there is a C_1 s.t. $A_1 \geq C_1$ and $\vartheta_V(C_1, \alpha_{A_1})$

$\vdots$

$\alpha_{A_n(x_1,\ldots,x_{n-1})}(\xi_1, \ldots, \xi_{n-1})$ is a well-defined object of τ_V for ξ_1 of $\tau_{A_1}, \ldots, \xi_{n-1}$ of $\tau_{A_{n-1}(x_1,\ldots,x_{n-2})}(\xi_1, \ldots, \xi_{n-2})$ and if $\vartheta_{A_1}(a_1, \xi_1), \ldots, \vartheta_{A_{n-1}(x_1,\ldots,x_{n-2})}(\xi_1, \ldots, \xi_{n-2})(a_{n-1}, \xi_{n-1})$,
then there is a C_n s.t. $A_n(a_1, \ldots, a_{n-1}) \geq C_n$ and $\vartheta_V(C_n, \alpha_{A_n(x_1,\ldots,x_{n-1})}(\xi_1, \ldots, \xi_{n-1}))$.

If one defines $\omega_1 = \omega_{A_1}, \ldots, \omega_n = \omega_{A_n(x_1,\ldots,x_{n-1})}(\omega_1, \ldots, \omega_{n-1})$ then $\vartheta_{A_1}(x_1, \omega_1), \ldots, \vartheta_{A_n(x_1,\ldots,x_{n-1})}(\omega_1, \ldots, \omega_{n-1})(x_n, \omega_n)$.

Conclude, by the preceding proof and the IH that there is an expression c s.t. $a(x_1, \ldots, x_n) \geq c$ and $\vartheta_{A(x_1,\ldots,x_n)}(\omega_1, \ldots, \omega_n)(c, \alpha_{a(x_1,\ldots,x_n)}(\omega_1, \ldots, \omega_n))$.

On the other hand, given a derivation of $a(x_1, \ldots, x_n) \in A(x_1, \ldots, x_n)$ there is also a derivation

$$\begin{array}{ccc} x_1 \in A_1 & \ldots\ldots & x_n \in A_n(x_1, \ldots, x_{n-1}) \\ \ddots & & \iddots \\ & A(x_1, \ldots, x_n) \in V & \end{array}$$

so that there is an expression C s.t. $A(x_1, \ldots, x_n) \geq C$ and $\vartheta_V(C, \alpha_{A(x_1,\ldots,x_n)}(\omega_1, \ldots, \omega_n))$; as was to be proved.

If one compares the normalization proof of version (2) with the one of version (1), the following differences appear: Φ of version (2) is a partial computability predicate defined for closed type expressions only, whereas ϑ of version (1) is a total computability predicate defined for any type expressions. Also, identity on Φ was seen to be extensional, whereas identity on ϑ is obviously definitional or intensional (cf. Lemma 3 on the compatibility with equality); this is due to another difference between the versions (2) and (1), namely that the computability predicate Φ_A is defined on canonical types while the computability predicate ϑ_A is defined on normal types. Finally, the normalization proof of version (2) is not formalized and proved in the theory itself, whereas the normalization proof of version (1) is. This is possible in version (1) only because of a reflection principle, according to which everything that can be done with types can be done inside V, and because of the inconsistent axiom $V \in V$. This axiom, however, was exactly the reason for abandoning version (1) and its normalization proof and for moving on to version (2), and in this case to version (3).

(3) As mentioned above, in version (3) an approach to type theory is adopted in which conversion is defined on well-typed terms. *a conv b* forms a judgement, and the rules for forming judgements of that form, i.e. the conversion rules, are such that it is immediately clear that parallel to a derivation of *a conv b* there run derivations of $a \in A$ and $b \in B$ for some type symbols A and B.

The notions of conversion and reduction are accordingly redefined by the following rules:

Π-conversion
Π-reduction

$$a_1 \in A_1, \ldots, a_k \in A_k[a_1, \ldots, a_{k-1}]$$
$$a \in A[a_1, \ldots, a_k]$$
f has been defined by the (ΠI)-scheme:
$$\frac{f(x_1, \ldots, x_k)(x) \overset{conv}{\underset{red}{}} b[x_1, \ldots, x_k, x]}{f(a_1, \ldots, a_k)(a) \overset{conv}{\underset{red}{}} b[a_1, \ldots, a_k, a]}$$

Σ-conversion
Σ-reduction

$$a_1 \in A_1, \ldots, a_k \in A_k[a_1, \ldots, a_{k-1}]$$
$$a \in A[a_1, \ldots, a_k] \quad b \in B[a_1, \ldots, a_k, a]$$
f has been defined by the (ΣE)-scheme:
$$\frac{f(x_1, \ldots, x_k, (x_1, \ldots, x_k, x, y)) \overset{conv}{\underset{red}{}} c[x_1, \ldots, x_k, x, y]}{f(a_1, \ldots, a_k, (a_1, \ldots, a_k, a, b)) \overset{conv}{\underset{red}{}} c[a_1, \ldots, a_k, a, b]}$$

+-conversion
+-reduction

$$a_1 \in A_1, \ldots, a_k \in A_k[a_1, \ldots, a_{k-1}]$$
$$a \in A[a_1, \ldots, a_k] \qquad b \in B[a_1, \ldots, a_k]$$
f has been defined by the $(+E)$-scheme:
$$\frac{\begin{cases} f(x_1, \ldots, x_k, i(x_1, \ldots, x_k, x)) \overset{conv}{\underset{red}{}} c[x_1, \ldots, x_k, x] \\ f(x_1, \ldots, x_k, j(x_1, \ldots, x_k, y)) \overset{conv}{\underset{red}{}} d[x_1, \ldots, x_k, y] \end{cases}}{\begin{cases} f(a_1, \ldots, a_k, i(a_1, \ldots, a_k, a)) \overset{conv}{\underset{red}{}} c[a_1, \ldots, a_k, a] \\ f(a_1, \ldots, a_k, j(a_1, \ldots, a_k, b)) \overset{conv}{\underset{red}{}} d[a_1, \ldots, a_k, b] \end{cases}}$$

N_n-conversion
N_n-reduction

$$a_1 \in A_1, \ldots, a_k \in A_k[a_1, \ldots, a_{k-1}]$$
f has been defined by the $(N_n E)$-scheme:
$$\frac{\begin{cases} f(x_1, \ldots, x_k, 1) \overset{conv}{\underset{red}{}} c_1[x_1, \ldots, x_k] \\ \vdots \\ f(x_1, \ldots, x_k, n) \overset{conv}{\underset{red}{}} c_n[x_1, \ldots, x_k] \end{cases}}{\begin{cases} f(a_1, \ldots, a_k, 1) \overset{conv}{\underset{red}{}} c_1[a_1, \ldots, a_k] \\ \vdots \\ f(a_1, \ldots, a_k, n) \overset{conv}{\underset{red}{}} c_n[a_1, \ldots, a_k] \end{cases}}$$

N-conversion $\quad a_1 \in A_1, \ldots, a_k \in A_k[a_1, \ldots, a_{k-1}] \quad a \in N$

N-reduction $\quad f$ has been defined by the (NE)-scheme:

$$\frac{\left\{\begin{array}{lcl} f(x_1, \ldots, x_k, 0) & \overset{conv}{\underset{red}{}} & c[x_1, \ldots, x_k] \\ f(x_1, \ldots, x_k, s(x)) & \overset{conv}{\underset{red}{}} & d[x_1, \ldots, x_k, x, f(x_1, \ldots, x_k, x)] \end{array}\right.}{\left\{\begin{array}{lcl} f(a_1, \ldots, a_k, 0) & \overset{conv}{\underset{red}{}} & c[a_1, \ldots, a_k] \\ f(a_1, \ldots, a_k, s(a)) & \overset{conv}{\underset{red}{}} & d[a_1, \ldots, a_k, a, f(a_1, \ldots, a_k)] \end{array}\right.}$$

General rules of conversion/reduction

$$\frac{\begin{array}{c} a \in A[a_1, \ldots, a_k] \qquad c \in A[a_1, \ldots, a_k] \\ b \in (\Pi x \in A[a_1, \ldots, a_k]) B[a_1, \ldots, a_k, x] \\ d \in (\Pi x \in A[a_1, \ldots, a_k]) B[a_1, \ldots, a_k, x] \\ a \overset{conv}{\underset{red}{}} c \qquad b \overset{conv}{\underset{red}{}} d \end{array}}{b(a) \overset{conv}{\underset{red}{}} d(c)}$$

$$\frac{\begin{array}{c} f \text{ is a function constant with arguments} \\ a_1 \in A_1, \ldots, a_k \in A_k[a_1, \ldots, a_{k-1}] \\ c_1 \in A_1, \ldots, c_k \in A_k[c_1, \ldots, c_{k-1}] \\ a_1 \overset{conv}{\underset{red}{}} c_1, \ldots, a_k \overset{conv}{\underset{red}{}} c_k \end{array}}{f(a_1, \ldots, a_k) \overset{conv}{\underset{red}{}} f(c_1, \ldots, c_k)}$$

$$\frac{a \in A}{a \overset{conv}{\underset{red}{}} a} \qquad\qquad \frac{a \overset{conv}{\underset{red}{}} b \quad b \overset{conv}{\underset{red}{}} c}{a \overset{conv}{\underset{red}{}} c}$$

but $\dfrac{a \text{ conv } b}{b \text{ conv } a}$ does not hold for reduction!

A function f is called an introductory constant if f has been introduced by one of the (Π-, Σ-, $+$-, N_n-, N- or V_n-) introduction rules. The type symbols of the arguments of an introductory constant f are always of the form $A_1, \ldots, A_k[x_1, \ldots, x_{k-1}]$ and the type constant of the value of f is always of the form $F(x_1, \ldots, x_k)$, where F is an introductory type constant, denoting one of the basic types (N_n, N, V_n) or one of the type forming (Π, Σ, $+$).

A term $c \in C$ is called a closed normal term with indicated type symbol if $c \in C$ can be derived by repeated applications of the following single rule of term formation ($CNTF$-rule):

$$\frac{\begin{array}{c} f \text{ is a (possibly 0-ary) introductory constant} \\ c_1 \in C_1, \ldots, c_k \in C_k \text{ are closed normal terms} \\ \text{there exist closed derivations of } A_1 \; red \; C_1, \ldots, A_k[c_1, \ldots, c_{k-1}] \; red \; C_k \end{array}}{f(c_1, \ldots, c_k) \in F(c_1, \ldots, c_k) \text{ is a closed normal term}}$$

Note: if $c \in C$ is a closed normal term with indicated type symbol, then C is called a closed normal type symbol (i.e. a closed normal term with type symbol V_n, for some $n \in \omega$) which is uniquely associated with c.

c is said to be a closed normal term if $c \in C$ is a closed normal term with indicated type symbol, for some C.

A term a is said to be normal (or irreducible) if for any term b, a *red* b implies $a = b$.

Clearly, a closed normal term is closed and normal. But the proof of the converse, i.e. that a closed and normal term is closed normal, requires the Church–Rosser property which will only be proved as a consequence of the normalization theorem.

Martin-Löf's proof of the normalization theorem (of version (3)) uses again an extension of the method of computability introduced by Tait. The proof consists of 1) the def. of a normal form a' for every closed term $a \in A$, 2) the def. of a computability predicate A'' which expresses the computability (reducibility) of a normal term a' of type A' together with the def. of a proof object a'' of the proposition $A''(a')$, i.e. the normal form a' of a term $a \in A$ is computable, 3) the proof that if $a \in A$ is derivable, then $a' \in A'$ is a closed normal term s.t. a *red* a', 4) the proof that if $a \in A$ is derivable, a'' is indeed a proof of $A''(a')$, 5) the proof that if a *conv* b is derivable, then $a' =_{\text{def}} b'$, and 6) the proof that if a *conv* b is derivable then $a'' =_{\text{def}} b''$.

As in version (2), the case of the closed terms can be extended to the one of open terms.

If one compares the normalization proof of version (2) with the one of version (3), one notices that step 1) above is new: now, steps 1), 3) and 5) serve to prove the Church–Rosser property as a corollary to the normalization theorem – whereas the Church–Rosser property in version (2) was proved before and independently of normalization in the standard combinatorial way. Moreover, in step 2) the introduction of a proof object for the verification that $A''(a')$ is new. Furthermore, if in step 6) the terms a and b are chosen to be A and B resp., step 6) becomes very much reminiscent of Lemma 3 in version (2): but in the case of version (3), if A *conv* B, then A'' and B'' are not extensionally, but intensionally or definitionally identical computability predicates. This is due to the fact that the computability predicate Φ_A in version (2) was defined on canonical types, while in version (3) the computability predicate A'' is defined on normal types. Ideally speaking, for each closed term $a \in A$, even a third object a''' should be defined which would be the proof object of a *red* a', but Martin-Löf renounces to the introduction of such an object and the same procedure will be followed here. Finally, the ξ-rule, that is conversion under an abstraction, is valid in version (2) but not in version (3).

1) **Definition of $'$ of a closed term** $a \in A$ (def. of the normal form of a)

1.1.1 $N'_n =_{\text{def}} N_n$
$m' =_{\text{def}} m$ where $m \in N_n$
$f'(m, m'') =_{\text{def}} c'_m$ where f is introduced by the $(N_n E)$-scheme
$f(m)\ conv\ c_m,\ m \in N_n$

1.1.2 $N' =_{\text{def}} N$
$0' =_{\text{def}} 0$
$s'(x', x'') =_{\text{def}} s(x')$ where $x' \in N$ and x'' is a proof of $N''(x')$

$$\begin{cases} f'(0, 0'') =_{\text{def}} c' \\ f'(s(x'), s''(x', x'')) =_{\text{def}} d'[x', x'', f'(x', x''), f''(x', x'')] \end{cases}$$

where f is introduced by the (NE)-scheme $\begin{cases} f(0)\ conv\ c \\ f(s(x))\ conv\ d[x, f(x)] \end{cases}$

1.2 $(\Pi x \in A)B[x]' =_{\text{def}} (\Pi x \in A)B[x]$
$f' =_{\text{def}} f$ where f is introduced by the(ΠI)-scheme
$f(x)\ conv\ b[x]$
$c(a)' =_{\text{def}} p'(c''(a', a''))$ where $c \in (\Pi x \in A)B[x]$, $a \in A$, p' is the first projection which takes a proof of the existential proposition: there exists a closed normal term $y' \in B'[a', a'']$ s.t. $B''[a', a''](y')$ and $c'(a', a'')\ red\ y'$, into the object y'

1.3 $(\Sigma x \in A)B[x]' =_{\text{def}} (\Sigma x \in A)B[x]$
$(x, y)' =_{\text{def}} (x', y')$
$f'((x', y'), (x', x'', y', y'')) =_{\text{def}} c'[x', x'', y', y'']$ where f is introduced by the (ΣE)-scheme $f((x, y))\ conv\ c[x, y]$

1.4 $(A + B)' =_{\text{def}} A + B$

$$\begin{cases} i'(x', x'') =_{\text{def}} i(x') \\ j'(y', y'') =_{\text{def}} j(y') \end{cases}$$

$$\begin{cases} f'(i(x'), i''(x', x'')) =_{\text{def}} c'[x', x''] \\ f'(j(y'), j''(y', y'')) =_{\text{def}} d'[y', y''] \end{cases}$$

where f is introduced by the $(+E)$-scheme

$$\begin{cases} f(i(x))\ conv\ c[x] \\ f(j(y))\ conv\ d[y] \end{cases}$$

1.5 $V'_n =_{\text{def}} V_n,\ n \in \omega$

2) **Definition of $''$ of a closed term** $a \in A$ (def. of the computability fct./predicate A'' and of the related proof objects a'')

2.1.1 N''_n is defined to be the least predicate s.t. if $m \in N_n$ is a closed normal term, then $N''_n(m)$ and m'' is a proof thereof.
If f is introduced by the $(N_n E)$-scheme $f(m)\ conv\ c_m$, for $m \in N_n$, then $f''(m, m'') =_{\text{def}} c''_m$.

2.1.2 N'' is defined to be the least predicate s.t.

1. if $0 \in N$ is a closed normal term, then $N''(0)$ and $0''$ is a proof thereof.

2. if $s(x') \in N'$, and $x' \in N'$ and x'' is a proof of $N''(x')$, then $N''(s(x'))$ and $s''(x', x'')$ is a proof thereof.

   Define

   $s'' =_{\text{def}}$ the function which to $x' \in N'$ and $x'' \in N''(x')$ assigns a proof of $N''(s(x'))$.

   If f is introduced by the (NE)-scheme $\begin{cases} f(0)\ \mathit{conv}\ c \\ f(s(x))\ \mathit{conv}\ d[x, f(x)] \end{cases}$

   then $\begin{cases} f''(0, 0'') =_{\text{def}} c'' \\ f''(s(x'), s''(x', x'')) =_{\text{def}} d''[x', x'', f'(x', x''), f''(x', x'')] \end{cases}$

2.2 $(\Pi x \in A)B[x]''$ is defined to be the least predicate s.t. if $f' \in (\Pi x \in A)B[x]'$ is a closed normal term, and for all closed normal terms $x' \in A'$ and for all proofs x'' of $A''(x')$, there is a closed normal term $y' \in B'[x', x'']$ s.t. $f'(x', x'')\ \mathit{red}\ y'$ and $B''[x', x''](y')$, then $(\Pi x \in A)B[x]''(f')$ and f'' is a proof thereof.

Define

$f'' =_{\text{def}}$ the fct. which to $x' \in A'$ and x'' is a proof of $A''(x')$ assigns the triple $(b'[x', x''], b''[x', x'']$, the proof of $f(x')\ \mathit{red}\ b'[x', x''])$

If f is introduced by the (ΠI)-scheme $f(x)\ \mathit{conv}\ b[x]$ then

$f(x)'' =_{\text{def}} p''(f''(x', x''))$ where p'' is the second projection which takes a proof of the existential proposition: there exists a closed normal term $z' \in B'[x', x'']$ s.t. $B''[x', x''](z')$ and $f'(x', x'')\ \mathit{red}\ z'$, into the proof of $B''[x', x''](z')$.

2.3 $(\Sigma x \in A)B[x]''$ is defined to be the least predicate s.t. if $(x', y') \in (\Sigma x \in A)B[x]'$ is a closed normal term, and $x' \in A'$ is a closed normal term, x'' is a proof of $A''(x')$, $y' \in B'[x', x'']$ is a closed normal term, y'' is a proof of $B''[x', x''](y')$, then $(\Sigma x \in A)B[x]''((x', y'))$ and (x', x'', y', y'') is a proof thereof.

Define

$$(x, y)'' =_{\text{def}} (x', x'', y', y'')$$

If f is introduced by the (ΣE)-scheme $f((x, y))\ \mathit{conv}\ c[x, y]$ then

$$f''((x', y'), (x', x'', y', y'')) =_{\text{def}} c''[x', x'', y', y''].$$

2.4 $(A + B)''$ is defined to be the least predicate s.t.

1. if $i(x') \in (A + B)'$ is a closed normal term, $x' \in A'$ is a closed normal term, x'' is a proof of $A''(x')$, then $(A + B)''(i(x'))$ and $i''(x', x'')$ is a proof thereof.
2. if $j(y') \in (A + B)'$ is a closed normal term, $y' \in B'$ is a closed normal term, y'' is a proof of $B''(y')$, then $(A + B)''(j(y'))$ and $j''(y', y'')$ is a proof thereof.

Define

$$i'' =_{\text{def}} \text{ the fct. which to } x' \in A' \text{ and } x'' \in A''(x') \text{ assigns a proof of } (A+B)''(i(x'))$$
$$j'' =_{\text{def}} \text{ the fct. which to } y' \in B' \text{ and } y'' \in B''(y') \text{ assigns a proof of } (A+B)''(j(y'))$$

If f is introduced by the $(+E)$-scheme $\begin{cases} f(i(x))\ conv\ c[x] \\ f(j(y))\ conv\ d[y] \end{cases}$

then $\begin{aligned} f''(i(x'), i''(x', x'')) &=_{\text{def}} c''[x', x''] \\ f''(j(y'), j''(y', y'')) &=_{\text{def}} d''[y', y''] \end{aligned}$

2.5 $V_n'' =_{\text{def}}$ the function which to a closed normal term $A' \in V_n' =_{\text{def}} V_n$ assigns the type of nth order species of closed normal terms with type symbol A'.

3) **Proof that if** $a \in A$ is a closed term, then $a' \in A'$ is a closed normal term s.t. $a\ red\ a'$

Exercise, or cf. version (3)

4) **Proof that if** $a \in A$ is a closed term, then a'' is a proof of $A''(a')$

4.1 (Ass. rule) If $x \in A$ is introduced, by IH, $A' \in V_n' =_{\text{def}} V_n$ is a closed normal term, and A'' has been defined and is an nth order species of closed normal terms with type symbol A'. Therefore one can let $x' \in A'$ be an arbitrary closed normal term with indicated type symbol, and x'' an arbitrary proof of $A''(x')$.

4.2 (Constant rule) By IH, $a_1' \in A_1', \ldots, a_k' \in A_k[a_1, \ldots, a_{k-1}]'$, $a_1'' \in A_1''(a_1')$, $\ldots, a_k'' \in A_k[a_1, \ldots, a_{k-1}]''(a_k')$, f is a function constant whose arguments and value have type symbols $A_1, \ldots, A_k[x_1, \ldots, x_{k-1}]$ and $A[x_1, \ldots, x_k]$, and $f'(x_1', x_1'', \ldots, x_k', x_k'') \in A'[x_1', x_1'', \ldots, x_k', x_k'']$ is to be a closed normal term s.t. $f(x_1', \ldots, x_k')\ red\ f'(x_1', x_1'', \ldots, x_k', x_k'')$, and $f''(x_1', x_1'', \ldots, x_k', x_k'')$ is to be a proof of $A''[x_1', x_1'', \ldots, x_k', x_k''](f'(x_1', x_1'', \ldots, x_k', x_k''))$, where $x_1' \in A_1'$ denotes an arbitrary closed normal term, x_1'' denotes an arbitrary proof of $A_1''(x_1')$, $\ldots, x_k' \in A_k'[x_1', x_1'', \ldots, x_{k-1}', x_{k-1}'']$ denotes an arbitrary closed normal term, and x_k'' denotes an arbitrary proof of $A_k''[x_1', x_1'', \ldots, x_{k-1}', x_{k-1}''](x_k')$.

Define

$$f(a_1, \ldots, a_k)' =_{\text{def}} f'(a_1', a_1'', \ldots, a_k', a_k'')$$
$$f(a_1, \ldots, a_k)'' =_{\text{def}} f''(a_1', a_1'', \ldots, a_k', a_k'')$$

Then by the IH and the substitution property, $f(a_1, \ldots, a_k)' \in A[a_1, \ldots, a_k]'$ so defined is a closed normal term s.t.

$$\begin{aligned} f(a_1, \ldots, a_k) \quad & \mathit{red}\ f(a_1', \ldots, a_k') \text{ by } IH \\ & \mathit{red}\ f'(a_1', a_1'', \ldots, a_k', a_k'') \\ & \mathit{red}\ f(a_1, \ldots, a_k)' \text{ by } =_{\mathrm{def}} \end{aligned}$$

and $f(a_1, \ldots, a_k)''$ is a proof of $A[a_1, \ldots, a_k]''(f(a_1, \ldots, a_k)')$.

4.3 (ΠI) By IH, $b[x]''$ is a proof of $B[x]''(b[x]')$, $x' \in A'$ and x'' is a proof of $A''(x')$.

$$f(x)'' =_{\mathrm{def}} p''(f''(x', x'')) =_{\mathrm{def}} b''[x', x''] =_{\mathrm{def}} b[x]''$$

is by IH a proof of

$$\begin{aligned} B[x]''(b[x]') & =_{\mathrm{def}} B[x]''(b'[x', x'']) \\ & =_{\mathrm{def}} B[x]''(p'(f''(x', x''))) \\ & =_{\mathrm{def}} B[x]''(f(x)') \end{aligned}$$

4.4 (ΠE)

$c(a)'' =_{\mathrm{def}}$ $p''(c''(a', a'')) =_{\mathrm{def}}$ the proof of $B''[a', a''](y')$ where $y' \in B'[a', a'']$ is a closed normal term s.t. $c'(a', a'')$ *red* y'. Let y' be $p'(c''(a', a''))$. Then

$$\begin{aligned} c(a)'' =_{\mathrm{def}} \text{ the proof of } B''[a', a''](p'(c''(a', a''))) & =_{\mathrm{def}} B[a]''(p'(c''(a', a''))) \\ & =_{\mathrm{def}} B[a]''(c(a)') \end{aligned}$$

4.5 (ΣI) By IH, $x' \in A'$, x'' is a proof of $A''(x')$, $y' \in B[x]'$, y'' is a proof of $B[x]''(y')$.

$(x, y)'' =_{\mathrm{def}} (x', x'', y', y'')$ which is a proof of

$$(\Sigma x \in A)B[x]''((x', y')) =_{\mathrm{def}} (\Sigma x \in A)B[x]''((x, y)')$$

4.6 (ΣE) By IH, $c[x, y]''$ is a proof of $C[(x, y)]''(c[x, y]')$, where $x' \in A'$, x'' is a proof of $A''(x')$, $y' \in B[x]'$, and y'' is a proof of $B[x]''(y')$.

$$\begin{aligned} f((x, y))'' =_{\mathrm{def}} f''((x, y)', (x, y)'') & =_{\mathrm{def}} f''((x', y'), (x', x'', y', y'')) \\ & =_{\mathrm{def}} c''[x', x'', y', y''] \\ & =_{\mathrm{def}} c[x, y]'' \text{ which by } IH \text{ is a proof of} \\ C[(x, y)]''(c[x, y]') & =_{\mathrm{def}} C[(x, y)]''(c'[x', x'', y', y'']) \\ & =_{\mathrm{def}} C[(x, y)]''(f'((x', y'), (x', x'', y', y''))) \\ & =_{\mathrm{def}} C[(x, y)]''(f'((x, y)', (x, y)'')) \\ & =_{\mathrm{def}} C[(x, y)]''(f((x, y)')) \end{aligned}$$

4.7 ($+I$) By IH, $x' \in A'$ and x'' is a proof of $A''(x')$.

$$\begin{aligned} i(x)'' =_{\mathrm{def}} i''(x', x'') \text{ which is a proof of } & (A + B)''(i(x')) \\ =_{\mathrm{def}} & (A + B)''(i'(x', x'')) \\ =_{\mathrm{def}} & (A + B)''(i(x)'). \end{aligned}$$

Similarly for $j(y)''$.

4.8 $(+E)$ By IH, $c[x]''$ is a proof of $C[i(x)]''(c[x]')$, and $d[y]''$ is a proof of $C[j(y)]''(d[y]')$ where $x' \in A'$, $y' \in B'$ etc.

$$\begin{aligned} f(i(x))'' &=_{\text{def}} f''(i(x)', i(x)'') \\ &=_{\text{def}} f''(i'(x', x''), i''(x', x'')) \\ &=_{\text{def}} f''(i(x'), i''(x', x'')) \\ &=_{\text{def}} c''[x', x''] \\ &=_{\text{def}} c[x]'' \text{ is by } IH \text{ a proof of } C[i(x)]''(c(x)') \\ &\qquad =_{\text{def}} C[i(x)]''(c'(x', x'')) \\ &\qquad =_{\text{def}} C[i(x)]''(f'(i'(x', x''), i''(x', x''))) \\ &\qquad =_{\text{def}} C[i(x)]''(f'(i(x)', i(x)'')) \\ &\qquad =_{\text{def}} C[i(x)]''(f(i(x))') \end{aligned}$$

4.9 $(N_n I)$ m'' is a proof of $N''_m(m')$, by definition of N''_n and $m', m \in N_n$.

4.10 $(N_n E)$ By IH, c''_m is a proof of $C[m]''(c'_m)$, $m \in N_n$.

$$\begin{aligned} f(m)'' =_{\text{def}} f''(m, m'') =_{\text{def}} c''_m \text{ is by } IH \text{ a proof of } & C[m]''(c'_m) \\ &=_{\text{def}} C[m]''(f'(m, m'')) \\ &=_{\text{def}} C[m]''(f(m)') \end{aligned}$$

4.11 (NI) $0''$ is a proof of $N''(0')$, by def. of N'' and $0'$.
By IH, x'' is a proof of $N''(x')$ and $x' \in N'$.

$$\begin{aligned} s(x)'' =_{\text{def}} s''(x', x'') \text{ which is a proof of } N''(s(x')) &=_{\text{def}} N''(s'(x', x'')) \\ &=_{\text{def}} N''(s(x)') \end{aligned}$$

4.12 (NE) By IH, c'' is a proof of $C[0]''(c')$.

$$\begin{aligned} f(0)'' =_{\text{def}} f''(0, 0'') =_{\text{def}} c'' \text{ which by } IH \text{ is a proof of } & C[0]''(c') \\ &=_{\text{def}} C[0]''(f'(0, 0'')) \\ &=_{\text{def}} C[0]''(f(0)') \end{aligned}$$

By IH, $d[x, f(x)]''$ is a proof of $C[s(x)]''(d[x, f(x)]')$

$$\begin{aligned} f(s(x))'' &=_{\text{def}} f''(s(x)', s(x)'') =_{\text{def}} f''(s'(x', x''), s''(x', x'')) \\ &=_{\text{def}} f''(s(x'), s''(x', x'')) =_{\text{def}} d''[x', x'', f'(x', x''), f''(x', x'')] \\ &=_{\text{def}} d[x, f(x)]'' \text{ which by } IH \text{ is a proof of } C[s(x)]''(d[x, f(x)]') \\ &\qquad =_{\text{def}} C[s(x)]''(d'[x', x'', f'(x', x''), f''(x', x'')]) \\ &\qquad =_{\text{def}} C[s(x)]''(f'(s(x'), s''(x', x''))) \\ &\qquad =_{\text{def}} C[s(x)]''(f(s(x))') \end{aligned}$$

4.13 (VI) For a closed type symbol A (e.g. $N_n, N, (\Pi x \in A)B[x], (\Sigma x \in A)B[x]$ or $A + B$) A'' is a nth order species of closed normal terms with type symbol A', for some $n \in \omega$. Hence, one clearly has $A'' \in V''_n(A')$, for a closed term $A \in V_n$.

5) **Proof that for closed terms** a, b **if** $a \; conv \; b$ **then** $a' =_{\text{def}} b'$
Exercise, or cf. version (3).

6) **Proof that for closed terms** a, b **if** $a \; conv \; b$**, then** $a'' =_{\text{def}} b''$

6.1 If $A \in V_n$, for some $n \in \omega$, one has to argue informally since no $(V_n E)$-rule is given in version (3) and hence no (V_n-conversion)-rule either.

6.1.1 If A is N_n or N and $A\,conv\,A$, then A'' is defined in 2.1 and $A'' =_{\text{def}} A''$.

6.1.2 If A and B have the forms $(\Pi x \in C)D[x]$ and $(\Pi x \in E)F[x]$ resp. and $A\,conv\,B$, then $C\,conv\,E$ and $D[x]\,conv\,F[x]$. By IH, $C =_{\text{def}} E$ and $D[c] =_{\text{def}} F[c]$ for all $c \in C$ s.t. $C''(c')$. Since $(\Pi x \in C)D[x]''$ and $(\Pi x \in E)F[x]''$ are both defined by 2.2, $A'' =_{\text{def}} B''$.

6.1.3 If A and B have the forms $(\Sigma x \in C)D[x]$ and $(\Sigma x \in E)F[x]$ resp. and $A\,conv\,B$, then $C\,conv\,E$ and $D[x]\,conv\,F[x]$. By IH, $C =_{\text{def}} E$ and $D[c] =_{\text{def}} F[c]$ for all $c \in C$ s.t. $C''(c')$. Since $(\Sigma x \in C)D[x]''$ and $(\Sigma x \in E)F[x]''$ are both defined by 2.3, $A'' =_{\text{def}} B''$.

6.1.4 If A and B have the forms $C + D$ and $E + F$ resp. and $A\,conv\,B$, then $C\,conv\,E$ and $D\,conv\,F$. By IH, $C =_{\text{def}} E$ and $D =_{\text{def}} F$. Since $(C + D)''$ and $(E + F)''$ are both defined by 2.4, $A'' =_{\text{def}} B''$.

6.2 (Π-conversion)

$$\begin{aligned} f(a)'' =_{\text{def}} p''(f''(a', a'')) &=_{\text{def}} b''[a', a''] \\ &=_{\text{def}} b[a]'' \end{aligned}$$

6.3 (Σ-conversion)

$$\begin{aligned} f((a, b))'' &=_{\text{def}} f''((a, b)', (a, b)'') \\ &=_{\text{def}} f''((a', b'), (a', a'', b', b'')) \\ &=_{\text{def}} c''[a', a'', b', b''] =_{\text{def}} c[a, b]'' \end{aligned}$$

6.4 ($+$–conversion)

$$\begin{aligned} f(i(a))'' &=_{\text{def}} f''(i(a)', i(a)'') \\ &=_{\text{def}} f''(i(a'), i''(a', a'')) \\ &=_{\text{def}} c''[a', a''] =_{\text{def}} c[a]'' \\ f(j(b))'' &=_{\text{def}} f''(j(b)', j(b)'') \\ &=_{\text{def}} f''(j(b'), j''(b', b'')) \\ &=_{\text{def}} d''[b', b''] =_{\text{def}} d[b]'' \end{aligned}$$

6.5 (N_n-conversion)

$$\begin{aligned} f(m)'' &=_{\text{def}} f''(m', m'') =_{\text{def}} f''(m, m'') \\ &=_{\text{def}} c''_m \end{aligned}$$

6.6 (N-conversion)

$$\begin{aligned} f(0)'' &=_{\text{def}} f''(0', 0'') =_{\text{def}} f''(0, 0'') =_{\text{def}} c'' \\ f(s(a))'' &=_{\text{def}} f''(s(a)', s(a)'') \\ &=_{\text{def}} f''(s(a'), s''(a', a'')) \\ &=_{\text{def}} d''[a', a'', f'(a', a''), f''(a', a'')] \\ &=_{\text{def}} d[a, f(a)]'' \end{aligned}$$

Notes

(i) As in the case of version (2) the normalization proof is carried out for closed terms only, but as Martin-Löf points out, it can readily be made to cover open terms as well.

Furthermore, in order to alleviate the already considerable notational burden, the normalization proof was carried out for the case of no parameters.

(ii) In many of the chains of definitional equality $=_{\text{def}}$ above, a certain property plays an important role, the so-called substitution property:
If one has derivations of $a_1 \in A_1, \ldots, a_k \in A_k[a_1, \ldots, a_{k-1}]$ to which the IH applies, i.e. s.t.

$$a_1' \in A_1', \ldots, a_k' \in A_k[a_1, \ldots, a_{k-1}]' =_{\text{def}} A_k'[a_1', a_1'', \ldots, a_{k-1}', a_{k-1}'']$$

and

$$\begin{aligned} a_1'' \in A_1''(a_1'), \ldots, a_k'' \in a_k'' \in \ & A_k[a_1, \ldots, a_{k-1}]''(a_k') \\ =_{\text{def}} \ & A_k''[a_1', a_1'', \ldots, a_{k-1}', a_{k-1}''](a_k') \end{aligned}$$

then

$$\begin{aligned} a[a_1, \ldots, a_k]' =_{\text{def}} a'[a_1', a_1'', \ldots, a_k', a_k''] \ &\in \ A[a_1, \ldots, a_k]' \\ &\qquad =_{\text{def}} A'[a_1', a_1'', \ldots, a_k', a_k''] \\ a[a_1, \ldots, a_k]'' =_{\text{def}} a''[a_1', a_1'', \ldots, a_k', a_k''] \ &\in \ A[a_1, \ldots, a_k]''(a[a_1, \ldots, a_k]') \\ &\qquad =_{\text{def}} A''[a_1', a_1'', \ldots, a_k' a_k''] \\ &\qquad\quad (a[a_1, \ldots, a_k]') \end{aligned}$$

In each case in which a derivation ends with $a[x_1, \ldots, x_k] \in A[x_1, \ldots, x_k]$ one has to verify this substitution property (which is used e.g. in the Proofs 4.2, 4.4, 4.6, ... , 6.2, 6.3, 6.4, ...).

(iii) The square brackets in $a'[x_1', x_1'', \ldots, x_k', x_k'']$ and $a''[x_1', x_1'', \ldots, x_k', x_k'']$ are used to indicate that these expressions depend on the (informal!) variables $x_1', x_1'', \ldots, x_k', x_k''$ which are explained in 4.2.

(iv) In order to emphasize the differences and similarities between the normalization proofs of version (2) and version (3), no attention was paid throughout the steps 1)–6) to the identity type I of version (3) which does not occur in version (2).

(4) Subsequently to version (3) Martin-Löf began to have doubts concerning normalization proofs and implicitly given intuitionistic term models of constructive type theory. The reasoning leading to his doubts is the following:[3] in model theory, when one models a formal language, one assigns to the syntactical entities of the formal language certain mathematical objects which are then referred to as the interpretations of the syntactical entities. If model theory is understood intuitionistically, then one proves that if a formula of the formal language is formally derivable, then the proof which is assigned to the formal derivation is a proof of the proposition assigned to the end formula of the derivation. This is no doubt one way of giving meaning to the various syntactical entities of a formal language.

But since it is not possible to deal with these mathematical objects, which have been modelled, in a language-free way – after all, one always needs a linguistic expression to refer to the object which is to serve as the interpretation of the syntactical entity – one can look at what is being interpreted (i.e. the syntactic entities of the formal language), as well as at what is the interpretation as linguistic expressions. And then, what appears as modelling to the model theorist is nothing but a translation from one formal language into another one: this is the way a proof theorist looks at models. But then it becomes equally clear that it is not always possible to give meaning to the various syntactical entities of a formal language in this way: it is possible only if there is another formal language in which to translate the former one.

However, if one is to give meaning to the syntactical entities of a formal language which is to contain the most primitive notions of a theory that one is dealing with ('primitive' meaning that they cannot be defined in terms of any other notions), then modelling or translating that formal language does not work anymore; these syntactical entities have to be given meaning in another way. One must proceed to a semantical or meaning-theoretical investigation of a radically new kind, which is, as Martin-Löf suggests, essentially philosophical or phenomenological.[4]

In his (1975) Martin-Löf writes that to understand a language is to understand its rules, and that to understand the rule of a language is to understand the conclusion under the assumption that the premisses have been understood. He also endorses Wittgenstein when he adds in the very same lectures that the meaning of an expression is determined by the rules that govern its use in the language of which it forms a part. And the difficulty in explaining the meaning of an expression of a language is to discern what are the rules of the language that determine its meaning.

In the lecture notes just mentioned Martin-Löf takes the language to be the one of prim. rec. functions and he singles out the following primitive notions of this language: natural number, numerical function, and definitional equality between two numerical functions. He then discerns the rules governing the use of these notions in the language of prim. rec. functions (where the rules governing the use of the notion of natural number include the so-called computation rules, and the rules governing the use of the notion of numerical function include the so-called denotation rules). And finally he explains each of these rules governing the use of any of the primitive notions just mentioned the way referred to above, i.e. by explaining the conclusion of the rule under the assumption that the premisses have already been explained.

In version (4) Martin-Löf chooses the language to be the one of general mathematical practice, and he singles out the following primitive concepts of this language: set, function, natural number, proposition, propositional function (property or relation), negation, conjunction, disjunction, implication, universal and existential quantification, and identity. He explicates these concepts in terms of type expression, object expression (of a type expression), object-valued functional expression, type-valued functional expression, and definitional identity between two object-valued functional expressions or between two type-valued functional expressions. (cf. the six basic forms of sentences (cf. Sect. i)) Then he discerns the rules governing the use of these latter concepts in their general form, as well as the rules governing the use of some particular forms of type expression (namely the familiar N, N_n, Π, Σ, $+$, I-type expressions), of some particular forms of object expression (namely the object expressions of the particular forms of type expression), of some particular forms of type-valued functional expressions (namely the ones corresponding to the particular forms of type expression), etc. (*Note*: these rules also include a whole lot of computation rules collected under the heading of mechanical or stipulatory part of the language.) But the semantical part of the project of version (4), namely the explanation of these rules the way it was explained above, is not carried out anymore.

However, Martin-Löf writes that version (4) is just 'a development and extension of our paper Syntax and Semantics of the language of primitive recursive functions',[5] and hence the direct semantics of that paper will provide an idea of what the semantics of version (4) may have looked like. Discussing the natural numbers in this paper he points to the fact that with each type of object there are always given some rules which show the form of an object of the resp. type. But he then warns not to fall into the trap of appearances, since an expression is an object of a certain type, not by virtue of its form, but by virtue of its function. The form of the object of a certain type is exhibited by the introduction rules of the type, whereas the function of these objects is actually determined by the elimination rule of the resp. type. Thus, an attempt is made in version (4) to explain the meaning of the possibly 0-ary type-theoretic operations not by the introduction, but by the elimination rules. But this attempt did and could not succeed. However, it is noteworthy that the semantical explanations in the notes on the language of prim. rec. functions and implicitly in version (4) show that the new direct style in which semantics is going to be developed in the following versions of constructive type theory has already emerged.

(5) In version (5) the number of primitive concepts of version (4) has shrunk from 6 to 4, and they are in the new version: type, definitional equality between types, object (of a type), and definitional equality between two objects (of a type) (cf. the four basic forms of judgement (cf. Sect. i)). The direct semantics of version (5) proceeds by explaining the meaning of each primitive concept by means of first, an explanation of each of the resp. basic forms of categorical judgement, and second, an explanation of each of the resp. basic forms of hypothetical judgement. The remainder of this version consists on the one hand of a lengthy enumeration of general rules, and of rules governing the use of the notions of cartesian product of a family of types (Π), of disjoint union of a family of types (Σ), of disjoint union of two types ($+$), of identity relation (I), of finite types (N_n), of natural numbers (N), of well-orderings (W), and of universes (U_n), and on the other hand of a considerable homework assignment for the reader, namely to try to supply all these rules with semantical explanations in the way which has been explained in version (4).

According to the way in which the meaning of the primitive concept of type has been explained, the meaning of the (possibly 0-ary) type-theoretic operations is explained in version (5) – other than in version (4) – by their resp. introduction rules. In version (5) the combinator formulation of constructive type theory of versions

(3) and (4) is abandoned and the earlier λ-abstraction formulation of version (2) is adopted again. And immediately the problem related on in the discussion of version (2) resurges (cf. part (1) of this section). Martin-Löf writes that the path to the new solution in version (5) of this old problem had been paved with the most serious difficulties. The new solution consists of replacing the traditional and honorable way of evaluating an expression from within, i.e. by evaluating the parts of the expression before evaluating the expression itself (also known as the applicative order or eager evaluation), by the modern way of evaluating an expression from without, i.e. by evaluating an expression only until it is on canonical or normal form (also called the normal order or lazy evaluation). Since e.g. $(\lambda x)b$ is an expression on canonical form, it is simply assigned itself as value, and it is not necessary – but also not possible – to evaluate its part b. It turns out to be insignificant that it no longer holds that an expression cannot have a value unless all its parts have values. What caused the troubles mentioned above was rather the task to make semantics of this new language work despite the loss of the principle according to which the value of an expression depends only on the values of its parts (a Fregean principle). From the semantical point of view, version (6) differs from version (5) only in being more reader-friendly: a large part of the homework assigned to the reader in version (5) is done and presented to the reader by Martin-Löf himself in version (6).

The direct semantics of version (6) is taken over and extended to the higher level type theory of version (7): the meaning of each of the primitive concepts of versions (5) or (6) is not only explained by means of first, an explanation of each of the resp. basic forms of categorical higher level judgement, second, an explanation of each of the resp. basic forms of hypothetical judgement, but also by means of third, an explanation of each of the resp. basic forms of categorical lower level judgement. Furthermore, version (7) contains a brief discussion of how correctness or soundness and completeness are to be understood in direct semantics – as opposed to soundness and completeness in metamathematical semantics or model theory (cf. Sect. iii., Chapt. 2). Version (8) contains the same direct semantics for higher level constructive type theory as version (7). But apart from these semantical explanations of the language of type theory, some extensive philosophical comments and reflections on the relationship between syntax and semantics and on the theory of indirect reference in constructive type theory can equally be found in version (8) (cf. Sect. viii., Chapt. 1). (Note that there are already quite a few scattered and sketchy philosophical observations and reflections on these topics in version (7).)

iv CONSTRUCTIVE TYPE THEORY AS A MATHEMATICAL LANGUAGE TO BE USED

Constructive type theory is a mathematical language which can be and is meant to be used to develop large parts of constructive mathematics in general and constructive analysis in particular. It is not a language inductively defined once and for all. To the contrary, it is a mathematical language which has to be extended or enriched as soon as achieving the aim just mentioned requires such an extension.

Versions (2), (3) and (5) consider adding the set of ordinals of the second number class and a well-ordering operator to the usual language of constructive type theory. Version (6) adds another set-operation *List* to the enriched language of version (5) and hints at yet another set-operation, namely the finite tree operation. Finally, Nordström et al. introduce a generalized well-ordering operation, called a general tree operation, into the language of constructive type theory in order to take care of families of mutually dependent inductive sets (which cannot be defined by means of the simple well-ordering operator).

In versions (2) and (3) a set U, called the universe, is introduced together with a reflection principle which roughly speaking states that whatever can be done with sets can be done inside the universe U. It is now possible of course to introduce a principle of transfinite induction over U expressing the idea that U is the smallest set which is closed w.r.t. the inductive clauses defining U.

However, Martin-Löf refrains from accepting such a principle of transfinite induction because he wants to keep the universe U open so as to be free to throw new ground sets into it or require it to be closed w.r.t. new set operations as the need for such sets or operations arises in the course of using constructive type theory as a mathematical language. He mentions two examples of sets or operations which one may like to add to U, namely the set $\mathcal{O}$ of ordinals of the second number class or the set-forming operation W for which $W(A)$: *set*, i.e. a set of well-founded trees, provided A : *set.*

In version (4) this topic is not touched upon. Version (5), however, demonstrates what such an enrichment of the basic collection of ground sets and set operations looks like: to the set operations of versions (2)–(4) is added a well-ordering operator W. It can be introduced in the style of Chapt. 1 as follows:

iv.1.1) What does a judgement of the form '$W(A, B)$: *set*' mean? or What does it mean to be a well-ordering of a family of sets?
To make a judgement of the form '$W(A, B)$: *set*' one has to know (i) what a canonical element of $W(A, B)$ is, and (ii) when two canonical elements of $W(A, B)$ are identical.

The formation axiom for the composite set $W(A, B)$ is

$$W : (X : set)(Y : (X)set)\ set$$

The formation axiom is tantamount to the following two rules

$$\frac{A : set \quad B : (A)\, set}{W(A, B) : set}\ (\text{WF } 1) \qquad \frac{A = C : set \quad B = D : (A)\, set}{W(A, B) = W(C, D) : set}\ (\text{WF } 2)$$

iv.1.2) What does a judgement of the form '$c : W(A, B)$' mean? or What does it mean to be a well-ordered set of a family of such sets?
To explain what a judgement of the form '$c : W(A, B)$' means, one has to know that (i) c has the form of an object formed by the element-formation rule of $W(A, B)$ (c has the form of a canonical element of $W(A, B)$) or (ii) c can be evaluated to that form (c is a non-canonical element of $W(A, B)$).

The element-formation or introduction axiom of $W(A, B)$ is

$$WI : (X : \mathit{set})(Y : (X)\ \mathit{set})(x : X)((Y(x))W(X, Y))W(X, Y)$$

which can be expressed alternatively by the introduction rule

$$\frac{\begin{array}{l} A = C : \mathit{set} \\ B = D : (A)\ \mathit{set} \\ a : A \\ b : (B(a))W(A, B) \end{array}}{WI(A, B, a, b) : W(C, D)} \qquad \text{(WI 1)}$$

iv.1.3) What does a judgement of the form '$c = d : W(A, B)$' mean? or What does it mean for two well-ordered sets of $W(A, B)$ to be identical?
To explain what a judgement of the form '$c = d : W(A, B)$' means, one has to know that (i) c and d have the form of two objects formed by the identity rule associated with the element-formation rule of $W(A, B)$ (c and d are identical canonical elements of $W(A, B)$) or (ii) c and d can be evaluated to that form (c and d are identical non-canonical elements of $W(A, B)$).

The identity rule associated with the element-formation rule or second introduction rule of $W(A, B)$ is

$$\frac{\begin{array}{l} A = C : \mathit{set} \\ B = D : (A)\ \mathit{set} \\ a = c : A \\ b = d : (B(a))W(A, B) \end{array}}{WI(A, B, a, b) = WI(C, D, c, d) : W(A, B)} \qquad \text{(WI 2)}$$

The W-elimination axiom states that the selector or eliminatory operator WE for the set $W(A, B)$ is an object of the type

$$\begin{array}{l} (X : \mathit{set})(Y : (X)\ \mathit{set})(Z : (W(X, Y))\ \mathit{set}) \\ \quad ((x : X)(v : (y : Y(x))W(X, Y))(u : (y : Y(x))Z(v(y)))Z(WI(X, Y, x, v))) \\ \qquad (z : W(X, Y))Z(z) : \mathit{type} \end{array}$$

i.e.

$$\begin{array}{l} WE : (X : \mathit{set})(Y : (X)\ \mathit{set})(Z : (W(X, Y))\ \mathit{set}) \\ \quad ((x : X)(v : (y : Y(x))W(X, Y))(u : (y : Y(x))Z(v(y)))Z(WI(X, Y, x, v))) \\ \qquad (z : W(X, Y))Z(z) \end{array}$$

The W-elimination axiom can be expressed alternatively by the two rules

$$\frac{\begin{array}{l} A : \mathit{set} \\ B : (A)\ \mathit{set} \\ C : (W(A, B))\ \mathit{set} \\ d : (x : A)(v : (y : B(x))W(A, B))(u : (y : B(x))C(v(y)))C(WI(A, B, x, v)) \\ c : W(A, B) \end{array}}{WE(A, B, C, d, c) : C(c)} \qquad \text{(WE 1)}$$

$$\frac{\begin{array}{l} A : set \\ B : (A)\ set \\ C : (W(A, B))\ set \\ d = f : (x : A)(v : (y : B(x))W(A, B))(u : (y : B(x))C(v(y)))C(WI(A, B, x, v)) \\ c = e : W(A, B) \end{array}}{WE(A, B, C, d, c) = WE(A, B, C, f, e) : C(c)} \quad (WE\ 2)$$

The meaning of the selector WE is explained as follows:

1. Evaluate c to canonical form.
2. If $c \Longrightarrow WI(A, B, a, b)$, then $WE(A, B, C, d, c) \Longrightarrow d(a, b, (x)WE(A, B, C, d, b(x)))$ and $WE(A, B, C, d, c) = d(a, b, (x)WE(A, B, C, d, b(x)))$.

This meaning explanation of WE justifies the two rules (WE 1) and (WE 2) as well as the following W-equality rule.

The W-equality axiom is

$$WE(A, B, C, d, WI(A, B, a, b)) = d(a, b, (x)WE(A, B, C, d, b(x))) : C(WI(A, B, a, b))$$

where $A :\ set$, $B : (A)\ set$, $C : (W(A, B))\ set$, $d : (x : A)(v : (y : B(x))W(A, B))(u : (y : B(x))C(v(y)))C(WI(A, B, x, v))$, $a : A$ and $b : (B(a))W(A, B)$.

The W-equality axiom can also be expressed by the following rule:

$$\frac{\begin{array}{l} A : set \\ B : (A)\ set \\ C : (W(A, B))\ set \\ d : (x : A)(v : (y : B(x))W(A, B))(u : (y : B(x))C(v(y)))C(WI(A, B, x, v)) \\ a : A \\ b : (B(a))W(A, B) \end{array}}{WE(A, B, C, d, WI(A, B, a, b)) = d(a, b, (x)WE(A, B, C, d, b(x))) : C(WI(A, B, a, b))} \quad (W\text{-equ})$$

C, n & o

(1) With the well-ordering operation one can construct many different sets of well-ordered sets (or many different sets of well-founded trees). In order to characterize a particular set, information has to be provided about two things: (i) the different ways the well-founded trees (or well-ordered sets) are formed, and (ii) for each way to form a well-founded tree (a well-ordered set) which parts it consists of. Accordingly, the well-ordered set operator W has two arguments, namely (i) the constructor set A, and (ii) the selector family B over A. Thus, given a constructor set A and a selector family B over A, one can form a well-ordering $W(A, B)$. As has just been explained, the elements of a well-ordering $W(A, B)$ are well-founded trees (or well-ordered sets); and to form a particular element one must indicate which way the tree is formed and what its parts are. If one has a constructor $a : A$, i.e. a particular form one wants the tree to have, and a function b from $B(a)$ to $W(A, B)$, i.e. a collection of subtrees

(elements of $W(A; B)$) constituting its parts, then one can form the well-founded tree $WI(A, B, a, b)$.[6]

(2) Now, ordinals can be intuitionistically constructed as well-founded trees. In this case they are no more totally ordered. But it is clear then that with a set A (the constructor set), a family of sets B over A (the selector family) and the well-ordering operator W, one can construct a set of ordinals $W(A, B)$. And $WI(A, B, a, b)$ can then be understood as the supremum of the ordinals $b(x)$ where x ranges over $B(a)$. Note that this interpretation is the reason why '*sup*' is used in version (5), version (6) and in Nordström et al. as a polymorphic notation of WI.

(3) WE is recursion over the set of ordinals $W(A, B)$. For, suppose $c : W(A, B)$; then $c \Longrightarrow WI(A, B, a, b)$ for some $a : A$ and $b : (B(a))W(A, B)$; now, for any $v : B(a)$, $b(v) : W(A, B)$ and $b(v) \Longrightarrow WI(A, B, a_1, b_1)$ for some $a_1 : A$ and $b_1 : (B(a_1))W(A, B)$; again, for any $v_1 : B(a_1)$, $b_1(v_1) : W(A, B)$ and $b_1(v_1) \Longrightarrow WI(A, B, a_2, b_2)$ for some $a_2 : A$ and $b_2 : (B(a_2))W(A, B)$, etc. until in any case one reaches a bottom element of $W(A, B)$ of the form $WI(A, B, a_n, b_n)$ where $B(a_n) = N_0$ s.t. no choice of an element in $B(a_n)$ is possible. b_n can then be chosen to be $(y)N_0E(W(A, B), y)$, and $WI(A, B, a_n, (y)N_0E(W(A, B), y)) : W(A, B)$ is a bottom element of the tree c.

That is why Nordström et al. refer to the polymorphic notation of WE by '*wrec*'. Martin-Löf uses in versions (5) and (6) 'T' instead of '*wrec*'.

(4) If one proceeds as in note (4) of C, n & o to ix.2.3), i.e. if one thinks of C in the rule (WE 1) in terms of a propositional function or a property of $W(A, B)$, and if one suppresses the proof objects of the occurring propositions and writes '*true*' instead behind the resp. propositions (and with a little handwaving), the rule (WE 1) is turned into the rule

$$\frac{\begin{array}{l} A : \mathit{set} \\ B : (A)\ \mathit{set} \\ C : (W(A, B))\mathit{prop} \\ (\forall x : A)(\forall y : (B(x))W(A, B))(\forall v : B(x))(C(y(v)) \supset C(WI(A, B, x, y)))\ \mathit{true} \\ c : W(A, B) \end{array}}{C(c)\ \mathit{true}}$$

which is nothing else but transfinite induction. The above transformation shows that transfinite recursion and transfinite induction essentially express the same concept.

(5) Several uses have been made of this new set operation W in constructive type theory to illustrate the expressibility of "old" and new mathematical objects:

(5.1) *Example 1*: The first number class (the natural numbers)

Instead of introducing the set of natural numbers as a ground set N, one could explicitly define it by means of the W operation.

There are two different ways of constructing a natural number, either to construct the leaf, i.e. 0, or to construct a thin compound tree. The constructor set must therefore contain two elements, and one can use the 2-element set $\{zero, succ\}$. The leaf does not have any parts, whence $B(zero) = N_0$, and a thin compound tree has one part, hence one can choose $B(succ) = S(N_0)$; hence let the set $(x : A)B(x)$ be

$$(x : \{zero, succ\})Set(\{zero, succ\}E(U, \hat{N}_0, UI1(\hat{N}_0), x)),$$

where $\{zero, succ\}E$ is like N_2E.

Then the rule $(W\text{F } 1)$ yields

$$\frac{\begin{array}{l}\{zero, succ\} : set \\ x : \{zero, succ\} \vdash Set(\{zero, succ\}E(U, \hat{N}_0, UI1(\hat{N}_0), x)) : set\end{array}}{(Wx : \{zero, succ\})Set(\{zero, succ\}E(U, \hat{N}_0, UI1(\hat{N}_0), x)) : set}$$

whence

$$\frac{\begin{array}{l}\{zero, succ\} : set \\ x : \{zero, succ\} \vdash Set(\{zero, succ\}E(U, \hat{N}_0, UI1(\hat{N}_0), x)) : set\end{array}}{\left\{\begin{array}{l} N : set \\ N = (Wx : \{zero, succ\})Set(\{zero, succ\}E(U, \hat{N}_0, UI1(\hat{N}_0), x)) : set\end{array}\right.}$$

which provides $(N\text{F } 1)$. $(N\text{F } 2)$ is, as usual, just an instance of set type identity. The rule $(W\text{I } 1)$ yields

$$\frac{\begin{array}{l}\{zero, succ\} : set \\ x : \{zero, succ\} \vdash Set(\{zero, succ\}E(U, \hat{N}_0, UI1(\hat{N}_0), x)) : set \\ zero : \{zero, succ\} \\ (y)N_0E(W(\{zero, succ\}, (x)Set(\{zero, succ\}E(U, \hat{N}_0, UI1(\hat{N}_0), x))), y) : \\ \qquad (Set(\{zero, succ\}E(U, \hat{N}_0, UI1(\hat{N}_0), zero))) \\ \qquad\qquad (Wx : \{zero, succ\})Set(\{zero, succ\}E(U, \hat{N}_0, UI1(\hat{N}_0), x))\end{array}}{\left\{\begin{array}{l} NI1 : (Wx : \{zero, succ\})Set(\{zero, succ\}E(U, \hat{N}_0, UI1(\hat{N}_0), x)) \\ NI1 = WI(\{zero, succ\}, (x)Set(\{zero, succ\}E(U, \hat{N}_0, UI1(\hat{N}_0), x)), zero, \\ \qquad (y)N_0E(W(\{zero, succ\}, (x)Set(\{zero, succ\}E(U, \hat{N}_0, UI1(\hat{N}_0), x))), y) : \\ \qquad\qquad (Wx : \{zero, succ\})Set(\{zero, succ\}E(U, \hat{N}_0, UI1(\hat{N}_0), x))\end{array}\right.}$$

The rule $(W\text{I } 1)$ yields also

$$\frac{\begin{array}{l}\{zero, succ\} : set \\ x : \{zero, succ\} \vdash Set(\{zero, succ\}E(U, \hat{N}_0, UI1(\hat{N}_0), x)) : set \\ succ : \{zero, succ\} \\ (y)x : (Set(\{zero, succ\}E(U, \hat{N}_0, UI1(\hat{N}_0), succ))) \\ \qquad (Wx : \{zero, succ\})Set(\{zero, succ\}E(U, \hat{N}_0, UI1(\hat{N}_0), x))\end{array}}{\left\{\begin{array}{l} NI2(x) : (Wx : \{zero, succ\})Set(\{zero, succ\}E(U, \hat{N}_o, UI1(\hat{N}_0), x)) \\ NI2(x) = WI(\{zero, succ\}, (x)Set(\{zero, succ\}E(U, \hat{N}_0, UI1(\hat{N}_0), x)), succ, (y)x) : \\ \qquad (Wx : \{zero, succ\})Set(\{zero, succ\}E(U, \hat{N}_0, UI1(\hat{N}_0), x))\end{array}\right.}$$

Finally, the eliminatory operator NE for N can be explicitly defined by

$$\begin{aligned} NE(C,d,e,c) = WE(&\{zero, succ\}, (x)Set(\{zero, succ\}E(U, \hat{N}_0, UI1(\hat{N}_0), x)),\\ &C, (z)(u)(y)\{zero, succ\}E(C, d, e(z(SI1(N_0)),\\ &u(SI1(N_0))), y), c) : C(c) \end{aligned}$$

Despite this definition, the rules (NE 2), (N-equ 1) and (N-equ 2) cannot be derived by means of the rules for W and intensional identity in constructive type theory, since there are more elements in the well-ordering $W(\{zero, succ\}, (x)Set(\{zero, succ\}E(U, \hat{N}_0, UI1(\hat{N}_0), x)))$ representing the natural numbers than there are natural numbers: an element $WI(A, B, a, b)$ of a well-ordering has a functional component b and two functions b and c are (intensionally) identical iff b *conv* c. Thus e.g. the two functions

$$(y)0 : N_0 \rightarrow (Wx : \{zero, succ\})Set(\{zero, succ\}E(U, \hat{N}_0, UI1(\hat{N}_0), x))$$

and

$$(y)1 : N_0 \rightarrow (Wx : \{zero, succ\})Set(\{zero, succ\}E(U, \hat{N}_0, UI1(\hat{N}_0), x))$$

are not intensionally identical although they are extensionally identical. Thus, with extensional identity this problem cannot occur. Observe also that the problem raised here is but the particular instance of a more general problem, namely: in note (3) above the question was asked how b_n should be determined in the case where $WI(A, B, a_n, b_n) : W(A, B)$ is s.t. $B(a_n) = N_0$. If one chooses an extensional approach, there is but one function and b_n was there defined to be $(y)N_0E(W(A, B), y)$. But b_n can be given in different ways, which accounts for the intensional differences of this function.[7]

(5.2) *Example 2*: The second number class

There are three ways of constructing an ordinal of the second number class: either to construct the leaf, i.e. 0, or to construct a thin successor tree from a given tree, or to construct a limit tree from (i.e. the least tree greater than each of the trees of) a sequence of trees. The constructor set A must therefore contain three elements, and one can use the 3-element set $\{zero, succ, lim\}$. As before, $B(zero) = N_0$, $B(succ) = S(N_0)$, but now $B(lim) = N$. Hence the set $(x : A)B(x)$ becomes

$$(x : \{zero, succ, lim\})Set(\{zero, succ, lim\}E(U, \hat{N}_0, UI1(\hat{N}_0), UI2, x))$$

where $\{zero, succ, lim\}E$ is like N_3E.

Then (WF 1) provides the definiens for the explicit definition of the second number class $\mathcal{O}$:

$$\begin{array}{l}
\{zero, succ, lim\} : set \\
x : \{zero, succ, lim\} \vdash Set(\{zero, succ, lim\}E(U, \hat{N}_0, UI1(\hat{N}_0), UI2, x)) : set \\
\hline
\left\{\begin{array}{l}
\mathcal{O} : set \\
\mathcal{O} = (Wx : \{zero, succ, lim\})Set(\{zero, succ, lim\}E(U, \hat{N}_0, UI1(\hat{N}_0), UI2, x)) : set
\end{array}\right.
\end{array}$$

Likewise, (WI 1) provides the definiens for the explicit definitions of $\mathcal{O}I1$, $\mathcal{O}I2(x)$ and $\mathcal{O}I3(b)$:

$$\begin{array}{l}
\{zero, succ, lim\} : set \\
x : \{zero, succ, lim\} \vdash Set(\{zero, succ, lim\}E(U, \hat{N}_0, UI1(\hat{N}_0), UI2, x)) : set \\
zero : \{zero, succ, lim\} \\
(y)N_0E(W(\{zero, succ, lim\}, (x)Set(\{zero, succ, lim\}E(U, \hat{N}_0, UI1(\hat{N}_0), UI2, x))), y) : \\
\quad (Set(\{zero, succ, lim\}E(U, \hat{N}_0, UI1(\hat{N}_0), UI2, zero))) \\
\qquad W(\{zero, succ, lim\}, (x)Set(\{zero, succ, lim\}E(\hat{N}_0, UI1(\hat{N}_0), UI2, x))) \\
\hline
\left\{\begin{array}{l}
\mathcal{O}I1 : (Wx : \{zero, succ, lim\})Set(\{zero, succ, lim\}E(U, \hat{N}_0, UI1(\hat{N}_0), UI2, x)) \\
\mathcal{O}I1 = WI(\{zero, succ, lim\}, (x)Set(\{zero, succ, lim\}E(U, \hat{N}_0, UI1(\hat{N}_0), \\
\qquad UI2, x)), zero, (y)N_0E(W(\{zero, succ, lim\}, (x)Set(\{zero, succ, lim\}E \\
\qquad\quad (U, \hat{N}_0, UI1(\hat{N}_0), UI2, x))), y)) : (Wx : \{zero, succ, lim\}) \\
\qquad\qquad Set(\{zero, succ, lim\}E(U, \hat{N}_0, UI1(\hat{N}_0), UI2, x))
\end{array}\right.
\end{array}$$

and

$$\begin{array}{l}
\{zero, succ, lim\} : set \\
x : \{zero, succ, lim\} \vdash Set(\{zero, succ, lim\}E(U, \hat{N}_0, UI1(\hat{N}_0), UI2, x)) : set \\
succ : \{zero, succ, lim\} \\
y(x) : (Set(\{zero, succ, lim\}E(U, \hat{N}_0, UI1(\hat{N}_0), UI2, succ))) \\
\qquad\qquad W(\{zero, succ, lim\}, (x)Set(\{zero, succ, lim\}E(U, \hat{N}_0, UI1(\hat{N}_0), \\
\qquad\qquad\quad UI2, x))) \\
\hline
\left\{\begin{array}{l}
\mathcal{O}I2(x) : (Wx : \{zero, succ, lim\})Set(\{zero, succ, lim\}E(U, \hat{N}_0, UI1(\hat{N}_0), \\
\qquad\qquad UI2, x)) \\
\mathcal{O}I2(x) = WI(\{zero, succ, lim\}, (x)Set(\{zero, succ, lim\}E(U, \hat{N}_0, UI1(\hat{N}_0), \\
\qquad\qquad UI2, x)), succ, (y)x) : (Wx : \{zero, succ, lim\})Set(\{zero, succ, lim\}E \\
\qquad\qquad\quad (U, \hat{N}_0, UI1(\hat{N}_0), UI2, x))
\end{array}\right.
\end{array}$$

and

$$\begin{array}{l}
\{zero, succ, lim\} : set \\
x : \{zero, succ, lim\} \vdash Set(\{zero, succ, lim\}E(U, \hat{N}_0, UI1(\hat{N}_0), UI2, x)) : set \\
lim : \{zero, succ, lim\} \\
b : (Set(\{zero, succ, lim\}E(U, \hat{N}_0, UI1(\hat{N}_0), UI2, lim)))W(\{zero, succ, lim\}, \\
\qquad (x)Set(\{zero, succ, lim\}E(U, \hat{N}_0, UI1(\hat{N}_0), UI2, x))) \\
\hline
\left\{\begin{array}{l}
\mathcal{O}I3(b) : (Wx : \{zero, succ, lim\})Set(\{zero, succ, lim\}E(U, \hat{N}_0, UI1(\hat{N}_0), UI2, x)) \\
\mathcal{O}I3(b) = WI(\{zero, succ, lim\}, (x)Set(\{zero, succ, lim\}E(U, \hat{N}_0, UI1(\hat{N}_0), \\
\qquad\qquad UI2, x)), lim, (x)b(x)) : (Wx : \{zero, succ, lim\})Set(\{zero, succ, lim\}E \\
\qquad\qquad\quad (U, \hat{N}_0, UI1(\hat{N}_0), UI2, x))
\end{array}\right.
\end{array}$$

Finally, the eliminatory operator $\mathcal{O}E$ for $\mathcal{O}$ can be explicitly defined by

$$\mathcal{O}E(C,d,e,b,c) = WE(\{zero, succ, lim\}, (x)Set(\{zero, succ, lim\}E(U, \hat{N}_0, UI1(\hat{N}_0), UI2, x)), C, (x)(y)(z)(u)(v)\{zero, succ, lim\}E (C, d, e(x,y), b(u,z), v), c) : C(c)$$

where $d : C(\mathcal{O}I1)$, $e : (x : W(\{zero, succ, lim\}, (x)Set(\{zero, succ, lim\}E(U, \hat{N}_0, UI1(\hat{N}_0), UI2, x)))(y : C(x))C(\mathcal{O}I2(x))$ and $b : (u : N \to W(\{zero, succ, lim\}, (x)Set(\{zero, succ, lim\}\ E(U, \hat{N}_0, UI1(\hat{N}_0), UI2, x)))(z : (\Pi x : N)C(u(x))) C(\mathcal{O}I3(u))$.

What has been explained about the rules (NE 2), (N-equ 1) and (N-equ 2) in *Example* (5.1), applies *mutatis mutandis* to ($\mathcal{O}$E 2), ($\mathcal{O}$-equ 1), ($\mathcal{O}$-equ 2) and ($\mathcal{O}$-equ 3). In order to be able to derive them in constructive type theory, one has to assume extensional identity.

(5.3) *Example* 3: The set of binary trees

There are two ways of constructing a binary tree of which one is to construct a leaf and one to construct a compound tree. Thus, the constructor set A must contain two elements and one can use the 2-element set {leaf, node}. Since a leaf has no parts, $B(leaf) = N_0$ and since a compound tree has two parts, $B(node) = S(S(N_0))$. Hence the set $(x : A)B(x)$ becomes $(x : \{leaf, node\})Set(\{leaf, node\}E(U, \hat{N}_0, UI1(UI1(\hat{N}_0)), x))$, where $\{leaf, node\}E$ is like N_2E.

Then (WF 1) provides the definiens for the explicit definition of the binary tree *Bintree*

$$\frac{\begin{array}{l}\{leaf, node\} : set \\ x : \{leaf, node\} \vdash Set(\{leaf, node\}E(U, \hat{N}_0, UI1(UI1(\hat{N}_0)), x)) : set\end{array}}{\left\{\begin{array}{l}Bintree : set \\ Bintree = (Wx : \{leaf, node\})Set(\{leaf, node\}E(U, \hat{N}_0, UI1(UI1(\hat{N}_0)), x)) : set\end{array}\right.}$$

The elements of *Bintree* are representations of all binary trees. They are provided by the following two rules:

$$\frac{\begin{array}{l}\{leaf, node\} : set \\ x : \{leaf, node\} \vdash Set(\{leaf, node\}E(U, \hat{N}_0, UI1(UI1(\hat{N}_0)), x)) : set \\ leaf : \{leaf, node\} \\ (y)N_0E(W(\{leaf, node\}, (x)Set(\{leaf, node\}E(U, \hat{N}_0, UI1(UI1(\hat{N}_0)), x))), y) : \\ \qquad (Wx : \{leaf, node\})Set(\{leaf, node\}E(U, \hat{N}_0, UI1(UI1(\hat{N}_0)), x))\end{array}}{\left\{\begin{array}{l}BintreeI1 : (Wx : \{leaf, node\})Set(\{leaf, node\}E(U, \hat{N}_0, UI1(UI1(\hat{N}_0)), x)) \\ BintreeI1 = WI(\{leaf, node\}, (x)Set(\{leaf, node\}E(U, \hat{N}_0, UI1(UI1(\hat{N}_0)), \\ \qquad x)), leaf, (y)N_0E(W(\{leaf, node\}, (x)Set(\{leaf, node\}E \\ \qquad (U, \hat{N}_0, UI1(UI1(\hat{N}_0)), x))), y)) : (Wx : \{leaf, node\}) \\ \qquad Set(\{leaf, node\}E(U, \hat{N}_0, UI1(UI1(\hat{N}_0)), x))\end{array}\right.}$$

and

$$\dfrac{\begin{array}{l}\{leaf, node\} : set \\ x : \{leaf, node\} \vdash Set(\{leaf, node\}E(U, \hat{N}_0, UI1(UI1(\hat{N}_0)), x)) : set \\ node : \{leaf, node\} \\ (u)(v)(y)\{leaf, node\}E(W(\{leaf, node\}, (x)Set(\{leaf, node\}E \\ \quad (U, \hat{N}_0, UI1(UI1(\hat{N}_0)), x))), u, v, y) : \\ \quad\quad (W(\{leaf, node\}, (x)Set(\{leaf, node\}E(U, \hat{N}_0, UI1(UI1(\hat{N}_0)), x)))) \\ \quad\quad\quad (W(\{leaf, node\}, (x)Set(\{leaf, node\}E(U, \hat{N}_0, UI1(UI1(\hat{N}_0)), x)))) \\ \quad\quad\quad\quad (Wx : \{leaf, node\})Set(\{leaf, node\}E(U, \hat{N}_0, UI1(UI1(\hat{N}_0)), x))\end{array}}{\left\{\begin{array}{l} BintreeI2(u, v) : (Wx : \{leaf, node\})Set(\{leaf, node\}E \\ \quad\quad\quad (U, \hat{N}_0, UI1(UI1(\hat{N}_0)), x)) \\ BintreeI2(u, v) = WI(\{leaf, node\}, (x)Set(\{leaf, node\}E \\ \quad\quad\quad (U, \hat{N}_0, UI1(UI1(\hat{N}_0)), x)), node, (y)\{leaf, node\}E \\ \quad\quad\quad (W(\{leaf, node\}, (x)Set(\{leaf, node\}E(U, \hat{N}_0, \\ \quad\quad\quad UI1(UI1(\hat{N}_0)), x))), u, v, y) : \\ \quad\quad\quad (Wx : \{leaf, node\})Set(\{leaf, node\}E(U, \\ \quad\quad\quad \hat{N}_0, UI1(UI1(\hat{N}_0)), x))\end{array}\right.}$$

And the eliminatory operator $BintreeE$ for $Bintree$ can be explicitly defined by

$$\begin{aligned} BintreeE(C, d, e, c) = WE(&\{leaf, node\}, (x)Set(\{leaf, node\}E(U, \\ &\hat{N}_0, UI1(UI1(\hat{N}_0)), x)), \\ &C, (y)(z)(x)\{leaf, node\}E(d, \\ &\quad e(y(SI1(N_0)), y(SI2(SI1(N_0))), \\ &\quad\quad z(SI1(N_0)), z(SI2(SI1(N_0))), x), c) : C(c)\end{aligned}$$

Again, the derivations of ($Bintree$E 2), ($Bintree$-equ 1) and ($Bintree$-equ 2) in constructive type theory presuppose extensional identity.[8]

(5.4) *Inductively defined sets*

In constructive type theory the well-ordering operator W can be used for representing many inductively defined sets.

But, as was remarked in the previous Exs. (5.1)–(5.3), one needs extensional identity E in order to get the correct elimination and equality rules.[9]

Version (6) not only takes over the extended collection of set operations from version (5), but it extends it even further by adding another set operation *List* to it and by hinting at yet another set operation, namely the finite tree operation *Fintree*. They both can be introduced in the style of Chapt. 1 as follows:

iv.2.1) What does a judgement of the form '$List(A) : set$', $A : set$, mean? or What does it mean to be a set of lists (i.e. finite sequences) of elements of a given set?

To make a judgement of the form '$List(A) : set$', for $A : set$, one has to know (i) what a canonical element of $List(A)$ is, and (ii) when two canonical elements of $List(A)$ are identical.

The formation axiom for the composite set $List(A)$ is

$$List : (\ set)\ set$$

The formation axiom is equivalent to the following two rules:

$$\frac{A:\ set}{List(A):\ set}\ (List\text{F }1) \qquad \frac{A = C:\ set}{List(A) = List(C):\ set}\ (List\text{F }2)$$

iv.2.2) What does a judgement of the form '$c : List(A)$' mean? or What does it mean to be a list of elements of a given set A?
To explain what a judgement of the form '$c : List(A)$' means, one has to know that (i) c has the form of an object formed by one of the element-formation rules of $List(A)$ (c has the form of a canonical element of $List(A)$), or (ii) c can be evaluated to one of those forms (c is a non-canonical element of $List(A)$).

The element-formation or introduction axioms of $List(A)$ are

$$\begin{aligned} &ListI1 : (X:\ set)List(X) \\ &ListI2 : (X:\ set)(X)(List(X))List(X) \end{aligned}$$

which can be expressed alternatively by the introduction rules

$$\frac{A = C:\ set}{ListI1(A) : List(C)}\ (List\text{I }1) \qquad \frac{A = C:\ set \quad a : A \quad b : List(A)}{ListI2(A, a, b) : List(C)}\ (List\text{I }2)$$

iv.2.3) What does a judgement of the form '$c = d : List(A)$', for $A:\ set$, mean? or What does it mean for two lists of elements of A to be identical?
To explain what a judgement of the form '$c = d : List(A)$' means, one has to know that (i) c and d have the form of two objects formed by one of the identity rules associated with the element-formation rules of $List(A)$ (c and d are identical canonical elements of $List(A)$) or (ii) c and d can be evaluated to those forms (c or d are identical non-canonical elements of $List(A)$).

The identity rules associated with the element-formation rules or third and fourth introduction rules of $List(A)$ are

$$\frac{A = C:\ set}{List\,I1(A) = List\,I1(C) : List(A)}\ (List\text{I }3)$$

$$\frac{\begin{array}{l} A = C:\ set \\ a = c : A \\ b = d : List(A) \end{array}}{List\,I2(A, a, b) = List\,I2(C, c, d) : List(A)}\ (List\text{I }4)$$

The *List*-elimination axiom states that the selector or eliminatory operator *ListE* for the set $List(A)$ is an object of the type

$$\begin{aligned} &(X:\ set)(Z : (List(X))\ set)(Z(List\,I1(X))) \\ &\quad ((x : X)(y : List(X))(Z(y))Z(ListI2(X, x, y))) \\ &\qquad (z : List(X))Z(z) : type \end{aligned}$$

i.e.

$$\begin{aligned}&\mathit{ListE} : (X : \mathit{set})(Z : (\mathit{List}(X))\ \mathit{set})(Z(\mathit{List}\,I1(X)))\\&\quad((x : X)(y : \mathit{List}(X))(Z(y))Z(\mathit{List}\,I2(X, x, y)))\\&\qquad(z : \mathit{List}(X))Z(z)\end{aligned}$$

This *List*-elimination axiom can be expressed alternatively by the two rules

$$\frac{\begin{array}{l}A : \mathit{set}\\ C : (\mathit{List}(A))\ \mathit{set}\\ d : C(\mathit{ListI1}(A))\\ e : (x : A)(y : \mathit{List}(A))(C(y))C(\mathit{ListI2}(A, x, y))\\ c : \mathit{List}(A)\end{array}}{\mathit{List}\,E(A, C, d, e, c) : C(c)}\ (\mathit{List}\mathrm{E}\ 1)$$

$$\frac{\begin{array}{l}A : \mathit{set}\\ C : (\mathit{List}(A))\ \mathit{set}\\ d = f : C(\mathit{List}\,I1(A))\\ e = g : (x : A)(y : \mathit{List}(A))(C(y))C(\mathit{ListI2}(A, x, y))\\ a = c : \mathit{List}(A)\end{array}}{\mathit{ListE}(A, C, d, e, a) = \mathit{ListE}(A, C, f, g, c) : C(a)}\ (\mathit{List}\mathrm{E}\ 2)$$

The meaning of the selector *ListE* is explained as follows:

1. Evaluate c to canonical form.

2a. If $c \Longrightarrow \mathit{ListI1}(A)$ then $\mathit{ListE}(A, C, d, e, c) \Longrightarrow d$ and $\mathit{ListE}(A, C, d, e, c) = d$.

2b. If $c \Longrightarrow \mathit{ListI2}(A, a, b)$, then $\mathit{ListE}(A, C, d, e, c) \Longrightarrow e(a, b, \mathit{ListE}(A, C, d, e, b))$ and $\mathit{ListE}(A, C, d, e, c) = e(a, b, \mathit{ListE}(A, C, d, e, b))$.

This meaning explanation of *ListE* justifies the two rules (*List*E 1) and (*List*E 2) as well as the two *List*-equality rules below. The *List*-equality axioms are

$$\begin{aligned}&\mathit{ListE}(A, C, d, e, \mathit{ListI1}(A)) = d : C(\mathit{ListI1}(A))\\&\mathit{ListE}(A, C, d, e, \mathit{ListI2}(A, a, b)) = e(a, b, \mathit{ListE}(A, C, d, e, b)) :\\&C(\mathit{ListI2}(A, a, b))\end{aligned}$$

where $A : \mathit{set}$, $C : (\mathit{List}(A))\ \mathit{set}$, $d : C(\mathit{ListI1}(A))$,
$e : (x : A)(y : \mathit{List}(A))(C(y))C(\mathit{ListI2}(A, x, y))$, $a : A$ and $b : \mathit{List}(A)$.

The *List*-equality axioms can also be expressed by the rules

$$\frac{\begin{array}{l}A : \mathit{set}\\ C : (\mathit{List}(A))\ \mathit{set}\\ d : C(\mathit{ListI1}(A))\\ e : (x : A)(y : \mathit{List}(A))(C(y))C(\mathit{ListI2}(A, x, y))\end{array}}{\mathit{ListE}(A, C, d, e, \mathit{ListI1}(A)) = d : C(\mathit{ListI1}(A))}\ (\mathit{List}\text{-equ}\ 1)$$

$$\begin{array}{ll}
A : set & \\
C : (List(A))\ set & \\
d : C(ListI1(A)) & \\
e : (x : A)(y : List(A))(C(y))C(ListI2(A, x, y)) & (List\text{-equ } 2)\\
a : A & \\
b : List(A) & \\
\hline
ListE(A, C, d, e, ListI2(A, a, b)) = e(a, b, ListE(A, C, d, e, b)) : C(ListI2(A, a, b)) &
\end{array}$$

C, n & o

(1) Martin-Löf uses in version (6) the polymorphic 'nil' instead of the monomorphic '*ListI1*' and the polymorphic '$(a.b)$' instead of the monomorphic '$ListI2(A, a, b)$' used here. Nordström et al. take over Martin-Löf's notation except for dropping the parentheses in '$(a.b)$'.

And instead of the monomorphic '*ListE*' Martin-Löf uses the polymorphic '$listrec$' (in order to express recursion on lists) and so do Nordström et al.

(2) Few applications of this new set operation *List* in constructive type theory can be found in the literature.

(2.1) *Example* 1: A function concatenating two lists (called $append$)

The rule (*List*E 1) yields

$$\begin{array}{l}
A : set\\
List(A) : set\\
b_2 : List(A)\\
(x)(y)(z)ListI2(A, x, z) : (x : A)(z : List(A))List(A)\\
b_1 : List(A)\\
\hline
ListE(A, List(A), b_2, (x)(y)(z)ListI2(A, x, z), b_1) : List(A)
\end{array}$$

This is the function concatenating two lists $b_1 : List(A)$ and $b_2 : List(A)$ which can be explicitly defined as follows:

$$\begin{array}{l}
u : List(A), v : List(A) \vdash ListE(A, List(A), v, (x)(y)(z)ListI2(A, x, z), u) : List(A)\\
b_1 : List(A)\\
b_2 : List(A)\\
\hline
append\ (A, List(A), b_1, b_2) : List(A)
\end{array}$$

and

$$\begin{array}{l}
u : List(A), v : List(A) \vdash ListE(A, List(A), v, (x)(y)(z)ListI2(A, x, z), u) : List(A)\\
b_1 : List(A)\\
b_2 : List(A)\\
\hline
append(A, List(A), b_1, b_2) = ListE(A, List(A), b_2, (x)(y)(z)ListI2(A, x, z), b_1) : List(A)
\end{array}$$

Applying *List*-equality to the definition of $append$ one gets the usual defining equations for a function concatenating two lists

$$append(A, List(A), b_2, ListI1(A)) = b_2 : List(A)$$

and

$$append(A, List(A), b_2, ListI2(A, a, b_1)) = ListI2(A, a, append(A, List(A), b_1, b_2)) : List(A)$$

Nordström et al. subsequently prove in constructive type theory the associativity of *append*.[10]

(2.2) *Example* 2 of Jan Smith's: course-of-values recursion on lists

Let $A :$ *set* and $C :$ *set* and assume the following:

$d : C(ListI1)$

$g : (x : A)(y : List(A))(z_1 : C) \ldots (z_k : C)C(ListI2(x, y))$

$d_i : (x : A)(y : List(A))List(A)$, for $i = 1, \ldots, k$

$r : (x : A)(y : List(A))length(d_i(x, y)) \leq length(y)$

The course-of-values equations to be solved are

$$f(ListI1) = d$$
$$f(ListI2(a, b)) = g(a, b, f(d_1(a, b)), \ldots, f(d_k(a, b)))$$

These equations are solved by deriving the following recursion rule in constructive type theory (a slightly simplified form will be given and derived in order to simplify the notation a little):

$$\dfrac{\begin{array}{l} A : \ set \\ C : (List(A))\ set \\ d : C(ListI1) \\ g : (x : A)(y : List(A))(z : C(d(x, y)))C(ListI2(x, y)) \\ r : (x : A)(y : List(A))length(d(x, y)) \leq length(y) \\ c : List(A) \end{array}}{f(c) : C(c)}$$

f is constructed via a proof of

$$(\forall y : List(A))(length(y) \leq n \supset C(y))$$

by induction (recursion) on n: assume the following judgements to hold

$$y : List(A) \vdash p : length(y) \leq NI1$$
$$y : List(A) \vdash t : length(y) \leq NI1 \supset I(List(A), y, ListI1)$$

then

$$\dfrac{\dfrac{\dfrac{\dfrac{y : List(A)}{app(t, p) : I(List(A), y, ListI1)} \quad d : C(ListI1)}{subst(app(t, p), d) : C(y)}}{\rightarrow I((p)subst(app(t, p), d)) : length(y) \leq NI1 \rightarrow C(y)}}{\Pi I((y) \rightarrow I((p)subst(app(t, p), d))) : (\Pi y : List(A))length(y) \leq NI1 \rightarrow C(y)}$$

$$\dfrac{\dfrac{\begin{array}{l} \dfrac{u : A \quad v : List(A) \quad z : N}{q : length(ListI2(u, v)) \leq NI2(z)} \\ h(q) : length(v) \leq z \\ r(u, v) : length(d(u, v)) \leq length(v) \end{array}}{trans(r(u, v), h(q)) : length(d(u, v)) \leq z} \quad \dfrac{\begin{array}{l} \dfrac{u : A \quad v : List(A)}{d(u, v) : List(A)} \\ z : N \\ w : (\Pi y : List(A))length(y) \leq z \rightarrow C(y) \end{array}}{app(w, d(u, v)) : length(d(u, v)) \leq z \rightarrow C(d(u, v))}}{app(app(w, d(u, v)), trans(r(u, v), h(q))) : C(d(u, v))}$$

$g(u, v, app(app(w, d(u, v)), trans(r(u, v), h(q)))) : C(ListI2(u, v))$

$\rightarrow I((q)g(u, v, app(app(w, d(u, v)), trans(r(u, v), h(q)))) : length(ListI2(u, v)) \leq NI2(z)$
$\rightarrow C(ListI2(u, v))$

$\rightarrow I((q)d) : length(ListI1) \leq NI2(z) \rightarrow C(ListI1)$

$ListE(\rightarrow I((q)d), (u)(v) \rightarrow I((q)g(u, v, app(app(w, d(u, v)), trans(r(u, v), h(q))))), y) :$
$length(y) \leq NI2(z) \rightarrow C(y)$

$\Pi I((y)(w)ListE(\rightarrow I((q)d), (u)(v) \rightarrow I((q)g(u, v, app(app(w, d(u, v)), trans(r(u, v),$
$h(q))))), y)) : (\Pi y : List(A))length(y) \leq NI2(z) \rightarrow C(y)$

$\Pi I((y) \rightarrow I((p)subst(app(t, p), d))) : (\Pi y : List(a))length(y) \leq NI1 \rightarrow C(y)$

$a : N$

$NE(\Pi I((y) \rightarrow I((p)subst(app(t, p), d))), \Pi I((y)(w)ListE(\rightarrow I((q)d), (u)(v) \rightarrow I((q)$
$g(u, v, app(app(w, d(u, v)), trans(r(u, v), h(q))))), y)), a) :$
$(\Pi y : List(A))length(y) \leq a \rightarrow C(y)$

for *length*$(y) : N$, $y : List(A)$ this yields

$NE(\Pi I((y) \rightarrow I((p)subst(app(t, p), d))), \Pi I((y)(w)ListE(\rightarrow I((q)d), (u)(v) \rightarrow I((q)$
$g(u, v, app(app(w, d(u, v)), trans(r(u, v), h(q))))), y)), length(y)) :$
$(\Pi y : List(A))length(y) \leq length(y) \rightarrow C(y)$

$c : List(A)$

$app(NE(\Pi I((y) \rightarrow I((p)subst(app(t, p), d))), \Pi I((y)(w)ListE(\rightarrow I((q)d), (u)(v) \rightarrow I((q)$
$g(u, v, app(app(w, d(u, v)), trans(r(u, v), h(q))))), y)), length(y)), c) :$
$length(c) \leq length(c) \rightarrow C(c)$

$s(c) : length(c) \leq length(c)$

$app(app(NE(\Pi I((y) \rightarrow I((p)subst(app(t, p), d))), \Pi I((y)(w)ListE(\rightarrow I((q)d),$
$(u)(v) \rightarrow I((q)g(u, v, app(app(w, d(u, v)), trans(r(u, v), h(q))))), y)),$
$length(y)), c), s(c)) : C(c)$

$f(c) = app(app(NE(\Pi I((y) \rightarrow I((p)subst(app(t, p), d))), \Pi I((y)(w)ListE(\rightarrow I((q)d),$
$(u)(v) \rightarrow I((q)g(u, v, app(app(w, d(u, v)), trans(r(u, v), h(q))))), y)),$
$length(y)), c), s(c)) : C(c)$

Note that *subst* is the substitution function defined in Nordström et al.[11]; as Jan Smith observes, if $q : length(ListI2(u, v)) \leq NI2(z)$, then, since *length* $(ListI2(u, v)) = NI2(length(v))$, there obviously exists a construction of the proposition $length(v) \leq z$ based on the proof object q, and this construction is here denoted by $h(q)$; *trans* is the function expressing the transitivity of $\leq$ (analogous to the homonymous function defined in Nordström et al.[12]). Finally, Smith uses dots instead of the function $s(c)$, for $c : List(A)$, since, as he points out, what the dots (or $s(c)$ here resp.) denote, depends on the particular definition one chooses of the proposition $m \leq n$.

Smith also proves that this method of solving the course-of-values equations is correct.[13]

iv.3.1) What does a judgement of the form '*Fintree*(A) : *set*', A : *set*, mean? or What does it mean to be a set of arbitrarily finitely branching trees decorated with elements of a given set?

To make a judgement of the form '*Fintree*(A) : *set*', for A : *set*, one has to know (i) what a canonical element of *Fintree*(A) is, and (ii) when two canonical elements of *Fintree*(A) are identical.

The formation axiom for the composite set $Fintree(A)$ is

$$Fintree : (\ set)\ set$$

The formation axiom is equivalent to the following two rules

$$\frac{A :\ set}{Fintree(A) :\ set}\ (Fintree\text{F } 1) \qquad \frac{A = C :\ set}{Fintree(A) = Fintree(C) :\ set}\ (Fintree\text{F } 2)$$

iv.3.2) What does a judgement of the form '$c : Fintree(A)$', $A : set$, mean? or What does it mean to be an arbitrarily finitely branching tree decorated with elements of a given set?
To explain what a judgement of the form '$c : Fintree(A)$' means, one has to know that (i) c has the form of an object formed by one of the element-formation rules of $Fintree(A)$ (c has the form of a canonical element of $Fintree(A)$), or (ii) c can be evaluated to one of those forms (c is a non-canonical element of $Fintree(A)$).

The element-formation or introduction axioms of $Fintree(A)$ are

$$FintreeI1 : (X :\ set)(X)Fintree(X)$$
$$FintreeI2 : (X :\ set)(Fintree(X))(Fintree(X))Fintree(X)$$

which can be expressed alternatively by the introduction rules

$$\frac{\begin{array}{l} A = C :\ set \\ a : A \end{array}}{FintreeI1(A, a) : Fintree(C)}\ (Fintree\text{I } 1)$$

$$\frac{\begin{array}{l} A = C :\ set \\ b : Fintree(A) \\ e : Fintree(A) \end{array}}{FintreeI2(A, b, e) : Fintree(C)}\ (Fintree\text{I } 2)$$

iv.3.3) What does a judgement of the form '$c = d : Fintree(A)$' mean? or What does it mean for two arbitrarily finitely branching trees decorated with elements of set A to be identical?
To explain what a judgement of the form '$c = d : Fintree(A)$' means, one has to know that (i) c and d have the form of two objects formed by one of the identity rules associated with the element-formation rules of $Fintree(A)$ (c and d are identical canonical elements of $Fintree(A)$) or (ii) c and d can be evaluated to those forms (c or d are identical non-canonical elements of $Fintree(A)$).

The identity rules associated with the element-formation rules or third and fourth introduction rules of $Fintree(A)$ are

$$\frac{A = C :\ set \qquad a = c : A}{FintreeI1(A, a) = FintreeI1(C, c) : Fintree(A)}\ (Fintree\text{I } 3)$$

$$\frac{\begin{array}{l} A = C : \ set \\ b = d : Fintree(A) \\ e = f : Fintree(A) \end{array}}{FintreeI2(A, b, e) = FintreeI2(C, d, f) : Fintree(A)} \quad (Fintree\mathrm{I}\ 4)$$

The *Fintree*-elimination axiom states that the eliminatory operator *FintreeE* for the set $Fintree(A)$ is an object of the type

$$\begin{array}{l} FintreeE : (X : \ set)(Z : (Fintree(X)) \ set)((x : X)Z(FintreeI1(X, x))) \\ \qquad ((v : Fintree(X))(w : Fintree(X))(Z(v))(Z(w))Z(FintreeI2(X, v, w))) \\ \qquad\quad (z : Fintree(X))Z(z) \end{array}$$

This *Fintree*-elimination axiom can be expressed alternatively by the two rules

$$\frac{\begin{array}{l} A : set \\ C : (Fintree(A)) \ set \\ d : (x : A)C(FintreeI1(A, x)) \\ e : (v : Fintree(A))(w : Fintree(A))(C(v))(C(w))C(FintreeI2(A, v, w)) \\ c : Fintree(A) \end{array}}{FintreeE(A, C, d, e, c) : C(c)} \quad (Fintree\mathrm{E}\ 1)$$

$$\frac{\begin{array}{l} A : \ set \\ C : (Fintree(A)) \ set \\ d = f : (x : A)C(FintreeI1(A, x)) \\ e = g : (v : Fintree(A))(w : Fintree(A))(C(v))(C(w))C(FintreeI2(A, v, w)) \\ a = c : Fintree(A) \end{array}}{FintreeE(A, C, d, e, a) = FintreeE(A, C, f, g, c) : C(a)} \quad (Fintree\mathrm{E}\ 2)$$

The meaning of the selector *FintreeE* is explained as follows:

1. Evaluate c to canonical form.
2a. If $c \Longrightarrow FintreeI1(A, a)$ then $FintreeE(A, C, d, e, c) \Longrightarrow d(a)$ and $FintreeE(A, C, d, e, c) = d(a)$.
2b. If $c \Longrightarrow FintreeI2(A, a, b)$ then $FintreeE(A, C, d, e, c) \Longrightarrow e(a, b, Fintree\ E(A, C, d, e, a),$ $FintreeE(A, C, d, e, b))$ and $FintreeE(A, C, d, e, c)$ $= e(a, b, FintreeE(A, C, d, e, a), FintreeE(A, C, d, e, b))$.

This meaning explanation of *FintreeE* justifies the two rules (*Fintree*E 1) and (*Fintree*E 2) as well as the following two *Fintree*-equality rules.

The *Fintree*-equality axioms are

$$\begin{array}{l} FintreeE(A, C, d, e, FintreeI1(A, a)) = d(a) : C(FintreeI1(A, a)) \\ FintreeE(A, C, d, e, FintreeI2(A, b, f)) = e(b, f, FintreeE(A, C, d, e, b), \\ \qquad\qquad FintreeE(A, C, d, e, f)) : C(FintreeI2(A, b, f)) \end{array}$$

where $A : \ set$, $C : (Fintree(A)) \ set$, $d : (x : A) \ C(FintreeI1(A, x))$, $e : (x : Fintree(A))(y : Fintree(A))(C(x))(C(y))C(FintreeI2(A, x, y))$, $a : A$, $b : Fintree(A)$ and $f : Fintree(A)$.

The *Fintree*-equality axioms can also be expressed by the rules

$$\frac{\begin{array}{l} A:\ set \\ C:(Fintree(A))\ set \\ d:(x:A)C(FintreeI1(A,x)) \\ e:(x:Fintree(A))(y:Fintree(A))(C(x))(C(y))C(FintreeI2(A,x,y)) \\ a:A \end{array}}{FintreeE(A,C,d,e,Fintree\ I1(A,a)) = d(a):C(FintreeI1(A,a))} \quad (Fintree\text{-equ }1)$$

$$\frac{\begin{array}{l} A:\ set \\ C:(Fintree(A))\ set \\ d:(x:A)C(FintreeI1(A,x)) \\ e:(x:Fintree(A))(y:Fintree(A))(C(x))(C(y))C(FintreeI2(A,x,y)) \\ b:Fintree(A) \\ f:Fintree(A) \end{array}}{\begin{array}{l} FintreeE(A,C,d,e,FintreeI2(A,b,f)) = \\ \quad e(b,f,FintreeE(A,C,d,e,b),FintreeE(A,C,d,e,f)):C(FintreeI2(A,b,f)) \end{array}} \quad (Fintree\text{-equ }2)$$

Note: The *Fintree*-operator is only made allusion to, and not fully written out in version (6).[14]

Finally, the well-ordering operation W is not easy to use when one wants to define a family of mutually dependent inductive sets or mutually dependent families of trees. Nordström et al. (originally Petersson and Synek) solved this problem by introducing a generalized well-ordering operation, which they call a general tree operation, into the basic collection of ground sets and set operations:

iv.4.1) What does a judgement of the form '$Tree(A, B, C, d, a) : set$' mean? or What does it mean to be a family of mutually dependent families of trees? To make a judgement of the form '$Tree(A, B, C, d, a) : set$', one has to know (i) what a canonical element of $Tree(A, B, C, d, a)$ is, and (ii) when two canonical elements of $Tree(A, B, C, d, a)$ are identical.

The formation axiom for the composite set $Tree(A, B, C, d, a)$ is

$$Tree:(X:\ set)(Y:(X)\ set)(Z:(X)(Y)\ set)((x:X)(y:Y(x))(Z(x,y))X)(X)\ set$$

The formation axiom is tantamount to the following two rules

$$\frac{\begin{array}{l} A:\ set \\ B:(A)\ set \\ C:(A)(B)\ set \\ d:(x:A)(y:B(x))(C(x,y))A \\ a:A \end{array}}{Tree(A,B,C,d,a):\ set} \quad (Tree\text{F }1)$$

$$\frac{\begin{array}{l} A=D:\ set \\ B=E:(A)\ set \\ C=F:(A)(B)\ set \\ d=f:(x:A)(y:B(x))(C(x,y))A \\ a=c:A \end{array}}{Tree(A,B,C,d,a)=Tree(D,E,F,f,c):\ set} \quad (Tree\text{F }2)$$

iv.4.2) What does a judgement of the form '$c : Tree(A, B, C, d, a)$' mean? or What does it mean to be a tree of a family of mutually dependent families of trees? To explain what a judgement of the form '$c : Tree(A, B, C, d, a)$' means, one has to know that (i) c has the form of an object formed by the element-formation rule of $Tree(A, B, C, d, a)$ (c has the form of a canonical element of $Tree(A, B, C, d, a)$) or (ii) c can be evaluated to that form (c is a non-canonical element of $Tree(A, B, C, d, a)$).

The element-formation or introduction axiom of $Tree(A, B, C, d, a)$ is

$$\begin{aligned}TreeI : (X : set)(Y : (X)\, set)(Z : (X)(Y)\, set)(x : X)(y : Y(x))\\ (u : (x : X)(y : Y(x))(z : Z(x, y))X)(v : (z : Z(x, y))Tree(X, Y, Z, u, u(x, y, z)))\\ Tree(X, Y, Z, u, x)\end{aligned}$$

which can be expressed alternatively by the introduction rule

$$\frac{\begin{array}{l}A = D : set\\ B = E : (A)\, set\\ C = F : (A)(B)\, set\\ a : A\\ b : B(a)\\ d : (x : A)(y : B(x))(C(x, y))A\\ h : (z : C(a, b))Tree(A, B, C, d, d(a, b, z))\end{array}}{Tree\ I(A, B, C, a, b, d, h) : Tree(D, E, F, d, a)}\qquad (TreeI\ 1)$$

iv.4.3) What does a judgement of the form '$c = e : Tree(A, B, C, d, a)$' mean? or What does it mean for two trees of $Tree(A, B, C, d, a)$ to be identical? To explain what a judgement of the form '$c = e : Tree(A, B, C, d, a)$' means, one has to know that (i) c and e have the form of two objects formed by the identity rule associated with the element-formation rule of $Tree(A, B, C, d, a)$ (c and e are identical canonical elements of $Tree(A, B, C, d, a)$), or (ii) c and e can be evaluated to that form (c or e are identical non-canonical elements of $Tree(A, B, C, d, a)$).

The identity rule associated with the element-formation rule or second introduction rule of $Tree(A, B, C, d, a)$ is

$$\frac{\begin{array}{l}A = D : set\\ B = E : (A)\, set\\ C = F : (A)(B)\, set\\ a = e : A\\ b = f : B(a)\\ d = i : (x : A)(y : B(x))(C(x, y))A\\ h = g : (z; C(a, b))Tree(A, B, C, d, d(a, b, z))\end{array}}{TreeI(A, B, C, a, b, d, h) = TreeI(D, E, F, e, f, i, g) : Tree(A, B, C, d, a)}\qquad (TreeI\ 2)$$

The *Tree*-elimination axiom states that the selector or eliminatory operator $TreeE$ for the set $Tree(A, B, C, d, a)$ is an object of the type

$$\begin{array}{l}(X: set)(Y: (X)\ set)(Z: (X)(Y)\ set)(u: (x: X)(y: Y(x))(Z(x, y))X) \\ \quad (V: (x: X)(t: Tree(X, Y, Z, u, x))\ set)(x: X) \\ \quad\quad (s: (x: X)(y: Y(x))(z: (v: Z(x, y))Tree(X, Y, Z, u, u(x, y, v))) \\ \quad\quad\quad (w: (v: Z(x, y))V(u(x, y, v), z(v))) \\ \quad\quad\quad\quad V(x, TreeI(X, Y, Z, x, y, u, z)))(t: Tree(X, Y, Z, u, x))V(x, t): type\end{array}$$

i.e.

$$\begin{array}{l}TreeE: (X: set)(Y: (X)\ set)(Z: (X)(Y)\ set)(u: (x: X)(y: Y(x))(Z(x, y))X) \\ \quad (V: (x: X)(t: Tree(X, Y, Z, u, x))\ set)(x: X)(s: (x: X)(y: Y(x)) \\ \quad\quad (z: (v: Z(x, y))Tree(X, Y, Z, u, u(x, y, v)))(w: (v: Z(x, y)) \\ \quad\quad\quad V(u(x, y, z), z(v)))V(x, TreeI(X, Y, Z, x, y, u, z))) \\ \quad\quad\quad\quad (t: Tree(X, Y, Z, u, x))V(x, t)\end{array}$$

The *Tree*-elimination axiom can be expressed alternatively by the two rules

$$\frac{\begin{array}{l}A: set \\ B: (A)\ set \\ C: (A)(B)\ set \\ d: (x: A)(y: B(x))(C(x, y))A \\ D: (x: A)(t: Tree(A, B, C, d, x))\ set \\ a: A \\ f: (x: A)(y: B(x))(z: (v: C(x, y))Tree(A, B, C, d, d(x, y, v))) \\ \quad (w: (v: C(x, y))D(d(x, y, v), z(v)))D(x, TreeI(A, B, C, x, y, d, z)) \\ c: Tree(A, B, C, d, a)\end{array}}{TreeE(A, B, C, d, D, f, c): D(a, c)} \quad (TreeE\ 1)$$

$$\frac{\begin{array}{l}A: set \\ B: (A)\ set \\ C: (A)(B)\ set \\ d = e: (x: A)(y: B(x))(C(x, y))A \\ D: (x: A)(t: Tree(A, B, C, d, x))\ set \\ a: A \\ f = g: (x: A)(y: B(x))(z: (v: C(x, y))Tree(A, B, C, d, d(x, y, v))) \\ \quad\quad\quad (w: (v: C(x, y))D(d(x, y, v), z(v)))D(x, TreeI(A, B, C, x, y, d, z)) \\ c = i: Tree(A, B, C, d, a)\end{array}}{TreeE(A, B, C, d, D, f, c) = TreeE(A, B, C, e, D, g, i): D(a, c)} \quad (TreeE\ 2)$$

The meaning of the eliminatory operator $TreeE$ is explained as follows:

1. Evaluate c to canonical form.
2. If $c \Longrightarrow TreeI(A, B, C, a, b, d, h)$ then $TreeE(A, B, C, d, D, f, c) \Longrightarrow f(a, b, h, (x)TreeE(A, B, C, d, D, f, h(x)))$ and $TreeE(A, B, C, d, D, f, c) = f(a, b, h, (x)TreeE(A, B, C, d, D, f, h(x)))$.

The meaning explanation of $TreeE$ justifies the two rules (*Tree*E 1) and (*Tree*E 2) as well as the *Tree*-equality rule below.

The *Tree*-equality axiom is

$$\begin{aligned}&TreeE(A, B, C, d, D, f, TreeI(A, B, C, a, b, d, h))\\ &\quad = f(a, b, h, (x)TreeE(A, B, C, d, D, f, h(x))) : D(a, TreeI(A, B, C, a, b, d, h))\end{aligned}$$

where $A:\ set$, $B : (A)\ set$, $C : (A)(B)\ set$, $d : (x : A)(y : B(x))(C(x, y))A$, $D : (x : A)(Tree(A, B, C, d, x))\ set$, $f : (x : A)(y : B(x))(z : (v : C(x, y))$ $Tree(A, B, C, d, d(x, y, v)))(w : (v : C(x, y))D(d(x, y, v), z(v)))$ $D(x, TreeI(A, B, C, x, y, d, z))$,
$a : A, b : B(a), h : (z : C(a, b))Tree(A, B, C, d, d(a, b, z))$.

The *Tree*-equality axiom can also be expressed by the following rule:

$$\frac{\begin{array}{l}A:\ set\\ B(A):\ set\\ C:(A)(B)\ set\\ d:(x:A)(y:B(x))(C(x,y))A\\ D:(x:A)(t:Tree(A,B,C,d,x))\ set\\ f:(x:A)(y:B(x))(z:(v:C(x,y))Tree(A,B,C,d,d(x,y,v)))\\ \qquad (w:(v:C(x,y))D(d(x,y,v),z(v)))D(x,TreeI(A,B,C,x,y,d,z))\\ a:A\\ b:B(a)\\ h:(z:C(a,b))Tree(A,B,C,d,d(a,b,z))\end{array}}{\begin{array}{l}TreeE(A,B,C,d,D,f,TreeI(A,B,C,a,b,d,h)\\ \quad =f(a,b,h,(x)TreeE(A,B,C,d,D,f,h(x))):D(a,TreeI(A,B,C,a,b,d,h))\end{array}}\quad (Tree\text{-equ})$$

C, n & o

(1) Nordström et al. use instead of the monomorphic '$TreeI$' and '$TreeE$' the polymorphic '*tree*' and '*treerec*', resp.

(2) The *Tree* operation produces not one inductive set, but a whole family of mutually dependent inductive sets. This is carried out by assuming a set A of names of the mutually dependent sets (called the name set by Nordström et al.). Instead of having one set of constructors (i.e. different ways the trees may be formed) as in the case of the well-ordering operation, one now has a constructor set $B(a)$ for each name $a : A$. And finally, instead of having a family of selectors over the set of constructors (i.e. for each way to form a tree its different parts) as in the case of the well-ordering operation, one now has a selector family $C(x, y)$ for each constructor $y : B$ and each name $x : A$. Moreover, since the parts of a tree may now come from different well-ordered sets, a function d is introduced, called the component set name by Nordström et al., which provides information about the set which a part of a tree belongs to: $d(x, y, z)$ is an element of A, if $x : A$, $y : B(x)$ and $z : C(x, y)$.

The family of sets $Tree(A, B, C, d, x)$ is a representation of the family of sets introduced by a collection A of (possibly mutually dependent) inductive definitions, i.e. s.t. $x : A$. Lastly, it should be emphasized that – as in the well-ordering case – one needs extensional identity in order to get the correct rules (*Tree*E 1), (*Tree*E 2) and (*Tree*-equ).[15]

(3) The *Tree*-operation was said to be a generalization of the well-ordering operation W, hence one can show the well-order set $W(B, C)$ to be an instance of a *Tree* set: one simply chooses a family of well-founded trees on a 1-element name set. (*Tree*F 1) yields

$$\frac{\begin{array}{l} X : \mathit{set}, Y : (X)\,\mathit{set}, Z : (x : X)(Y(x))\,\mathit{set}, u : (x : X)(y : Y(x))(z : Z(x, y))X, \\ x : X \vdash \mathit{Tree} : \mathit{set} \\ S(N_0) : \mathit{set} \\ B : (S(N_0))\,\mathit{set} \\ C : (S(N_0))(B)\,\mathit{set} \\ (x)(y)(z)SI1 : (x : S(N_0))(y : B)(z : C(y))S(N_0) \\ SI1 : S(N_0) \end{array}}{\left\{ \begin{array}{l} W(B, C) : \mathit{set} \\ W(B, C) = \mathit{Tree}(S(N_0), B, C, (x)(y)(z)SI1, SI1) : \mathit{set} \end{array} \right.}$$

where $((x : S(N_0))B)(x) = B : \mathit{set}$ and $((x : S(N_0))(y : B)C)(x, y) = (B)C : \mathit{set}$.

(*Tree*I 1) yields

$$\frac{\begin{array}{l} X : \mathit{set}, Y : (X)\,\mathit{set}, Z : (x : X)(Y(x))\,\mathit{set}, x : X, y : Y(x), \\ u : (x : X)(y : Y(x))(z : Z(x, y))X, v : (z : Z(x, y))\mathit{Tree}(X, Y, Z, u(x, y, z), x) \vdash \\ \quad \mathit{TreeI} : \mathit{Tree}(X, Y, Z, u, x) \\ S(N_0) : \mathit{set} \\ B : (S(N_0))\,\mathit{set} \\ C : (S(N_0))(B)\,\mathit{set} \\ SI1 : S(N_0) \\ b : B \\ (x)(y)(z)SI1 : (x : S(N_0))(y : B)(z : C(y))S(N_0) \\ c : (C(b))\mathit{Tree}(S(N_0), B, C, (x)(y)(z)SI1, SI1) \end{array}}{\left\{ \begin{array}{l} WI(B, C, b, c) : \mathit{Tree}(S(N_0), B, C, (x)(y)(z)SI1, SI1) = W(B, C) \\ WI(B, C, b, c) = \mathit{TreeI}(S(N_0), B, C, SI1, b, (x)(y)(z)SI1, c) : W(B, C) \end{array} \right.}$$

and (*Tree*E 1) yields

$$\frac{\begin{array}{l} X : \mathit{set}, Y : (X)\,\mathit{set}, Z : (x : X)(Y(x))\,\mathit{set}, u : (x : X)(y : Y(x)) \\ \quad (z : Z(x, y))X, V : (x : X)(t : \mathit{Tree}(X, Y, Z, u, x))\,\mathit{set}, \\ x : X, s : (x : X)(y : Y(x))(v : (z : Z(x, y))\mathit{Tree}(X, Y, Z, u(x, y, z), x)) \\ (w : (z : Z(x, y))V(u(x, y, z), v(z)))V(x, \mathit{TreeI}(X, Y, Z, \\ \quad x, y, u, v)), t : \mathit{Tree}(X, Y, Z, u, x) \vdash \mathit{TreeE} : V(x, t) \\ S(N_0) : \mathit{set} \\ B : (S(N_0))\,\mathit{set} \\ C : (S(N_0))(B)\,\mathit{set} \\ (x)(y)(z)SI1 : (x : S(N_0))(y : B)(z : C(y))S(N_0) \\ D : (S(N_0))(\mathit{Tree}(S(N_0), B, C, (x)(y)(z)SI1, SI1))\,\mathit{set} \\ SI1 : S(N_0) \\ f : (x : S(N_0))(y : B)(v : (z : C(y))\mathit{Tree}(S(N_0), B, C, (x)(y)(z)SI1, SI1)) \\ \quad (w : (z : C(y))D(SI1, v(z)))D(x, \mathit{TreeI}(S(N_0), B, C, x, y, SI1, z)) \\ e : \mathit{Tree}(S(N_0), B, C, (x)(y)(z)SI1, SI1) \end{array}}{\left\{ \begin{array}{l} WE(B, C, D, f, c) : D(SI1, e) = D(e) \\ WE(B, C, D, f, c) = \mathit{TreeE}(S(N_0), B, C, (x)(y)(z)SI1, D, SI1, f, e) : D(e) \end{array} \right.}$$

(*Tree*-equ) yields

$$
\begin{aligned}
&WE(B, C, D, f, WI(B, C, b, c))\\
&\quad = TreeE(S(N_0), B, C, (x)(y)(z)SI1, D, SI1, f, WI(B, C, b, c))\\
&\quad = TreeE(S(N_0), B, C, (x)(y)(z)SI1, D, SI1, f,\\
&\quad\quad TreeI(S(N_0), B, C, SI1, b, (x)(y)(z)SI1, c))\\
&\quad = f(SI1, b, c, (x)TreeE(S(N_0), B, C, SI1, d, f, c(x)))\\
&\quad = f(b, c, (x)WE(B, C, D, f, c(x))) : D(WI(B, C, b, c))
\end{aligned}
$$

(4) Nordström et al. show how to define the set of even and the set of odd natural numbers by means of the *Tree* operation. They also show how to define a set of lists with exactly n-elements from a set A by means of *Tree*; note that the latter is an infinite family of inductive sets, since $n \in \omega$.

These are a few examples of how constructive type theory can be and has been enriched by adding new ground sets or set operations to its basic collection of such sets and operations, and of how these new sets and operations have increased the expressibility of new mathematical objects and states of affairs in constructive type theory. Versions (7) and (8) add in this respect nothing to version (6).

v KINDS OF EQUALITY/IDENTITY

Equality or identity relations are not only one of the more tricky ingredients of constructive type theory (cf. Sect. i., Chapt. 2), they also underwent a considerable development from the origin to the present state of this theory. Version (1) has an informal as well as a formal notion of definitional or intensional equality which are both notions regarding the linguistic level. A third notion on the linguistic level is called identity and it is basically syntactical identity. In version (2), the formal notion of definitional equality is again supposed to formalize the informal one and it consists in a new def. of the convertibility relation; moreover, the informal notion is obtained as a consequence of the formal one and a formal rule of inference called type conversion. But it turned out that although convertibility was supposed to be a correct formalization of (informal) definitional equality, it actually was not. The troubles with the convertibility relation of version (2) are taken care of in version (3) by defining a new convertibility relation on a universe of well-typed terms and by abolishing bound variables altogether. This new def. led to a distinction between general rules of conversion and the particular rules of conversion for the various set-theoretic operations of version (3). Apart from the syntactical identity relation preserved through version (2) also in version (3), a new identity relation appears in (3), called identity, which is like an internal form of the convertibility relation between terms, a relation on which, unlike the convertibility relation, one can operate with the logical operations. There are still two concepts of definitional equality in version (4), namely an informal and a formal one, but since version (4)'s main concern is a new sort of semantics, these concepts merely concern the kind of language, ordinary or formal resp., in which they occur and not what they mean. Therefore, the formal concept of definitional equality is not expressed any longer by the convertibility relation. Furthermore, definitional equality is, according to the new semantical approach, no more a relation regarding the linguistic level, but rather a relation between abstract entities, i.e. a relation regarding the object level. Version (4) also takes over from version (3) the new relation of identity I. Version (5) achieves what version (4) tried to achieve, namely to provide a new direct semantics for the language of constructive type theory. Thus, unlike in version (4), in version (5) there is a meaning explanation of its well-known forms of judgement. But like in version (4), definitional equality is in version (5) an equivalence relation on the level of abstract entities and not on the linguistic level. Version (5) also takes Quine's slogan "No entity without equality" (in version (5) terminology) radically seriously and equips each type and each object of some type with equality. Finally, version (5) contains two elimination rules for the no longer new relation of identity I, a weak and a strong one. Now, the I-elimination rule clearly introduces extensional identity into constructive type theory. The equality or identity relations of version (6) are essentially the same as the ones of version (5), except that in version (6) the ambiguity of I-elimination of version (5) is dissolved in favor of strong I-elimination, i.e. of extensional identity. And version (6) reintroduces a relation called intensional equality or equality of meaning, which regards the linguistic level and corresponds to the informal definitional equality relation of version (3). As far as version (7) is concerned, all the equality or identity relations on the object level are taken over from version (6), except that all the equality rules for the various set-theoretic operations are now in a monomorphic and higher-level form. Version (7) resembles version (2) insofar as the relation of identity I is missing in both these versions. As for the equality or identity relations on the linguistic level, not only are the two of intensional equality of version (6) taken over in version (7), but the one of syntactical identity of versions (1) and (2) is warmed up in version (7) too. As for version (8), so much is revealed in advance: version (8) carries on with the philosophical and historical elucidation of constructive type theory started in version (7) and it contains the most differentiated and sophisticated account of matters concerned with equality and identity and the like in this theory.

In version (1) an informal and a formal notion of definitional or intensional equality occur; both are denoted by =, and both are relations between expressions.

The informal notion is defined as follows: Two linguistic expressions a and b are said to be definitionally or intensionally equal if they denote the same object.[16]

Thus, the concept of sameness of abstract entities (types or objects of some type) is implicitly introduced, but nowhere explicated or discussed.

The formal notion of definitional or intensional equality is defined as follows: Two expressions a and b are definitionally or intensionally equal if they reduce to a common expression, i.e. if there is an expression c s.t. $a \geq c$ and $b \geq c$. (For the full def., cf. Sect. iii., Subsect.(2).) The formal relation $=$ is then proved to be an equivalence relation.[17]

On the same linguistic level the notion of identity is introduced as follows: Two expressions a and b are said to be identical if they are syntactically identical modulo naming of bound variables.[18]

These are the only prim. notions of equality or identity in this version of constructive type theory, and it is pointed out that other notions of equality, such as equality between real numbers or extensional equality between species of objects of some given type, ought to be defined.[19]

The only notion of equality/identity on the object level is the notion of identity denoted by I between two objects denoted by a and b of some type denoted by A, and defined following Russell as follows: $I(a, b) = (\Pi X \in A \to V)\,(Xa \to Xb) \in V$ which can readily be shown to be an equivalence relation.[20]

This identity relation on the object level is related to the notion of definitional or intensional equality as follows: it should first be mentioned that one of the rules of inference of version (1) is the so-called equality rule $\dfrac{a \in A \quad B \in V}{a \in B}\quad A = B$; then

i) if $a \in A, b \in A, a = b$ then $I(a, b)$
because of the equality rule

ii) if $a \in A, b \in A, I(a, b)$ then $a = b$
by the normalization theorem (Cor. 5.5)[21]

Finally, it may be noted that the (formal) definitional or intensional equality relation = is decidable as a consequence of the normalization theorem (Cor. 5.7).[22]

Due to the discovery of the inconsistency of version (1), a new version (2) of constructive type theory followed which differs from version (1) as far as the equality/identity issue is concerned in the following ways:

There are again two notions of definitional equality: The informal notion is denoted by = and defined as follows: Two type expressions A and B are said to be definitionally equal if they both denote the same (abstract) type. Two terms a and b are called definitionally equal if they both denote the same object of the (abstract) type denoted by their type expressions.[23]

The formal notion is defined by: Two type expressions A and B are said to be definitionally equal if A *conv* B. Two terms $a \in A$ and $b \in B$ are said to be definitionally equal if a *conv* b and A *conv* B.[24] The formal notion of convertibility *conv* is defined in Sect. iii., Subsect. (1). *conv* is then proved to be an equivalence relation (Cor. 2.4.4 to the Church–Rosser property).[25] Note that the relation of convertibility between terms is defined on a universe of untyped or raw terms.

The relation on the linguistic level called identity in version (1) is called (syntactical) equality in version (2).[26]

Note that the informal notion of definitional equality is introduced as a consequence of the formal one and a rule of inference called type conversion (which parallels the equality rule of version (1)) $\dfrac{a \in A}{a \in B}\quad A\ \mathit{conv}\ B$. By means of this rule one can show that two terms a and b are of definitionally equal type expressions iff the objects denoted by a and b are of the same (abstract) type. Moreover, since definitional equality is a relation between (linguistic) expressions and not between (abstract) entities which they denote, one can easily see why the formal rule of type conversion has (unlike all the other rules of term formation) no informal counterpart.[27]

As is the case in version (1), the definitional equality between two terms or between two type expressions is decidable as a consequence of the normalization theorem (Cor. 4.3).[28] Although this equality between two type expressions is intensional, the term model obtained from the normalization proof has an extensional equality on the interpretation of the types (extensional equality understood in the set-theoretical sense) (cf. Sect. iii., Subsect. (1)). Furthermore, for this term model it does not hold that $a\ conv\ b$ implies $\overline{a} = \overline{b}$, which of course points to a serious flaw in the theory, since the convertibility relation *conv* is supposed to formalize the informal notion of definitional equality =.[29]

After version (2) it became clear that although convertibility was supposed to be a correct formalization of definitional equality, it actually was not. This reason and a couple of other changes lead to version (3) of constructive type theory (cf. Sect. iii, Subsect. (1)).

An analysis of the informal notion of definitional equality denoted by $=_{\text{def}}$ provided the following result: the definitional equality relation is a relation between expressions generated by exactly the following three principles:

(i) definiendum $=_{\text{def}}$ definiens,

(ii) $\dfrac{a =_{\text{def}} c}{b[a] =_{\text{def}} b[c]}$, i.e. definitional equality is preserved under substitution, and

(iii) $a =_{\text{def}} a$, $\dfrac{a =_{\text{def}} b}{b =_{\text{def}} a}$, $\dfrac{a =_{\text{def}} b \quad b =_{\text{def}} c}{a =_{\text{def}} c}$ i.e. definitional equality is an equivalence relation.

Moreover, it turned out that the only way in which definitional equality entered into a mathematical construction except for definitional schematas themselves, was in inferences of the following form $\dfrac{a \in A \quad A =_{\text{def}} B}{a \in B}$ where either *A set* and *B set* or *A prop* and *B prop*. This inference or principle is called the principle of replacing a set (proposition) by a definitionally equal set (proposition). It is justified by reference to the fact that $A =_{\text{def}} B$ implies that the expressions A and B are mere notational variants of one and the same set or proposition. (Note that contrary to version (2) the justification of this principle is here based on the definition of the informal notion of definitional equality of versions (1) and (2).)[30]

If convertibility is to be the correct formal counterpart of the informal definitional equality, then it is determined by the following conversion rules:

(i) redex *conv* contractum, where redex and contractum belong to the same type,

(ii) $\dfrac{a_1\ conv\ c_1 \ldots a_k\ conv\ c_k}{b[a_1, \ldots, a_k]\ conv\ b[c_1, \ldots, c_k]}$ where $a_1 \in A_1, \ldots, a_k \in A_k[a_1, \ldots, a_{k-1}]$, $c_1 \in A_1, \ldots, c_k \in A_k[c_1, \ldots, c_{k-1}]$ and $b[x_1, \ldots, x_k]$ is a term which depends on no other variables than $x_1 \in A_1, \ldots, x_k \in A_k[x_1, \ldots, x_{k-1}]$,

(iii) if $a \in A$ has been derived for some A, then $a\ conv\ a$; if $a \in A, b \in A$ have been derived for some A then $\dfrac{a\ conv\ b}{b\ conv\ a}$; and if $a \in A, b \in A, c \in A$ have been derived for some A then $\dfrac{a\ conv\ b \quad b\ conv\ c}{a\ conv\ c}$.

To these rules is added a formalization of the principle of replacing a set (proposition) by a definitionally equal set (proposition) called the rule of conversion of a type symbol $\dfrac{a \in A \quad A\ conv\ B}{a \in B}$. (*Note*: this rule formally corresponds to the equality rule of version (1) and the type conversion rule of version (2).)[31]

Now, the problems with the convertibility relation of version (2) arose from unrestricted conversions under the λ sign and other variable binding operations. The following rules which are valid under the definition of convertibility of version (2) are not valid anymore under the new definition of convertibility of version (3):[32]

$$(\lambda x)(b(x))\ conv\ b \qquad (\eta\text{-conversion})$$
$$\text{where } x \text{ does not occur free in } b$$
$$(E(c, (\lambda x)(\lambda y)x),\ E(c, (\lambda x)(\lambda y)y))\ conv\ c$$
$$\left.\begin{array}{l} i(D(c, (\lambda x)x, (\lambda y)y)) \\ j(D(c, (\lambda x)x, (\lambda y)y)) \end{array}\right\}\ conv\ c$$

One of the positive effects of the new definition of convertibility is that one has a natural correspondence between λ-calculus and combinatory logic:[33] there is a natural transformation denoted by the superscript $^\circ$ from λ-terms to terms constructed from constants or combinators, and there is a natural transformation denoted by the superscript $^\bullet$ from terms constructed from constants and combinators to λ-terms, s.t. the following holds:

i) for λ-terms a and b, $a\ conv\ b$ implies $a^\circ\ conv\ b^\circ$
ii) for terms a and b constructed from constants and combinators, $a\ conv\ b$ implies $a^\bullet\ conv\ b^\bullet$

where $^\circ$ is defined by

$$x^\circ =_{\text{def}} x$$
$$(b(a))^\circ =_{\text{def}} b^\circ(a^\circ)$$
$$((\lambda x)b[a_1, \ldots, a_n, x])^\circ =_{\text{def}} f(a_1^\circ, \ldots, a_n^\circ)$$

where $a_1, \ldots, a_n$ are the (necessarily disjoint) maximal subterms of $b[a_1, \ldots, a_n, x]$ which do not contain any free occurrences of x and f is the function constant introduced by the schema $f(a_1, \ldots, a_n, a)\ conv\ b^\circ[a_1, \ldots, a_n, a]$

and where $^\bullet$ is defined by

$$x^\bullet =_{\text{def}} x$$
$$(b(a))^\bullet =_{\text{def}} b^\bullet(a^\bullet)$$
$$f^\bullet =_{\text{def}} (\lambda x_1)\ldots(\lambda x_n)b^\bullet[x_1, \ldots, x_n]$$

where f is a function constant introduced by the schema $f(a_1, \ldots, a_n)\ conv\ b[a_1, \ldots, a_n]$

The problems with the convertibility relation of version (2) are solved in version (3) by defining the new convertibility relation *conv* on a universe of well-typed terms (hence the conversions are subject to type restrictions) and by abolishing bound variables altogether.[34] The latter is achieved by giving a combinator formulation of the theory rather than a λ-calculus one (as had been the case in versions (1) and (2)). And this can be done quite naturally if use is made of the natural correspondence between λ-calculus and combinatory logic explained above. Thus, analogous to version (2), a now typed function, the eliminatory operator, is informally explicitly defined for each set and set forming operation.[35] These definitions are formally given by corresponding conversion rules always in a combinator language. Hence in version (3) – unlike the preceding versions – a distinction is made between the general rules of conversion and the particular rules of Π-conversion, Σ-conversion, +-conversion, etc. (the only exception being the structural rule of conversion of a type symbol and its analogues in versions (1) and (2)) (for details, cf. Sect. iii., Subsect. (3)).

The relation on the linguistic level called (syntactical) equality in version (2) is called syntactical identity in version (3) and denoted by =.[36] Unlike versions (1) and (2), version (3) contains a new equality/identity relation called identity and denoted by 'I'. The identity relation I is like an internal form of the relation *conv* between terms: if a and b are terms of the same type symbol A, then $I(a, b)$ is the proposition that the objects denoted by a and b are identical or the same. Unlike *conv*, I is a relation on which one can operate with the logical operations.[37]

Finally, the following relations can be established between the different notions of equality/identity of version (3):

i) if $a \in A, b \in A, a =_{\text{def}} b$ then there is a term c s.t. $c \in I(a, b)$
(by the principle of replacing a set (proposition) by a definitionally equal set (proposition))[38]
ii) $a \in A, b \in A, a\ conv\ b$ iff there exists a closed term c s.t. $c \in I(a, b)$
(Thm. 3.14 as a corollary to the normalization theorem;[39] *note*: this theorem does not hold for open terms c)
iii) if $a \in A, b \in A$ are closed normal terms s.t. $a\ conv\ b$ then $a = b$
(Thm. 3.4 as a corollary to the normalization theorem)[40]
iv) for the term model obtained from the normalization proof one has[41]
for $a \in A, b \in A, a\ conv\ b$ iff $\overline{a} =_{\text{def}} \overline{b}$.

The convertibility and hence definitional equality relation between two terms is decidable as a consequence of the normalization theorem (Cor. 3.8,[42] and remarks concerning a strengthening of the normalization theorem[43]). (Note that type expressions are in version (3) also terms, they are type expressions for some terms and terms for some other type expressions.[44]) And, not only is the definitional equality between two type expressions intensional, but the term model obtained from the normalization proof has also an intensional equality on the interpretation of the types (cf. Sect. iii., Subsect. (3)).[45]

The main drive of version (4) as opposed to version (3) is to find a new semantics for the language or syntax of constructive type theory (cf. Sect. iii., Subsect. (4)). As far as the equality/identity issue is concerned, version (4) compares with version (3) as follows:

There are again two concepts of definitional equality (also called definitional equivalence) in version (4), namely a formal and an informal one, which however regard only the kind of language, formal or ordinary, in which they occur and not what they mean. Definitional equality is formally expressed by judgements of the form $\overline{A} \ni \overline{a} = \overline{b} \in \overline{B}$ and $\overline{A} = \overline{B}$; and these forms of judgement are informally read as: $a \equiv b$ or $a =_{\text{def}} b$ or $a = b\ Df$ or the $\overline{A}$-valued function a and the $\overline{B}$-valued function b are definitionally equal or equivalent, and, $A \equiv B$ or $A =_{\text{def}} B$ or $A = B\ Df$ or the type-valued functions A and B are definitionally equal or equivalent, resp.[46] Thus, the informal readings of the preceding formal judgements must not be mistaken as meaning explanations; they only show how to read these judgements in ordinary mathematical language or in English. Whether they are expressed in the formal language of constructive type theory or in ordinary English, the meanings of the judgements concerning the definitional equality between two object-valued functions or between two type-valued functions have yet to be explained.[47] Note that definitional equality is according to the new semantical approach no more a relation between expressions, but rather a relation between abstract entities, i.e. between functions. However, since the specific attempt of version (4) at a new meaning explanation of the various forms of judgement failed in general, the meaning explanations of the forms of judgement on definitional equality in particular do not work and will not be given here (cf. Sect. iii., Subsect. (4)).

The general rules of definitional equality in version (4) are:[48]

$$\frac{\overline{a} \in \overline{A} \quad \overline{A} = \overline{B}}{\overline{A} \ni \overline{a} = \overline{a} \in \overline{B}} \qquad\qquad \frac{\overline{A} \text{ is a type}}{\overline{A} = \overline{A}}$$

$$\frac{\overline{A} \ni \overline{a} = \overline{b} \in \overline{B}}{\overline{B} \ni \overline{b} = \overline{a} \in \overline{A}} \qquad\qquad \frac{\overline{A} = \overline{B}}{\overline{B} = \overline{A}}$$

$$\frac{\overline{A} \ni \overline{a} = \overline{b} \in \overline{B} \quad \overline{B} \ni \overline{b} = \overline{c} \in \overline{C}}{\overline{A} \ni \overline{a} = \overline{c} \in \overline{C}} \qquad\qquad \frac{\overline{A} = \overline{B} \quad \overline{B} = \overline{C}}{\overline{A} = \overline{C}}$$

and the rules called preservation under composition:

$$\frac{\begin{array}{c} \overline{A}_1 \in \overline{a}_1 = \overline{b}_1 \in \overline{A}_1 \\ \vdots \\ \overline{A_k(a_1, \ldots, a_{k-1})} \ni \overline{a}_k = \overline{b}_k \in \overline{A_k(b_1, \ldots, b_{k-1})} \end{array}}{\overline{A(a_1, \ldots, a_k)} \ni \overline{a(a_1, \ldots, a_k)} = \overline{a(b_1, \ldots, b_k)} \in \overline{A(b_1, \ldots, b_k)}}$$

provided that $\begin{array}{c} \overline{u}_1 \in \overline{A}_1, \ldots, \overline{u}_k \in \overline{A_k(u_1, \ldots, u_{k-1})} \\ \cdots\cdots\cdots\cdots\cdots\cdots\cdots\cdots\cdots\cdots \\ \overline{a(u_1, \ldots, u_k)} \in \overline{A(u_1, \ldots, u_k)} \end{array}$ and

$$\frac{\begin{array}{c} \overline{A}_1 \ni \overline{a}_1 = \overline{b}_1 \in \overline{A}_1 \\ \vdots \\ \overline{A_k(a_1, \ldots, a_{k-1})} \ni \overline{a}_k = \overline{b}_k \in \overline{A_k(b_1, \ldots, b_{k-1})} \end{array}}{\overline{A(a_1, \ldots, a_k)} = \overline{A(b_1, \ldots, b_k)}}$$

provided that $\begin{array}{c} \overline{u}_1 \in \overline{A}_1, \ldots, \overline{u}_k \in \overline{A_k(u_1, \ldots, u_{k-1})} \\ \cdots\cdots\cdots\cdots\cdots\cdots\cdots\cdots\cdots\cdots \\ \overline{A(u_1, \ldots, u_k)} \text{ is a type} \end{array}$

(Note that preservation under composition is nothing but preservation of definitional equality under substitution in version (3).)

To these rules is added the rule linking the rules of function formation with the general rules of definitional equality, i.e. the rule of replacing the range of a function by one with which it is definitionally equal[49] $\frac{\overline{a} \in \overline{A} \quad \overline{A} = \overline{B}}{\overline{a} \in \overline{B}}$ (called the principle of replacing a set (proposition) by a definitionally equal set (proposition) in version (3).)

With the exception of the Π-operation, there is an explicit definition of an eliminatory operator for each set and set-forming operation of version (4) in the language of combinatory logic – and these definitions are almost identical with the corresponding formal definitions of version (3). The new definition of the eliminatory operator for Π differs from the one of version (3) by introducing a new symbol for this operator, namely '*ap*', and by using a λ-notation rather than a combinator one.[50]

Version (4) takes over from version (3) the new relation called identity and denoted by 'I'. Whereas identity is formally expressed in constructive type theory by '$I_A(x, y)$', it is informally expressed in ordinary mathematical language or English by $x =_A y$ or $x = y$ or the proposition that x and y are identical objects of the type which is the value of A.[51]

($x, y, z, \ldots$ are used for arbitrary object expressions, $X, Y, Z, \ldots$ for arbitrary type expressions, $a, b, c, \ldots$ for arbitrary object-valued functional expressions, and $u, v, w, \ldots$ as variables to indicate how these functional expressions take arguments.[52] And the phrases $\overline{a(u_1, \ldots, u_k) = x \text{ for } \overline{u}_1 = x_1, \ldots, \overline{u}_k = x_k}$ and, $\overline{A(u_1, \ldots, u_k) = X \text{ for } \overline{u}_1 = x_1, \ldots, \overline{u}_k = x_k}$ in constructive type theory express informally: x (X) is the value of $a(u_1, \ldots, u_k)$ ($A(u_1, \ldots, u_k)$) when the variables $u_1, \ldots, u_k$ are assigned values $x_1, \ldots, x_k$ resp.[53] Note that $=$ expresses here the identity of evaluation or computation, and it means that $a(u_1, \ldots, u_k)$ is definitionally equal to x if u_1 is definitionally equal to $x_1, \ldots, u_k$ is definitionally equal to x_k.)

Version (5) realizes what version (4) tried to achieve, namely to provide a new direct semantics for the language or syntax of constructive type theory (cf. Sect. iii., Subsect. (5)).

What has been called 'definitional equality' in version (4) is called 'equality' in version (5). The inferential part of the language of constructive type theory consists of rules for making judgements – among others – of the following form: $A = B$, which is read informally as A and B are equal types, and $a = b \in A$, which is informally read as a and b are equal objects of type A. In their more general, i.e. hypothetical form, these judgements are: $A = B\,(x_1 \in A_1, \ldots, x_n \in A_n)$ and $a = b \in A\,(x_1 \in A_1, \ldots, x_n \in A_n)$.[54] Unlike in version (4), in version (5) there is a meaning explanation of the just mentioned forms of judgement, which for $A = B$ is the familiar explanation that each canonical object a of A is a canonical object of B and vice versa, and two identical canonical objects $a = b$ of A are identical canonical objects of B and vice versa, provided A and B are canonical types; o.w. the judgement means that A and B have equal canonical types as values. And for $a = b \in A$ it is the familiar meaning explanation that a and b have equal canonical objects of the canonical type denoted by A as values.[55] As is the case in version (4), equality is an equivalence relation on the level of abstract entities and not on the linguistic level (as used to be the case in versions (1)–(3)).

The general rules of definitional equality of version (4) are taken over in version (5) and a second rule of equality of types (called the rule of replacing the range of a function by one to which it is definitionally equal, in version (4)), is added: $\dfrac{a = b \in A \quad A = B}{a = b \in B}$.

Finally, the rule called preservation under composition in version (4) is replaced by the following more general substitution rules in version (5):[56]

$$\frac{a = c \in A \qquad \begin{matrix}(x \in A)\\ B = D\end{matrix}}{B(a/x) = D(c/x)} \quad \text{and} \quad \frac{a = c \in A \qquad \begin{matrix}(x \in A)\\ b = d \in B\end{matrix}}{b(a/x) = d(c/x) \in B(a/x)}$$

Version (5) takes Quine's slogan 'No entity without equality' (in version (5) terminology) radically seriously and endows each type (canonical or non-canonical) and each canonical as well as non-canonical object of some type with an equality. The non-dependent types and the non-dependent objects all have an intensional equality whereas the dependent types (families of types) and the dependent objects (functions in the oldfashioned sense) all have an extensional equality. This manifests itself in that each formation-rule, each introduction- and elimination-rule for each type and type forming operation has an equality rule associated with it. As far as the Π-equality, Σ-equality, $+$-equality, N_n-equality and N-equality rules explicitly defining the eliminatory operators for Π, Σ, $+$, N_n and N resp. are concerned, version (5) returns on the one hand to the λ-notation of version (2), preserving on the other hand the conversion being defined on well-typed terms approach of versions (3) and (4) in order to correct the formalization of equality of version (2) (equality is here a relation defined on a universe of well-typed objects).[57]

Like versions (3) and (4), this version also has an identity relation denoted by 'I'. The formal judgement $I(A, a, b)$ *type* is informally read as a and b are identical objects of type A, and $I(A, a, b)$ is a proposition or set. But the rules associated with and defining identity are different from those of version (4). According to the I-introduction rule $\dfrac{a = b \in A}{r \in I(A, a, b)}$ all identity sets (identity propositions) have one and the same canonical element r. This is clearly not the case in version (4) and it is an equally clear sign that version (5) is a polymorphic version of constructive type theory (cf. Subsect. ix. 8, Chapt. 1). Furthermore, there are two kinds of I-elimination rules. The one kind is similar to the I-elimination rule of version (4) and is here called weak I-elimination rule, and the other one, i.e. $\dfrac{c \in I(A, a, b)}{a = b \in A}$ introduces a symmetry with the I-introduction rule and is here called the strong I-elimination rule. If propositional identities are expressed by means of the strong I-elimination rule, then equality can no longer be understood as convertibility, since it is then possible to prove a judgement expressing an equality by means of an argument using propositions. Moreover, equality is, unlike in versions (1)–(3), no longer decidable. One can say that the strong I-elimination rule introduces extensional equality into constructive type theory. – Note that the I-equality rule is still based on the weak I-elimination rule.[58] Version (5) also has a second Π-equality rule which can be derived by means of the strong I-elimination rule.

Version (6) is the first extensive presentation of constructive type theory after its evolution through the various preceding stages/versions. Version (6) distinguishes four kinds of equality/identity relations.

The first two are called 'equality' as in version (5): they are equality of sets, formally expressed in constructive type theory by the form of judgement $A = B$, and equality of elements of a set, formally expressed by the form of judgement $a = b \in A$. They are both equivalence relations on the level of abstract entities (sets or elements of a set) and denoted by the same symbol $=$. The meaning explanations of these two forms of judgement are essentially the same as in version (5) (except that in version (6), unlike version (5), non-canonical sets are not explicitly taken into account).[59]

All the general rules of equality of version (5) are taken over into version (6), the substitution rules included.[60] The radically "equalitarian approach" of version (5) is likewise preserved in version (6): to each rule of set formation, introduction and elimination there corresponds an equality rule which allows for the substitution of equals for equals.[61] The Π-, Σ-, $+$-, N_n- and N-equality rules of version (5) appear almost identically in version (6). Version (6) finally not only contains the second Π-equality rule of version (5), but also a second Σ-equality rule derived again by means of the strong I-elimination rule (of version (5)).[62]

The third notion is the one of propositional equality or identity, denoted as in the three preceding versions by 'I'. $I(A, a, b)$ asserts that a and b are equal elements of the set A in such a way that one can operate on it with logical operations. It also is a relation on the level of abstract entities and an internal form of the notion of equality of elements of a set. Its formation and introduction rules are the same as those of version (5). Hence version (6) is also a polymorphic version of constructive type theory. From the I-elimination rules of version (5) only the strong I-elimination rule survives in version (6). Thus, the I-equality rule of version (6) is unlike the one of version (5) based on the strong I-elimination rule.[63] This points to the fact that extensional identity plays an even more important role in version (6) than it does in version (5).

The fourth and last notion of equality in version (6) is the one of definitional equality, also called intensional equality or equality of meaning, and denoted by $\equiv$ or $=_{\mathrm{def}}$. Definitional equality is a relation between expressions (hence on the linguistic level) satisfying the three principles of version (3).[64]

These different notions of equality of version (6) can be compared with each other as follows:

i) the relation $\equiv$ is intensional, while the relations $=$ and I are extensional.[65]

ii) the relation $\equiv$ is decidable, while the relations $=$ and I are not decidable. But the decidability of $\equiv$ cannot be expressed in the theory itself: $a \equiv b$ is not a proposition of constructive type theory.[66]

iii) the second notion of $=$ and the notion I are coextensional, but they are not cointensional.[67]

iv) the relations $=$ serve to form judgements, the relation I to form propositions, and the relation $\equiv$ to form mere linguistic conventions or stipulations.[68]

Version (7) is concerned with a philosophical elucidation of constructive type theory developed up to version (6) and beyond.

What has been called 'equality' in versions (5) and (6) is here called 'identity' or 'definitional identity (equality)' or 'intensional identity (equality)' and denoted by =.[69] The inferential part of the language of constructive type theory consists of rules for making judgements – among others – of the form $\alpha = \beta : type$, $a = b : \alpha$, $\alpha = \beta : type\ (x_1 : \alpha_1, \ldots, x_n : \alpha_n)$, $a = b : \alpha\ (x_1 : \alpha_1, \ldots, x_n : \alpha_n)$ as in versions (5) and (6) as well as new, higher-level judgements of the form $(x : \alpha)\beta = (x : \gamma)\delta : type$, $(x)b = (x)d : (x : \alpha)\beta$ and finally categorical and hypothetical judgements of the form $\alpha = \beta :\ seq$.[70] The meanings of all these forms of judgement are explained in Chapt. 1. This definitional or intensional identity relation is a relation on the level of abstract entities (objects, functions in the oldfashioned sense, functions in the modern sense, non-dependent non-function types, dependent types, function types, etc.) and following version (6) one could distinguish between a definitional identity of types and a definitional identity of objects of some type.

Version (7) contains all the general rules of definitional identity of version (6) (called general rules of equality there).[71] And as in the two preceding versions there is an identity rule associated with each rule of set or proposition formation, introduction and elimination. The Π- and N-equality rules of version (6) appear now as Π- and N-identity rules in a monomorphic and higher-level form and there is also a second formulation of the Π-identity rule which follows the same pattern as the identity rules of all the other set forming operations.[72] Version (7) has moreover $\forall$-, $\supset$-, $\exists$- &- and $\vee$-identity rules.[73]

In version (7) definitional or intensional identity, denoted by =, is opposed to material equivalence, denoted by $\supset\subset$. Material equivalence can be conceived of as extensional identity of propositions or identity of truth values. Extensional identity of propositions is obviously weaker than the corresponding intensional identity, and the relationship between the two notions of identity is such that intensional identity entails extensional identity but not vice versa.[74]

The identity or equality relations so far related on all hold on the level of abstract entities. The following three notions of identity all hold on the linguistic level.

The first two notions of identity on the linguistic level are called identity of meaning, also sameness of meaning or synonymy. They are the relations between expressions induced by the relations of definitional identity between the entities denoted by these expressions: two expressions are identical with regards to their meaning, i.e. they are synonymous, if their meanings, i.e. the entities they stand for, are definitionally identical.[75] These notions of identity are also called definitional identity and denoted by $. = .Df$ or $=_{\mathrm{Def}}$ or $\equiv$. They are the same notions as the homonymous one in version (6), and although they are briefly discussed, no use is made of them in version (7) contrary to version (6).[76] The definitional identities of version (6) are transformed into definitional identities on the level of abstract entities in version (7) and thus become expressible in constructive type theory itself.

The third notion of identity on the linguistic level is syntactical identity.[77] It is the same notion as the one called identity in version (1) and (syntactical) equality in version (2).

Observe that version (7) resembles version (2) insofar as what has been called propositional equality or identity I in version (6) is missing in both these versions; and the definitional identity relation $=$ of both versions is decidable. Version (7) contains a sketch of a definitional identity checking algorithm.[78] Finally, version (7) resembles versions (3) and (4) insofar as it is also monomorphic.

Version (8) continues the philosophical and historical elucidation of constructive type theory started in version (7). It contains e.g. the most elaborate and differentiated account of matters concerned with equality and identity in this theory.

In version (8) the identities of types and the identities of objects of some type are always conceived as definitional identities, denoted by $=$.[79] The inferential part of the language of constructive type theory version (8) is basically the same as the one of version (7) (the only differences being that 1) sequents do not occur in version (8) and hence nor do judgements of the form $\alpha : seq$, etc., and 2) the rule of abstraction of version (7) being replaced by two other rules of function object formation).[80] As in version (7), this relation of definitional identity is a relation on the level of abstract entities.[81] Version (8) contains all the general rules of definitional identity of version (7). As with the three preceding versions each rule of set formation, introduction and elimination comes together with an identity rule which is exemplified by the Π- and I-operation.[82] The Π- and I-identity rules are presented as in version (7) in a monomorphic and higher-level notation following the same pattern as the identity rules of all the other set forming operations.[83] Accordingly, as was explained in version (5), each non-dependent (function) type and each non-dependent (function) object has an intensional identity, whereas each dependent (function) type and each dependent (function) object has an extensional identity, i.e. each type and each object has an identity.

The method for making explicit definitions inside constructive type theory used implicitly in version (7) is made explicit in version (8) by way of the following rule schemas:[84]

$$\frac{\begin{array}{l} a_1 : \alpha_1 \\ \quad\vdots \\ a_n : \alpha_n(a_1/x_1, \ldots, a_{n-1}/x_{n-1}) \\ x_1 : \alpha_1, \ldots, x_n : \alpha_n \vdash \beta : type \end{array}}{\begin{cases} \varphi(a_1)\ldots(a_n) : type \\ \varphi(a_1)\ldots(a_n) = \beta(a_1/x_1, \ldots, a_n/x_n) : type \end{cases}} \quad \text{on the type level}$$

$$\frac{\begin{array}{l} a_1 : \alpha_1 \\ \quad\vdots \\ a_n : \alpha_n(a_1/x_1, \ldots, a_{n-1}/x_{n-1}) \\ x_1 : \alpha_1, \ldots, x_n : \alpha_n \vdash b : \beta \end{array}}{\begin{cases} f(a_1)\ldots(a_n) : \beta(a_1/x_1, \ldots, a_n/x_n) \\ f(a_1)\ldots(a_n) = b(a_1/x_1, \ldots, a_n/x_n) : \beta(a_1/x_1, \ldots, a_n/x_n) \end{cases}} \quad \text{on the object level}$$

There is also a notion of definitional identity with regard to the linguistic level, called identity of meaning or synonymy and denoted, similarly to version (7), by $. = .Df$, or $=_{\mathrm{Def}}$ or $\equiv$ or $\stackrel{\Delta}{=}$ or simply by $=$. What has been said in version (7) about this notion applies equally well to version (8).[85]

A second notion of identity on the linguistic level is the notion of syntactical identity, denoted by $\equiv$. (cf. also version (7))[86]

There is a second notion of identity on the level of abstract entities called identity of expression or syntactic identity and denoted by $\equiv$. It is a relation induced by the relation of syntactical identity, which holds between two entities if the expressions standing for these two entities are syntactically identical.[87]

There is even a third notion of identity on the level of abstract entities which is the notion of extensional identity. Two objects a and b of some type A are extensionally identical if $I(A, a, b)$ *true*, i.e. if there is an element c s.t. $c : I(A, a, b)$.[88]

The identity relations introduced above are all judgemental identities, i.e. identity relations forming a judgement. But the notion of extensional identity shows that there is a propositional identity involved, simply called identity in version (8) and denoted as usual by 'I'. It is, as was said in version (3), an internal form of the definitional identity relation $=$, but this time, unlike version (3), on the level of abstract entities.

The I-formation and I-introduction rules are like those of versions (3) and (4), i.e. monomorphic, and thus different from those of versions (5) and (6).

The I-elimination and I-equality rules of version (8) are likewise like those of versions (3) and (4); this I-elimination rule has been called weak I-elimination rule in version (5) and it preserves intensional identity in constructive type theory.[89]

Finally, the following relationship can be established between the different notions of judgemental identity on the level of abstract entities:[90]

i) if $a : A$, $b : A$, $a \equiv b : A$ then $a = b : A$, but not conversely
ii) if $a : A$, $b : A$, $a = b : A$ then $I(A, a, b)$ *true*, but not conversely.

Note: the relation of definitional equality $=$ is decidable as in version (7) and since the new I-rules are monomorphic and preserve the intensionality of identity.

vi UNIVERSES

Universes are present in constructive type theory from the very beginning. They serve at least the two purposes of enabling type theory to go beyond the structure of finite types and to formulate at least parts of category theory. Version (1) of constructive type theory starts off with an allembracing universe, that is a type of all types, which thus is an object of itself. However, Girard showed that this assumption ends in a logical paradox, so that Martin-Löf put forward another version (version (2)) in which this assumption is dropped. But since the types and type-valued functions Π, Σ, $+$, N, and N_n do not allow to define enough types and type-valued functions, in particular no transfinite types, a universe is introduced together with a reflection principle saying: whatever can be done with types, can be done inside this universe. An object of this universe is said to be small, whereas the universe itself as well as any type defined by means of it, is said to be large. This universe is thus the type of small types and not the type of all types anymore. Since Martin-Löf was interested from scratch in the possibility of formalizing category theory in constructive type theory, he assumed the allembracing universe in version (1) to deal with the construction of the category of all categories in category theory, and he adopts the universe of small types and makes the distinction of small and large types in version (2) in order to be able to formulate the basic notions and constructions of category theory (but of course not the construction of the category of all categories anymore). In version (3), the universe of version (2) is extended to a whole sequence of universes which provides this new version of constructive type theory with more natural closure properties. The adoption of such a sequence of universes entails that every type is at the same time an object of some type. This is the full realization of the doctrine of types: No entity without type. Whereas version (3) is a monomorphic version of constructive type theory, version (5) is polymorphic, which can clearly be read off the rules for the universes of the sequence of universes. The novelty of version (6) lies in the distinction of the two different ways of formulating the universes, namely Tarski's way and Russell's way. Universes make their last appearance in this history in version (6).

In version (1) two assumptions are made, namely

(i) every mathematical object is of a certain type, and
(ii) the types themselves are mathematical objects.

From these assumptions follows that there is a type of types, symbolized by 'V'. Now, V is itself a type and hence, according to (ii), a mathematical object, and thus, according to (i), of a certain type which is the type of (all) types, i.e. an object of V. $V \in V$ is the basic and only axiom of the formal system of version (1) with which every derivation begins.

In (1971B) Martin-Löf mentions Kreisel's drawing attention to the intuitionistic significance of the impredicative theory of species. What motivated Martin-Löf to formulate version (1) of constructive type theory was his insight, following Kreisel, into the intuitionistic significance of the full intuitionistic theory of finite types, and, because of some unnatural limitations and shortcomings of such a theory, into the significance of a general intuitionistic theory of types. In this paper he shows how the idea of a type of types can also be seen to arise from the three assumptions: (a) quantification over propositions is meaningful (as in second order logic), (b) ranges of significance of propositional functions form types, and in particular, it is only meaningful to quantify over objects of some type (Russell's doctrine of types), (c) identification of propositions and types. Now, from (a) and (b) follows that there is a type of (all) propositions, and from this and (c) follows that there is a type of (all) types. Finally, Martin-Löf refers to Gödel (1964) who pointed out that the type of types V introduces a strong kind of self-reference which transcends the cumulative hierarchy notion of set and may seem to verge on the logical paradoxes. But this kind of strong self-reference is actually used in category theory, notably in the construction of the category of all categories.

Version (2) abandons the strongly impredicative axiom of version (1) that there is a type of all types, i.e. $V \in V$, because it does indeed lead to a logical paradox, namely to Girard's paradox (cf. C, n & o (6) to ix.9.3) Chapt. 1).

Now, Π, Σ, $+$, N and N_n do not allow to define enough types and type-valued functions – e.g. equality on natural numbers $E(m, n)$ or transfinite types such as $(\Pi x \in N)F(x)$ where $F(0) = N$ and $F(s(n)) = F(n) \to N$ – and that is why a universe, i.e. a type denoted by 'V' is introduced, whose objects are types, together with the reflection principle which roughly speaking says: whatever one can do with types can be done inside the universe V. More precisely, it means that V is closed w.r.t. the following inductive clauses: if A is an object of V and B is a function which to an arbitrary object of A assigns an object of type V, then $(\Pi x \in A)B(x)$ and $(\Sigma x \in A)B(x)$ are objects of type V, and if A and B are objects of type V then so is $A + B$ an object of V, and $N_0, N_1, \ldots, N$ are objects of type V. Observe that this reflection principle does not justify the axiom $V \in V$ of version (1) which Girard (1972) showed to be contradictory (because then V would have had to be in V before it was introduced into V).

No principle of (transfinite) induction over V is adopted in version (2) expressing the idea that V is the least type closed under the inductive clauses defining V. The reason for refraining from adopting such a principle is that V should be kept open so as to be free to throw new types into it or require it to be closed w.r.t. new type forming operations (cf. beginning of Sect. iv.)

Borrowing terminology from category theory, an object of type V is said to be small, whereas V and all types formed by means of V are said to be large. The universe V is thus the type of small types (and not the type of all types anymore). With this distinction, version (2) of constructive type theory is adequate for the formulation of the basic notions and constructions of category theory – however, it does not legitimize the construction of the category of all categories anymore (which in view of Girard's paradox has turned out to be a highly dubious category).

The reflection principle of version (2) hence serves a double purpose:

1) to overcome the unnatural limitations of finite types (i.e. the finite type structure over the basic types);
2) to render possible the formalization of category theory in constructive type theory.

The incoherence of the idea of a type of all types whatsoever made it necessary in version (2) to distinguish between small and large types. This makes the types well-founded and the theory predicative.

In version (3) the universe V of version (2) has been extended to a whole sequence $V =_{\text{def}} V_0 \in V_1 \in V_2 \in \cdots \in V_n \in \cdots$ of universes which provides this new version of constructive type theory with more natural closure properties. The reasons for introducing universes are the same as in version (2).

As has just been mentioned, the idea of introducing a universe has been iterated in version (3) so as to obtain a sequence of universes $V =_{\text{def}} V_0, V_1, \ldots, V_n, \ldots$ As before, $V =_{\text{def}} V_0$ is the type of small types, V_1 is the type of large types, V_2 is the type of extralarge types, etc.; a type A which is an object of type V_n is said to be a type of order n, and a function from type A into the universe V_n is an nth order property (or species) of objects of type A.

The axioms governing the sequence of universes are obtained by extending in the natural way those for small and large types of version (2); thus, for every n, V_n is a type of order $n+1$, i.e. $V_n \in V_{n+1}$ (which is V_n-reflection). If A is a type of order m, i.e. $A \in V_m$, and B is a function from A into V_n then $(\Pi x \in A)B(x)$ and $(\Sigma x \in A)B(x)$ are of order $\max(m, n)$, i.e. $(\Pi x \in A)B(x)$, $(\Sigma x \in A)B(x) \in V_{\max(m,n)}$. If $A \in V_m$ and $B \in V_n$, then $A + B \in V_{\max(m,n)}$, etc. and $N_0, N_1, \ldots, N \in V_0$ (V_n-introduction).

A consequence of the adoption of a sequence of universes is that every type is at the same time an object and as such it has itself got a type. Note that the difference between having one universe V in version (2) and a whole countable sequence of universes $V_0, V_1, \ldots, V_n, \ldots$ in version (3) is like the difference in set theory between the simple axiom that there exists a universe, and Grothendieck's axiom that every set is contained in a universe. Finally, it may be pointed out that the axioms for universes serve to link the generation of objects and types (as they already do in version (2)), and that they play a role for constructive type theory similar to the one of the replacement axiom for ZF set theory.

Universes do not feature in version (4).

Although version (5) also has a countable sequence of universes $U_0, U_1, \ldots, U_n, \ldots$, the rules for these universes differ considerably from those of version (3) and that is why the two sets of rules will be given here in comparison:

Version (3)	Version (5)
V_n-formation rule(s)	*U_n-formation rules*
———	*U_n type* $\quad U_n = U_n$ for every $n = 0, 1, \ldots$
V_n-introduction rules	*U_n-introduction rules*
$\dfrac{\begin{array}{l} x_1 \in A_1, \ldots, x_k \in A_k[x_1, \ldots, x_{k-1}] \\ x \in A[x_1, \ldots, x_k] \\ A[x_1, \ldots, x_k] \in V_m,\ B[x_1, \ldots, x_k, x] \in V_n \end{array}}{F(x_1, \ldots, x_k) \in V_{\max(m,n)}}$ where F is a function constant uniquely associated with $A[x_1, \ldots, x_k]$, $B[x_1, \ldots, x_k, x]$ and Π	$\dfrac{A \in U_n \qquad \begin{array}{c}(x \in A)\\ B \in U_n\end{array}}{(\Pi x \in A)B \in U_n}$ $\dfrac{A = C \in U_n \qquad \begin{array}{c}(x \in A)\\ B = D \in U_n\end{array}}{(\Pi x \in A)B = (\Pi x \in C)D \in U_n}$
$\dfrac{\begin{array}{l} x_1 \in A_1, \ldots, x_k \in A_k[x_1, \ldots, x_{k-1}] \\ x \in A[x_1, \ldots, x_k] \\ A[x_1, \ldots, x_k] \in V_m,\ B[x_1, \ldots, x_k, x] \in V_n \end{array}}{F(x_1, \ldots, x_k) \in V_{\max(m,n)}}$ where F is a function constant uniquely associated with $A[x_1, \ldots, x_k]$, $B[x_1, \ldots, x_k, x]$ and Σ	$\dfrac{A \in U_n \qquad \begin{array}{c}(x \in A)\\ B \in U_n\end{array}}{(\Sigma x \in A)B \in U_n}$ $\dfrac{A = C \in U_n \qquad \begin{array}{c}(x \in A)\\ B = D \in U_n\end{array}}{(\Sigma x \in A)B = (\Sigma x \in C)D \in U_n}$
$\dfrac{\begin{array}{l} x_1 \in A_1, \ldots, x_k \in A_k[x_1, \ldots, x_{k-1}] \\ A[x_1, \ldots, x_k] \in V_m,\ B[x_1, \ldots, x_k] \in V_n \end{array}}{F(x_1, \ldots, x_k) \in V_{\max(m,n)}}$ where F is a function constant uniquely associated with $A[x_1, \ldots, x_k]$, $B[x_1, \ldots, x_k]$ and $+$	$\dfrac{A \in U_n \quad B \in U_n}{A + B \in U_n}$ $\dfrac{A = C \in U_n \quad B = D \in U_n}{A + B = C + D \in U_n}$

$\dfrac{x_1 \in A_1, \ldots, x_k \in A_k[x_1, \ldots, x_{k-1}] \quad x \in A[x_1, \ldots, x_k] \in V_n}{I(x_1, \ldots, x_k, x, x) \in V_n}$	$\dfrac{A \in U_n \quad a \in A \quad b \in A}{I(A, a, b) \in U_n}$ $\dfrac{A = C \in U_n \quad a = c \in A \quad b = d \in A}{I(A, a, b) = I(C, c, d) \in U_n}$
$N_n \in V_0$ for $n = 0, 1, \ldots$	$N_0 \in U_n \quad N_0 = N_0 \in U_n$ $N_1 \in U_n \quad N_1 = N_1 \in U_n$ $\vdots$
$N \in V_0$	$N \in U_n \quad N = N \in U_n$
———	$\dfrac{A \in U_n \quad \begin{matrix}(x \in A)\\ B \in U_n\end{matrix}}{(Wx \in A)B \in U_n}$
	$\dfrac{A = C \in U_n \quad \begin{matrix}(x \in A)\\ B = D \in U_n\end{matrix}}{(Wx \in A)B = (Wx \in C)D \in U_n}$
V_n-reflection rules $V_n \in V_{n+1}$ for $n = 0, 1, \ldots$	*U_n-reflection rules* $U_0 \in U_n \quad U_0 = U_0 \in U_n$ $\vdots$ $U_{n-1} \in U_n \quad U_{n-1} = U_{n-1} \in U_n$
V_n-elimination rule(s) ———	*U_n-elimination rules* $\dfrac{A \in U_n}{A \text{ type}} \quad \dfrac{A = B \in U_n}{A = B}$ $\dfrac{A \in U_n}{A \in U_{n+1}} \quad \dfrac{A = B \in U_n}{A = B \in U_{n+1}}$

In version (3) each type is of a unique order or of a unique type, i.e. each type belongs to exactly one universe (and the system is of course monomorphic), whereas in version (5) each type belongs to a minimal universe as well as to each universe higher than the minimal one (e.g. $N \in U_0$, but then also by U_n-elimination, $N \in U_1, N \in U_2, \ldots$) (hence, this version of constructive type theory is polymorphic). Note that in a certain sense one can say (abstraction being made of e.g. the combinator formulation of version (3) and the λ-abstraction formulation of version (5)) that all the V_n-rules of version (3) are contained in the U_n-rules of version (5) but not vice versa.

As has been remarked already under version (2), Π, Σ, $+$, I, *List* and W only allow to build up a finite type structure, because one can iterate the given set forming operations starting from $I(A, a, b)$, N_0, $N_1, \ldots, N$ only a finite number of times. To strengthen the language of constructive type theory, one can introduce transfinite types which are obtained by introducing universes into the theory. Moreover, if constructive type theory is to deal with the foundational problems of category theory, it has to be able to deal with sets of sets. The idea then is to define a universe as the (least) set closed under certain specified set forming operations.

All of this is stated also in version (6). Version (6), however, distinguishes two possible ways of building a universe, i.e. to obtain closure under possibly transfinite iterations of such operations:

The Tarski way It is so called by Martin-Löf because of the similarity between the family of sets $T(x)$ $(x \in U)$ below and Tarski's truth def. To build up the canonical elements of $U (= U_0)$, new symbols are used which reflect $\Pi, \Sigma, +, I \ldots$ (cf. reflection principle). Then U consists of indices of sets (like in recursion theory). The rules are $U(= U_0)$ formation

$$U_0\ set \qquad \frac{a \in U_0}{T^0(a)\ set}$$

U_0 and $T^0(x)$ $(x \in U_0)$ are defined by a simultaneous transfinite induction which can be read off the following introduction rules:

$U(= U_0)$ introduction

$$\frac{a \in U_0 \quad \begin{matrix}(x \in T^0(a))\\ b(x) \in U_0\end{matrix}}{\pi(a, (x)b(x)) \in U_0} \qquad \frac{a \in U_0 \quad \begin{matrix}(x \in T^0(a))\\ b(x) \in U_0\end{matrix}}{T^0(\pi(a, (x)b(x))) = (\Pi x \in T^0(a))T^0(b(x))}$$

$$\frac{a \in U_0 \quad \begin{matrix}(x \in T^0(a))\\ b(x) \in U_0\end{matrix}}{\sigma(a, (x)b(x)) \in U_0} \qquad \frac{a \in U_0 \quad \begin{matrix}(x \in T^0(a))\\ b(x) \in U_0\end{matrix}}{T^0(\sigma(a, (x)b(x))) = (\Sigma x \in T^0(a))T^0(b(x))}$$

$$\frac{a \in U_0 \quad b \in U_0}{a + b \in U_0} \qquad \frac{a \in U_0 \quad b \in U_0}{T^0(a + b) = T^0(a) + T^0(b)}$$

$$\frac{a \in U_0 \quad b \in T^0(a) \quad c \in T^0(a)}{i(a, b, c) \in U_0} \qquad \frac{a \in U_0 \quad b \in T^0(a) \quad c \in T^0(a)}{T^0(i(a, b, c)) = I(T^0(a), b, c)}$$

$$n_0 \in U_0, n_1 \in U_0, \ldots \qquad T^0(n_0) = N_0, T^0(n_1) = N_1, \ldots$$

$$n \in U_0 \qquad T^0(n) = N$$

$$\frac{a \in U_0 \quad \begin{matrix}(x \in T^0(a))\\ b(x) \in U_0\end{matrix}}{w(a, (x)b(x)) \in U_0} \qquad \frac{a_0 \in U_0 \quad \begin{matrix}(x \in T^0(a))\\ b(x) \in U_0\end{matrix}}{T^0(w(a, (x)b(x))) = (Wx \in T^0(a))T^0(b(x))}$$

One can iterate the process thus obtaining a second universe U_1 with the two U_1 formation rules

U_1-formation

$$U_1\ set \qquad \frac{a \in U_1}{T^1(a)\ set}$$

and the four new U_1 introduction rules

U_1-introduction

$$u_0 \in U_1 \qquad T^1(u_0) = U_0$$

$$\frac{a \in U_0}{t^0(a) \in U_1} \qquad \frac{a \in U_0}{T^1(t^0(a)) = T^0(a)}$$

U_2-formation

$$U_2\ set \qquad \frac{a \in U_2}{T^2(a)\ set}$$

U_2-introduction

$$u_1 \in U_2 \qquad\qquad T^2(u_1) = U_1$$

$$\frac{a \in U_1}{t^1(a) \in U_2} \qquad\qquad \frac{a \in U_1}{T^2(t^1(a)) = T^1(a)}$$

etc. or generally a universe U_n, $n > 0$, with the rules

U_n-formation

$$U_n\ set \qquad\qquad \frac{a \in U_n}{T^n(a)\ set}$$

U_n-introduction

$$u_{n-1} \in U_n \qquad\qquad T^n(u_{n-1}) = U_{n-1}$$

$$\frac{a \in U_{n-1}}{t^{n-1}(a) \in U_n} \qquad\qquad \frac{a \in U_{n-1}}{T^n(t^{n-1}(a)) = T^{n-1}(a)}$$

where t^n is obviously the coding function, coding all sets resp. their indices of the universe U_n in the universe U_{n+1}, and T^n is the decoding function which when applied to the code of set a of universe U_n simply yields the resp. set A.

The Russell way Consider Π, Σ, $+$, I, W, N_0, $N_1, \ldots, N$ all as set forming operations and at the same time as operations to form canonical elements of the set $U(= U_0)$, of the set U_1, of the set $U_2, \ldots$, i.e. all the universes. This is like ramified type theory and that is why Martin-Löf calls it Russell's way. T above disappears (since there is no coding) and the rules above are turned into:

U_0-formation

$$U_0\ set \qquad\qquad \frac{A \in U_0}{A\ set}$$

U_0-introduction

$$\frac{\begin{array}{cc} & (x \in A) \\ A \in U_0 & B(x) \in U_0 \end{array}}{(\Pi x \in A)B(x) \in U_0} \qquad\qquad \frac{\begin{array}{cc} & (x \in A) \\ A \in U_0 & B(x) \in U_0 \end{array}}{(\Sigma x \in A)B(x) \in U_0}$$

$$\frac{A \in U_0 \quad B \in U_0}{A + B \in U_0} \qquad\qquad \frac{A \in U_0 \quad b \in A \quad c \in A}{I(A, b, c) \in U_0}$$

$$N_0 \in U_0,\ N_1 \in U_0, \ldots \qquad\qquad N \in U_0$$

$$\frac{\begin{array}{cc} & (x \in A) \\ A \in U_0 & B(x) \in U_0 \end{array}}{(Wx \in A)B(x) \in U_0}$$

Here too one can iterate the process obtaining a second universe U_1 with U_1-formation

$$U_1 \ set \qquad \frac{A \in U_1}{A \ set}$$

U_1-introduction

$$U_0 \in U_1 \qquad \frac{A \in U_0}{A \in U_1}$$

and a third universe U_2 with the rules
U_2 formation

$$U_2 \ set \qquad \frac{A \in U_2}{A \ set}$$

U_2-introduction

$$U_1 \in U_2 \qquad \frac{A \in U_1}{A \in U_2}$$

or generally a universe U_n, $n > 0$, with the rules
U_n-formation

$$U_n \ set \qquad \frac{A \in U_n}{A \ set}$$

U_n-introduction

$$U_{n-1} \in U_n \qquad \frac{A \in U_{n-1}}{A \in U_n}$$

Now, obviously, U_0 is not itself an element of U_0, and in general $U_n \in U_n$ does not hold; it would lead to a contradiction (Girard's paradox).

A set A is said to be small or a U_0-set if it has a code $a \in U_0$, i.e. if there is an element $a \in U_0$ s.t. $T^0(a) = A$. And generally, a family of sets $A(x_1, \ldots, x_n)$ $(x_1 \in A_1, \ldots, x_n \in A_n(x_1, \ldots, x_{n-1}))$ is said to be small provided $A(x_1, \ldots, x_n) = T^0(a(x_1, \ldots, x_n))$ for some indexing function $a(x_1, \ldots, x_n) \in U_0$ $(x_1 \in T^0(a_1), \ldots, x_n \in T^0(a_n(x_1, \ldots, x_{n-1})))$ where $A_1 = T^0(a_1), \ldots, A_n(x_1, \ldots, x_{n-1}) = T^0(a_n(x_1, \ldots, x_n))$ for $a_1 \in U_0, \ldots, a_n(x_1, \ldots, x_{n-1}) \in U_0(x_1 \in T^0(a_1), \ldots, x_{n-1} \in T^0(a_{n-1}(x_1, \ldots, x_{n-2})))$. So the set of small sets is closed under the operations Π, Σ, $+$, I and W, and U_0 is a perfectly good set, but it is not small. Using U_0 one can form transfinite types (using e.g. a recursion with value in U_0). *Example* The large set $V \equiv (Wx \in U_0)T^0(x) \in U_1$ has been used by Aczel (1978) to give meaning to a constructive version of ZF set theory via constructive type theory (cf. C, n & o (1)iii) to vi.1) of Chapt. 1).

In version (6) the set formation rule $U_n = U_n$ of version (5) is missing, but it is a simple consequence of the first set equality rule of version (6).

All the identity rules associated with the U_n-introduction and the U_n-elimination rules of version (5) are missing in version (6). However, they are simply taken for granted in version (6).

The U_n-elimination rules of version (5) are called U_n-formation or U_n-introduction rules in version (6).

The formulation of the rules of U_n-introduction and U_n-elimination of version (5) is simplified in version (6) in the sense that instead of using the countable sequence of universes $(U_n)_{n\in\omega}$ one universe U_i, $i \geq 1$, is introduced at a time and then two U_i-formation and four U_i-introduction rules are indicated with which to eventually obtain the whole generality of version (5). In this sense also, the universe rules of version (5) and (6) are essentially the same.

The novelty of version (6) lies in the distinction of the two formulations of the universes, the Tarski and the Russell one.

Universes have their last appearance in version (6), i.e. no trace of them can be found in either version (7) or version (8).

vii VARIATIONS ON THE BASIC STOCK RULES $\Pi, \Sigma, +, N_n, N$

The changes in the basic stock rules of Π, Σ, $+$, N_n and N reflect to a large extent the various (major) innovations of the different versions of constructive type theory. Hence the differences in the formulations of the basic stock rules of versions (2) and (3) are due to the innovations of version (3): the new conversion-as-judgement approach, the combinator language and the use of a parametrized form. The basic stock rules of version (4) are basically the same as those of version (3) except for a couple of elimination- and equality rules of version (4) which are the same as those of version (2). Due to the return to the λ-abstraction formulation of versions (5) and (6), the introduction- and elimination rules of their basic stock set-theoretical operations are just about identical with those of version (2). The homonymous rules of versions (7) and (8) are given in a higher level and monomorphic notation.

Dyckhoff has shown how the elimination rules of versions (7) and (8) can be strengthened in order to suppress the occurrence of spurious and yet necessary components in certain derivations of constructive type theory. This story ends with an outlook on Swaen's characterization of what he calls the strong elimination rules of the basic stock set-theoretical operations (that is the polymorphic form of the corresponding elimination rules of versions (7) and (8)) in terms of his exhaustion principles.

In versions (2) and (3) it is observed that since the language of constructive type theory is richer than the ones of first order predicate logic and traditional intuitionistic systems (in the sense that proofs can appear as parts of propositions), the elimination rules for absurdity, disjunction and existential quantification can be strengthened in type theory. The first one to have realized the possibility of strengthening the usual elimination rule for existential quantification seems to have been Howard [1980]. He proposed existential elimination axioms of the following form:[91]

$$\begin{aligned} P_1 F^{\exists y\alpha(y)} &: 0 \\ P_2 F^{\exists y\alpha(y)} &: \alpha(P_1 F^{\exists y\alpha(y)}) \end{aligned}$$

which in Martin-Löf's notation are

$$\frac{c : (\Sigma x : 0)\alpha(x)}{p(c) : 0} \qquad \frac{c : (\Sigma x : 0)\alpha(x)}{q(c) : \alpha(p(c))}$$

Now, Howard's existential elimination axioms are obviously a special case of the Σ-elimination rule of version (2)

$$\frac{c \in (\Sigma x \in A)B[x] \quad \begin{array}{c} x \in A \quad y \in B[x] \\ d[x,y] \in C[(x,y)] \end{array}}{E(c,(\lambda x)(\lambda y)d[x,y]) \in C[c]}$$

(cf. C, n & o (2) to ix.6.3), Chapt. 1) which in version (3) is formulated in a combinator language with parameters as follows:

$$\frac{\begin{array}{l} x_1 \in A_1, \ldots, x_k \in A_k[x_1, \ldots, x_{k-1}] \\ x \in A[x_1, \ldots, x_k],\, y \in B[x_1, \ldots, x_k, x] \\ c[x_1, \ldots, x_k, x, y] \in C[x_1, \ldots, x_k, (x_1, \ldots, x_k, x, y)] \end{array}}{f(x_1, \ldots, x_k, (x_1, \ldots, x_k, x, y))\ conv\ c[x_1, \ldots, x_k, x, y]}$$

where $f(x_1, \ldots, x_k, (x_1, \ldots, x_k, x, y)) \in C[x_1, \ldots, x_k, (x_1, \ldots, x_k, x, y)]$. The Π-, Σ-, $+$-, N_n- and N-introduction and elimination rules of version (2) are:

Π – introduction (version (2))

$$\frac{\begin{array}{c} x \in A \\ b[x] \in B[x] \end{array}}{(\lambda x)b[x] \in (\Pi x \in A)B[x]}$$

Π – elimination (version (2))

$$\frac{b \in (\Pi x \in A)B[x] \quad a \in A}{b(a) \in B[a]}$$

Σ – introduction (version (2))

$$\frac{a \in A \qquad b \in B[a]}{(a,b) \in (\Sigma x \in A)B[x]}$$

$+$ – introduction (version (2))

$$\frac{a \in A}{i(a) \in A + B} \qquad \frac{b \in B}{j(b) \in A + B}$$

$+$ – elimination (version (2))

$$\frac{c \in A + B \quad \begin{array}{c} x \in A \\ d[x] \in C[i(x)] \end{array} \quad \begin{array}{c} y \in B \\ e[y] \in C[j(y)] \end{array}}{D(c,(\lambda x)d[x],(\lambda y)e[y]) \in C[c]}$$

N_n – introduction (version(2))

$$1 \in N_n \ \ldots\ n \in N_n$$

N_n – elimination (version(2))

$$\frac{c \in N_n \quad c_1 \in C[1] \ldots c_n \in C[n]}{R_n(c, c_1, \ldots, c_n) \in C[c]}$$

N – introduction (version(2))

$$0 \in N \qquad \frac{a \in N}{s(a) \in N}$$

N – elimination (version(2))

$$\frac{c \in N \quad d \in C[0] \quad \begin{array}{c} x \in N \quad y \in C[x] \\ e[x,y] \in C[s(x)] \end{array}}{R(c,d,(\lambda x)(\lambda y)e[x,y]) \in C[c]}$$

The same kind of reformulation of the $+$-, Π-, N_n- and N-elimination rules of version (2) in the language of combinators is carried out in version (3).

The N_n- and N-introduction rules of version (3) are identical with those of version (2), and the Π-, Σ- and $+$-introduction rules as well as the Π-elimination rule of version (3) are just parametrized forms of the corresponding rules of version (2).

The $+$-, N_n- and N-elimination rules of version (3) are parametrized rules standing so to speak in between the corresponding elimination- and equality- or conversion rules of version (2) (due to the new conversion-as-judgement approach of version (3)) (cf. Sect. iii. or Sect. v.).

$+$ – elimination (version (3))

$$\frac{\begin{array}{l} x_1 \in A_1, \ldots, x_k \in A_k[x_1, \ldots, x_{k-1}] \\ \left\{ \begin{array}{l} x \in A[x_1, \ldots, x_k] \qquad y \in B[x_1, \ldots, x_k] \\ c[x_1, \ldots, x_k, x] \in C[x_1, \ldots, x_k, i(x_1, \ldots, x_k, x)] \\ d[x_1, \ldots, x_k, y] \in C[x_1, \ldots, x_k, j(x_1, \ldots, x_k, y)] \end{array} \right. \end{array}}{\left\{ \begin{array}{l} f(x_1, \ldots, x_k, i(x_1, \ldots, x_k, x)) \; conv \; c[x_1, \ldots, x_k, x] \\ f(x_1, \ldots, x_k, j(x_1, \ldots, x_k, y)) \; conv \; d[x_1, \ldots, x_k, y] \end{array} \right.}$$

N_n – elimination (version (3))

$$\frac{\begin{array}{l} x_1 \in A_1, \ldots, x_k \in A_k[x_1, \ldots, x_{k-1}] \\ \left\{ \begin{array}{l} c_1[x_1, \ldots, x_k] \in C[x_1, \ldots, x_k, 1] \\ \quad \vdots \\ c_n[x_1, \ldots, x_k] \in C[x_1, \ldots, x_k, n] \end{array} \right. \end{array}}{\left\{ \begin{array}{l} f(x_1, \ldots, x_k, 1) \; conv \; c_1[x_1, \ldots, x_k] \\ \quad \vdots \\ f(x_1, \ldots, x_k, n) \; conv \; c_n[x_1, \ldots, x_k] \end{array} \right.}$$

N – elimination (version (3))

$$\frac{\begin{array}{l} x_1 \in A_1, \ldots, x_k \in A_k[x_1, \ldots, x_{k-1}] \\ \left\{ \begin{array}{l} c[x_1, \ldots, x_k] \in C[x_1, \ldots, x_k, 0] \\ x \in N \qquad y \in C[x_1, \ldots, x_k, x] \\ d[x_1, \ldots, x_k, x, y] \in C[x_1, \ldots, x_k, s(x)] \end{array} \right. \end{array}}{\left\{ \begin{array}{l} f(x_1, \ldots, x_k, 0) \; conv \; c[x_1, \ldots, x_k] \\ f(x_1, \ldots, x_k, s(x)) \; conv \; d[x_1, \ldots, x_k, x, f(x_1, \ldots, x_k, x)] \end{array} \right.}$$

'{' holds together two or more alternatives; e.g. in the case of $+$-elimination: accordingly to whether $x \in A[x_1, \ldots, x_k]$ and $c[x_1, \ldots, x_k, x] \in C[x_1, \ldots, x_k, i(x_1, \ldots, x_k, x)]$, or $y \in B[x_1, \ldots, x_k]$ and $d[x_1, \ldots, x_k, y] \in C[x_1, \ldots, x_k, j(x_1, \ldots, x_k, y)]$, $f(x_1, \ldots, x_k, i(x_1, \ldots, x_k, x)) \; conv \; c[x_1, \ldots, x_k, x]$, or $f(x_1, \ldots, x_k, j(x_1, \ldots, x_k, y)) \; conv \; d[x_1, \ldots, x_k, y]$ resp.

The Π-, Σ-, $+$-, N_n- and N-introduction and -elimination rules of version (4) are essentially the same as those of version (3) except that the sharper distinction between $+$-, N_n- and N-elimination and $+$-, N_n- and N-equality or conversion rules of version (2) is – contrary to version (3) – taken over in version (4) (note that the $+$-, N_n- and N-conversion rules of version (3) are but substitution instances of the resp. $+$-, N_n- and N-elimination rules):

$+$ – elimination (version (4))

$$\frac{\begin{array}{c} x_1 \in A_1, \ldots, x_k \in A_k[x_1, \ldots, x_{k-1}] \\ w \in A[x_1, \ldots, x_k] + B[x_1, \ldots, x_k] \end{array}}{f(x_1, \ldots, x_k, w) \in C[x_1, \ldots, x_k, w]}$$

N_n – elimination (version (4))

$$\frac{\begin{array}{l} x_1 \in A_1, \ldots, x_k \in A_k[x_1, \ldots, x_{k-1}] \\ z \in N_n \end{array}}{f(x_1, \ldots, x_k, z) \in C[x_1, \ldots, x_k, z]}$$

N – elimination (version (4))

$$\frac{\begin{array}{l} x_1 \in A_1, \ldots, x_k \in A_k[x_1, \ldots, x_{k-1}] \\ z \in N \end{array}}{f(x_1, \ldots, x_k, z) \in C[x_1, \ldots, x_k, z]}$$

(A few minor notational differences between versions (3) and (4) are: in version (4) the symbol 'ap' is introduced for the eliminatory operator for Π – this is due to the recognition that there are not only different sorts of functions, but correspondingly different ways of applying these functions to their arguments. Moreover, the introductory operator for Σ as well as the introductory operators for $+$ are not parametrized in version (4).)

The rules of Π-, Σ-, $+$-, N_n- and N-introduction and -elimination of version (5) are almost identical with those of version (2) (signalling the return from the combinator formulation of versions (3) and (4) to the λ-abstraction formulation of version (2)). The only differences are i) another notation for substitution in the two versions, ii) the introduction of indices for the elements of N_n, i.e. N_n-introduction of version (5) is $m_n \in N_n$ $(m = 0, \ldots, n-1)$, iii) the systematic shift of abstraction from the "main body" to the "head" of the eliminatory operators for Σ, $+$, N_n and N in version (5); iv) the use of prime instead of 's' for the successor function. The rules of version (6) are even closer to those of version (2) than are those of version (5). The only novelties of version (5) w.r.t. version (2) retained in version (6) are the points ii) and iv). (Some tiny notational differences between (6) and (2) are the absence of brackets and the absence of λ-symbols inside the eliminatory operators for Σ, $+$, N_n and N; and the symbol 'Ap' is used for the eliminatory operator for Π.)

As far as the Π-, Σ, …, N-introduction and -elimination rules are explicitly given in versions (7) and (8), they are given in a higher level and monomorphic notation. (In version (7) only the rules for Π and N (i.e. plus formation and equality rules) are all explicitly given, and in version (8) only the rules for Π (and I).) Note that the higher level notation seems to make its first appearance in version (6a). The Π-rules of versions (7) and (8) are:

Π-introduction (versions (7) and (8))

$$\frac{\begin{array}{l} A = C : set \\ B = D : (A)\, set \\ b : (x : A)B(x) \end{array}}{\Pi I(A, B, b) : \Pi(C, D)}$$

Π-elimination (versions (7) and (8))

$$\frac{\begin{array}{l} A : set \\ B : (A)\, set \\ b : (x : A)B(x) \\ a : A \end{array}}{\Pi E(A, B, b, a) : B(a)} \qquad \text{(weak form)}$$

$$\frac{\begin{array}{l} A : \ set \\ B : (A)\ set \\ C : (\Pi(A, B))\ set \\ d : (y : (x : A)B(x))C(\Pi I(A, B, y)) \\ c : \Pi(A, B) \end{array}}{\Pi E(A, B, C, d, c) : C(c)} \qquad \text{(strong form)}$$

The strong form of Π-elimination also makes its first appearance in a polymorphic form in version (6a), and it makes Π-elimination follow the same pattern as the elimination rules for all the other set forming operations.

Nordström et al. relate on the possibility of strengthening the rules (ΠE), (ΣE), $(+E)$, $(N_n E)$ and (NE) of versions (7) and (8) (cf. Sect. ix, Chapt. 1) through weakening their premisses.[92]

These new rules are

strong Π-elimination $(\Pi E')$

$$\frac{\begin{array}{l} A : \ set \\ B : (A)\ set \\ C : (\Pi(A, B))\ set \\ c : \Pi(A, B) \\ y : (x : A)B(x), I(\Pi(A, B), c, \Pi I(A, B, y))\ true \vdash d(y) : C(\Pi I(A, B, y)) \end{array}}{\Pi E'(A, B, C, c, d) : C(c)}$$

strong Σ-elimination $(\Sigma E')$

$$\frac{\begin{array}{l} A : \ set \\ B : (A)\ set \\ C : (\Sigma(A, B))\ set \\ c : \Sigma(A, B) \\ x : A, y : B(x), I(\Sigma(A, B), c, \Sigma I(A, B, x, y))\ true \vdash d(x, y) : C(\Sigma I(A, B, x, y)) \end{array}}{\Sigma E'(A, B, C, c, d) : C(c)}$$

strong +-elimination $(+E')$

$$\frac{\begin{array}{l} A : \ set \\ B : \ set \\ C : (A + B)\ set \\ c : A + B \\ x : A, I(A + B, c, +I1(A, B, x))\ true \vdash d(x) : C(+I1(A, B, x)) \\ y : B, I(A + B, c, +I2(A, B, y))\ true \vdash e(y) : C(+I2(A, B, y)) \end{array}}{+E'(A, B, C, c, d, e) : C(c)}$$

strong N_n-elimination $(N_n E')$

$$\frac{\begin{array}{l} C : (N_n)\ set \\ c : N_n \\ I(N_n, c, N_n I0)\ true \vdash b_0 : C(N_n I0) \\ \qquad\vdots \\ I(N_n, c, N_n In - 1)\ true \vdash b_{n-1} : C(N_n In - 1) \end{array}}{N_n E'(C, c, b_0, \ldots, b_{n-1}) : C(c)}$$

strong N-elimination (NE')

$$\frac{\begin{array}{l} C : (N)\, set \\ c : N \\ I(N, c, NI1)\, true \vdash d : C(NI1) \\ x : N, y : C(x), I(N, c, NI2(x))\, true \vdash e(x, y) : C(NI2(x)) \end{array}}{NE'(C, c, d, e) : C(c)}$$

It should be noticed that the last premiss of the strong Π-elimination rule $(\Pi E')$ is weaker than the second last premiss of (the strong form of) the ordinary Π-elimination rule (ΠE) in that the assumption $I(\Pi(A, B), c, \Pi I(A, B, y))\, true$ is added. Likewise, the last premiss of the strong Σ-elimination rule $(\Sigma E')$ is weaker than the second last premiss of the ordinary Σ-elimination rule (ΣE) (cf. Sect. ix.6), Chapt. 1) in that the assumption $I(\Sigma(A, B), c, \Sigma I(A, B, x, y))\, true$ is added; etc. By applying (ΠE) to $c : \Pi(A, B)$ and $C'(u) = I(\Pi(A, B), u, c) \to C(u)$, or (ΣE) to $c : \Sigma(A, B)$ and $C'(u) = I(\Sigma(A, B), u, c) \to C(u)$ etc. one derives $(\Pi E')$ or $(\Sigma E')$ resp., etc.

E.g. $(\Pi E')$ is derived by means of (ΠE) as follows:

$$\frac{\dfrac{\dfrac{\dfrac{\begin{array}{l}(y : (A)B) \\ (z : I(\Pi(A, B), \Pi I(A, B, y), c))\end{array}}{d(y) : C(\Pi I(A, B, y))}}{\begin{array}{l}\to I(I(\Pi(A, B), \Pi I(A, B, y), c), C(\Pi I(A, B, y)), (z)d(y)) : \\ \qquad I(\Pi(A, B), \Pi I(A, B, y), c) \to C(\Pi I(A, B, y))\end{array}}}{\begin{array}{l}\Pi E(A, B, C, (y) \to I(I(\Pi(A, B), \Pi I(A, B, y), c), C(\Pi I(A, B, y)), (z)d(y)), c) : \\ \qquad I(\Pi(A, B), c, c) \to C(c) \\ II(\Pi(A, B), c) : I(\Pi(A, B), c, c)\end{array}}}{\begin{array}{l}\to E(I(\Pi(A, B), c, c), C(c), \Pi E(A, B, C, (y) \to I(I(\Pi(A, B), \Pi I(A, B, y), c), \\ \qquad C(\Pi I(A, B, y)), (z)d(y)), c), II(\Pi(A, B), c)) : C(c)\end{array}}}{\begin{array}{l}\Pi E'(A, B, C, c, d) =\to E(I(\Pi(A, B), c, c), C(c), \Pi E(A, B, C, (y) \\ \qquad \to I(I(\Pi(A, B), \Pi I(A, B, y), c), C(\Pi I(A, B, y)), \\ \qquad (z)d(y), c), II(\Pi(A, B), c)) : C(c)\end{array}}$$

Analogously, if $z : I(\Sigma(A, B), c, \Sigma I(A, B, x, y))$ or, $z : I(A + B, c, + I1(A, B, x))$ or $z : I(A + B, c, +I2(A, B, y))$, etc.

$$\begin{array}{ll}
\Sigma E'(A, B, C, c, d) = & \to E(I(\Sigma(A, B), c, c), C(c), \Sigma E(A, B, C, \\
& \quad (x)(y) \to I(I(\Sigma(A, B), \Sigma I(A, B, x, y), c), C(\Sigma I(A, B, x, y)), \\
& \quad (z)d(x, y)), c), II(\Sigma(A, B), c)) : C(c) \\
+E'(A, B, C, c, d, e) = & \to E(I(A + B, c, c), C(c), +E(A, B, C, \\
& \quad (x) \to I(I(A + B, c, +I1(A, B, x)), C(+I1(A, B, x)), \\
& \quad (z)d(x)), (y) \to I(I(A + B, c, +I2(A, B, y)), \\
& \quad C(+I2(A, B, y)), (z)e(y))), II(A + B, c)) : C(c) \\
N_n E'(C, c, b_0, \ldots, b_{n-1}) = & \to E(I(N_n, c, c), C(c), N_n E(C, \to I(I(N_n, c, N_n I0), \\
& \quad C(N_n I0), (z)b_0), \ldots, \to I(I(N_n, c, N_n In - 1), \\
& \quad C(N_n In - 1), (z)b_{n-1})), II(N_n, c)) : C(c) \\
NE'(C, c, d, e) = & \to E(I(N, c, c), C(c), NE(C, \to I(I(N, c, NI1), C(NI1), (z)d), \\
& \quad (x)(y) \to I(I(N, c, NI2(x)), C(NI2(x)), (z)e(x, y))), II(N, c)) : \\
& \quad C(c)
\end{array}$$

The strong Π-, Σ-, $+$-, N_n- and N-elimination rules have been introduced by Dyckhoff in order to avoid the appearance of spurious and yet necessary elements/components in certain derivations.[93] Furthermore, these strong elimination rules can alternatively be expressed by means of 'hypothetical hypotheses'.[94]

As far as weak and strong elimination rules are concerned, there is a certain risk of confusion for the following reason: the following rules in polymorphic form have been called weak elimination rules by Swaen and others:[95]

weak Π-elimination ($\Pi E^{\sim}$)

$$\frac{\begin{array}{l} c : \Pi(A, B) \\ d : (y : (A)B)C \end{array}}{\Pi E^{\sim}(c, d) : C}$$

weak Σ-elimination ($\Sigma E^{\sim}$)

$$\frac{\begin{array}{l} c : \Sigma(A, B) \\ d : (x : A)(y : B(x))C \end{array}}{\Sigma E^{\sim}(c, d) : C}$$

weak +-elimination ($+E^{\sim}$)

$$\frac{\begin{array}{l} c : A + B \\ d : (A)C \\ e : (B)C \end{array}}{+E^{\sim}(c, d, e) : C}$$

weak N_n-elimination ($N_n E^{\sim}$)

$$\frac{\begin{array}{l} c : N_n \\ b_0 : C \\ \vdots \\ b_{n-1} : C \end{array}}{N_n E^{\sim}(c, b_0, \ldots, b_{n-1}) : C}$$

weak N-elimination ($NE^{\sim}$)

$$\frac{\begin{array}{l} c : N \\ d : C \\ e : (N)(C)C \end{array}}{NE^{\sim}(c, d, e) : C}$$

Note that in ($\Pi E^{\sim}$) C should not contain y and in ($\Sigma E^{\sim}$) C should not contain x and y as free variables.

The same logicians call (the polymorphic forms of) (ΠE) strong Π-elimination, (ΣE) strong Σ-elimination, ($+E$) strong +-elimination, etc. Now, Swaen characterizes his strong Σ-elimination (ΣE) and strong +-elimination ($+E$) by what he calls exhaustion principles for Σ sets and for + sets resp.[96] Analogous principles can be found for the Π sets, the N_n sets and the set N.

The various exhaustion principles are pairs of rules.

The exhaustion principle for Π sets

Π-exhaustion (ΠEXH)

$$\frac{c : \Pi(A, B)}{d(c) : (\Sigma y : (A)B)E(\Pi(A, B), \Pi I(y), c)}$$

Πd-equality (Πd – equ)

$$\frac{b : (A)B}{d(\Pi I(b)) = \Pi I(\Sigma I(b, EI)) : (\Sigma y : (A)B)E(\Pi(A, B), \Pi I(y), \Pi I(b))}$$

The exhaustion principle for Σ sets

Σ-exhaustion (ΣEXH)

$$\frac{c : \Sigma(A, B)}{d(c) : (\Sigma x : A)(\Sigma y : B)E(\Sigma(A, B), \Sigma I(x, y), c)}$$

Σd-equality (Σd-equ)

$$\frac{a : A \qquad b : B(a/x)}{d(\Sigma I(a, b)) = \Sigma I(a, \Sigma I(b, EI)) : (\Sigma x : A)(\Sigma y : B)E(\Sigma(A, B), \Sigma I(x, y), \Sigma I(a, b))}$$

The exhaustion principle for + sets

+-exhaustion (+EXH)
$$\frac{c : A + B}{d(c) : (\Sigma x : A)E(A + B, +I1(x), c) + (\Sigma x : B)E(A + B, +I2(x), c)}$$

+d-equality (+d-equ 1)
$$\frac{a : A}{d(+I1(a)) = +I1(\Sigma I(a, EI)) : (\Sigma x : A)E(A + B, +I1(x), +I1(a)) + (\Sigma x : B)E(A + B, +I2(x), +I1(a))}$$

+d-equality (+d-equ 2)
$$\frac{b : B}{d(+I2(b)) = +I2(\Sigma I(b, EI)) : (\Sigma x : A)E(A + B, +I1(x), +I2(b)) + (\Sigma x : B)E(A + B, +I2(x), +I2(b))}$$

The exhaustion principle for N_n sets

N_n-exhaustion (N_nEXH)
$$\frac{c : N_n}{d(c) : E(N_n, N_nI0, c) + \cdots + E(N_n, NIn - 1, c)}$$

N_nd-equality (N_nd-equ i)
$$d(N_nIi) = \Sigma I(N_nIi, EI) : E(N_n, N_nI0, N_nIi) + \cdots + E(N_n, N_nIn - 1, N_nIi)$$

The exhaustion principle for the set N

N-exhaustion ($NEXH$)
$$\frac{c : N}{d(c) : E(N, NI1, c) + (\Sigma x : N)E(N, NI2(x), c)}$$

Nd-equality (Nd-equ 1)
$$d(NI1) = \Sigma I(NI1, EI) : E(N, NI1, NI1) + (\Sigma x : N) E(N, NI2(x), NI1)$$

(Nd-equ 2)
$$\frac{n : N}{d(NI2(n)) = NI2(\Sigma I(n, EI)) : E(N, NI1, NI2(n)) + (\Sigma x : N) E(N, NI2(x), NI2(n))}$$

Swaen then proves that[97]

- from ($\Sigma E^{\sim}$) plus (Σ-equ$^{\sim}$) plus (ΣEXH) plus (Σd-equ) are derivable (the polymorphic forms of) (ΣE) and (Σ-equ)
- and from (the polymorphic forms of) (ΣE) plus (Σ-equ) are derivable (ΣEXH) and (Σd-equ),

as well as[98]

- from ($+E^{\sim}$) plus (+-equ$^{\sim}$) plus ($+EXH$) plus ($+d$-equ) are derivable (the polymorphic forms of) ($+E$) and (+-equ)
- and from (the polymorphic forms of) ($+E$) plus (+-equ) are derivable ($+EXH$) and ($+d$-equ).

Similar proofs can be given for the Π sets, the finite sets N_n and the set N.

However, the weakness of Swaen's characterization of his strong elimination rules by means of the exhaustion principles is, that it obviously uses extensional identity E which is not available in constructive type theory. (Note that already in version (5) Martin-Löf mentions a second Π-equality rule and in version (6) he derives a second Π-equality rule and a second Σ-equality rule which correspond to Swaen's exhaustion principles for the Π and the Σ sets resp. *Comme il faut*, versions (5) and (6)' propositional identity is actually extensional identity E.)

As far as the main use or uses of the basic stock type-theoretical operations is/are concerned

- the main use of N, i.e. of the rules of N, is to provide for number theory; these rules yield at once the principle of complete induction and the "thm" of finite recursion. Or if number theory is taken in the form of PA these rules provide the PA axioms 1, 2 and 5 directly and PA axiom 3 indirectly (i.e. it can be derived).
- N_n can almost be "defined away": if N_0 and N_1 are assumed, N_n, for $n > 1$, can be defined by means of N_1, $+$ and N_0.[99]
- the main logical use of $+$ is disjunction $\vee$ and the main set-theoretical use is the disjoint union or coproduct of two sets.
- the main logical use of Σ are existential quantification $\exists$ and conjunction &, and the main set-theoretical uses of Σ are the disjoint union of a family of sets, the cartesian product $\times$ of two sets, and finally comprehension (i.e. the such-that-operation).
- the main logical uses of Π are universal quantification $\forall$ and implication $\supset$, and the main set-theoretical uses of Π are the cartesian product of a family of sets and the power operation $\rightarrow$ of two sets.

viii MONOMORPHISM / POLYMORPHISM AND DECIDABILITY / UNDECIDABILITY OF TYPE CHECKING

In version (2) every term is defined in such a way that it is a term of some type expression, but many terms are also defined s.t. they can be terms of many different type expressions (e.g. the terms $(\lambda x)b[x]$, $b(a)$, (a, b), $E(c, (\lambda x)(\lambda y)d[x, y])$, $i(a)$, $j(b)$, etc.). Hence version (2) is obviously polymorphic(cf. also Subsubsect. ii.3.3, Chapt. 2).

Version (3), however, is completely different in the sense that a new term is introduced with each application of a rule. This entails that each term is of one and only one type expression, whence version (3) is monomorphic. It should be observed though that the notation is not strictly monomorphic since no type information is explicitly contained in the terms. Moreover, some terms even seem to be terms of many different type expressions, namely $1 \in N_1, 1 \in N_2, \ldots, 2 \in N_2, 2 \in N_3, \ldots, 3 \in N_3, 3 \in N_4, \ldots$, etc., but this is merely an imprecision of notation. It is then proved that the judgements A *type*, A *conv* B, $a \in A$ and a *conv* b, where $a \in A$ and $b \in B$, are all decidable (cf. Cor. 3.8 and Cor. 3.13 to the Normalization theorem). These comments concerning the monomorphism of version (3) also hold for version (4).

In version (5) the picture changes again. To dwell on the obvious, every object is still defined in such a way that it is an object of some type. But due to the definition of the canonical objects of identity sets/types, some objects are now unmistakably objects of different types. Thus, not only is the appearance, i.e. notation, of this version of constructive type theory polymorphic, but the same holds true for the theory itself. Note that the imprecision of notation pointed out for versions (3) and (4) is corrected in version (5). This polymorphism also means that the judgements A *set*, $A = B$,

$a \in A$ and $a = b \in A$ are in general not decidable. All of this applies equally well to version (6).

Version (7) not only returns to the "monomorphic reality" of versions (3) and (4), but it introduces for the first time also a corresponding "monomorphic appearance", i.e. notation, by including full type information in all of its objects. Accordingly all the judgements $x_1 : \alpha_1, \ldots, x_n : \alpha_n \vdash \alpha : \mathit{set}$, $x_1 : \alpha_1, \ldots, x_n : \alpha_n \vdash \alpha = \beta : \mathit{set}$, $x_1 : \alpha_1, \ldots, x_n : \alpha_n \vdash a : \alpha$ and $x_1 : \alpha_1, \ldots, x_n : \alpha_n \vdash a = b : \alpha$ are decidable, and a type checking as well as a definitional identity checking algorithm are briefly sketched in version (7). Essentially the same thing also holds for version (8).

ix APPENDIX. THE BATTLE WITH THE DIRECT SEMANTICAL TURN: FROM TYPE THEORY TO LOGICAL THEORY AND BACK

There are two phases during the 25 years of constructive type theory, namely from 1973 to 1979 and from 1982 to 1984–85, where Martin-Löf considered reversing the order of things, that is founding his type theory on a logical theory rather than the other way round. In 1978 Smith demonstrated how to interpret a crossing between versions (2) and (3) of constructive type theory in a standard logical theory based on ordinary first order logic, and in 1984 he demonstrated how to interpret version (5) without well-orderings in a non-standard first order logical theory.

Already in 1977 Aczel provided a sketch of an interpretation of a variant of version (2) of constructive type theory in a theory similar to Smith's standard first order logical theory. This interpretation led Aczel to the notion of Frege structure which can be conceived of as a model of these logical theories. In 1980 Aczel tried indeed to show that there is a foundational typefree language for constructive mathematics whose semantics would be given by something like a Frege structure – but he soon came to realize that this attempt was doomed to fail. What this means and what the story which has just been told looks like in detail is the subject of the following explanations.

During the years 1973 to 1979, when Martin-Löf was battling with the problem of finding a radically new, direct semantics with which to replace the old meta-mathematical or model-theoretical semantics (cf. Sect. iii.), he considered founding (constructive) type theory on another theory, namely a logical theory. In his thesis (1978) Smith shows exactly how to interpret constructive type theory in a standard logical theory based on ordinary predicate logic. (Note that Smith leaves the question as to which theory is more fundamental open, by also showing the converse, namely how to interpret this logical theory in constructive type theory.) This interpretation of constructive type theory in the logical theory consists of associating with each type A a formula $A^*[z]$ (expressing in the logical theory the condition for an object to be an element of A) s.t. if $a \in A$ is derivable in constructive type theory, then $A^*[a]$ is derivable in the logical theory.

Smith's version of constructive type theory CTT is a mix of Martin-Löf's versions (2) and (3) without universe(s): it combines the λ-abstraction formulation of type theory of version (2) with the conversion as judgement defined only on well-typed terms approach of version (3). The first group of definitional axioms of CTT is the clearest manifestation of this combination. The second group of rules of CTT are the general rules of conversion of version (3), and the third group are the rules of compatibility of epsilon with equality to the left and to the right of version (3) as well. The assumption rule of CTT is the one of version (2) or (3), and the Π-, Σ-, $+$-, N_n- and N-introduction and -elimination rules are clearly those of version (2). To these are added in CTT the I-introduction and -elimination rules of version (3) in the style of version (2), and a V-introduction axiom $a \in V$ which is neither from version (2) nor from version (3) (what is here denoted by 'V' is not the usual universe of constructive type theory).

Smith's logical theory LT is the following: its expressions are the same as those of CTT, and its formulas are given by the rules:

1. If $B[x]$ is a formula, then $(\forall x)B[x]$, $(\exists x)B[x]$ are formulas.
2. If A, B are formulas, then $(A \supset B)$, $(A\&B)$, $(A \vee B)$ are formulas.
3. $\perp$ is a formula.
4. If a, b are arbitrary expressions, then $a = b$ is a formula.
5. If a is an arbitrary expression, then $N(a)$ is a formula.

Finally, the axioms and rules of LT are

1. the same definitional axioms as CTT
2. $\forall$-introduction $\quad \dfrac{B[x]}{(\forall x)B[x]}$

   Critical condition: $B[x]$ must be derivable from assumptions in which x does not occur free.
3. $\forall$-elimination $\quad \dfrac{(\forall x)B[x]}{B[a]}$
4. $\exists$-introduction $\quad \dfrac{B[a]}{(\exists x)B[x]}$
5. $\exists$-elimination $\quad \dfrac{(\exists x)B[x] \quad \begin{matrix}[B[x]]\\ C\end{matrix}}{C}$

   Critical condition: x must not occur free neither in C nor in any assumption on which the upper occurrence of C depends other than $B[x]$.
6. $\supset$-introduction $\quad \dfrac{\begin{matrix}[A]\\ B\end{matrix}}{A \supset B}$
7. $\supset$-elimination $\quad \dfrac{A \supset B \qquad A}{B}$
8. &-introduction $\quad \dfrac{A \quad B}{A\&B}$
9. &-elimination $\quad \dfrac{A\&B}{A} \qquad \dfrac{A\&B}{B}$
10. $\vee$-introduction $\quad \dfrac{A}{A \vee B} \qquad \dfrac{B}{A \vee B}$
11. $\vee$-elimination $\quad \dfrac{A \vee B \quad \begin{matrix}[A]\\ C\end{matrix} \quad \begin{matrix}[B]\\ C\end{matrix}}{C}$

12. $\bot$-elimination $\dfrac{\bot}{C}$

13. $=$-introduction $a = a$

14. $=$-elimination $\dfrac{a = b \qquad C[a]}{C[b]}$

15. N-introduction $N(0) \qquad \dfrac{N(a)}{N(s(a))}$

16. N-elimination $\dfrac{N(c) \quad C[0] \quad \begin{matrix}[C[x]]\\ C[s(x)]\end{matrix}}{C[c]}$

Critical condition: x must not occur free in any assumption on which $C[s(x)]$ depends other than $C[x]$.

17. Assumption A

The interpretation $*$ of CTT in LT is s.t. to each type A in CTT a formula $A^*[z]$ is associated s.t. $A^*[z]$ will have as free variables the variables free in A and a new variable z. $A^*[z]$ will be constructed by induction on its formation according to type formation in CTT:

$(\Pi x \in A)B[x]$ is interpreted as the species of elements c s.t. for all elements a of the species $A^*[x]$, $c(a)$ is an element of the species $B^*[a, y]$.
$((\Pi x \in A)B[x])^*[z] \equiv (\forall x)(A^*[x] \supset B^*[x, z(x)])$
$(\Sigma x \in A)B[x]$ is interpreted as the species of pairs (a, b) s.t. a is an element of the species $A^*[x]$ and b is an element of the species $B^*[a, y]$.
$((\Sigma x \in A)B[x])^*[z] \equiv (\exists x)(\exists y)(A^*[x] \,\&\, B^*[x, y] \,\&\, z = (x, y))$
$A + B$ is interpreted as the disjoint union of the species $A^*[x]$ and $B^*[y]$.
$(A + B)^*[z] \equiv (\exists x)(A^*[x] \,\&\, z = i(x)) \vee (\exists y)(B^*[y] \,\&\, z = j(y))$
$I(a, b)$ is interpreted as the species which contains the single element r provided $a = b$
$(I(a, b))^*[z] \equiv (a = b \,\&\, z = r)$
N_0 is interpreted as absurdity
$N_0^*[z] \equiv \bot$
N_n, for $n = 1, 2, \ldots,$ is interpreted as the species whose elements are $1_n, \ldots, n_n$
$N_n^*[z] \equiv (z = 1_n \vee \cdots \vee z = n_n)$
N is interpreted as the species of natural numbers
$N^*[z] \equiv N(z)$
V is interpreted as the species of all objects
$V^*[z] \equiv (z = z)$
At last $a = b$ is interpreted by defining
$(a = b)^* \equiv (a = b)$

It is now easily verified by induction on the construction of $A^*[z]$ that the substitution property $A[a]^*[z] = A^*[a, z]$ is satisfied. Smith finally proves by induction on the length of the derivation in CTT that if $a \in A$ is derivable from $x_1 \in A_1, \ldots, x_n \in A_n$ in CTT, then $A^*[a]$ is derivable from $A_1^*[x_1], \ldots, A_n^*[x_n]$ in LT. This means verifying that the $*$-interpretation of each axiom of CTT is an axiom of LT, and showing that if the $*$-interpretations of the premisses of a rule of CTT are derivable in

LT (IH), then the $*$-interpretation of the conclusion of the resp. rule is derivable in LT.[100]

Martin-Löf's (1983A) is a witness to his occupation with or search for a direct semantics for a logical theory which could serve as a foundation for constructive type theory. In the paper just mentioned Martin-Löf presents a direct semantics for a subtheory of Smith's LT, namely for LT without the definitional axioms and without the $=$- and N-introduction and -elimination rules. (This statement is actually not entirely precise, since Martin-Löf's $\supset$-formation rule is non-standard, whereas Smith's $\supset$-formation rule is standard.) The direct semantical explanations of the $\forall$-, $\exists$-, $\supset$-, &-, $\vee$-, $\bot$-formation rules are called the explanations of the meanings of the logical constants, and the direct semantical explanations of the $\forall$-, $\exists$-, $\supset$-, &-, $\vee$-introduction and -elimination rules as well as of the $\bot$-elimination rule are called the justifications of the logical laws.

G. Sambin drew my attention to the fact that one should distinguish two logical interludes in the development of constructive type theory. The first one lasted, as already mentioned, from 1973 to 1979, and the second one roughly from 1982 to 1984–85. Martin-Löf's (1983) and (1983A) are witnesses of this second logical framework phase rather than of the first one.

Version (5) is the first version of constructive type theory which takes the identity issue radically seriously in the sense of endowing each type and each object of some type with an (intensional or extensional) identity (cf. Sect. v.). This was an important step in the development of constructive type theory towards its present form. In (1984) Smith shows how to interpret version (5) of constructive type theory without well-orderings in a non-standard first order logical theory (also called a formal theory of propositions and combinator terms by Smith). This interpretation is based on the direct semantics of version (5) and can as a matter of fact be regarded as a metamathematical formalization of this direct semantics in the logical theory.
Let version (5) of constructive type theory be denoted by 'CTT5' and the logical theory, in which CTT5 is going to be interpreted, by 'LT5'. Then the expressions of LT5 not involving reflection are the same as those of CTT5. The expressions of LT5 concerning reflection are $(\dot\forall x)a$, $(\dot\exists x)a$, $a\dot\supset b$, $a\dot\& b$, $a\dot\vee b$, $\dot\bot$, $\dot\top$, $a\dot= b$, $\dot N(a)$. And the axioms and rules of LT5 are the same as in LT except for the $\supset$-formation rule which in LT5 is non-standard: If A is a formula (proposition) and B is a formula (proposition) provided A is true, then $A \supset B$ is a formula (proposition). Moreover, LT5 has the following additional formula, or as they are called in LT5, proposition formation rules:

6. $\top$ *prop*
7. If a is an arbitrary expression, then $P(a)$ *prop* and if $P(a)$ is true, then $T(a)$ *prop*

and the additional axioms or rules:

18. Symmetry $\dfrac{a = b}{b = a}$
19. Transitivity $\dfrac{a = b \qquad b = c}{a = c}$
20. $\top$-introduction $\quad \top$

21. Reflection introduction

$P(a \dot{=} b) \qquad T(a \dot{=} b) \supset\subset (a = b)$

$P(\dot{\top}) \qquad T(\dot{\top})$

$P(\dot{\bot}) \qquad T(\dot{\bot}) \supset\subset \bot$

$$\frac{P(a) \quad P(b)}{P(a \,\dot{\&}\, b)} \qquad \frac{P(a) \qquad P(b)}{T(a \,\dot{\&}\, b) \supset\subset (T(a) \,\&\, T(b))}$$

$$\frac{P(a) \qquad P(b)}{P(a \dot{\vee} b)} \qquad \frac{P(a) \qquad P(b)}{T(a \dot{\vee} b) \supset\subset (T(a) \vee T(b))}$$

$$\frac{P(a) \quad \begin{matrix}(T(a))\\ P(b)\end{matrix}}{P(a \dot{\supset} b)} \qquad \frac{P(a) \qquad \begin{matrix}(T(a))\\ P(b)\end{matrix}}{T(a \dot{\supset} b) \supset\subset (T(a) \supset T(b))}$$

$$\frac{P(b)}{P((\dot{\forall} x)b)} \qquad \frac{P(b)}{T((\dot{\forall} x)b) \supset\subset (\forall x)T(b)}$$

$$\frac{P(b)}{P((\dot{\exists} x)b)} \qquad \frac{P(b)}{T((\dot{\exists} x)b) \supset\subset (\exists x)T(b)}$$

$P(\dot{N}(a)) \qquad T(\dot{N}(a)) \supset\subset N(a)$

where the two unary predicates $P(.)$ and $T(.)$ express A *prop* and A *true* resp. *Note*: Smith puts '*true*' behind every axiom and every premiss and conlusion of all the rules of inference of LT5.

The interpretation of CTT5 in LT5 is s.t. to each type A in CTT5 will be associated two expressions $A'_{\in}[z']$ and $A''_{=}[z', z'']$ of LT5, and to each object a of type A one object a' in LT5. $A'_{\in}[z']$ will be a proposition for unrestricted z', expressing the membership relation on the type A, and $A''_{=}[z', z'']$ will be a proposition on the assumption that both propositions $A'_{\in}[z']$ and $A'_{\in}[z'']$ are true, expressing the equality relation on the type A, and $A''_{=}[z', z'']$ will have as free variables the variables free in A and the two new variables z' and z'' (while a' must not have any free variables other than those of a). This yields, if $a \in A$ then $A'_{\in}[a]$ is true, and if $a = b \in A$ then $A''_{=}[a, b]$ is true, and if $A = B$ then $A'_{\in}[z'] \supset\subset B'_{\in}[z']$ and $(A'_{\in}[z'] \,\&\, A'_{\in}[z'']) \supset (A''_{=}[z', z''] \supset\subset B''_{=}[z', z''])$. But Smith deviates from this pattern by giving $a \in A$ the same interpretation as $a = a \in A$, i.e. $A'_{\in}[z']$ is defined by: $A'_{\in}[z'] =_{\text{def}} A'_{=}[z', z']$.

The interpretation $'$ will be constructed by first associating to each term a of CTT5 an expression a' of LT5:

$$x' =_{\text{def}} x$$

$$((\lambda x)b)' =_{\text{def}} (\lambda x')b' \qquad c(a)' =_{\text{def}} c'(a')$$

$$(a, b)' =_{\text{def}} (a', b') \qquad E(c, (x, y)d)' =_{\text{def}} E(c', (x', y')d')$$

$$i(a)' =_{\text{def}} i(a') \qquad D(c, (x)d, (y)e)' =_{\text{def}} D(c', (x')d', (y')e')$$

$$j(b)' =_{\text{def}} j(b')$$

$$r' =_{\text{def}} r \qquad J(c, d)' =_{\text{def}} J(c', d')$$

$$m'_n =_{\text{def}} m_n \qquad R_n(c, c_0, \ldots, c_{n-1}) =_{\text{def}} R_n(c', c'_0, \ldots, c'_{n-1})$$

$$0' =_{\text{def}} 0 \qquad R(c, d, (x, y)e)' =_{\text{def}} R(c', d', (x', y')e')$$

$$s(a)' =_{\text{def}} s(a')$$

The forms of expression above have been interpreted as themselves, but the remaining expressions all involve reflection and their interpretation reflects the interpretation of the corresponding types given below (cf. the def. of $(\Pi x \in A)B'_{=}[z', z'']$, etc.).

$$\dot{\Pi}(a, (x)b)' =_{\text{def}} (\lambda z'z'')[(\dot{\forall} x'x'')[a'(x', x'')\dot{\supset} b'(z'(x'), z''(x''))] \\ \dot{\&}\, [z' \dot{=} (\lambda x')(z'(x'))]\,\dot{\&}\, [z'' \dot{=} (\lambda x'')(z''(x''))]]$$

$$\dot{\Sigma}(a, (x)b)' =_{\text{def}} (\lambda z'z'')(\dot{\exists} x'x'')(\dot{\exists} y'y'')[[a'(x', x'')\,\dot{\&} \\ (a'(x', x'')\dot{\supset} b'(y', y''))]\,\dot{\&}\, [z' \dot{=} (x', y')]\,\dot{\&}\, [z'' \dot{=} (x'', y'')]]$$

$$(a \dot{+} b)' =_{\text{def}} (\lambda z'z'')[(\dot{\exists} x'x'')(a'(x', x'')\,\dot{\&}\, [z' \dot{=} i(x')] \\ \dot{\&}\, [z'' \dot{=} i(x'')])\dot{\vee}(\dot{\exists} y'y'')(b'(y', y'')\,\dot{\&}\, [z' \dot{=} j(y')]\,\dot{\&}\, [z'' \dot{=} j(y'')])]$$

$$i(a, b, c)' =_{\text{def}} (\lambda z'z'')((a'(b'))(c')\,\dot{\&}\, [z' \dot{=} r]\,\dot{\&}\, [z'' \dot{=} r])$$

$$\dot{N}_0{}' =_{\text{def}} (\lambda z'z'')(\dot{\bot}\,\dot{\&}\, [z' \dot{=} z''])$$

$$\dot{N}_n{}' =_{\text{def}} (\lambda z'z'')[([z' \dot{=} 0_n]\dot{\vee} \cdots \dot{\vee}[z' \dot{=} (n-1)_n])\,\dot{\&}\, [z' \dot{=} z'']]$$

$$\dot{N}' =_{\text{def}} (\lambda z'z'')(\dot{N}(z')\dot{\&}[z' \dot{=} z''])$$

Note that if $b[x]$ has been built up from expressions only and if $b[x]' =_{\text{def}} b'[x']$, then the substitution property $b[a]' =_{\text{def}} b'[a']$ holds.

$A'_{=}[z', z'']$ will be constructed by induction on its formation according to type formation in CTT5; in order to motivate the definition of the interpretation of all the types A not involving reflection, the equality relation on that type will be semantically explained:

$a = b \in (\Pi x \in A)B[x]$ means that a, b both have values of λ-form and that if they are applied to equal objects $c = d \in A$ they yield equal objects of type $B[c]$

$$(\Pi x \in A)B'_{=}[z', z''] =_{\text{def}} (\forall x'x'')((A'_{=}[x', x''] \supset B'_{=}[z'(x'), z''(x'')]) \\ \& [z' = (\lambda x')(z'(x'))] \& [z'' = (\lambda x'')(z''(x''))])$$

$a = b \in (\Sigma x \in A)B[x]$ means that a, b have values (c, d), (e, f) resp. where $c = e \in A$ and $d = f \in B[c]$.

$$(\Sigma x \in A)B'_{=}[z', z''] =_{\text{def}} (\exists x'x'')(\exists y'y'')[(A'_{=}[x', x''] \\ \& (A'_{=}[x', x''] \supset B'_{=}[y', y''])) \& [z' = (x', y')] \\ \& [z'' = (x'', y'')]]$$

$a = b \in A + B$ means that a, b have values $i(c)$ and $i(d)$ resp. where $c = d \in A$, or that a, b have values $j(e)$ and $j(f)$ resp. where $e = f \in B$

$$(A + B)'_{=}[z', z''] =_{\text{def}} (\exists x'x'')(A'_{=}[x', x''] \& [z' = i(x')] \\ \& [z'' = i(x'')]) \vee (\exists y'y'')(B'_{=}[y', y''] \\ \& [z' = j(y')] \& [z'' = j(y'')])$$

$d = e \in I(A, a, b)$ means that d, e both have value r and that $a = b \in A$.

$$I(A, a, b)'_{=}[z', z''] =_{\text{def}} (A'_{=}[a', b'] \& [z' = r] \& [z'' = r])$$

There are no equal objects of type N_0.

$$N'_{0=}[z', z''] =_{\text{def}} (\bot \ \& [z' = z''])$$

$a = b \in N_n$, for $n > 0$, means that a, b both have value m_n for some $m = 0, \ldots, m = (n-1)$.

$$N'_{n=}[z', z''] =_{\text{def}} [([z' = 0_n] \vee \ldots \vee [z' = (n-1)_n]) \& [z' = z'']]$$

$a = b \in N$ means that a, b both have value 0 or that there are objects $c = d \in N$ s.t. a, b have values $s(c)$ and $s(d)$ resp.

$$N'_{=}[z', z''] =_{\text{def}} (N(z') \& [z' = z''])$$

The interpretation of U does not follow the semantical explanation of $a = b \in U$, but it rather reflects the conception of the notion of type (according to which types are α-convertible and equality on types is an equivalence relation):

$$U'_{=}[z', z''] =_{\text{def}} [(\forall u'u'')P(z'(u', u'')) \& (\forall u'u'')P(z''(u', u''))] \\ \& ([(\forall u'u'')P(z'(u', u''))\& (\forall u'u'')P(z''(u', u''))] \\ \supset [(\forall u'u'')[T(z'(u', u'')) \supset\subset T(z''(u', u''))] \& \\ (\forall u'u'')[T(z'(u', u'')) \supset\subset T(z'(u', u''))] \\ \& (\forall u'u''u''')(T(z'(u', u'')) \& T(z'(u'', u'''))] \\ \supset T(z'(u', u''')))])$$

$$T(a)'_{=}[z', z''] =_{\text{def}} T(a'(z', z''))$$

Analogously to the interpretation of the expressions, the following substitution property, $B[a]'_{=}[z', z''] = B'_{=}[a'][z', z'']$ is readily seen to hold.

Finally, if a' is the interpretation of an expression a and has the free variables $x'_1, \ldots, x'_n$, then a'' denotes the result of simultaneously substituting the new variables $x''_1, \ldots, x''_n$ for $x'_1, \ldots, x'_n$ in a', and if $A'_{=}[z', z'']$ is the interpretation of a type A and has the free variables $x'_1, \ldots, x'_n, z', z''$, then $A''_{=}[z', z'']$ denotes the result of simultaneously substituting $x''_1, \ldots, x''_n$ for $x'_1, \ldots, x'_n$ in $A'_{=}[z', z'']$.

Smith then proves by induction on the length of a derivation in CTT5 that

(I) If $A\ type$ is derivable in CTT5 from the assumptions $x_1 \in A_1, \ldots, x_n \in A_n$ then

(Ia) $A'_{=}[z', z'']\ prop$
(Ib) $A'_{=}[z', z''] \supset\subset A''_{=}[z', z'']$
(Ic) $[(A'_{=}[z', z''] \supset A'_{=}[z'', z']) \,\&$
$((A'_{=}[z', z''] \,\&\, A'_{=}[z'', z''']) \supset A'_{=}[z', z'''])]$

are all derivable in LT5 from the assumptions $A'_{1=}[x'_1, x''_1], \ldots, A'_{n=}[x'_n, x''_n]$.

(II) If $A = B$ is derivable in CTT5 from the assumptions $x_1 \in A_1, \ldots, x_n \in A_n$ then $A'_{=}[z', z''] \supset\subset B''_{=}[z', z'']$ is derivable in LT5 from the assumptions $A'_{1=}[x'_1, x''_1], \ldots, A'_{n=}[x'_n, x''_n]$.

(III) If $a \in A$ is derivable in CTT5 from the assumptions $x_1 \in A_1, \ldots, x_n \in A_n$ then $A'_{=}[a', a'']$ is derivable in LT5 from the assumptions $A'_{1=}[x'_1, x''_1], \ldots, A'_{n=}[x'_n, x''_n]$.

(IV) If $a = b \in A$ is derivable in CTT5 from the assumptions $x_1 \in A_1, \ldots, x_n \in A_n$ then $A'_{=}[a', b'']$ is derivable in LT5 from the assumptions $A'_{1=}[x'_1, x''_1], \ldots, A'_{n=}[x'_n, x''_n]$

in a way analogous to the one explained for the interpretation of CTT in LT.[101]

Smith's interpretation $*$ of CTT in LT is a detailed working out of a λ-calculus interpretation, sketched by Aczel in his (1977), of a variant of constructive type theory version (2) in a logical theory similar to LT (Aczel's logical theory is a first order theory of combinators containing the Peano axioms and a form of reflection on this theory). Aczel's interpretation led him to the notion of Frege structure, where a Frege structure consists of a logical system in Aczel's sense together with a lambda system and can be constructed from arbitrary models of λ-calculus. Frege structures (or, more precisely, N-standard Frege structures) are models of these logical theories and hence they can also be considered to be models of the corresponding constructive type theories.

Are Frege structures thus the sought after semantics and foundation of constructive type theory? To show that there is a foundational typefree language – which treats the notion of proposition as primitive rather than reducing it to that of type as in the theory of types – for constructive mathematics whose semantics would be given by something like a Frege structure, was Aczel's original motivation for writing his (1980). But Aczel answers in this paper the above question in a careful negative way: he refers to the weaknesses in attempts to formulate an adequate semantics for such a language (such a logical theory) and questions the foundational significance of the notion of Frege structure.[102] And Smith indirectly affirms the foundational priority of constructive type theory (e.g. in its version (5) or any of its later versions) when he points out that also a logical theory (such as LT5) needs a direct semantics which can presumably be given even less easily than the one for constructive type theory e.g. in its version (5). This in all likelihood also explains why Martin-Löf after all has stuck to constructive type theory as the foundational language for constructive mathematics.

NOTES

1 Version (1), pp. 2, 3.
2 The definition of 1-step reducibility and n-step reducibility are taken from C. Svensson's thesis (1990), pp. 7, 8, since Martin-Löf's own definition of reducibility in version (2) makes it difficult to justify the reflexivity of *conv*.
3 Martin-Löf (1985), pp. 407, 408.
4 pp. 407, 408.
5 Version (4), p. 11.
6 These explanations are taken from Nordström et al. (1990), pp. 109, 110.
7 This example is based on Nordström et al. (1990), pp. 112, 113.
8 Cf. Nordström et al. (1990), pp. 110–112.
9 Cf. Nordström et al. (1990), p. 113 and Dybjer (1994).
10 Cf. Nordström et al. (1990), pp. 76–79.
11 p. 64.
12 p. 63.
13 Cf. Smith (1983), p. 454.
14 Cf. version (6), p. 78. *Fintree* was here written out with the aid of Per Martin-Löf.
15 Cf. Nordström et al. (1990), pp. 115–117.
16 Version (1), p. 5.
17 p. 8ff.
18 p. 12.
19 p. 5.
20 pp. 25, 26.
21 p. 48.
22 p. 49.
23 Version (2), p. 31.
24 pp. 30, 31.
25 p. 29.
26 p. 30.
27 p. 31.
28 p. 76.
29 Cf. Martin-Löf (1972A), point 2.1.1, p. 97.
30 Version (3), p. 85.
31 pp. 96, 97.
32 Martin-Löf (1972A), p. 100.
33 pp. 98,99.
34 Version (3), p. 74.
35 pp. 78–82.
36 p. 113.
37 p. 81.
38 p. 86.
39 p. 116.

40 p. 113.
41 Martin-Löf (1972A), p. 97.
42 Version (3), p. 114.
43 p. 101.
44 Cf. p. 87.
45 Observe that in (1972A) Martin-Löf uses = instead of 'I'.
46 Version (4), pp. 32–34.
47 Cf. Martin-Löf (1975), p. 4.
48 Version (4), pp. 54, 55.
49 p. 53.
50 pp. 38–51.
51 p. 50.
52 pp. 16, 17, 20.
53 p. 21.
54 Version (5), p. 161.
55 p. 163.
56 pp. 166, 167.
57 pp. 167–171.
58 pp. 169, 170, and Nordström et al. (1990), p. 66.
59 Version (6), p. 59.
60 pp. 14–20.
61 p. 25.
62 pp. 62, 63.
63 pp. 59–61.
64 p. 31.
65 p. 60.
66 p. 31.
67 pp. 59, 60.
68 p. 59.
69 Version (7), p. 9.
70 pp. 6, 7, 11, 16, 17, 19–21, 34, 35, 98.
71 pp. 6, 7, 10,11, 16–18.
72 pp. 75, 76.
73 pp. 65–75.
74 pp. 10, 54, 55, 81, 82.
75 p. 31.
76 pp. 22–24.
77 p. 32.
78 p. 105.
79 Version (8), p. 19.
80 pp. 36, 37.
81 pp. 49, 51.
82 pp. 95–99, pp. 102–104.
83 pp. 97, 104.
84 pp. 41, 42, 45.
85 pp. 21, 58, 61.
86 pp. 58, 61.
87 pp. 58, 61.
88 pp. 82, 83, 104.
89 pp. 103, 104.
90 pp. 104, 105.
91 Howard (1980), pp. 488, 489.
92 Nordström et al. (1990), pp. 180–182.
93 For an example, cf. Thompson (1991), pp. 287–289.
94 For an example, cf. p. 290.
95 Swaen (1989), p. 112.

[96] pp. 113, 114.
[97] Thm. 1.4 in Swaen (1989), p. 114.
[98] Thm. 1.5, pp. 116, 117.
[99] Cf. Nordström et al. (1990), p. 102.
[100] Smith (1978), pp. 33–47.
[101] Smith (1984), pp. 743–752.
[102] Aczel (1980), pp. 33, 34.

CHAPTER 4
PHILOSOPHICAL AND TECHNICAL PREHISTORY OF CONSTRUCTIVE TYPE THEORY (1880–1970)

This chapter intends to introduce some of the philosophical logical ideas and preconceptions as well as some of the mathematical logical instruments and techniques which led up to or influenced the development of constructive type theory. The first section discusses the place of constructive type theory w.r.t. two different approaches to or traditions of logic. It is of a more philosophical flavor. The second section presents in a chronological and systematic way some of the important technical prerequisites of constructive type theory. In this chronology are inserted some of Martin-Löf's works on normalization which can be seen to give rise to the first version of his type theory in 1970. This second section is by its nature more mathematical. Both sections mean to cast some light onto the logical history prior to the establishment of constructive type theory, that is onto the prehistory of this theory.

i PHILOSOPHICAL SOURCES OF CONSTRUCTIVE TYPE THEORY: TWO APPROACHES TO LOGIC

In 1967, van Heijenoort published an important paper in which he explains how two traditions or strands in logic around the turn of this century led in the 20s to the formation of standard mathematical logic as it is well-known today. Hintikka elaborated and modified van Heijenoort's insight into the history of logic in various publications,[1] and Goldfarb also contributed an influential paper on this topic.[2] The two conceptions will here be called logic as universal language and logic as calculus. (Van Heijenoort calls the first one just 'logic as language' and Hintikka calls the two 'language as universal medium' and 'language as calculus'; furthermore, van Heijenoort refers to the first tradition as the axiomatic one and to the second as the set-theoretic one, whereas Hintikka calls the first the universalist and the second the model-theoretic tradition.) The following explanations are based on the papers just mentioned.[3] They will show that logic as calculus has since the 20s been the dominant paradigm of logic, but that with the rise of constructive type theory at the beginning of the 70s the paradigm of logic as universal language has been revitalized if not revived. To be more precise, a shift from the conception of logic as calculus to the one of logic as universal language can be observed as part of the development of constructive type theory.

i.1 *Logic as Universal Language*

The main representatives of this tradition are Frege, Peano, Russell, Whitehead and the early Wittgenstein. Later adherents to this conception are (the late Wittgenstein) Quine and the Vienna Circle in the hayday 1930–32.

The conception can be characterized by two main theses and various consequences of these theses:

(1) Thesis of the universality of logic (in the sense of inescapability), i.e. the logical system deals with everything there is (no metalanguage).

E.g. in Frege's case the variable binding quantifiers range over all objects, in Russell's case the variable binding quantifiers range over all individuals or over all propositional functions of some particular order.

Van Heijenoort and Goldfarb link the universality thesis also to the central role played by the universal quantifier in this strand of logic.

1. Consequence: the universe underlying the logical system consists of all there is, it is one and it is fixed. E.g. in Russell's *PM* Frege's universe is stratified by the introduction of types, but it is still the only one and fixed universe that there is.

2. Consequence: the laws of logic have content: they are the most general truths about the logical furniture of the universe.

3. Consequence: e.g. in Frege's logic, functions must be defined for all objects.

(2) Thesis of the inexpressibility of metalogical considerations, formulated also as: it is impossible to say anything outside the formal system, to describe it from outside etc. (no metalogic)

Consequence: thesis of the inexpressibility of semantics (and the concept of truth), i.e. it is impossible to meaningfully and significantly say in logic what its semantical relations are, since any attempt to do so must already presuppose them and thus becomes non-sensical or tautological.

Hintikka points out that it is wrong to call this thesis the thesis of the impossibility of semantics, since Frege clearly had a theory about the meaning of expressions in intensional contexts (the sense–reference theory), and he had a definite set of ideas about the semantics of ordinary extensional logic, but Frege does not and cannot incorporate them into his 'official' logical system (Hintikka calls Frege a semanticist without semantics):

1. Consequence to this consequence: e.g. in Wittgenstein's *Tractatus* (1922) a distinction is made between what can be said and what can only be shown: all the semantical relations of logic are exactly what can only be shown.

2. Consequence: since the semantics cannot be expressed in formulating the system of logic, the emphasis is on formalization, on a purely formal system in logic (a logical syntax of language;

cf. Carnap (1934)): the point of formalization is to render mathematics in such a way that all principles used in mathematical reasoning are explicit.

3. Consequence: thesis of the impossibility of varying the interpretation of the language of logic; every logical formula has a fixed meaning and there are no schematic placeholders which may be assigned one value or another, there is no question of reinterpreting any sign.

Consequence to this 3. Consequence: thesis of the impossibility of all model theory (no model theory).

The only question of completeness that may arise in this strand of logic is, as mentioned by van Heijenoort, an experimental one: Can the intuitive modes of reasoning actually used in science be exhausted by the resp. logical system? Frege, Russell and Peano tried to answer this question in their own way with their resp. enterprise. Logic as the universal formal language was meant to replace natural language (in science).

i.2 *Logic as Calculus*

The main representatives of this tradition in logic are Boole, Schröder, Löwenheim, and later on Gödel, the later Carnap and Tarski.

This conception of logic can be characterized as follows:

(1) Thesis of the changeability of the universe, i.e. the universe comprises only what is agreed upon to be considered at a certain time in a certain context and can be changed *ad libitum*. E.g. de Morgan calls it 'the universe of discourse'.
(2) Thesis of the reinterpretability of the language of logic. Hintikka stresses that this is the true meaning of logic as calculus and not the conception of logic as an uninterpreted calculus, a mere game with characters.

This thesis (2) implies

1. the thesis of the expressibility of semantics, i.e. the possibility to talk meaningfully and non-trivially about semantical relations,
2. the thesis of the variability of the semantical relations in logic,
3. the thesis of the expressibility of metalogical considerations.

Moreover, to such a conception of logic belong equally

- the use of the concept of truth
- reliance on metalanguage
- every use of intuitive model-theoretical ideas in the 'official' logical system under consideration
- the use of algebraic methods and attempts to find parallels between logic and algebra.

The logic as calculus tradition of logic gives rise to the idea that the relation of a well-formed formula to its models is the cornerstone of all logical semantics or model theory.

i.3 *Where Does Metamathematics Come in?*

Van Heijenoort and Goldfarb conceive of the metamathematics of the 20s as an amalgamation of the Frege–Russell–Peano tradition and the Boole–Schröder–Löwenheim tradition of logic. Goldfarb puts it in the following terms:

> To arrive at metamathematics from the Frege–Russell–Peano approach (logic as universal language) one must add the 'meta', i.e. the possibility to examine a logical system from an external standpoint.
>
> To arrive at metamathematics from the Boole–Schröder–Löwenheim approach (logic as calculus) one must add the 'mathematics', i.e. an accurate appreciation of how the logical system can be used to encode/formalize mathematics and hence of how the metalogical considerations can be taken to be about mathematics.

Hintikka, however, regards metamathematics as a continuation of the Boole–-Schröder–Löwenheim approach (logic as calculus) which picks up a few ingredients from the opposite approach. This seems to be the right view in view of the fact that in metamathematics both main theses of the logic as calculus conception are still valid, whereas both main theses of the logic as universal language conception are dumped.

Goldfarb and Hintikka finally emphasize the pivotal role played by Hilbert in the transformation of the Boole–Schröder–Löwenheim approach to metamathematics: (a) the system of *Principia Mathematica* (1910–13) provides empirical evidence that all mathematical results can be captured by appropriate formal systems. Hilbert then suggests that one ought to think of the content of much of mathematics as given by its formal representation, thereby adopting an un-Russellian external standpoint; thus, it becomes possible to investigate the system metalogically or metamathematically. (b) The modern axiomatic approach to scientific theories, including mathematical theories, and esp. Hilbert's *Grundlagen der Geometrie* (1899) constituted a formidable ally for the early (Boole–Schröder–Löwenheim) model-theoretic (or set-theoretic) tradition. (c) Although the introduction of the notion of quantifier was one of Frege's main logical achievements, it was Hilbert who gave the first satisfactory semantical analysis of this notion in terms of quantifier-dependence and choice functions. (It is interesting to notice that although Hintikka appears to postulate a revision or reassessment of Hilbert's formalism in terms of model theory, he does not count Hilbert among the representatives of the language as calculus or model-theoretic tradition.)

i.4 *Universal Language Vs. Foundational Language*

Set theory in general, and ZF set theory in particular, is commonly considered to be a theory which can serve as a universal language for the formalization of mathematics.

First order predicate logic conceived as a general schematism or underlying logic, a logic applicable to any particular area of mathematics (the way described by Goldfarb) can be and has been considered another universal language for the formalization of mathematics.

The modern conception of first order logic as well as set theory merely borrow the phrase 'universal language' from the conception of logic as universal language, while they belong essentially to the conception of logic as calculus. To avoid confusing them with real universal languages it may be better to call them following frequent practice 'foundational languages'.

i.5 *Where does Constructive Type Theory Come in?*

In his (1971B) commenting on version (1) of constructive type theory, Martin-Löf compares constructive type theory with the foundational language of set theory. He writes that constructive type theory is intended to be a full scale system for the formalization of intuitionistic mathematics (as developed e.g. by Bishop (1967)) in somewhat the same way (e.g. ZF or NBG) that set theory is intended to be a full scale system for the formalization of classical mathematics. In versions (2) and (3) of constructive type theory, the first half of this statement is repeated, the comparison with set theory is dropped.

In version (4) the comparison with foundational languages is taken up again. It is stated that one of the unresolved problems of mathematics is 'to decide the relative merits of the multiplying Babel of artificial languages put up as analysis of mathematical practice'. As examples of such artificial languages are listed: ramified type theory, ZF set theory, NBG set theory, Quine's set theories, MK set theory, Scott's set theory, various languages inspired by category theory, Automath on the classical side, and systems of Heyting, Kleene-Vesley, Troelstra, Myhill, Kreisel, Scott, Feferman, Bishop and Martin-Löf on the intuitionistic side.[4]

Martin-Löf also reformulates what he wrote in versions (1)–(3) about what constructive type theory is all about: the language he is going to present is intended as a full scale system which accounts for essentially everything one does in mathematics. He will not only introduce but also explain the rules which constitute that language and will thus engage in what is usually called syntax and semantics. The language is a development of constructive type theory. All these explanations could be considered an elaboration of the corresponding comment of (1971B).

In addition and apart from a little terminological hint (cf. 'language') version (4) contains another clearer terminological indication as to how constructive type theory relates to the conceptions of logic as universal language and logic as calculus. Martin-Löf points out that semantics as he understands this term and is going to get engaged in, namely as the theory which regulates the explanation of mathematics, has nothing to do with semantics in the sense of model theory, which according to him is a part of metamathematics and is nothing about, but simply a branch of mathematics. (This addition is typical for the direct semantical turn beginning with version (4). cf. Sect. iii,

Chapt. 3.) Version (5) just summarizes one part of the explanations of version (4), according to which constructive type theory has been developed in order to clarify the syntax and semantics of intuitionistic mathematics.

As far as the relation of constructive type theory to the conceptions of logic as universal language and logic as calculus is concerned, version (6) is more outspoken and explicit than any of the previous versions.

First, Martin-Löf expresses his interest in and his intention to do mathematical logic in the sense of foundation (or philosophy) of mathematics and also in the sense of logic using mathematical symbolism, but not in the sense of logic studied by mathematical methods as a branch of mathematics (which can easily be associated with metamathematics).

Second, he proceeds to compare once again constructive type theory with other foundational languages, but unlike earlier such comparisons, he refers here to advantages of his theory over other foundational languages: Ramified type theory had to be reduced to simple type theory because it was not sufficient for deriving even elementary parts of analysis; and the official justification of simple type theory does not justify the quantifier laws if quantification is extended to infinite domains. ZF set theory cannot adequately deal with the foundational problems of category theory where the category of all sets or the category of all groups etc. are considered. First order predicate logic is not rich enough to be suitable as a higher level programming language. Now, constructive type theory is predicative (ramified) and sufficient for deriving elementary parts of mathematics such as the theory of real numbers, it can adequately deal with the foundational problems of category theory and it is suitable as a higher level programming language.

Third and last, Martin-Löf compares a standard presentation of first order predicate logic with a presentation of constructive logic/set theory as a part of constructive type theory. In a standard i.e. metamathematical introduction to mathematical logic, the presentation of first order predicate logic proceeds in 4 steps:[5]

(i) inductive def. of individual terms
(ii) inductive def. of formulas
(iii) inductive def. of deductions (or derivations) and theorems (or derivable formulas)
(iv) semantical interpretation of (i)–(iii) following Tarski and assuming set theory

In constructive type theory

to the individual terms (i)	correspond the (dependent or non-dependent) elements (of some set) which are inductively defined only as far as they belong to their resp. set
to the formulas (ii)	correspond the propositions which are not defined inductively (but form an open concept)
to the deductions and theorems (iii)	correspond the deductions of constructive type theory and their resp. judgements both of which are not defined inductively either.

The open concepts of element/proof-object, set/proposition, and deduction of constructive type theory and resp. judgement are very reminiscent of the experimental completeness, which from another point of view can be regarded as openess, of the conception of logic as universal language.

In the Frege–Russell approach to logic it was customary to distinguish between propositions and judgements (this does not appear to apply to Peano), and the same distinction can be found in constructive type theory (cf. Subsect. ii.2, Chapt. 2). Furthermore, semantics is not the last step (iv) of a presentation of constructive logic/set theory. Constructive type theory avoids separating form and meaning (content) by explaining semantically each form of judgement and each rule of inference at the same time as they are introduced (and this in a direct, intuitive and not in a recursive, set-theoretical way).

Finally, in the metamathematical tradition judgements and rules of inference are only partially formally represented in the so-called object language, whereas they are implicitly used in the so-called metalanguage. In constructive type theory there is only one language and the judgements and rules of inference are to be completely and formally represented in that language.

Martin-Löf's aim with constructive type theory is to build a system of formal rules representing in the best possible way informal (mathematical) reasoning. Frege expressed his aim with his *Begriffsschrift* (1879) in the following way: 'Ich habe versucht, die mathematische Formelsprache durch Zeichen für die logischen Verhältnisse zu ergänzen, sodass daraus zunächst für das Gebiet der Mathematik eine Begriffsschrift hervorgehe, wie ich sie als wünschenswert dargestellt habe.'[6] It is now quite plain that Martin-Löf's approach to logic belongs to the tradition of logic as universal language rather than the tradition of logic as calculus.

The last doubt – had there possibly been any left from version (6) – as to the association of constructive type theory with one of two approaches to logic would be definitively eradicated in version (8). There, as in version (6) but more pointedly, Martin-Löf distinguishes between logic as the study of a formal system of rules which the mathematician uses or applies in his proving or reasoning, i.e. logic in the sense of philosophy or foundations of mathematics – he calls this the Frege–Russell–Peano approach to logic which has earlier on been called logic as universal language – and logic as the study of logical systems by mathematical means, i.e. logic in the sense of metamathematics and being a branch of mathematics – he calls this the Hilbert–Gödel–Tarski approach to logic which has earlier on been associated with the logic as calculus conception. He expresses his interest in developing the Frege–Peano–Russell line of logic for the following reasons: he believes that the greatest challenge and the greatest stimulation for contemporary logic stem from computer science, and that computer science is interested in new formalisms, new languages formalizing mathematics. This interest is closely linked with the logic as universal language tradition. (Note that this shift of interest from the foundations of intuitionistic mathematics to computer science is already hinted at in version (5).) Martin-Löf thus explicitly associates himself with this latter tradition and conception of logic. He also believes that the metamathematical tradition of logic is basically exhausted and finished.

ii TECHNICAL SOURCES OF CONSTRUCTIVE TYPE THEORY

In this section the technical work of some logicians will be outlined, work which provided the main tools for Martin-Löf's work on normalization which not only precedes but also leads to his first formulation of constructive type theory. This more technical prehistory of constructive type theory starts with Gentzen and Prawitz on natural deduction and normalization, it goes on to Curry and Howard's isomorphism, and it combines these two methods in Martin-Löf's normalization for infinitary propositional logic. This is followed by Tait's computability technique which was subsequently applied by Martin-Löf (in combination with other methods) in his normalization for the intuitionistic theory of iterated inductive definitions. Next comes Girard's generalization of Tait's technique which was finally used by Martin-Löf – again in connection with the other methods – in his normalization for intuitionistic 2nd order logic and for intuitionistic simple type theory both of which are forerunners to a general intuitionistic theory of types, in other terms, of constructive type theory.

ii.1 *Gentzen and Prawitz on Natural Deduction*

The question as to how the notion of logical proof within first order languages is to be explained, was answered by Gentzen in 2 steps:[7]

1) Gentzen showed how this notion of logical proof could be defined in terms of certain formal systems, called systems of natural deduction, which he constructed.
2) Gentzen showed by a deeper analysis of the structure of these logical proofs that they could be written in a special form.

This led to a very satisfactory understanding of the notion of logical proof of first order languages. Now, 2) was carried out by Gentzen for certain formal systems called calculi of sequents. Prawitz's important contribution to proof theory was to carry out 2) for Gentzen's systems of natural deduction.[8]

Gentzen's main idea is beautifully expressed by Prawitz in the following terms: 'Gentzen's systems of natural deduction arise from a particular analysis of deductive inferences by which the deductive role of the different logical constants are separated. The inferences are broken down into atomic steps in such a way that each step involves only one logical constant. The steps are of two kinds: steps that allow the *introduction* of the logical constant (i.e., the conclusion of the inference has the constant as outermost symbol) and steps that allow the *elimination* of the logical constant (i.e., the premiss or one of the premisses of the inference has the constant as outermost symbol). The proofs start from *assumptions*, which at certain steps in the proof may be discharged or *closed*; typically, an assumption A is closed at an introduction of an implication $A \supset B$'.[9]

The logical proofs are formalized by derivations in tree form, and the inference rules formalize the atomic inferences mentioned above. The top formulas of the tree are the assumptions, and the other formulas of the tree are to follow from the one(s) immediately above by one of the inference rules. A formula A is said to depend on the assumptions standing above A if these assumptions have not been closed by some inference preceding A. The open assumptions of a derivation are the assumptions on which the formula at the bottom of the tree depends.[10]

Gentzen's inference rules of his systems of natural deduction are:

$$\&I\ \frac{A \quad B}{A\&B} \qquad \&E1\ \frac{A\&B}{A} \qquad \&E2\ \frac{A\&B}{B}$$

$$\vee I1\ \frac{A}{A \vee B} \qquad \vee I2\ \frac{B}{A \vee B} \qquad \vee E\ \frac{A \vee B \quad \begin{array}{c}[A]\\ C\end{array} \begin{array}{c}[B]\\ C\end{array}}{C}$$

$$\supset I\ \frac{\begin{array}{c}[A]\\ B\end{array}}{A \supset B} \qquad \supset E\ \frac{A \quad A \supset B}{B}$$

$$\forall I\ \frac{A(a)}{\forall x A(x)} \qquad \forall E\ \frac{\forall x A(x)}{A(t)}$$

$$\exists I\ \frac{A(t)}{\exists x A(x)} \qquad \exists E\ \frac{\exists x A(x) \quad \begin{array}{c}[A(a)]\\ B\end{array}}{B}$$

plus $\bot E$ $\frac{\bot}{A}$ for intuitionistic logic

or plus $\bot E$ $\frac{\begin{array}{c}[\sim A]\\ \bot\end{array}}{A}$ for classical logic.

Restrictions: in either rule $\bot E$, A is to be atomic and different from $\bot$.

1. Definition The major premiss of an elimination inference is the premiss whose outermost logical sign is eliminated by the inference. The other premisses of the inference are called minor premisses. The deduction of the major premiss is called major deduction and the deductions of the minor premisses are called minor deductions of the elimination inference.

The most interesting features of Gentzen's systems of natural deduction appear to be (1) the analysis of deductive inferences into atomic steps by which the deductive role of the various logical operators is separated, and (2) the discovery that these atomic steps are of two sorts, i.e. introductions and eliminations, standing in a certain symmetrical relation to each other. Gentzen's analysis may be taken as an attempt to characterize the notion of logical proof (for first order languages) in the sense that it is an attempt (a) to isolate the essential deductive operations, and (b) to break them down as far as possible (i.e. until they become atomic).[11]

Prawitz took a special interest in point (2). He observed that the proof of the conclusion of an elimination is already 'contained' in the proofs of the premisses when the major premiss is inferred by an introduction. He puts this observation into the formula that the pairs of corresponding introductions and eliminations satisfy the inversion principle. The inversion principle says in other terms that nothing new is obtained by an elimination immediately following an introduction (of the major premiss of the elimination) and hence such sequences of inferences can be dispensed with. *Note:* only the rules not expressing any special properties of $\perp$ are governed by the inversion principle.[12]

2. Definition Martin-Löf calls such pairs of corresponding introductions and eliminations cuts. Prawitz calls a formula occurrence in a derivation which stands at the same time as the conclusion of an introduction and as the major premiss of an elimination a maximum formula, and Martin-Löf calls it a cut formula. And Prawitz calls a sequence of formula occurrences $A_1, \ldots, A_n$ which is s.t. they all are of the same form, A_{i+1} stands immediately below A_i, A_1 is the conclusion of an introduction, and A_n the major premiss of an elimination ($A_i, i < n$, is then the minor premiss of a $\vee E$ or $\exists E$) a maximum segment in a derivation.[13] The five reductions corresponding to the five possible forms of a maximum or cut formula are called proper reductions, and the two reductions which make it possible to remove the two possible forms of maximal segments are called permutative reductions by Prawitz.[14]

For the proper reductions, cf. Chapt. 1, Sect. x, C, n & o (1) to x.3.3), C, n & o (3) to x.4.3), C, n & o (3) to x.5.3), C, n & o (1) to x.6.3), C, n & o (3) to x.7.3).

The permutative reductions are:

$\vee E$-reduction

$$\dfrac{\dfrac{\begin{matrix}\Sigma_1\\ A\vee B\end{matrix}\qquad \begin{matrix}\Sigma_2\\ C\end{matrix}\quad \begin{matrix}\Sigma_3\\ C\end{matrix}}{C}\qquad \Sigma_4}{D} \qquad \geq \qquad \dfrac{\begin{matrix}\Sigma_1\\ A\vee B\end{matrix}\qquad \dfrac{\begin{matrix}\Sigma_2\\ C\end{matrix}\quad \Sigma_4}{D}\qquad \dfrac{\begin{matrix}\Sigma_3\\ C\end{matrix}\quad \Sigma_4}{D}}{D}$$

$\exists E$-reduction

$$\dfrac{\dfrac{\begin{matrix}\Sigma_1\\ \exists x A(x)\end{matrix}\qquad \begin{matrix}\Sigma_2\\ C\end{matrix}}{C}\qquad \Sigma_4}{D} \qquad \geq \qquad \dfrac{\begin{matrix}\Sigma_1\\ \exists x A(x)\end{matrix}\qquad \dfrac{\begin{matrix}\Sigma_2\\ C\end{matrix}\qquad \Sigma_4}{D}}{D}$$

3. Definition A derivation Π is said to reduce immediately to a derivation Π' if Π' is obtained from Π by replacing a subtree of Π by a reduction of it. A derivation Π is said to reduce to a derivation Π' if there is a sequence $\Pi_1, \Pi_2, \ldots, \Pi_n, n \geq 1$, where $\Pi_1 = \Pi$, Π_i reduces immediately to Π_{i+1}, for $i < n$, and $\Pi_n = \Pi'$. And a derivation or deduction in any of the systems of natural deduction considered above is defined to be normal or in normal form (n.f.), if it contains no maximum or cut formula and no maximum segment. A derivation that reduces to n.f. is said to be normalizable.[15]

Prawitz then proves a

4. Normal form theorem If A is derivable in any of the systems of natural deduction considered above from assumptions Γ, then there is a normal derivation in the system in question of A from Γ.

5. Normalization theorem Every derivation in any of the systems of natural deduction considered above is normalizable (i.e., reduces to n.f.).[16]

Prawitz thus achieves for systems of natural deduction what Gentzen achieved for calculi of sequents, i.e. he shows, concerning the structure of first order logical proofs, that a proof formalized by a normal derivation is direct in the sense that 'it proceeds from assumptions to the conclusions by first only using the meaning of the assumptions by breaking them down to their components (the analytical part), and then only verifying the meaning of the conclusions by building them up from their components (the synthetical part)'.[17]

Note on the so-called 'Complete Discharge Convention' CDC. Gentzen in (1934) as well as e.g. Curry (1977) thought of natural deduction on the basis of the CDC which says: Assumptions are always discharged as soon as possible. In his discussion of normalization for a natural deduction system for the modal logic $S4$, Prawitz (1965) introduces a more liberal convention concerning the closure of open assumptions:

- When $\supset I$ is applied, the set $[A]^u$ of open assumptions of the form A becomes closed.
- When $\vee E$ is applied, the set $[A]^u$ of open assumptions of the form A and the set $[B]^v$ of open assumptions of the form B become closed.
- When $\exists E$ is applied, the set $[A]^u$ of open assumptions of the form A becomes closed.
- When $\bot E$ for classical logic is applied, the set $[\sim A]^u$ of open assumptions of the form $\sim A$ becomes closed.

All other assumptions, not covered by the cases just mentioned, stay open.[18]

ii.2 *The Curry–Howard Isomorphism*

The type structure of the subsystem $HA(\supset, \&, \forall, \exists)$ of Heyting Arithmetic obtained by omitting $\vee$ is defined as follows:

1. Definition of the type structure T_{HAH} (HA-Howard)

1) $0 \in T_{HAH}$ (0 is the type of natural numbers)
2) every equation of $HA(\supset, \&, \forall, \exists)$ denotes a type of T_{HAH}
3) if $\sigma, \tau \in T_{HAH}$ then $\sigma \supset \tau, \sigma\&\tau \in T_{HAH}$
4) if $\sigma \in T_{HAH}$ and x is a number variable, then $\forall x\sigma, \exists x\sigma \in T_{HAH}$.

2. Definition The system $HA(\supset, \&, \forall, \exists)$ **formulated by Howard as a sequent** calculus is defined as follows:

axioms

(a) $\sigma \rightarrow \sigma$

(b) $x = x$
(c) $x = y \supset t(x) = t(y)$, for $t(x) \in Tm_0^{HAH}$
(d) $x + 0 = x$
(e) $x + (y + 1) = (x + y) + 1$
(f) $x \cdot 0 = 0$
(g) $x \cdot (y + 1) = x \cdot y + x$
(h) $\tau(y) \equiv \sigma(0) \supset [\forall x(\sigma(x) \supset \sigma(x + 1)) \supset \sigma(y)]$

3. Definition The class of terms Tm^{HAH} of $HA(\supset, \&, \forall, \exists)$ is defined as follows:

prime terms

constants:	$0 \in Tm_0^{HAH}$
	$1 \in Tm_0^{HAH}$
	$+ \in Tm_{0\supset(0\supset 0)}^{HAH}$
	$\cdot \in Tm_{0\supset(0\supset 0)}$
number variables:	$x, y, \ldots \in Tm_0^{HAH}$
type variables:	$X^\sigma, Y^\sigma, \ldots \in Tm_\sigma^{HAH}$

(a) a term of type $x = x$
(b) a term of type $x = y \supset t(x) = t(y)$
(c) a term of type $x + 0 = x$
(d) a term of type $x + (y + 1) = (x + y) + 1$
(e) a term of type $x \cdot 0 = 0$
(f) a term of type $x \cdot (y + 1) = x \cdot y + x$
(g) a term $R^{\forall y \tau(y)}$ for each $\tau(y)$
(h) a term $R^{\tau(n)}$ for each numeral n

What is actually meant here is that each closed substitution instance of one of the types mentioned above is 'inhabited' by a closed term (in fact, one may simply take constant terms for all these arithmetical axioms).

$\sigma \supset (\tau \supset \sigma \& \tau)$	a term $B_1^{\sigma\supset(\tau\supset\sigma\&\tau)}$
$\sigma \& \tau \supset \sigma$	a term $B_2^{\sigma\&\tau\supset\sigma}$
$\sigma \& \tau \supset \tau$	a term $B_3^{\sigma\&\tau\supset\tau}$
Rules of inference	Rules of term formation

$$\frac{\Gamma, \sigma \to \tau}{\Gamma \to \sigma \supset \tau} \qquad \frac{t^\tau}{(\lambda X^\sigma . t^\tau)^{\sigma\supset\tau}}$$

$$\frac{\Gamma \to \sigma}{\Gamma \to \forall x \sigma} \qquad \frac{t^\sigma}{(\lambda x . t^\sigma)^{\forall x \sigma}}$$

$$\frac{\Gamma \to \sigma t}{\Gamma \to \exists x \sigma} \; t \in Tm_0^{HAH} \qquad \frac{t'^{\sigma t}}{\langle t, t' \rangle^{\exists x \sigma}}$$

$$\frac{\Gamma \to \sigma \quad \Delta \to \sigma \supset \tau}{\Gamma, \Delta \to \tau} \qquad \frac{s^\sigma \quad t^{\sigma\supset\tau}}{(ts)^\tau}$$

$$\frac{\Gamma \to \forall x \sigma x}{\Gamma \to \sigma t} \quad t \in Tm_0^{HAH} \qquad \frac{s^{\forall x\sigma} \quad t^0}{(st)^{\sigma t}}$$

$$\frac{\Gamma \to \exists x\sigma \quad \Delta, \sigma \to \tau}{\Gamma, \Delta \to \tau} \qquad \frac{t^{\exists x\sigma}}{p_2 t^{\sigma(p_1 t)}}$$

The axioms and prime terms on the same line correspond to each other and so do the rules of inference and the rules of term formation on the same line.

4. Definition A construction of a sequent $\Gamma \to \tau$ is defined to be a term t^τ of type τ s.t. for every free type variable X^σ occurring in t^τ there is a corresponding occurrence of σ in Γ (whereby it is understood that the existence of k distinct free variables of the same type σ in t^τ is reflected by at least k occurrences of σ in Γ).

After all these preliminary definitions Howard's theorem 5 is quite straightforward:[19]

5. Theorem Given any derivation of a sequent $\Gamma \to \tau$ in $HA(\supset, \&, \forall, \exists)$ one can find a construction of $\Gamma \to \tau$ and conversely.

Consider $HA(\supset, \&, \forall, \exists)$ to be a logical system. Then it can be interpreted in set theory as follows:

(i) set-theoretical interpretation $*$ of the closed formulas σ of $HA(\supset, \&, \forall, \exists)$
- σ is a true closed equation: $\sigma^* = \{1\}$
- σ is a false closed equation: $\sigma^* = \{0\}$
- $\sigma \equiv \tau \supset \varrho$: $\sigma^* = \varrho^{*^{\tau^*}}$, or the set of all functions whose domain is a superset of τ^* and whose range is a subset of ϱ^*
- $\sigma \equiv \tau \& \varrho$: $\sigma^* = \tau^* \times \varrho^*$
- $\sigma \equiv \forall x \tau(x)$: σ^* is the set of all functions f s.t. $f(n) \in \tau(n)^*$, provided $\tau(n)^*$ has been defined for every n, i.e. $\sigma^* = (\Pi x \in N)\tau(x)^*$ (in constructive type theory notation)
- $\sigma \equiv \exists x \tau(x)$: $\sigma^* = \{\langle n, Z\rangle \mid Z \in \tau(n)^*\}$, i.e. $\sigma^* = (\Sigma x \in N)\tau(x)^*$ (in constructive type theory notation)

(ii) set-theoretical interpretation $*$ of the terms t of $HA(\supset, \&, \forall, \exists)$
Note that when t is a closed term of $HA(\supset, \&, \forall, \exists)$, then the type symbol of t is a closed formula σ of $HA(\supset, \&, \forall, \exists)$ and $t^* \in \sigma^*$.
- $t \equiv s(x_1, \ldots, x_k)$ where $x_1, \ldots, x_k$ are all the free number variables of t; t^* is the mapping which sends each k-tuple $n_1, \ldots, n_k$ into $s(n_1, \ldots, n_k)^*$ provided $s(n_1, \ldots, n_k)^*$ has been defined for all numerals $n_1, \ldots, n_k$, i.e. $t^* \in (\Pi x_1 \in N) \ldots (\Pi x_k \in N)s(x_1, \ldots, x_k)^*$ or $t^* \in (\Pi z \in \underbrace{N \times \cdots \times N}_{k})s(z)^*$ (in constructive type theory notation)
- suppose t has no free number variables. If $t \equiv s(X^{\tau_1}, \ldots, X^{\tau_k})$ where $X^{\tau_1}, \ldots, X^{\tau_k}$ are all the free type variables of t; then t^* is a mapping from $\tau_1^* \times \cdots \times \tau_k^*$ into $s(X^{\tau_1}, \ldots, X^{\tau_k})^*$ where $\tau_1^*, \ldots, \tau_k^*$ have already been defined by (i), i.e. $t^* \in (\Pi Z \in \tau_1^* \times \cdots \times \tau_k^*)s(Z)^*$ (in constructive type theory notation)

If $t \equiv \lambda X^{\tau}. s$, then t^* is a mapping from τ^* into s^*, i.e. $t^* \in (\Pi X \in \tau^*)s(X)^*$ (in constructive type theory notation)

– $t \equiv R_n^{\tau(n)}$; then t^* is defined by prim. recursion on n.
  $t \equiv R^{\forall y \tau(y)}$; then t^* is the mapping which takes each n into $(R_n^{\tau(n)})^*$, i.e.
  $t^* \in (\Pi x \in N)(R_x^{\tau(x)})^*$ (in constructive type theory notation)

Thus, there is a bijection between the deductions in Howard's $HA(\supset, \&, \forall, \exists)$ and the corresponding terms in the suitable typed lambda calculus (as stated in Thm. 5). Since, as Howard also proves in his paper, there is moreover a perfect correspondence between the notions of reduction and normal form for the two resp. theories, there exists as a matter of fact an isomorphism between $HA(\supset, \&, \forall, \exists)$ and the corresponding typed lambda calculus. Howard's set-theoretical interpretation of the 'logical theory' $HA(\supset, \&, \forall, \exists)$ also shows: there exists a most interesting and precise correspondence, often called isomorphism, between the axioms and rules for the logical operators on the one hand and the set-theoretical operations on the other hand.

Note that, as pointed out by Howard, the logical operations $\sim$ and $\vee$ can be defined

$$\begin{array}{lll} \sim \sigma & \text{by} & \sigma \supset 0 = 1 \\ \sigma \vee \tau & \text{by} & \exists x((x = 0 \supset \sigma)\&(\sim x = 0 \supset \tau)). \end{array}$$

Note that the Curry–Howard isomorphism applies only to the notion of deduction with Prawitz' general assumption discharge mechanism, not to deductions based on the CDC. The isomorphism is called Curry–Howard isomorphism because Curry discovered in (1958) the bijection between the axioms of positive implicational propositional logic and the corresponding combinators of a suitable system of combinatory logic (called theory of functionality by Curry). For more on the history of the Curry–Howard isomorphism, cf. Troelstra and Schwichtenberg (1996), Note 2.5.6.

ii.3 *Martin-Löf's Normalization for Infinitary Propositional Logic*

Martin-Löf's first Hauptsatz-paper,[20] i.e. his normalization for infinitary propositional logic constitutes essentially an application of the Curry–Howard isomorphism to this logic.

At first a typed lambda calculus containing infinite terms will be defined.

1. Definition of the type structure T_∞

1) $0 \in T_\infty$ (0 is just an atomic type)
2) if $\sigma, \tau \in T_\infty$ then $(\sigma)\tau \in T_\infty$
3) if $\tau_0, \tau_1, \ldots, \tau_n, \ldots$ is a countable sequence of types of T_∞ then $\Pi\tau_n \in T_\infty$.

Next, the class of terms Tm^∞ of this calculus is defined by $Tm^\infty =_{\text{def}} \bigcup\{Tm_\sigma^\infty \mid \sigma \in T_\infty\}$.

2. Definition of the class of terms of type σ as the smallest class Tm_σ^∞, $\sigma \in T_\infty$, s.t.

1) $x^\sigma, y^\sigma, \ldots \in Tm_\sigma^\infty$
2) if x^σ is a variable, $t(x) \in Tm_\tau^\infty$ then $\lambda x t(x) \in Tm_{(\sigma)\tau}^\infty$

3) if $t_n \in Tm^{\infty}_{\tau_n}$, for $n = 0, 1, \ldots$ then $(t_0, t_1, \ldots, t_n, \ldots) \in Tm^{\infty}_{\Pi\tau_n}$
4) if $t \in Tm^{\infty}_{(\sigma)\tau}$, $t' \in Tm^{\infty}_{\sigma}$, then $tt' \in Tm^{\infty}_{\tau}$
5) if $t \in Tm^{\infty}_{\Pi\tau_n}$ and $n \in \omega$ then $tn \in Tm^{\infty}_{\tau_n}$.

Next, a relation of reduction which holds between objects of Tm^{∞} will be defined:

3. Definition of reduction for terms of the infinitary typed lambda calculus. The relation of reduction, denoted by $\succeq$, can be defined as a binary relation between objects of Tm^{∞}, inductively generated by the following closure conditions:

1) $x \succeq x$, x is a variable
2) $\lambda xt(x)t' \succeq t(t')$ (λ-contraction)
3) $(t_0, t_1, \ldots, t_n, \ldots)n \succeq t_n$ (projection)
4) if $t(x) \succeq t'(x)$ then $\lambda xt(x) \succeq \lambda xt'(x)$
5) if $t_n \succeq t'_n$ for $n = 0, 1, \ldots$ then $(t_0, t_1, \ldots, t_n, \ldots) \succeq (t'_0, t'_1, \ldots, t'_n, \ldots)$
6) if $t \succeq t'$ then $t''t \succeq t''t'$
7) if $t \succeq t'$ then $tt'' \succeq t't''$
8) if $t \succeq t'$ then $tn \succeq t'n$
9) if $t \succeq t'$ and $t' \succeq t''$ then $t \succeq t''$

4. Definition A term t of Tm^{∞} is said to be in n.f. or simply normal if for every $t' \in Tm^{\infty}$, if $t \succeq t'$ then $t = t'$. And a term $t \in Tm^{\infty}$ is said to be normalizable or to reduce to n.f. if there exists a normal term $t' \in Tm^{\infty}$ s.t. $t \succeq t'$.

The language of infinitary propositional logic consists of a fragment of the standard language of propositional logic, namely of at least one atomic formula A and of the logical operators $\supset$ and &, where & is infinitary, i.e. non-standard.

5. Definition of the class of formulas of infinitary propositional logic

1) A is a formula (atomic formula)
2) if A, B are formulas then $A \supset B$ is a formula
3) if $A_0, A_1, \ldots, A_n, \ldots$ is a countable sequence of formulas, then $\& A_n$ is a formula

6. Definition of the class of deductions of infinitary propositional logic as the smallest class $\mathcal{D}$ s.t.

1) assumption $B \in \mathcal{D}$, B is a formula
2) introductions

2.1) if $\begin{matrix}\Sigma \\ B \\ C\end{matrix} \in \mathcal{D}$ then $\dfrac{\begin{matrix}[B] \\ C\end{matrix}}{B \supset C} \in \mathcal{D}$

2.2) if $\begin{matrix}\Sigma_0 \\ A_0\end{matrix} \in \mathcal{D}, \begin{matrix}\Sigma_1 \\ A_1\end{matrix} \in \mathcal{D}, \ldots, \begin{matrix}\Sigma_n \\ A_n\end{matrix} \in \mathcal{D}, \ldots$ then

$$\dfrac{\begin{matrix}\Sigma_0 & \Sigma_1 & \ldots & \Sigma_n & \ldots \\ A_0 & A_1 & & A_n & \end{matrix}}{\& A_n} \in \mathcal{D}$$

3) eliminations

3.1) if $\genfrac{}{}{0pt}{}{\Sigma_1}{B} \in \mathcal{D}$ and $\genfrac{}{}{0pt}{}{\Sigma_2}{B \supset C} \in \mathcal{D}$ then $\dfrac{\genfrac{}{}{0pt}{}{\Sigma_1}{B} \qquad \genfrac{}{}{0pt}{}{\Sigma_2}{B \supset C}}{C} \in \mathcal{D}$

3.2) if $\genfrac{}{}{0pt}{}{\Sigma}{\& A_n} \in \mathcal{D}$ then $\dfrac{\genfrac{}{}{0pt}{}{\Sigma}{\& A_n}}{A_n} \in \mathcal{D}$

Next, a relation of reduction is going to be defined which holds between deductions of $\mathcal{D}$:

7. Definition of reduction for deductions of $\mathcal{D}$. The relation of reduction, denoted by $\succeq$, can be defined as a relation between objects of $\mathcal{D}$, inductively generated by the following closure conditions:

1) $\Sigma \succeq \Sigma, \qquad \Sigma \in \mathcal{D}$
2) $\supset$-reduction (cf. C, n & o (3) to x.5.3), Sect. x., Chapt. 1)
3) $$\dfrac{\dfrac{\genfrac{}{}{0pt}{}{\Sigma_0}{A_0} \quad \genfrac{}{}{0pt}{}{\Sigma_1}{A_1} \quad \ldots \quad \genfrac{}{}{0pt}{}{\Sigma_n}{A_n} \quad \ldots}{\& A_n}}{A_n} \succeq \genfrac{}{}{0pt}{}{\Sigma_n}{A_n}$$
4) if $\Sigma \succeq \Sigma'$ and $\Sigma' \succeq \Sigma''$ then $\Sigma \succeq \Sigma''$.

A normal deduction – Martin-Löf calls it a cut-free deduction – is defined as in Subsect. ii.1, Def. 3.

Inspired by the Curry–Howard isomorphism for $HA(\supset, \&, \forall, \exists)$, Martin-Löf developed the infinitary typed lambda calculus and the infinitary propositional logic in such a way as to obtain the following isomorphism:

Infinitary propositional logic	*Infinitary typed lambda calculus*
atomic formula	atomic type
formula	type
assumption	variable
deduction rule	rule of term formation
$\supset$-reduction	λ-contraction
&-reduction	projection
cut-free deduction	normal term

He then proves on the basis of this isomorphism a

8. Normalization theorem for infinitary propositional logic All deductions of infinitary propositional logic are normalizable (i.e. reduce to n.f.).

By proving:

All terms of the infinitary typed lambda calculus are normalizable (i.e. more precisely, a term of length α and cut degree β reduces to a term in n.f. of length $\leq \varphi_\beta(\alpha)$).

ii.4 *Tait's Normalization for Intuitionistic Arithmetic in All Finite Types by Means of the Notion of Computability*

With his notion of computability, Tait introduced an important new instrument which was used in many subsequent normalization proofs.[21]

Following Troelstra, the basic system of intuitionistic arithmetic in all finite types will be denoted by $N-HA^{\omega}$ ('N' for 'neutral'). The finite type structure is inductively defined as follows:

1. Definition of the finite type structure T

1) $0 \in T$
2) if $\sigma, \tau \in T$ then $(\sigma)\tau \in T$.

Next, the class of terms Tm of $N - HA^{\omega}$ is defined by $Tm =_{\text{def}} \cup\{Tm_{\sigma} \mid \sigma \in T\}$, where Tm_{σ} is the class of terms of type σ inductively defined as follows:

2. Definition of the class of terms of type σ as the smallest class $Tm_{\sigma}, \sigma \in T$, s.t.

1) $x^{\sigma}, y^{\sigma}, z^{\sigma}, \ldots \in Tm_{\sigma}$
2) $0 \in Tm_0$, $S \in Tm_{(0)0}$, $\Pi_{\varrho,\sigma} \in Tm_{(\varrho)(\sigma)\varrho}$, $\Sigma_{\varrho,\sigma,\tau} \in Tm_{((\varrho)(\sigma)\tau)((\varrho)\sigma)(\varrho)\tau}$, $R_{\sigma} \in Tm_{(\sigma)((\sigma)(0)\sigma)(0)\sigma}$.
3) if $t \in Tm_{(\sigma)\tau}$, $t' \in Tm_{\sigma}$ then $(tt') \in Tm_{\tau}$.

Next, a relation of reduction will be defined which holds between objects of Tm:

3. Definition of reduction for terms of $N-HA^{\omega}$.The relation of reduction, denoted by $\succeq$, can be defined as a relation between objects of Tm, inductively generated by the following closure conditions:

1) $t \succeq t$
2) $\Pi tt' \succeq t$
3) $\Sigma tt't'' \succeq tt''(t't'')$
4) $Rtt'0 \succeq t$
5) $Rtt'(St'') \succeq t'(Rtt't'')t''$
6) if $t \succeq t'$ then $t''t \succeq t''t'$
7) if $t \succeq t'$ then $tt'' \succeq t't''$
8) if $t \succeq t'$ and $t' \succeq t''$ then $t \succeq t''$.

'Normal term' and 'normalizable term' are defined as in Subsect. ii.1, Def. 3.
Finally, the computability predicate *Comp* is defined by $Comp =_{\text{def}} \cup\{Comp_{\sigma} \mid \sigma \in T\}$, where $Comp_{\sigma}$ is defined by induction over T as follows:

4. Definition of the computability predicate $Comp_{\sigma}, \sigma \in T$.

1) $Comp_0(t) =_{\text{def}} t \in 0$ and t is normalizable
2) $Comp_{(\sigma)\tau}(t) =_{\text{def}}$ for all t', if $Comp_{\sigma}(t')$ then $Comp_{\tau}(tt')$, and t is normalizable.

5. Normalization theorem for $N - HA^{\omega}$ All terms t of $N - HA^{\omega}$ satisfy *Comp* and hence are normalizable.

Note that Tait's notion of reduction is not the one defined above, but rather something called "strict reduction" by Troelstra. Moreover, Tait proves normalization for closed terms of $N - HA^{\omega}$ only, but the proof extends to open terms as well — that is why the theorem is formulated the way it is formulated above.[22]

6. Appendix: Gödel's theory T What sparked off Tait's invention of the computability method and indirectly Martin-Löf's normalization for infinitary propositional logic was Gödel's theory T in (1958). In fact, T led to a whole series of extensions of Gödel's main result as well as to a series of investigations of extensions of Gödel's T, both relevant in this context.
Gödel's theory T – denoted by '$qf - I - HA^{\omega}$' by Troelstra – can be described as follows:[23]

1) axioms and rules for (many-sorted) intuitionistic propositional logic
2) axioms for equality
   $x^{\sigma} = x^{\sigma}$
   $x^{\sigma} = y^{\sigma} \;\&\; z^{\sigma} = y^{\sigma} \supset x^{\sigma} = z^{\sigma}$
   $x^{\sigma} = y^{\sigma} \supset z^{(\sigma)\tau}x^{\sigma} = z^{(\sigma)\tau}y^{\sigma}$
   $x^{(\sigma)\tau} = y^{(\sigma)\tau} \supset x^{(\sigma)\tau}z^{\sigma} = y^{(\sigma)\tau}z^{\sigma}$
   equality axioms for successor
   $x^{0} \neq 0$
   $x^{0} = y^{0} \supset\subset Sx^{0} = Sy^{0}$
   where $t^{\sigma} = s^{\sigma}$ is to be conceived as an abbreviation for $t^{\sigma}x_1 \ldots x_n = s^{\sigma}x_1 \ldots x_n$, $x_1, \ldots, x_n$ variables not occuring in t^{σ}, s^{σ} or (open) assumptions of the deduction s.t. $t^{\sigma}x_1 \ldots x_n \in 0$.
3) induction rule
   $$\frac{A(0) \quad A(x^0) \supset A(Sx^0)}{A(x^0)}$$
   for A qf (i.e. quantifier-free), x^0 not occurring free in assumptions of the deduction.
4) $$\frac{A(x^{\sigma})}{A(t^{\sigma})}$$
   if x^{σ} does not occur in (open) assumptions of the deduction.
5) I- (i.e. intensional) equality axioms
   $E_{\sigma}x^{\sigma}y^{\sigma} = 0 \supset\subset x^{\sigma} = y^{\sigma}$
   $E_{\sigma}x^{\sigma}y^{\sigma} = 0 \vee E_{\sigma}x^{\sigma}y^{\sigma} = 1$
   where $E_{\sigma} \in (\sigma)(\sigma)0$, $\sigma \in T$

ii.5 *Martin-Löf's Normalization for the Intuitionistic Theory of Iterated Inductive Definitions*

Martin-Löf's second Hauptsatz paper,[24] i.e. his normalization for the intuitionistic theory of iterated inductive definitions, makes essential use of Tait's notion of computability and it also relies on the Curry–Howard isomorphism.

The language of the intuitionistic theory of iterated inductive definitions is the standard language of first order logic with an arbitrary finite number of function symbols and with predicate symbols with each of which is associated a place index and a non-negative integer called its level and denoted by '$lev(P_k)$' for 'P_k' a predicate symbol with index k. (In the sequel, the place index will often be skipped.) If $P_i, \ldots, Q_j$ are the predicate symbols occurring in a formula A of the language of the theory, then the level of the formula A is defined by

$$lev(A) = \max\big(lev(P_i), \ldots, lev(Q_j)\big).$$

The rules of inference of the intuitionistic theory of iterated inductive definitions are Gentzen's inference rules for his system of natural deduction: $\&I$, $\vee I1$, $\vee I2$, $\supset I$, $\forall I$, $\exists I$, $\&E1$, $\&E2$, $\vee E$, $\supset E$, $\forall E$, $\exists E$ as well as the three rules of the form

$$PI1 \quad \frac{Q\ \overrightarrow{q(x)} \ldots R\ \overrightarrow{r(x)}}{P\ \overrightarrow{p(x)}} \quad \text{called ordinary production}$$

with 0 or more atomic formulas as premisses, where $\vec{x}$ denotes the totality of all variables that occur in the production, and with $lev(P) \geq \max(lev(Q), \ldots, lev(R))$;

$$PI2 \quad \frac{C(\vec{x}) \supset Q\ \overrightarrow{q(x)}}{P\ \overrightarrow{p(x)}} \quad \text{called } \supset\text{-production}$$

where $lev(C(\vec{x})) + 1 \leq lev(P) \geq lev(Q)$;

$$PI3 \quad \frac{\forall y Q\ \overrightarrow{q(x, y)}}{P\ \overrightarrow{p(x)}} \quad \text{called } \forall\text{-production}$$

where $0 < lev(P) \geq lev(Q)$, and the variable y must not occur neither in t nor free in any assumption of the deduction of $Q\ \overrightarrow{q(t, y)}$. The productions $PI2$ and $PI3$ are also called generalized productions. The level of a production PIi, $i = 1, 2, 3$, is defined by $lev(PIi) = lev(P)$, where P is the predicate symbol occuring in the conclusion of PIi.

$$PE \quad \frac{P\vec{t} \qquad \text{minor deductions}}{A(\vec{t})}$$

To understand the rule PE one has to understand what it means for a predicate symbol to be linked with a predicate symbol. This is defined as follows:

1. Definition

i) Every predicate symbol is linked with itself.
ii) If P occurs in the conclusion of $PI1$ then P is linked with every predicate symbol which is linked with one of $Q, \ldots, R$.
iii) If P occurs in the conclusion of $PI2$ or $PI3$, then P is linked with every predicate symbol which is linked with Q.

An instance of PE for an inductively defined predicate P is obtained as follows: associate with every predicate symbol which is linked with P an abstraction term, i.e. a formula and as many variables as indicated by the place index of the resp. predicate symbol

$$P\vec{x} \qquad Q\vec{y} \qquad R\vec{z} \qquad \ldots$$

$$\lambda\vec{x}A(\vec{x}) \quad \lambda\vec{y}B(\vec{y}) \quad \lambda\vec{z}C(\vec{z}) \quad \ldots$$

For every $PI1$ which in its conclusion has a predicate symbol which is linked with P, there should among the minor deductions of PE be one of the form

$$\begin{array}{ccc} B(\overrightarrow{q(x)}) & \ldots & C(\overrightarrow{r(x)}) \\ \ddots & & \vdots \\ & A(\overrightarrow{p(x)}) & \end{array}$$

satisfying the restriction that a variable in the sequence $\vec{x}$, whose free occurrences in the minor deduction become bound by the elimination inference, must not occur free in any assumption other than $B(\overrightarrow{q(x)}), \ldots, C(\overrightarrow{r(x)})$ (which are all closed at the inference under consideration).

For every $PI2$ or $PI3$ which in their conclusion have a predicate symbol which is linked with P, there should among the minor deductions of PE be one of the form

$$\begin{array}{ccc} C(\vec{x}) \supset B(\overrightarrow{q(x)}) & \text{or} & \forall\vec{y}B(\overrightarrow{q(x,y)}) \quad \text{resp.} \\ \vdots & & \vdots \\ A(\overrightarrow{p(x)}) & & A(\overrightarrow{p(x)}) \end{array}$$

satisfying the restriction that a variable in the sequence $\vec{x}$, whose free occurrences in the minor deduction become bound by the elimination inference, must not occur free in any assumption other than $C(\vec{x}) \supset B(\overrightarrow{q(x)})$ and $\forall\vec{y}B(\overrightarrow{q(x,y)})$ resp. (which are closed at the inference under consideration).

Example. 1 $\bot E$, i.e. the intuitionistic rule of absurdity can be introduced as follows: Let $\bot$ be an 0-ary predicate symbol s.t. $lev(\bot) = 0$ which does not occur in the conclusion of any production. Thus, there is no introduction rule for $\bot$. The elimination rule PE described above takes the form $\bot E \ \dfrac{\bot}{A}$.

Example. 2 The class of inductive trees $\mathcal{JT}$ defined as follows:

(i) $0 \in \mathcal{JT}$ (code singleton tree)
(ii) $\forall x(e \cdot x \in \mathcal{JT}) \supset j(1, e) \in \mathcal{JT}$, where $j(x, y) =_{\text{def}} 2^x \cdot (2y + 1) \dot{-} 1$, and $e \cdot x$ is short for $\{e\}(x)$.

$\mathcal{JT}$ is then defined by the following productions:

(i) $\dfrac{}{\mathcal{JT}(0)}$ (ordinary production)

(ii) $\dfrac{\forall x(\mathcal{JT}(e \cdot x))}{\mathcal{JT}(j(1,e))}$ ($\forall$-production).

Example. 3 The recursive second number class

(a) $\dfrac{}{<_{\mathcal{O}} (0, j(1,0))}$ (b) $\dfrac{<_{\mathcal{O}} (t,s)}{<_{\mathcal{O}} (t, j(1,s))}$ (c) $\dfrac{<_{\mathcal{O}} (t,s)}{<_{\mathcal{O}} (s, j(1,s))}$

(d) $\dfrac{<_{\mathcal{O}} (t,t') \quad <_{\mathcal{O}} (t',t'')}{<_{\mathcal{O}} (t,t'')}$ (e) $\dfrac{\forall z\big(<_{\mathcal{O}} (s \cdot z, s \cdot (z+1))\big)}{<_{\mathcal{O}} (s \cdot t, j(2,s))}$

where j and $s \cdot t$ are defined as in Ex. 2, and $\mathcal{O} =_{\text{def}} \{x \mid x = 0 \vee 0 <_{\mathcal{O}} x\}$.

(a)–(d) are ordinary productions, (e) is a $\forall$-production. The minimum level of $<_{\mathcal{O}}$ is 0, i.e. $lev(<_{\mathcal{O}}) \geq 0$.

Example. 4 The recursive third number class
(a)–(e) as in Ex. 3 plus

(A) $\dfrac{}{<_{\mathcal{O}'} (0, j(1,0))}$ (B) $\dfrac{<_{\mathcal{O}'} (t,s)}{<_{\mathcal{O}'} (t, j(1,s))}$ (C) $\dfrac{<_{\mathcal{O}'} (t,s)}{<_{\mathcal{O}'} (s, j(1,s))}$

(D) $\dfrac{<_{\mathcal{O}'} (t,s) \quad <_{\mathcal{O}'} (s,r)}{<_{\mathcal{O}'} (t,r)}$ (E) $\dfrac{\forall z\big(<_{\mathcal{O}'} (t \cdot z, t \cdot (z+1))\big)}{<_{\mathcal{O}'} (t \cdot s, j(2,t))}$

(F) $\dfrac{<_{\mathcal{O}} (t,t') \supset <_{\mathcal{O}'} (s \cdot t, s \cdot t')}{Q(t,t',s)}$ (F′) $\dfrac{\forall x \forall y Q(x,y,s)}{Q'(s)}$

(F″) $\dfrac{<_{\mathcal{O}} (t,t') \quad Q'(s)}{<_{\mathcal{O}'} (t', j(3,s))}$

where Q, Q' are predicate letters, j and $s \cdot t$ are defined as in Ex. 2, and $\mathcal{O}' =_{\text{def}} \{x \mid x = 0 \vee 0 <_{\mathcal{O}'} x\}$.
(A)-(D) plus (F″) are ordinary productions, (E) and (F′) are $\forall$-productions, and (F) is a $\supset$-production. The minimum level of $<_{\mathcal{O}'}$, Q and Q' is 1, i.e. $lev(<_{\mathcal{O}'})$, $lev(Q)$, $lev(Q') \geq 1$.[25]

2. Definition of the class of deductions of the intuitionistic theory of iterated inductive definitions as the smallest class $\mathcal{D}$, s.t.

1) assumption $B \in \mathcal{D}$, B is a formula
2) introductions

2.1) if $\begin{matrix}\Sigma_1\\ A\end{matrix} \in \mathcal{D}$, $\begin{matrix}\Sigma_2\\ B\end{matrix} \in \mathcal{D}$ then $\dfrac{\begin{matrix}\Sigma_1 & \Sigma_2\\ A & B\end{matrix}}{A \mathbin{\&} B} \in \mathcal{D}$

2.2) if $\begin{matrix}\Sigma\\ A\end{matrix} \in \mathcal{D}$ then $\dfrac{\begin{matrix}\Sigma\\ A\end{matrix}}{A \vee B} \in \mathcal{D}$; if $\begin{matrix}\Sigma\\ B\end{matrix} \in \mathcal{D}$ then $\dfrac{\begin{matrix}\Sigma\\ B\end{matrix}}{A \vee B} \in \mathcal{D}$

2.3) if $\begin{array}{c}\Sigma\\ A\\ B\end{array} \in \mathcal{D}$ then $\dfrac{\begin{array}{c}[A]\\ B\end{array}}{A \supset B} \in \mathcal{D}$

2.4) if $\begin{array}{c}\Sigma\\ A(a)\end{array} \in \mathcal{D}$ then $\dfrac{\begin{array}{c}\Sigma\\ A(a)\end{array}}{\forall x A(x)} \in \mathcal{D}$

2.5) if $\begin{array}{c}\Sigma\\ A(t)\end{array} \in \mathcal{D}$ then $\dfrac{\begin{array}{c}\Sigma\\ A(t)\end{array}}{\exists x A(x)} \in \mathcal{D}$

2.6) if $\begin{array}{c}\Sigma\\ Q\ \overrightarrow{q(t)}\end{array} \in \mathcal{D}, \ldots, \begin{array}{c}\Pi\\ R\ \overrightarrow{r(t)}\end{array} \in \mathcal{D}$ then $\dfrac{\begin{array}{c}\Sigma\\ Q\ \overrightarrow{q(t)}\end{array} \quad \ldots \quad \begin{array}{c}\Pi\\ R\ \overrightarrow{r(t)}\end{array}}{P\ \overrightarrow{p(t)}} \in \mathcal{D}$

2.7) if $\begin{array}{c}\Sigma\\ C(\vec{t})\\ Q\ \overrightarrow{q(t)}\end{array} \in \mathcal{D}$ then $\dfrac{\begin{array}{c}[C(\vec{t})]\\ Q\ \overrightarrow{q(t)}\end{array}}{P\ \overrightarrow{p(t)}} \in \mathcal{D}$

2.8) if $\begin{array}{c}\Sigma\\ Q\ \overrightarrow{q(t,a)}\end{array} \in \mathcal{D}$ then $\dfrac{\begin{array}{c}\Sigma\\ Q\ \overrightarrow{q(t,a)}\end{array}}{P\ \overrightarrow{p(t)}} \in \mathcal{D}$

3) eliminations

3.1) if $\begin{array}{c}\Sigma\\ A\&B\end{array} \in \mathcal{D}$ then $\dfrac{\begin{array}{c}\Sigma\\ A\&B\end{array}}{A}, \quad \dfrac{\begin{array}{c}\Sigma\\ A\&B\end{array}}{B} \in \mathcal{D}$

3.2) if $\begin{array}{c}\Sigma_1\\ A \vee B\end{array} \in \mathcal{D}, \begin{array}{c}\Sigma_2\\ A\\ C\end{array} \in \mathcal{D}, \begin{array}{c}\Sigma_3\\ B\\ C\end{array} \in \mathcal{D}$, then $\dfrac{\begin{array}{c}\Sigma_1\\ A \vee B\end{array} \quad \begin{array}{c}[A]\\ C\end{array} \quad \begin{array}{c}[B]\\ C\end{array}}{C} \in \mathcal{D}$

3.3) if $\begin{array}{c}\Sigma_1\\ A\end{array} \in \mathcal{D}, \begin{array}{c}\Sigma_2\\ A \supset B\end{array} \in \mathcal{D}$, then $\dfrac{\begin{array}{c}\Sigma_1\\ A\end{array} \quad \begin{array}{c}\Sigma_2\\ A \supset B\end{array}}{B} \in \mathcal{D}$

3.4) if $\begin{array}{c}\Sigma\\ \forall x A(x)\end{array} \in \mathcal{D}$ then $\dfrac{\begin{array}{c}\Sigma\\ \forall x A(x)\end{array}}{A(t)} \in \mathcal{D}$

3.5) if $\begin{array}{c}\Sigma_1\\ \exists x A(x)\end{array} \in \mathcal{D}, \begin{array}{c}\Sigma_2\\ A(a)\\ B\end{array} \in \mathcal{D}$ then $\dfrac{\begin{array}{c}\Sigma_1\\ \exists x A(x)\end{array} \quad \begin{array}{c}[A(a)]\\ B\end{array}}{B} \in \mathcal{D}$

3.6) if $\begin{array}{c}\Sigma\\ P\vec{t}\end{array} \in \mathcal{D}$ then $\dfrac{\begin{array}{cc}\Sigma & \text{minor}\\ P\vec{t} & \text{deductions}\end{array}}{A(\vec{t})} \in \mathcal{D}$

Next, a relation of reduction is going to be defined which holds between deductions of $\mathcal{D}$:

3. Definition of reduction for deductions of $\mathcal{D}$

The relation of reduction, denoted by $\succeq$, can be defined as a binary relation between objects of $\mathcal{D}$, inductively generated by the following closure conditions:

1) $\Sigma \succeq \Sigma, \quad \Sigma \in \mathcal{D}$
2) &-reduction
3) $\vee$-reduction
4) $\supset$-reduction
5) $\forall$-reduction
6) $\exists$-reduction
   Cf. Chapt. 1, C, n & o (3) to x.7.3), C, n & o (1) to x.3.3), C, n & o (3) to x.5.3), C, n & o (3) to x.4.3) and C, n & o (1) to x.6.3).
7) $P1$-reduction (ordinary production-reduction)

$$\dfrac{\dfrac{\begin{array}{ccc}\Sigma & & \Pi\\ Q\,\overrightarrow{q(t)} & \dots & R\,\overrightarrow{r(t)}\end{array}}{P\,\overrightarrow{p(t)}} \qquad \begin{array}{c}\text{minor}\\ \text{deductions}\end{array}}{A(\overrightarrow{p(t)})} \quad \succeq$$

$$\begin{array}{ccc} \dfrac{\begin{array}{cc}\Sigma & \text{minor}\\ Q\,\overrightarrow{q(t)} & \text{deductions}\end{array}}{B(\overrightarrow{q(t)})} & \dots & \dfrac{\begin{array}{cc}\Pi & \text{minor}\\ R\,\overrightarrow{r(t)} & \text{deductions}\end{array}}{C(\overrightarrow{r(t)})}\\ & \ddots \qquad \cdot^{\cdot^{\cdot}} & \\ & A(\overrightarrow{p(t)}) & \end{array}$$

8) $P2$-reduction ($\supset$-production-reduction)

$$\dfrac{\dfrac{\begin{array}{c}[C(\vec{t})]\\ Q\,\overrightarrow{q(t)}\end{array}}{P\,\overrightarrow{p(t)}} \qquad \begin{array}{c}\text{minor}\\ \text{deductions}\end{array}}{A(\overrightarrow{p(t)})} \quad \succeq \quad \begin{array}{c}\dfrac{\dfrac{\begin{array}{cc}[C(\vec{t})] & \text{minor}\\ Q\,\overrightarrow{q(t)} & \text{deductions}\end{array}}{B(\overrightarrow{q(t)})}}{C(\vec{t}) \supset B(\overrightarrow{q(t)})}\\ \vdots\\ A(\overrightarrow{p(t)})\end{array}$$

9) $P3$-reduction ($\forall$-production-reduction)

$$\dfrac{\dfrac{\begin{matrix}\Sigma\\ Q\,\vec{q}(t,a)\end{matrix}}{P\,\vec{p}(t)} \qquad \begin{matrix}\text{minor}\\ \text{deductions}\end{matrix}}{A(\vec{p}(t))} \quad \succeq \quad \begin{matrix} \dfrac{\dfrac{\begin{matrix}\Sigma\\ Q\,\vec{q}(t,a)\end{matrix} \qquad \begin{matrix}\text{minor}\\ \text{deductions}\end{matrix}}{B(\vec{q}(t,a))}}{\forall\vec{y}B(\vec{q}(t,y))}\\ \vdots \\ A(\vec{p}(t))\end{matrix}$$

10) if $\Sigma \succeq \Sigma'$ and $\Sigma' \succeq \Sigma''$ then $\Sigma \succeq \Sigma''$.

4. Definition Note that each elimination inference has exactly one major premiss. Thus, if a deduction Σ ends with an elimination inference, it is possible to proceed in a unique way upwards in the deduction from the end formula, by always choosing the major premiss of an elimination inference. Two cases can occur: either one hits upon a cut, which will then be called the main cut, or one reaches a top formula of Σ, in which case the branch one proceeded up along will be called the main branch. If Σ ends with an elimination inference E, this will be written $\Sigma(E)$; and analogously for $\Sigma(I)$. If $\Sigma(E)$ has a main cut, this will be written $\Sigma(E)^{mc}$ o.w. $\Sigma(E)^{mcf}$, i.e. $\Sigma(E)$ is main cut free. If a deduction $\Sigma(E)^{mc}$ has a main cut, the result of eliminating this main cut will be denoted by $\Sigma(E)^*$.

A normal (or cut free) deduction is defined as in Subsect. ii.1, Def. 3. But the term 'normalizable' has a different definition:

5. Definition A deduction Σ of $\mathcal{D}$ is said to be normalizable if

1) $\Sigma \equiv A$, $\quad A$ is formula
2) $\Sigma(I)$ is normalizable if the deductions of the premisses of I are all normalizable
3.1) $\Sigma(E)^{mc}$ is normalizable if $\Sigma(E)^*$ is normalizable
3.2) $\Sigma(E)^{mcf}$ is normalizable if the minor deductions of the eliminations on the main branch are all normalizable.

It is readily verified that if a deduction Σ of $\mathcal{D}$ is normalizable then it reduces to n.f. Finally, the computability predicate *Comp* is defined by induction on $\omega n + m$ where m is the number of logical signs and n the level of the endformula of the deduction:

6. Definition of the computability predicate *Comp*

1) *Comp(A)* provided $\Sigma \equiv A$

2.1) $Comp\,(\Sigma(\&I)) =_{\text{def}} Comp\left(\begin{matrix}\Sigma_1\\ A\end{matrix}\right)$ and $Comp\left(\begin{matrix}\Sigma_2\\ B\end{matrix}\right)$

2.2) $Comp\,(\Sigma(\vee I1)) =_{\text{def}} Comp\left(\begin{matrix}\Sigma_1\\ A\end{matrix}\right) \qquad Comp\big(\Sigma(\vee I2)\big) =_{\text{def}} Comp\left(\begin{matrix}\Sigma_2\\ B\end{matrix}\right)$

2.3) $Comp\,(\Sigma(\supset I)) =_{\text{def}} \underset{\sim}{\forall}\;\begin{matrix}\Sigma'\\ A\end{matrix}\left(\text{if } Comp\left(\begin{matrix}\Sigma'\\ A\end{matrix}\right) \text{ then } Comp\left(\begin{matrix}\Sigma'\\ A\\ B\end{matrix}\right)\right)$

2.4) $Comp\,(\Sigma(\forall I)) =_{\text{def}} \underset{\sim}{\forall t}\; Comp \begin{pmatrix} \Sigma'(t) \\ A(t) \end{pmatrix}$

2.5) $Comp\,(\Sigma(\exists I)) =_{\text{def}} Comp \begin{pmatrix} \Sigma' \\ A(t) \end{pmatrix}$

2.6) $Comp\,(\Sigma(PI1)) =_{\text{def}} Comp \begin{pmatrix} \Sigma' \\ \overrightarrow{Q\,q(t)} \end{pmatrix}, \ldots, Comp \begin{pmatrix} \Pi' \\ \overrightarrow{R\,r(t)} \end{pmatrix}$

2.7) $Comp\,(\Sigma(PI2)) =_{\text{def}} \underset{\sim}{\forall} \begin{matrix} \Sigma' \\ C(\vec{t}) \end{matrix} \left(\text{if } Comp \begin{pmatrix} \Sigma' \\ C(\vec{t}) \end{pmatrix} \text{ then } Comp \begin{pmatrix} \Sigma' \\ C(\vec{t}) \\ \overrightarrow{Q\,q(t)} \end{pmatrix} \right)$

2.8) $Comp\,(\Sigma(PI3)) =_{\text{def}} \underset{\sim}{\forall \vec{u}}\; Comp \begin{pmatrix} \Sigma'(\vec{u}) \\ \overrightarrow{Q\,q(t,u)} \end{pmatrix}$

3.1) $Comp\,(\Sigma(E)^{mc}) =_{\text{def}} Comp\big(\Sigma(E)^{*}\big)$

3.2) $Comp\,\big(\Sigma(E)^{mcf}\big) =_{\text{def}}$ the minor deductions of the eliminations on the main branch are all computable.

For all Σ of $\mathcal{D}$, if $Comp(\Sigma)$ then Σ is normalizable. This is seen by comparing clause for clause the Def. 5 of normalizability with the Def. 6 of *Comp*, and by considering that for a deduction Σ s.t. $\Sigma = A$, $Comp(\Sigma)$.
One then proves the

7. Normalization theorem for the intuitionistic theory of iterated inductive definitions. All deductions Σ of the intuitionistic theory of iterated inductive definitions satisfy *Comp* and hence are normalizable.

ii.6 *Girard's Normalization for his System F by Means of the Notion of Parametric Computability*

Girard extended Tait's notion of computability to parametric computability (also called reducibility predicate) and thus introduced a fruitful new technique, which was again applied in various later normalization proofs.[26]

How does one arrive at the idea of parametric computability? Well, straight computability does not work, since it is by no means clear, what computability has to be assigned to a type (set) variable, when something rather complex may be substituted for it later on. The notion of computability has to reflect this change of situation. Thus, one also introduces 'variable computability predicates', also called 'computability or reducibility candidates'. These are not completely arbitrary subsets of terms, but have to satisfy a few closure conditions (cf. Def. 4 below). The interesting point is that this device allows one to show that every computability predicate relative to an assignment of variable computability predicates to type variables (Def. 5 below) satisfies those closure conditions (cf. Lemma 6 below). This device also permits one to show that computability predicates can be defined impredicatively (use of strong comprehension on the metalevel).

F's variable type structure is inductively defined as follows:

1. Definition of the variable type structure T_F

1) $0 \in T_F$
2) $\kappa, \kappa', \kappa'', \ldots$ belong to T_F (they are called type variables)
3) if $\sigma, \tau \in T_F$ then $(\sigma)\tau, \sigma \times \tau \in T_F$
4) if $\sigma \in T_F$ then $\forall\kappa\sigma, \exists\kappa\sigma \in T_F$

Next, the class of terms Tm^F of F is defined by $Tm^F =_{\text{def}} \cup\{Tm^F_\sigma \mid \sigma \in T_F\}$, where Tm^F_σ is the class of terms of variable type σ inductively defined as follows:

2. Definition of the class of terms of variable type σ as the smallest class Tm^F_σ, $\sigma \in T_F$, s.t.

1) functional constants: $0_\sigma \in Tm^F_\sigma$, $S \in Tm^F_{(0)0}$, $\lambda x^\sigma.t \in Tm^F_{(\sigma)\tau}$ for $t \in \tau$, $R_\sigma \in Tm^F_{(\sigma)((\sigma)(0)\sigma)(0)\sigma}$, $D_{\sigma,\tau} \in Tm^F_{(\sigma)(\tau)\sigma\times\tau}$, $D'_{\sigma,\tau} \in Tm^F_{(\sigma\times\tau)\sigma}$, $D''_{\sigma,\tau} \in Tm^F_{(\sigma\times\tau)\tau}$, two injection functionals $I_{\forall\kappa.\sigma,\tau} \in Tm^F_{(\forall\kappa.\sigma[\kappa])(\sigma[\tau])}$, $I'_{\exists\kappa.\sigma,\tau} \in Tm^F_{(\sigma[\tau])(\exists\kappa.\sigma[\kappa])}$, and finally two operators DT, ST which are not functionals but schemata to introduce functionals s.t. $DT\kappa t \in Tm^F_{\forall\kappa.\sigma}$ for $t \in \sigma$, $ST\kappa t \in Tm^F_{(\exists\kappa\sigma[\kappa])\tau}$ for $t \in (\sigma[\kappa])\tau$
2) variables: $x^\sigma, y^\sigma, z^\sigma, \ldots$
3) if $t \in Tm^F_{(\sigma)\tau}$, $t' \in Tm^F_\sigma$ then $(tt') \in Tm^F_\tau$.

Next, a relation of reduction will be defined which holds between objects of Tm^F:

3. Definition of reduction for terms of F

The relation of reduction, denoted by $\succeq$, can be defined as a relation between objects of Tm^F, inductively generated by the following closure conditions:

1) $t \succeq t, \quad t \in Tm^F$
2) $0_{(\sigma)\tau}(t) \succeq 0_\tau$
3) $D'0_{\sigma\times\tau} \succeq 0_\sigma, \quad D''0_{\sigma\times\tau} \succeq 0_\tau$
4) $I_{\forall\kappa\sigma[\kappa],\tau}0_{\forall\kappa.\sigma[\kappa]} \succeq 0_{\sigma[\tau]}$
5) $ST\kappa t(0_{\exists\kappa.\sigma[\kappa]}) \succeq t(0_{\sigma[\kappa]}) \quad (t \in (\sigma[\kappa]))$
6) $R_\sigma tt'0_0 \succeq t$
7) $R_\sigma tt'(St'') \succeq t'(R_\sigma tt't'')t''$
8) $(\lambda x^\sigma.t[x^\sigma])t'^\sigma \succeq tt' \quad$ (t' free for x in t)
9) $D'(Dt', t'') \succeq t', \quad D''(Dt', t'') \succeq t''$
10) $I_{\forall\kappa\sigma[\kappa],\tau}(DT\kappa t^{\sigma[\kappa]}) \succeq t^{\sigma[\tau]} \quad (t \in \sigma)$
11) $ST\kappa t\,(I'_{\exists\kappa.\sigma[\kappa],\varrho}t'^{\sigma[\varrho]}) \succeq t^{(\sigma[\varrho])}t'^{\sigma[\varrho]}$
12) if $t \succeq t'$ and $t' \succeq t''$ then $t \succeq t''$

Note: this is not Girard's notion of reduction, since his is a strict notion of reduction. The definition of n.f. and normalizability is the same as Def. 4, Subsect. ii.3.

Finally, the typed computability predicate $Comp_\sigma$ (also called reducibility candidate by Girard) is defined.

4. Definition of the computability candidate predicate $Comp_\sigma$, $\sigma \in T_F$, as a predicate satisfying the following closure conditions:

1) if $Comp_\sigma(t)$ then $t \in \sigma$
2) $Comp_\sigma(0_\sigma)$
3) if $Comp_\sigma(t')$ and $t \succeq t'$ then $Comp_\sigma(t)$
4) if $Comp_\sigma(t)$ then t is normalizable

5. Definition of a parametric computability (or reducibility) predicate $\mathcal{C}_\sigma(Comp_{\vec{\tau}})$ of terms of variable type σ by induction over T_F.

Let $\vec{\kappa} = \kappa_1, \cdots, \kappa_n$ be a sequence of type variables, $\vec{\tau} = \tau_1, \ldots, \tau_n$ a sequence of variable types without type variables, and $Comp_{\vec{\tau}} = Comp_{\tau_1}, \ldots, Comp_{\tau_n}$ a sequence of computability candidates of variable types $\tau_1, \ldots, \tau_n$ resp.

If $\sigma[\vec{\kappa}] = \lambda\vec{x}\varrho[\vec{x}, \vec{\kappa}]$ is a variable type, all of whose free type variables occur in $\vec{\kappa}$, a new predicate $\mathcal{C}_{\sigma[\vec{\kappa}]}$ $(Comp_{\vec{\tau}})$ can be introduced which is defined on terms t of the form $s(\vec{t}, \vec{\tau})$. Since such a term t is to satisfy $\mathcal{C}_{\sigma[\vec{\kappa}]}$ $(Comp_{\vec{\tau}})$ iff t satisfies $\mathcal{C}_{\varrho(\vec{t},\vec{\tau})}$ $(Comp_{\vec{\tau}})$ for all sequences of computable terms $\vec{t}$, it suffices to define $\mathcal{C}_{\sigma[\vec{\kappa}]}$ $(Comp_{\vec{\tau}})$ for arbitrary $\sigma[\vec{\kappa}]$.

1) $\mathcal{C}_0(Comp_{\vec{\tau}})(t) =_{\text{def}} t$ is normalizable
2) $\mathcal{C}_\kappa(Comp_\tau, Comp_{\vec{\tau}})(t) =_{\text{def}} Comp_\tau(t)$
3) $\mathcal{C}_{\sigma\times\kappa}(Comp_{\vec{\tau}})(t) =_{\text{def}} \mathcal{C}_\sigma(Comp_{\vec{\tau}})(D't)$ and $\mathcal{C}_\tau(Comp_{\vec{\tau}})(D''t)$
4) $\mathcal{C}_{(\sigma)\tau}(Comp_{\vec{\tau}})(t) =_{\text{def}} \underset{\sim}{\forall} t'$ (if $\mathcal{C}_\sigma(Comp_{\vec{\tau}})(t')$ then $\mathcal{C}_\tau(Comp_{\vec{\tau}})(tt')$)
5) $\mathcal{C}_{\forall\kappa.\sigma[\kappa,\vec{\kappa}]}(Comp_{\vec{\tau}})(t) =_{\text{def}} \underset{\sim}{\forall}\tau\underset{\sim}{\forall}Comp_\tau, \mathcal{C}_\sigma(Comp_\tau, Comp_{\vec{\tau}})(I_{\forall\kappa.\sigma[\kappa,\vec{\tau}],\tau}t)$
6) $\mathcal{C}_{\exists\kappa.\sigma[\kappa,\vec{\kappa}]}(Comp_{\vec{\tau}})(t) = {}_{\text{def}}\underset{\sim}{\exists}\tau\underset{\sim}{\exists}t'\underset{\sim}{\exists}Comp_\tau(t \succeq I'_{\exists\kappa.\sigma[\kappa,\vec{\tau}],\sigma(\kappa,\vec{\tau})}t'$ and
$\mathcal{C}_\sigma(Comp_\tau, Comp_{\vec{\tau}})(t'))$ or $t \succeq 0_{\exists\kappa.\sigma[\kappa,\vec{\tau}]}$

By means of the two lemmas

6. Lemma (Girard's lemma 2)
$\mathcal{C}_\sigma(Comp_{\vec{\tau}})$ is a computability predicate of type $\sigma[\vec{\tau}]$
and

7. Substitution lemma
$\mathcal{C}_{\sigma[\vec{\kappa},\vec{\kappa}']}(Comp_{\vec{\tau}}, \mathcal{C}_{\sigma'[\vec{\kappa}']}(Comp_{\vec{\tau}'})) = \mathcal{C}_{\sigma[\vec{\kappa},\sigma'[\vec{\kappa}']]}(Comp_{\vec{\tau}}, Comp_{\vec{\tau}'})$

Girard proves the normalization theorem for F. A term $t[\vec{\kappa}] \in \sigma$ is said to be computable if $\underset{\sim}{\forall}\vec{\tau}\underset{\sim}{\forall}Comp_{\vec{\tau}}, \mathcal{C}_{\sigma[\vec{\kappa}]}(Comp_{\vec{\tau}})(t[\vec{\tau}])$.

8. Normalization theorem for F All terms $t[\vec{\kappa}]$ of F satisfy $\mathcal{C}_{\sigma[\vec{\kappa}]}(Comp_{\vec{\tau}})$, for all $\vec{\tau}$ and $Comp_{\vec{\tau}}$, and hence are normalizable.

ii.7 *Martin-Löf's Normalization for the Theory of Species (i.e. Intuitionistic Second Order Logic)*

Martin-Löf's third Hauptsatz paper,[27] i.e. his normalization for the theory of species, essentially applies Girard's notion of parametric computability (or reducibility predicate), and it is also based on the Curry–Howard isomorphism. This Hauptsatz paper constitutes the first concrete step in the direction of constructive type theory. (Cf. Martin-Löf's own comments on intuitionistic second order logic in his (1971B).)

The language of the theory of species is the standard language of first order logic with species variables, possibly species constants, and the second order quantifiers $\forall^2$ and $\exists^2$.

If $\vec{x}$ is a sequence of n individual variables and $A(\vec{x})$ is a formula, then $T = \lambda\vec{x}A(\vec{x})$ is a n-ary species term. If $\vec{t}$ is a sequence of n individual terms, then $T\vec{t}$ denotes the formula $A(\vec{t})$. And if T is a n-ary species term, X is a n-ary species variable and $A(X)$ is a formula, then $A(T)$ is the formula obtained by replacing every part of $A(X)$ of the form $X\vec{t}$ for which X is free by $T\vec{t}$ (whereby it is assumed that undesired ties are avoided). $\bot$ is defined as in Prawitz 1965 by $\forall^2 Y\, Y$, where Y is an 0-ary species variable.

The rules of inference of the theory of species are Gentzen's inference rules for his system of natural deduction: $\&I$, $\vee I1$, $\vee I2$, $\supset I$, $\forall I$, $\exists I$, $\&E1$, $\&E2$, $\vee E$, $\supset E$, $\forall E$, $\exists E$ as well as the four rules of the form

$$\forall^2 I \quad \frac{A(X)}{\forall^2 X A(X)} \qquad\qquad \forall^2 E \quad \frac{\forall^2 X A(X)}{A(T)}$$

$$\exists^2 I \quad \frac{A(T)}{\exists^2 X A(X)} \qquad\qquad \exists^2 E \quad \frac{\exists^2 X A(X) \qquad \begin{matrix}[A(X)]\\ C\end{matrix}}{C}$$

1. Definition of the class of deductions of the theory of species as the smallest class $\mathcal{D}$ s.t.

1), 2.1)–2.5) of the def. of $\mathcal{D}$ of the intuitionistic theory of iterated inductive definitions (i.e. 2. Def. of Subsect. ii.5) hold

2.6) if $\begin{matrix}\Sigma\\ A(X)\end{matrix} \in \mathcal{D}$ then $\dfrac{\begin{matrix}\Sigma\\ A(X)\end{matrix}}{\forall^2 X A(X)} \in \mathcal{D}$

2.7) if $\begin{matrix}\Sigma\\ A(T)\end{matrix} \in \mathcal{D}$ then $\dfrac{\begin{matrix}\Sigma\\ A(T)\end{matrix}}{\exists^2 X A(X)} \in \mathcal{D}$

3.1)–3.5) of the def. of $\mathcal{D}$ of the intuitionistic theory of iterated inductive definitions hold

3.6) if $\begin{matrix}\Sigma\\ \forall^2 X A(X)\end{matrix} \in \mathcal{D}$, then $\dfrac{\begin{matrix}\Sigma\\ \forall^2 X A(X)\end{matrix}}{A(T)} \in \mathcal{D}$

3.7) if $\begin{matrix}\Sigma_1\\ \exists^2 X A(X)\end{matrix} \in \mathcal{D}$, $\begin{matrix}\Sigma_2\\ A(X)\\ B\end{matrix} \in \mathcal{D}$, then $\dfrac{\begin{matrix}\Sigma_1\\ \exists^2 X A(X)\end{matrix} \qquad \begin{matrix}[A(X)]\\ B\end{matrix}}{B} \in \mathcal{D}$

As usual, a relation of reduction is going to be defined next, which holds between deductions of $\mathcal{D}$:

2. Definition of reduction for deductions of $\mathcal{D}$
The relation of reduction, denoted by $\succeq$, can be defined as a binary relation between

objects of $\mathcal{D}$, inductively generated by the following closure conditions:

1)–7) as in the def. of $\succeq$ of the intuitionistic theory of iterated inductive definitions (i.e. 3. Def. of Subsect. ii.5).

8) $\forall^2$-reduction

$$\dfrac{\dfrac{\overset{\Sigma}{A(X)}}{\forall^2 X A(X)}}{A(T)} \succeq \begin{array}{c}\Sigma(T)\\ A(T)\end{array}$$

9) $\exists^2$-reduction

$$\dfrac{\dfrac{\overset{\Sigma_1}{A(T)}}{\exists^2 X A(X)} \qquad \begin{array}{c}[A(X)]\\ B\end{array}}{B} \succeq \begin{array}{c}\Sigma_1\\ A(T)\\ B\end{array}$$

The definitions of 'main cut', 'main branch', and 'normalizable deduction' as well as the corresponding notation can all be taken over (literally) from Defs. 4 and 5 resp. of Subsect. ii.5.

The computability predicate $Comp_T$ of type T, where T is a species term (also called reducibility candidate) is defined as follows:

3. Definition of $Comp_T$ for deductions Σ of the form $\begin{array}{c}\Sigma'\\ T\vec{t}\end{array}$ as a candidate satisfying the following conditions:

1) $Comp_T(T\vec{t})$ provided $\Sigma \equiv T\vec{t}$
2) If $Comp_T(\Sigma)$ then Σ is normalizable
3.1) $Comp_T(\Sigma(E)^{mc}) =_{\text{def}} Comp_T(\Sigma(E)^*)$
3.2) $Comp_T(\Sigma(E)^{mcf}) =_{\text{def}}$ the minor deductions of the eliminations on the main branch are all normalizable.

4. Definition of a parametric computability (or reducibility) predicate $\mathcal{C}_{A(\vec{X})}(Comp_{\vec{T}})$, by induction on $A(\vec{X})$.

Let $\vec{X} = X_1, \ldots, X_n$ be a sequence of species variables, $\vec{T} = T_1, \ldots, T_n$ and $Comp_{\vec{T}} = Comp_{T_1}, \ldots, Comp_{T_n}$ be corresponding sequences of species terms and computability predicates for $T_1, \ldots, T_n$ resp. If $T(\vec{X}) = \lambda\vec{x} A(\vec{x}, \vec{X})$ is a species term all of whose free variables occur in $\vec{X}$, a new predicate $\mathcal{C}_{T(\vec{X})}(Comp_{\vec{T}})$ can be introduced which is defined on deductions Σ of the form $\begin{array}{c}\Sigma'\\ A(\vec{t}, \vec{T})\end{array}$. Since such a deduction Σ is to satisfy $\mathcal{C}_{T(\vec{X})}(Comp_{\vec{T}})$ iff Σ satisfies $\mathcal{C}_{A(\vec{t},\vec{X})}(Comp_{\vec{T}})$, it suffices to define $\mathcal{C}_{A(\vec{X})}(Comp_{\vec{T}})$ for arbitrary $A(\vec{X})$.

1.1) $\mathcal{C}_{C\vec{t}}(Comp_{\vec{T}})\left(\begin{array}{c}\Sigma'\\ C\vec{t}\end{array}\right) =_{\text{def}} \begin{array}{c}\Sigma'\\ C\vec{t}\end{array}$ is normalizable, C is a species constant

1.2) $\mathcal{C}_{X\vec{t}}(Comp_T, Comp_{\vec{T}})\left(\begin{array}{c}\Sigma'\\ T\vec{t}\end{array}\right) =_{\text{def}} Comp_T\left(\begin{array}{c}\Sigma'\\ T\vec{t}\end{array}\right)$

2.1) $\mathcal{C}_{A(\vec{X})}(Comp_{\vec{T}})(A(\vec{T}))$ provided $\Sigma \equiv A(\vec{T})$

2.2) $\mathcal{C}_{A(\vec{X})\&B(\vec{X})}(Comp_{\vec{T}})(\Sigma(\&I)) =_{\mathrm{def}} \mathcal{C}_{A(\vec{X})}(Comp_{\vec{T}})\begin{pmatrix}\Sigma_1\\ A(\vec{T})\end{pmatrix}$ and

$$\mathcal{C}_{B(\vec{X})}(Comp_{\vec{T}})\begin{pmatrix}\Sigma_2\\ B(\vec{T})\end{pmatrix}$$

2.3) $\mathcal{C}_{A(\vec{X})\vee B(\vec{X})}(Comp_{\vec{T}})(\Sigma(\vee I1)) =_{\mathrm{def}} \mathcal{C}_{A(\vec{X})}(Comp_{\vec{T}})\begin{pmatrix}\Sigma_1\\ A(\vec{T})\end{pmatrix}$

$\mathcal{C}_{A(\vec{X})\vee B(\vec{X})}(Comp_{\vec{T}})(\Sigma(\vee I2)) =_{\mathrm{def}} \mathcal{C}_{B(\vec{X})}(Comp_{\vec{T}})\begin{pmatrix}\Sigma_2\\ B(\vec{T})\end{pmatrix}$

2.4) $\mathcal{C}_{A(\vec{X})\supset B(\vec{X})}(Comp_{\vec{T}})\ (\Sigma(\supset I)) =_{\mathrm{def}} \underset{\sim}{\forall}\ \begin{matrix}\Sigma'\\ A(\vec{T})\end{matrix}\ \left(\text{if } \mathcal{C}_{A(\vec{X})}(Comp_{\vec{T}})\begin{pmatrix}\Sigma'\\ A(\vec{T})\end{pmatrix}\right.$

$$\left.\text{then } \mathcal{C}_{B(\vec{X})}(Comp_{\vec{T}})\begin{pmatrix}\Sigma'\\ A(\vec{T})\\ B(\vec{T})\end{pmatrix}\right)$$

2.5) $\mathcal{C}_{\forall x A(x,\vec{x})}(Comp_{\vec{T}})(\Sigma(\forall I)) =_{\mathrm{def}} \underset{\sim}{\forall} t\ \mathcal{C}_{A(t,\vec{x})}(Comp_{\vec{T}})\begin{pmatrix}\Sigma'\\ A(t,\vec{T})\end{pmatrix}$

2.6) $\mathcal{C}_{\exists x A(x,\vec{x})}(Comp_{\vec{T}})(\Sigma(\exists I)) =_{\mathrm{def}} \mathcal{C}_{A(t,\vec{x})}(Comp_{\vec{T}})\begin{pmatrix}\Sigma'\\ A(t,\vec{T})\end{pmatrix}$

2.7) $\mathcal{C}_{\forall^2 X A(X,\vec{X})}(Comp_{\vec{T}})(\Sigma(\forall^2 I)) =_{\mathrm{def}} \underset{\sim}{\forall} T\ \underset{\sim}{\forall} Comp_T$

$$\mathcal{C}_{A(X,\vec{X})}(Comp_T, Comp_{\vec{T}})\begin{pmatrix}\Sigma'\\ A(T,\vec{T})\end{pmatrix}$$

2.8) $\mathcal{C}_{\exists^2 X A(X,\vec{X})}(Comp_{\vec{T}})(\Sigma(\exists^2 I)) =_{\mathrm{def}} \mathcal{C}_{A(X,\vec{X})}(Comp_T, Comp_{\vec{T}})\begin{pmatrix}\Sigma'\\ A(T,\vec{T})\end{pmatrix}$

3.1) $\mathcal{C}_{A(\vec{X})}(Comp_{\vec{T}})(\Sigma(E)^{mc}) =_{\mathrm{def}} \mathcal{C}_{A(\vec{X})}(Comp_{\vec{T}})(\Sigma(E)^{*})$

3.2) $\mathcal{C}_{A(\vec{X})}(Comp_{\vec{T}})(\Sigma(E)^{mcf}) =_{\mathrm{def}}$ the minor deductions of the eliminations on the main branch are all normalizable.

The normalization theorem for the theory of species is then readily proved by means of the following two lemmas:

5. Lemma $\mathcal{C}_{T(\vec{X})}(Comp_{\vec{T}})$ is a computability predicate of type $T(\vec{T})$.

6. Lemma Consider a deduction $\begin{matrix}A_1(\vec{X}) & \dots & A_n(\vec{X})\\ \ddots & & \cdot^{\cdot^{\cdot}}\\ & A(\vec{X}) & \end{matrix}$ whose free variables form the sequence $\vec{X}$ and whose assumptions are $A_1(\vec{X}), \dots, A_n(\vec{X})$. If for all $\vec{t}$ that are substituted for $\vec{x}$, for all $\vec{T}$ that are substituted for $\vec{X}$ and for all $Comp_{\vec{T}}$,

$$\mathcal{C}_{A_1(\vec{X})}(Comp_{\vec{T}})\begin{pmatrix}\Sigma_1\\ A_1(\vec{T})\end{pmatrix}, \dots, \mathcal{C}_{A_n(\vec{X})}(Comp_{\vec{T}})\begin{pmatrix}\Sigma_n\\ A_n(\vec{T})\end{pmatrix},$$

$$\text{then } \mathcal{C}_{A(\vec{X})}(Comp_{\vec{T}}) \left(\begin{array}{ccc} \Sigma_1 & & \Sigma_n \\ A_1(\vec{T}) & \cdots & A_n(\vec{T}) \\ \ddots & & \cdot^{\cdot^{\cdot}} \\ & A(\vec{T}) & \end{array} \right).$$

7. Normalization theorem for the theory of species All deductions Σ (of the form $\genfrac{}{}{0pt}{}{\Sigma'}{A(\vec{X})}$) of the theory of species satisfy $\mathcal{C}_{A(\vec{X})}(Comp_{\vec{T}})$, for all $\vec{T}$ and $Comp_{\vec{T}}$, and hence are normalizable.

Note: The same theorem was proved at the same time but independently by Girard and Prawitz.

ii.8 *Martin-Löf's Normalization for Intuitionistic Simple Type Theory*

Martin-Löf's fourth Hauptsatz-paper,[28] i.e. his normalization for intuitionistic simple type theory, again applies Girard's notion of parametric computability (or reducibility predicate) in an essential way and is likewise based on the Curry–Howard isomorphism. This Hauptsatz paper constitutes the second and last concrete step towards constructive type theory. (For more on this point, cf. end of this section.)

The simple type structure is inductively defined as follows:

1. Definition of the simple type structure T_{ST}

1) $0 \in T_{ST}$ (the type of individuals)
2) if $\sigma_1, \ldots, \sigma_n \in T_{ST}$ then $[\sigma_1, \ldots, \sigma_n] \in T_{ST}$, where $n = 0, 1, \ldots$

Note that [ ] is the type of propositions.

(This is Russell's simple type structure minus Frege's type structure in Dummett's formulation; cf. C,n&o (2) to iii.1), Sect. iii Chapt. 1).

The language of intuitionistic simple type theory is the standard language of first order logic with variables of each type and with predicate constants, each of which has a definite type of the form $[\sigma_1, \ldots, \sigma_n]$.

Next, the class of terms Tm^{ST} of intuitionistic simple type theory is defined by $Tm^{ST} =_{\text{def}} \bigcup\{Tm^{ST}_{\sigma} \mid \sigma \in T_{ST}\}$, where Tm^{ST}_{σ} is the class of terms of type σ inductively defined as follows:

2. Definition of the class of terms of type σ as the smallest class Tm^{ST}_{σ}, $\sigma \in T_{ST}$, s.t.

1) If $A \in [\,]$ then $A \in Tm^{ST}_{[\,]}$
2) If $x \in 0$ then $x \in Tm^{ST}_0$
3) If f is a function constant whose arguments are of type 0 and $\vec{t}$ is a sequence of terms of type 0, then $f\vec{t} \in Tm^{ST}_0$.
4) $c \in [\sigma_1, \ldots, \sigma_n]$ and $\vec{t}$ is a sequence of terms of types $\sigma_1, \ldots, \sigma_n$ resp. then $c\vec{t} \in Tm^{ST}_{[\,]}$ (c is a predicate constant)
5) If $x \in [\sigma_1, \ldots, \sigma_n]$ and $\vec{t}$ is a sequence of terms of types $\sigma_1, \ldots, \sigma_n$ resp. then $x\vec{t} \in Tm^{ST}_{[\,]}$

6) If $A, B \in Tm^{ST}_{[\,]}$ then $A \supset B, A\&B, A \vee B \in Tm^{ST}_{[\,]}$
7) If $A(x) \in Tm^{ST}_{[\,]}$, then $\forall x A(x), \exists x A(x) \in Tm^{ST}_{[\,]}$
8) If $\vec{x}$ is a sequence of variables of types $\sigma_1, \ldots, \sigma_n$ resp. and $A(\vec{x}) \in Tm^{ST}_{[\,]}$, then $\lambda\vec{x}A(\vec{x}) \in Tm^{ST}_{[\sigma_1,\ldots,\sigma_n]}$

Note: A variable $x \in \sigma \equiv [\sigma_1, \ldots, \sigma_n]$ is not a term if $n > 0$. But x will be used as a notation for the term $t(x)$ which is defined inductively as follows:

$t(x) = x$ if $\sigma \equiv 0$ or $[\,]$
$t(x) = \lambda\vec{x}x\vec{t}(\vec{x})$ if $\sigma \equiv [\sigma_1, \ldots, \sigma_n]$ where $\vec{x} = x_1, \ldots, x_n$ is a sequence of variables of types $\sigma_1, \ldots, \sigma_n$ resp. and $\vec{t}(\vec{x}) = t_1(x_1), \ldots, t_n(x_n)$.

The rules of inference of intuitionistic simple type theory are Gentzen's inference rules for his system of natural deduction: $\&I$, $\vee I1$, $\vee I2$, $\supset I$, $\forall I$, $\exists I$, $\&E1$, $\&E2$, $\vee E$, $\supset E$, $\forall E$ and $\exists E$.

3. Definition of the class of deductions of intuitionistic simple type theory as the smallest class $\mathcal{D}$ s.t.

1), 2.1)–2.5) of the definition of $\mathcal{D}$ of the intuitionistic theory of iterated inductive definitions (i.e. Def. 2 of Subsect. ii.5) hold.
3.1)–3.5) of the definition of $\mathcal{D}$ of the intuitionistic theory of iterated inductive definitions hold.

4. Definition of reduction for deductions of $\mathcal{D}$
The relation of reduction, denoted by $\succeq$, can be defined as a binary relation between objects of $\mathcal{D}$, inductively generated by the following closure conditions:

1)–7) as in the definition of $\succeq$ of the intuitionistic theory of iterated inductive definitions (3. Def. of Subsect. ii.5).

As was the case with the theory of species, the definitions of 'main cut', 'main branch' and 'normalizable deduction' as well as the resp. notation can all be taken over (literally) from Defs. 4 and 5 resp. of Subsect. ii.5.

The computability candidate $Comp_t$ of term t of type σ (also called reducibility candidate) is defined as follows:

5. Definition of $Comp_t$, $t \in Tm^{ST}_{\sigma}$, by induction on $\sigma \in T_{ST}$

1) if $t \in 0$ then $Comp_t =_{\text{def}} t$
2) if $A \in [\,]$ then $Comp_A$ is a unary predicate which is defined for deductions Σ of the form $\dfrac{\Sigma'}{A}$ as follows

2.1) $Comp_A(A)$ provided $\Sigma \equiv A$
2.2) If $Comp_A(\Sigma)$ then Σ is normalizable
2.3) $Comp_A(\Sigma(E)^{mc}) =_{\text{def}} Comp_A(\Sigma(E)^*)$
2.4) $Comp_A(\Sigma(E)^{mcf}) =_{\text{def}}$ the minor deductions of the eliminations on the main branch are all normalizable

3) if $t \in \sigma \equiv [\sigma_1, \ldots, \sigma_n]$ then $Comp_t$ is an $(n+1)$-ary relation whose first

n arguments are computability predicates $Comp_{t_1}, \ldots, Comp_{t_n}$ of terms $t_1 \in \sigma_1, \ldots, t_n \in \sigma_n$ resp. and whose last argument is a deduction $\begin{matrix}\Sigma' \\ t\vec{t}\end{matrix}$ where $\vec{t} = t_1, \ldots, t_n$.

Moreover, for every sequence of computability predicates $Comp_{\vec{t}} = Comp_{t_1}, \ldots, Comp_{t_n}$, the section $Comp_t(Comp_{\vec{t}})$ of $Comp_t$ at $Comp_{\vec{t}}$ is to be a computability predicate of term $t\vec{t}$ of type [ ]. $Comp_t$ acts as a function transforming a sequence $Comp_{\vec{t}}$ into a computability predicate $Comp_{t\vec{t}}$.

6. Definition of the computability candidate $Comp_\sigma$, $\sigma \in T_{ST}$, as a predicate satisfying the following conditions:

1) $Comp_0 =_{\text{def}} 0$
2) $Comp_{[\,]}$ is a unary predicate which is defined for deductions Σ of the form $\begin{matrix}\Sigma' \\ A\end{matrix}$ as follows: $Comp_{[\,]} =_{\text{def}} Comp_A$, for $A \in [\,]$
3) $Comp_{[\sigma_1,\ldots,\sigma_n]} =_{\text{def}} Comp_t$, for any $t \in \sigma \equiv [\sigma_1, \ldots, \sigma_n]$. $Comp_{[\sigma_1,\ldots,\sigma_n]}$ is a $(n+1)$-ary relation whose first n arguments are computability predicates $Comp_{\sigma_1}, \ldots, Comp_{\sigma_n}$ and whose last argument is any deduction $\begin{matrix}\Sigma' \\ t\vec{t}\end{matrix}$ s.t. $\vec{t} = t_1, \ldots, t_n$ and $t_1 \in \sigma_1, \ldots, t_n \in \sigma_n$.

7. Definition of a parametric computability (or reducibility) predicate $\mathcal{C}_{u(\vec{x})}(Comp_{\vec{t}})$, by induction on $u(\vec{x})$, $\sigma \in T_{ST}$. Let $u(\vec{x})$ be a term, all of whose free variables occur in $\vec{x} = x_1, \ldots, x_n$, and for $t_1 \in \sigma_1, \ldots, t_n \in \sigma_n$. Let $Comp_{\vec{t}} = Comp_{t_1}, \ldots, Comp_{t_n}$ be a sequence of computability predicates of terms $t_1, \ldots, t_n$ resp.

1) $u(\vec{x}) \in 0$. Then $\mathcal{C}_{u(\vec{x})}(Comp_{\vec{t}}) =_{\text{def}} u(\vec{t})$

2.1) $\mathcal{C}_{c\overrightarrow{u(x)}}(Comp_{\vec{t}})\left(\begin{matrix}\Sigma' \\ c\,\overrightarrow{u(t)}\end{matrix}\right) =_{\text{def}} \begin{matrix}\Sigma' \\ c\,\overrightarrow{u(t)}\end{matrix}$ is normalizable

2.2) $\mathcal{C}_{x_i\overrightarrow{u(x)}}(Comp_{\vec{t}}) =_{\text{def}} Comp_{t_i}\big(\mathcal{C}_{\overrightarrow{u(x)}}(Comp_{\vec{t}})\big)$ where by IH $\mathcal{C}_{\overrightarrow{u(x)}}(Comp_{\vec{t}}) = \mathcal{C}_{u_1(\vec{x})}(Comp_{\vec{t}}), \ldots, \mathcal{C}_{u_n(\vec{x})}(Comp_{\vec{t}})$ is a sequence of reducibility predicates of types $u_1(\vec{t}), \ldots, u_n(\vec{t})$ resp. and $1 \leq i \leq n$.

3.1) $\mathcal{C}_{A(\vec{x})\&B(\vec{x})}(Comp_{\vec{t}})$ is defined by the following clauses

3.1.1) $\mathcal{C}_{A(\vec{x})\&B(\vec{x})}(Comp_{\vec{t}})(A(\vec{t})\&B(\vec{t}))$ provided $\Sigma \equiv A(\vec{t})\&B(\vec{t})$

3.1.2) $\mathcal{C}_{A(\vec{x})\&B(\vec{x})}(Comp_{\vec{t}})(\Sigma(E)^{mc}) =_{\text{def}} \mathcal{C}_{A(\vec{x})\&B(\vec{x})}(Comp_{\vec{t}})(\Sigma(E)^*)$

3.1.3) $\mathcal{C}_{A(\vec{x})\&B(\vec{x})}(Comp_{\vec{t}})(\Sigma(E)^{mcf}) =_{\text{def}}$ the minor deductions of the eliminations on the main branch are all normalizable

3.1.4) $\mathcal{C}_{A(\vec{x})\&B(\vec{x})}(Comp_{\vec{t}})(\Sigma(\&I)) =_{\text{def}} \mathcal{C}_{A(\vec{x})}(Comp_{\vec{t}})\left(\begin{matrix}\Sigma_1 \\ A(\vec{t})\end{matrix}\right)$ and $\mathcal{C}_{B(\vec{x})}(Comp_{\vec{t}})\left(\begin{matrix}\Sigma_2 \\ B(\vec{t})\end{matrix}\right)$

3.2) $\mathcal{C}_{A(\vec{x})\vee B(\vec{x})}(Comp_{\vec{t}})$ is defined by the following clauses

3.2.1) $\mathcal{C}_{A(\vec{x})\vee B(\vec{x})}(Comp_{\vec{t}})(A(\vec{t}) \vee B(\vec{t}))$ provided $\Sigma \equiv A(\vec{t}) \vee B(\vec{t})$

3.2.2) $\mathcal{C}_{A(\vec{x})\vee B(\vec{x})}(Comp_{\vec{t}})(\Sigma(E)^{mc}) =_{\text{def}} \mathcal{C}_{A(\vec{x})\vee B(\vec{x})}(Comp_{\vec{t}})(\Sigma(E)^{*})$

3.2.3) $\mathcal{C}_{A(\vec{x})\vee B(\vec{x})}(Comp_{\vec{t}})(\Sigma(E)^{mcf}) =_{\text{def}}$ the minor deductions of the eliminations on the main branch are all normalizable

3.2.4) $\mathcal{C}_{A(\vec{x})\vee B(\vec{x})}(Comp_{\vec{t}})(\Sigma(\vee I1)) =_{\text{def}} \mathcal{C}_{A(\vec{x})}(Comp_{\vec{t}})\begin{pmatrix}\Sigma_1\\ A(\vec{t})\end{pmatrix}$

Similarly for $\vee I2$.

3.3) $\mathcal{C}_{A(\vec{x})\supset B(\vec{x})}(Comp_{\vec{t}})$ is defined by the clauses:

3.3.1) $\mathcal{C}_{A(\vec{x})\supset B(\vec{x})}(Comp_{\vec{t}})(A(\vec{t}) \supset B(\vec{t}))$ provided $\Sigma \equiv A(\vec{t}) \supset B(\vec{t})$

3.3.2) $\mathcal{C}_{A(\vec{x})\supset B(\vec{x})}(Comp_{\vec{t}})(\Sigma(E)^{mc}) =_{\text{def}} \mathcal{C}_{A(\vec{x})\supset B(\vec{x})}(Comp_{\vec{t}})(\Sigma(E)^{*})$

3.3.3) $\mathcal{C}_{A(\vec{x})\supset B(\vec{x})}(Comp_{\vec{t}})(\Sigma(E)^{mcf}) =_{\text{def}}$ the minor deductions of the eliminations on the main branch are all normalizable

3.3.4) $\mathcal{C}_{A(\vec{x})\supset B(\vec{x})}(Comp_{\vec{t}})(\Sigma(\supset I)) =_{\text{def}} \underset{\sim}{\forall} \begin{matrix}\Sigma'\\ A(\vec{t})\end{matrix} \left(\text{if } \mathcal{C}_{A(\vec{x})}(Comp_{\vec{t}})\begin{pmatrix}\Sigma'\\ A(\vec{t})\end{pmatrix}\right.$

$\left.\text{then } \mathcal{C}_{B(\vec{x})}(Comp_{\vec{t}})\begin{pmatrix}\Sigma'\\ A(\vec{t})\\ B(\vec{t})\end{pmatrix}\right)$

3.4) $\mathcal{C}_{\forall x A(x,\vec{x})}(Comp_{\vec{t}})$ is defined by the clauses:

3.4.1) $\mathcal{C}_{\forall x A(x,\vec{x})}(Comp_{\vec{t}})(\forall x A(x, \vec{t}))$ provided $\Sigma \equiv \forall x A(x, \vec{t})$

3.4.2) $\mathcal{C}_{\forall x A(x,\vec{x})}(Comp_{\vec{t}})(\Sigma(E)^{mc}) =_{\text{def}} \mathcal{C}_{\forall x A(x,\vec{x})}(Comp_{\vec{t}})(\Sigma(E)^{*})$

3.4.3) $\mathcal{C}_{\forall x A(x,\vec{x})}(Comp_{\vec{t}})(\Sigma(E)^{mcf}) =_{\text{def}}$ the minor deductions of the eliminations on the main branch are all normalizable

3.4.4) $\mathcal{C}_{\forall x A(x,\vec{x})}(Comp_{\vec{t}})(\Sigma(\forall I)) =_{\text{def}} \underset{\sim}{\forall} Comp_t$ of type σ,

$$\mathcal{C}_{A(x,\vec{x})}(Comp_t, Comp_{\vec{t}})\begin{pmatrix}\Sigma'\\ A(t, \vec{t})\end{pmatrix}$$

3.5) $\mathcal{C}_{\exists x A(x,\vec{x})}(Comp_{\vec{t}})$ is defined by the clauses:

3.5.1) $\mathcal{C}_{\exists x A(x,\vec{x})}(Comp_{\vec{t}})(\exists x A(x, \vec{t}))$ provided $\Sigma \equiv \exists x A(x, \vec{t})$

3.5.2) $\mathcal{C}_{\exists x A(x,\vec{x})}(Comp_{\vec{t}})(\Sigma(E)^{mc}) =_{\text{def}} \mathcal{C}_{\exists x A(x,\vec{x})}(Comp_{\vec{t}})(\Sigma(E)^{*})$

3.5.3) $\mathcal{C}_{\exists x A(x,\vec{x})}(Comp_{\vec{t}})(\Sigma(E)^{mcf}) =_{\text{def}}$ the minor deductions of the eliminations on the main branch are all normalizable

3.5.4) $\mathcal{C}_{\exists x A(x,\vec{x})}(Comp_{\vec{t}})(\Sigma(\exists I)) =_{\text{def}} \mathcal{C}_{A(x,\vec{x})}(Comp_t, Comp_{\vec{t}})\begin{pmatrix}\Sigma'\\ A(t, \vec{t})\end{pmatrix}$

4) Define $\mathcal{C}_{\lambda\vec{y}A(\vec{x},\vec{y})}(Comp_{\vec{t}})(Comp_{\vec{u}}) =_{\text{def}} \mathcal{C}_{A(\vec{x},\vec{y})}(Comp_{\vec{t}}, Comp_{\vec{u}})$

The normalization theorem for intuitionistic simple type theory is then proved by means of the following three lemmas:

8. Lemma

$\mathcal{C}_{u(\vec{x})}(Comp_{\vec{t}})$ is a computability predicate of type $u(\vec{t})$.

9. Substitution lemma

$\mathcal{C}_{u(x,\vec{x})}\big(\mathcal{C}_{t(\vec{x})}(Comp_{\vec{t}}), Comp_{\vec{t}}\big) = \mathcal{C}_{u(t(\vec{x}),\vec{x})}(Comp_{\vec{t}})$.

10. Lemma

Consider a deduction $\begin{array}{ccc} A_1(\vec{x}) & \dots & A_n(\vec{x}) \\ \ddots & & \cdot^{\cdot^{\cdot}} \\ & A(\vec{x}) & \end{array}$ whose free variables form the sequence $\vec{x} = x_1, \dots, x_n$ of types $\sigma_1, \dots, \sigma_n$ resp. and whose assumptions are $A_1(\vec{x}), \dots, A_n(\vec{x})$. Then for all sequences $Comp_{\vec{t}} = Comp_{t_1}, \dots, Comp_{t_n}$ of computability predicates of types $\sigma_1, \dots, \sigma_n$ resp., if $\mathcal{C}_{A_1(\vec{x})}(Comp_{\vec{t}})\begin{pmatrix} \Sigma_1 \\ A_1(\vec{t}) \end{pmatrix}, \dots, \mathcal{C}_{A_n(\vec{x})}(Comp_{\vec{t}})\begin{pmatrix} \Sigma_n \\ A_n(\vec{t}) \end{pmatrix}$,

$$\text{then } \mathcal{C}_{A(\vec{x})}(Comp_{\vec{t}})\begin{pmatrix} \Sigma_1 & & \Sigma_n \\ A_1(\vec{t}) & \dots & A_n(\vec{t}) \\ \ddots & & \cdot^{\cdot^{\cdot}} \\ & A(\vec{t}) & \end{pmatrix}.$$

11. Normalization theorem for intuitionistic simple type theory All deductions Σ (of the form $\begin{array}{c}\Sigma' \\ u(\vec{x})\end{array}$) of intuitionistic simple type theory satisfy $\mathcal{C}_{u(\vec{x})}(Comp_{\vec{t}})$, for all $\vec{t}$ and $Comp_{\vec{t}}$, and hence are normalizable.

Note: As has been seen, Girard showed how to prove normalization for second order theories. This was readily adapted to natural deduction for second order logic by Girard, Martin-Löf and Prawitz. Then, both Girard and Martin-Löf discovered how to extend it to higher order.

The first version of constructive type theory is an intuitionistic theory of types in which intuitionistic simple type theory can be embedded as a subsystem. Martin-Löf actually shows how this can be done. The conception of a general intuitionistic theory of types, as Martin-Löf called the first version of constructive type theory, grew out of the desire on the one hand to overcome some unnatural limitations of intuitionistic simple type theory (which e.g. allows only for finite iterations of the power operation), and on the other hand to be able to talk about arbitrary sets and not just about sets of natural numbers, sets of sets of natural numbers etc., which seems to be necessary for the formalization of certain parts of mathematics dealing with arbitrary sets. According to Martin-Löf, this general intuitionistic theory of types or constructive type theory, as it will be called in the sequel, was to serve as a logical basis for intuitionistic mathematics in somewhat the same way as set theory has served as a basis for classical mathematics. Here then is the starting point of the history proper of constructive type theory.[29]

NOTES

[1] Hintikka (1988), Hintikka and Hintikka (1986), Chapt. 1, and other papers by Hintikka.

[2] Goldfarb (1979).

[3] Cf. also Sundholm (1988); there the first tradition is called objectivistic epistemological and the second one structuralistic ontological.

[4] Version (4), pp. 4, 5.
[5] Cf. e.g. Shoenfield (1967), Chapt. 2.
[6] Frege (1882), pp. 55, 56.
[7] Cf. Gentzen (1934).
[8] Prawitz (1971), p. 239; cf. also Prawitz (1965).
[9] p. 240.
[10] p. 240.
[11] Prawitz (1971), pp. 243, 244.
[12] pp. 246,247.
[13] p. 248.
[14] p. 251.
[15] pp. 248, 255.
[16] pp. 255,256.
[17] p. 258.
[18] Cf. Troelstra and Schwichtenberg (1996), pp. 31, 36, 37.
[19] Howard (1980), p. 486.
[20] Martin-Löf (1969).
[21] Tait (1967).
[22] For this whole Subsect. ii.4, cf. Troelstra (1973), pp. 39, 40, 101–104, 113, 114.
[23] Cf. Troelstra (1973), pp. 41, 43, 45, 49, and also Troelstra (1990).
[24] Martin-Löf (1970).
[25] Ex. (2)–(4) were communicated to me by A.S. Troelstra.
[26] Girard (1971). Cf. also Troelstra (1973), p. 84 and Troelstra and Schwichtenberg (1996), pp. 291–297.
[27] Martin-Löf (1970A).
[28] Martin-Löf (1971).
[29] Cf. Martin-Löf (1971B), pp. 2, 3.

CONCLUSION

This work is in many ways essentially incomplete or open.

'I believe that there also is a place and a function in our ongoing intellectual life for a less complete work, containing unfinished presentations, conjectures, open questions and problems, leads, side connections, as well as a main line of argument. There is room for words on subjects other than last words.' (R. Nozick)

One of the main subjects of this work is the historical development of constructive type theory. Now, this subject is itself open-ended since neither the history of constructive type theory nor history in general has reached its end yet – contrary to what certain false prophets nowadays try to make us believe. Thus, as the development of constructive type theory continues beyond 1995, so does its historiography , i.e. the sort of work pursued here. But this is only one aspect of the story.

Another one is that I far from claim to have fully grasped the philosophical ideas inherent in constructive type theory. To the contrary, it is my conviction that the present work has barely laid the foundation for such an endeavor. It appears that the most important philosophical sources for a better understanding as well as for a future inspiration of constructive type theory, are phenomenology, above all Husserl, and analytical philosophy, above all Frege. Each of them, that is, constructive type theory and Husserl (including phenomenology), and constructive type theory and Frege (including analytical philosophy) deserve an in-depth study in their own right. At the beginning of version (8), Martin-Löf declares that his main concern is to construct a bridge between analytical philosophy and phenomenology within the philosophy of mathematics.[1] Other important philosophical figures to be further scrutinized within the context of constructive type theory are Bolzano, Brentano, Brouwer and Kant. (One may be tempted to further add Aristotle, Quine, Russell, Wittgenstein, etc. etc. but one will end up quoting the index of names of a pocket history of philosophy!)

Yet another aspect of the unfinished story are certain recent and very stimulating ideas put forward by Sambin and Valentini which are of equally philosophical and mathematical interest. Sambin and Valentini characterize the spirit of constructive type theory or rather of constructive set theory by the principle that only those notions and rules are to be admitted which keep total control of the amount of information contained in the different forms of judgement.[2] This boils down to saying that constructive set theory is the logic of analytic judgements. (Cf. Subsect. ii.3, Chapt. 2.) Now, on the one hand Sambin and Valentini adhere to the principle just mentioned. On the other hand they observe that according to their experience in developing pieces of 'actual mathematics' within constructive set theory, this set theory is not suitable as it stands, since the control of the amount of information is too strict. To bridge this gap between constructive set theory and 'real mathematics' they have invented a toolbox: to make constructive set theory more handy they have fabricated some suitable tools, i.e. definitions and rules, which 'forget' some information in order to achieve a higher level of abstraction. The main still constructive tool so far has turned out to be a more liberal notion of subset than could possibly be defined (strictly) inside constructive set theory. For Sambin and Valentini, adherents to the spirit of constructive set theory, a tool is constructive or reliable only when it allows us to forget some information in such a way that it can be restored *ad libitum*.[3] This opens a whole new philosophical and mathematical research program.

A last aspect of the open story – which actually could be the subject of a new historiography – is the formal topologies introduced by Sambin as a constructive approach to topology based on constructive type theory.[4] The keynote in the formal approach to topology is the study of the properties of a topological space $\langle X, \Omega(X)\rangle$, where $\Omega(X)$ is the family of the open subsets of the set X which can be expressed without reference to the points, i.e. elements of the set X. That is why formal topology is sometimes also referred to as pointfree topology. Constructive type theory permits via formal topology to constructivize classical non-constructive results in analysis and topology and to pursue the project of a machine-implemented formalization of analysis. I thus finish off where Martin-Löf's work in constructive mathematics (analysis) started (1968).

Still, as Beeson dixit at the end of his book: 'This is not the end of the story'.[5]

NOTES

1 Version (8), p. 2.
2 Sambin and Valentini (1995), p. 3.
3 pp. 4, 5.
4 Cf. Sambin (1987, 1989) and other papers.
5 Beeson (1985), p. 438.

REFERENCES

Aczel, P. (1977): The strength of Martin-Löf's intuitionistic type theory with one universe, in: Miettinen, S. and Väänänen, S. (eds.), *Proceedings of Symposia in Mathematical Logic*, Oulu 1974, and Helsinki 1975. Report No. 2, Department of Philosophy, University of Helsinki, Helsinki, pp. 1–32.

Aczel, P. (1978): The type theoretic interpretation of constructive set theory, in: Macintyre, A. et al. (eds.), *Logic Colloquium '77*, North-Holland, Amsterdam, pp. 55–66.

Aczel, P. (1980): Frege structures and the notions of proposition, truth and set, in: Barwise, J. et al. (eds.), *The Kleene Symposium*, North-Holland, Amsterdam, pp. 31–59.

Bar-Hillel, Y. (1957): Husserl's conception of a purely logical grammar, *Philos. Phenomen. Research* 17, 362–369.

Beeson, M. (1985): *Foundations of Constructive Mathematics. Metamathematical Studies*, Springer, Berlin.

Berg, J. (1962): *Bolzano's Logic*, Almqvist & Wiksell, Stockholm.

Bishop, E. (1967): *Foundations of Constructive Analysis*, McGraw-Hill, New York.

Bocheński, J.M. (1970): *Formale Logik*, Karl Alber, Freiburg i. Br. 1970, 3rd edn.

Boolos, G. (1971): The iterative conception of set, *Journal of Philosophy* 68, 215–232.

Boolos, G. (1982): Extremely undecidable sentences, *JSL* 47, 191–196.

Broadie, A. (1993): *Introduction to Medieval Logic*, Clarendon Press, Oxford, 2nd edn.

Bruijn, N.G. de (1970): The mathematical language AUTOMATH, its usage, and some of its extensions, in: Laudet, M. et al. (eds.), *Symposium on Automatic Demonstration*, December 1980, Rocquencourt, Springer, Berlin, pp. 29–61.

Bruijn, N.G. de (1980): A survey of the project AUTOMATH, in: Seldin, J.P. and Hindley, J.R. (eds.), *To H.B. Curry: Essays on Combinatory Logic, Lambda Calculus and Formalism*, Academic Press, New York, pp. 579–606.

Carnap, R. (1931): Die logizistische Grundlegung der Mathematik, *Erkenntnis* 2, 91–105.

Carnap, R. (1934): *Logische Syntax der Sprache*, Springer, Wien.

Chisholm, R.M. (1967): Brentano, Franz, in: Edwards, P. (ed.), *The Encyclopedia of Philosophy*, Vol. I/II, Macmillan, New York, pp. 365–368.

Chisholm, R.M. (1982): Correct and incorrect emotion, in: Chisholm, R.M. (ed.), *Brentano and Meinong Studies*, Rodopi, Amsterdam, pp. 68–79.

Curry, H.B. and Feys, R. (1958): *Combinatory Logic I*, North-Holland, Amsterdam.

Curry, H.B. (1977): *Foundations of Mathematical Logic*, Dover, New York.

Dalen, D. van et al. (1978): *Sets: Naïve, Axiomatic and Applied*, Pergamon, Oxford.

Dubislav, W. (1931): *Die Definition*, Meiner, Hamburg.

Dummett, M. (1975): The philosophical basis of intuitionistic logic, in: Rose, H.E. and Shepherdson, J. (eds.), *Logic Colloquium '73*, North-Holland, Amsterdam, pp. 5–40.

Dummett, M. (1976): What is a theory of meaning? (II), in: Evans, G. and McDowell, J. (eds.), *Truth and Meaning. Essays in Semantics*, Clarendon Press, Oxford, pp. 67–137.

Dummett, M. (1977): *Elements of Intuitionism*, Clarendon Press, Oxford.

Dummett, M. (1981): *Frege. Philosophy of Language*, Duckworth, London, 2nd edn.

Dybjer, P. (1988): Representing inductively defined sets by wellorderings in Martin-Löf's type theory, in: Avron, A. et al. (eds.), *Proceedings of the Workshop on General Logic*, Edinburgh, February 1987, Report ECS-LFCS-88-52, Department of Computer Science, University of Edinburgh, Edinburgh.

Føllesdal, D. (1991): The justification of logic and mathematics in Husserl's phenomenology, in: Seebohm, Th.M. et al. (eds.), *Phenomenology and The Formal Sciences*, Kluwer, Dordrecht, pp. 25–34.

Fraenkel, A.A. et al. (1973): *Foundations of Set Theory*, North-Holland, Amsterdam, 2nd rev. edn.

Frede, M. (1987): Categories in Aristotle, in: Frede, M. (ed.), *Essays in Ancient Philosophy*, Clarendon Press, Oxford, pp. 29–48.

Frege, G. (1879): *Begriffsschrift, eine der arithmetischen nachgebildete Formelsprache des reinen Denkens*, Louis Nebert, Halle A/S.

Frege, G. (1882): Über die wissenschaftliche Berechtigung einer Begriffsschrift, *Zeitschrift für Philosophie und philosophische Kritik* 81, 48–56.

Frege, G. (1893,1903): *Grundgesetze der Arithmetik, begriffsschriftlich abgeleitet*, Jena I: 1893; II: 1903.

Geach, P.T. (1967): Identity, *The Review of Metaphysics* 21, 3–12.

Geach, P.T. (1973): Ontological relativity and relative identity, in: Munitz, M.K. (ed.), *Logic and Ontology*, New York UP, New York, pp. 287–302.

Gentzen, G. (1934): Untersuchungen über das logische Schliessen, *Mathemat. Zeitschrift* 39 176–210; 405–431.

Girard, J.-Y. (1971): Une extension de l'interprétation de Gödel à l'analyse, et son application à l'élimination des coupures dans l'analyse et la théorie des types, in: Fenstad, J.E. (ed.), *Proceedings of the Second Scandinavian Logic Symposium*, North-Holland, Amsterdam, pp. 63–92.

Girard, J.-Y. (1972): Interprétation fonctionnelle et élimination des coupures de l'arithmétique d'ordre supérieur, Ph.D. thesis, Université Paris VII, Paris.

Gödel, K. (1958): Über eine bisher noch nicht benützte Erweiterung des finiten Standpunktes, *Dialectica* 12, 280–287.

Gödel, K. (1964): What is Cantor's continuum problem? in Benacerraf, P. and Putnam, H. (eds.), *Philosophy of Mathematics: Selected Readings*, Prentice-Hall, Englewood Cliffs N.J., pp. 258–272.

Goldfarb, W.D. (1979): Logic in the twenties: The nature of the quantifier, *JSL* 44, 351–368.

Gottlieb, D. (1979): No entity without identity, in: Shahan, R.W. and Swoyer, Ch. (eds.), *Essays on the Philosophy of W.V. Quine*, University of Oklahoma Press, Norman, pp. 79–96.

Griffin, N. (1977): *Relative Identity*, Clarendon Press, Oxford.

Hallett, M. (1984): *Cantorian Set Theory and Limitation of Size*, Clarendon Press, Oxford.

Hatcher, W.S. (1982): *The Logical Foundations of Mathematics*, Pergamon, Oxford.

Heijenoort, J. van (1967): Logic as calculus and logic as language, *Synthese* 17, 324–330.

Heyting, A. (1931): Die intuitionistische Grundlegung der Mathematik, *Erkenntnis* 2, 106–115.

Heyting, A. (1934): *Mathematische Grundlagenforschung. Intuitionismus. Beweistheorie*, Springer, Berlin.

Heyting, A. (1971): *Intuitionism, an Introduction*, North-Holland, Amsterdam, 3rd rev. edn.

Hilbert, D. (1899): *Grundlagen der Geometrie*, Teubner, Leipzig.

Hintikka, M.B. and Hintikka, J. (1986): *Investigating Wittgenstein*, Basil Blackwell, Oxford.

Hintikka, J. (1988): On the development of the model-theoretic viewpoint in logical theory, *Synthese* 77, 1–36.

Howard, W.A. (1980): The formulae-as-types notion of construction, in: Seldin, J.P. and Hindley, J.R. (eds.), *To H.B. Curry: Essays on Combinatory Logic, Lambda Calculus and Formalism*, Academic Press, New York, pp. 480–490; circulated as preprint since 1969.

Husserl, E. (1900,1901): *Logische Untersuchungen*, 2 vols., Max Niemeyer, Halle a.d. Saale I: 1900; II: 1901 (Husserliana XVIII, XIX/1, XIX/2).

Husserl, E. (1913): *Ideen zu einer reinen Phänomenologie und phänomenologischen Philosophie. Erstes Buch: Allgemeine Einführung in die reine Phänomenologie*, Max Niemeyer, Halle a.d. Saale (Husserliana III, III/1).

Husserl, E. (1950): *Cartesianische Meditationen und Pariser Vorträge*, Martinus Nijhoff, Den Haag (Husserliana I).

Husserl, E. (1954): *Die Krisis der europäischen Wissenschaften und die transzendentale Phänomenologie. Eine Einleitung in die phänomenologische Philosophie*, Martinus Nijhoff, Den Haag (Husserliana VI).

Jacobs, B. (1989): The inconsistency of higher order extensions of Martin-Löf's type theory, *Journal of Philosophical Logic* 18, 399–422.

Kneale, W. and Kneale, M. (1978): *The Development of Logic*, Clarendon Press, Oxford.

Kolmogorov, A.N. (1932): Zur Deutung der intuitionistischen Logik, *Mathemat. Zeitschrift* 35, 58–65.

Körner, S. (1955): *Kant*, Penguin, Harmondsworth.

Körner, S. (1960): *The Philosophy of Mathematics. An Introductory Essay*, Hutchinson, London.

Kreisel, G. (1968): Functions, ordinals, species, in: Rootselaar, B. van and Staal, J.F. (eds.), *Logic, Methodology and Philosophy of Science III*, North-Holland, Amsterdam, pp. 145–159.

Kreisel, G. (1970): Church's thesis: A kind of reducibility axiom for constructive mathematics, in: Kino, A. et al. (eds.), *Intuitionism and Proof Theory. Proceedings of the Summer Conference at Buffalo N.Y. 1968*, North-Holland, Amsterdam, pp. 121–150.
Magnusson, L. (1994): The implementation of ALF – a proof editor based on Martin-Löf's monomorphic type theory with explicit substitution, Ph.D. thesis, University of Göteborg, Göteborg.
Neumann, J. von (1931): Die formalistische Grundlegung der Mathematik, *Erkenntnis* 2, 116–121.
Nordström, B. et al. (1990): *Programming in Martin-Löf's Type Theory*, Clarendon Press, Oxford.
Palmgren, E. (1991): On fixed point operators, inductive definitions and universes in Martin-Löf's type theory, Ph.D. thesis, University of Uppsala, Uppsala.
Parsons, Ch. (1967): Foundations of Mathematics, in: Edwards, P. (ed.), *The Encyclopedia of Philosophy*, V/VI, Macmillan, New York, pp. 188–213.
Patzig, G. (1973): Bemerkungen zu den Kategorien des Aristoteles, in: Scheibe, E. and Süssmann, G. (eds.), *Einheit und Vielheit. Festschrift für Carl Friedrich v. Weizsäcker zum 60. Geburtstag*, Vandenhoeck & Ruprecht, Göttingen, pp. 60–76.
Patzig, G. (1973A): Kritische Bemerkungen zu Husserls Thesen über das Verhältnis von Wahrheit und Evidenz, *Neue Hefte für Philosophie* 1, 12–32.
Pfeiffer, J. and Dahan-Dalmedico, A. (1986): *Une histoire des mathématiques*, Ed. du Seuil, Paris.
Pietersma, H. (1977): Husserl's views on the evident and the true, in: Elliston, F.A. and Mc Cormick, P. (eds.), *Husserl. Expositions and Appraisals*, University of Notre Dame Press, Notre Dame, pp. 38–53.
Prawitz, D. (1965): *Natural Deduction. A Proof-Theoretical Study*, Almqvist & Wiksell, Stockholm.
Prawitz, D. (1971): Ideas and results in proof theory, in Fenstad, J.E. (ed.), *Proceedings of the Second Scandinavian Logic Symposium*, North-Holland, Amsterdam, pp. 235–307.
Prawitz, D. (1980): Intuitionistic logic: A philosophical challenge, in: von Wright, G.H. (ed.), *Logic and Philosophy*, Martinus Nijhoff, The Hague, pp. 1–10.
Prawitz, D. (1987): Dummett on a theory of meaning and its impact on logic, in: Taylor, B.M. (ed.), *Michael Dummett. Contributions to Philosophy*, Martinus Nijhoff, Dordrecht, pp. 117–165.
Putnam, H. (1981): *Reason, Truth and History*, CUP, Cambridge.
Quine, W.V. (1960): *Word and Object*, M.I.T. Press, Cambridge Mass.
Quine, W.V. (1969): Speaking of objects, in: Quine, W.V. (ed.), *Ontological Relativity and other Essays*, Columbia UP, New York, pp. 1–25.
Ranta, A. (1988): Propositions as games as types, *Synthese* 76, 377–395.
Ranta, A. (1991): Intuitionistic categorial grammar, *Linguistics and Philosophy* 14, 203–239.
Ranta, A. (1994): *Type-Theoretical Grammar*, Clarendon Press, Oxford.
Russell, B. (1903): *The Principles of Mathematics*, George Allen & Unwin, London.
Rüthing, D. (1984): Some definitions of the concept of function from Joh. Bernoulli to N. Bourbaki, *The Mathematical Intelligencer* 6, 72–77.
Salvesen, A. (1989): Polymorphism and monomorphism in Martin-Löf's type theory, in: On information discharging and retrieval in Martin-Löf's type theory, Ph.D. thesis, University of Oslo, Oslo pp. 62–88.
Sambin, G. (1987): Intuitionistic formal spaces – a first communication, in: Skordev, D. (ed.), *Mathematical Logic and its Applications. Proceedings of an Advanced International Summer School and Conference on Mathematical Logic and its Applications, in Honor of the 80th Anniversary of Kurt Gödel's Birth*, held in September 24 – October 4, 1986, in Druzhba, Bulgaria, Plenum Press, New York, pp. 187–204.
Sambin, G. (1989): Intuitionistic formal spaces and their neighborhood, in: Ferro, R. et al. (eds.), *Logic Colloquium '88*, North-Holland, Amsterdam, pp. 261–285.
Sambin, G. and Valentini, S. (1995): Building up a tool-box for Martin-Löf's type theory I: subset theory, preprint, Department of Mathematics, University of Padova, Padova.
Scholz, H. (1930): Die Axiomatik der Alten, in: Scholz, H. (ed.), *Mathesis Universalis. Abhandlungen zur Philosophie als strenger Wissenschaft*, Benno Schwabe, Basel 1961, pp. 27–44.
Schroeder-Heister, P. (1984): Generalized rules for quantifiers and the completeness of the intuitionistic operators &, $\vee$, $\supset$, $\perp$, $\forall$, $\exists$, in: Richter, M.M. et al. (eds.), *Computation and Proof Theory. Proc. Logic Colloquium Aachen '83*, Springer, Berlin, pp. 399–426.
Schroeder-Heister, P. (1984A): A natural extension of natural deduction, *JSL* 49, 1284–1300.
Shoenfield, J.R. (1967): *Mathematical Logic*, Addison-Wesley, Reading Mass.
Sluga, H. (1980): *Gottlob Frege*, Routledge & Kegan Paul, Boston.

Smith, J. (1978): On the relation between a type theoretic and a logical formulation of the theory of constructions, Ph.D. thesis, University of Göteborg, Göteborg.

Smith, J. (1983): The identification of propositions and types in Martin-Löf's type theory, in: Karpinski, M. (ed.), *Foundations of Computation Theory: Proceedings of the 1983 International FCT-Conference*; Borgholm, August 21–27, 1983, Springer, Berlin, pp. 445–456.

Smith, J. (1984): An interpretation of Martin-Löf's type theory in a typefree theory of propositions, *JSL* 49, 730–753.

Solomon, R.C. (1970): Sense and essence: Frege and Husserl, *International Philosophical Quarterly* 10, 378–401.

Stegmüller, W. (1956,1957): Das Universalienproblem einst und jetzt, *Archiv für Philosophie*, 6 (1956) 192–225; 7 (1957) 45–81.

Stegmüller, W. (1969): *Metaphysik Skepsis Wissenschaft*, Springer, Berlin 1969, 2nd improved edn.

Stegmüller, W. (1978): Einleitung, in: Stegmüller, W. (ed.), *Das Universalien-Problem*, Wissenschaft. Buchgesellschaft, Darmstadt, pp. 1–19.

Sundholm, G. (1988): Oordeel en gevolgtrekking. Bedreigde species? Rede, uitgesproken bij de aanvaarding van het ambt van gewoon hoogleraar in de Wijsbegeerte, in het bijzonder de Wijsbegeerte van de Logica, aan de Rijksuniversiteit te Leiden op vrijdag 9 september 1988 door dr. B.G. Sundholm, Leiden University, Leiden.

Sundholm, G. (1990): Sätze der Logik: an alternative conception, in: Haller, R. and Brandl, J. (eds.), *Wittgenstein - Eine Neubewertung. Akten des 14. Internationalen Wittgenstein-Symposiums. Feier des 100. Geburtstages. 13. bis 20. August 1989 Kirchberg am Wechsel (Österreich)*, Hölder-Pichler-Tempsky, Wien, pp. 59–61.

Sundholm, G. (1993): Questions of proof, *Manuscrito* 16, 47–70.

Sundholm, G. (1994): Existence, proof and truth-making: A perspective on the intuitionistic conception of truth, *Topoi* 13, 117–126.

Svensson, C. (1990): A normalization proof for Martin-Löf's type theory, Masters thesis, Department of Computer Science, University of Göteborg, Göteborg.

Swaen, M. (1989): Weak and strong sum-elimination in intuitionistic type theory, Ph.D. thesis, University of Amsterdam, Amsterdam.

Tait, W.W. (1965): Infinitely long terms of transfinite type, in: Crossley, J.N. and Dummett, M.A.E. (eds.), *Formal Systems and Recursive Functions*, North-Holland, Amsterdam, pp. 176–185.

Tait, W.W. (1967): Intensional interpretation of functionals of finite type I, *JSL* 32, 198–212.

Tait, W.W. (1968): Constructive reasoning, in: Rootselaar, B. van and Staal, J.F. (eds.), *Logic, Methodology and Philosophy of Science III*, North-Holland, Amsterdam, pp. 185–199.

Tasistro, A. (1993): Formulation of Martin-Löf's theory of types with explicit substitutions, Masters thesis, Department of Computer Science, University of Göteborg, Göteborg.

Thompson, M. (1967): Categories, in: Edwards, P. (ed.), *The Encyclopedia of Philosophy*, Vol. I/II, Macmillan, New York, pp. 46–55.

Thompson, S. (1991): *Type Theory and Functional Programming*, Addison-Wesley, Wokingham.

Troelstra, A.S. (1969): *Principles of Intuitionism*, Springer, Berlin.

Troelstra, A.S. (ed.) (1973): *Metamathematical Investigation of Intuitionistic Arithmetic and Analysis*, Springer, Berlin.

Troelstra, A.S. (1990): Introductory note to 1958 and 1972, in: Gödel, K., *Collected Works*, II, OUP, New York, pp. 217–241.

Troelstra, A.S. (1991): Remarks on intuitionism and the philosophy of mathematics, in: *Atti del Congresso* Nuovi problemi della logica e della filosofia della scienza. *Viareggio, 8–13 gennaio 1990*, II, CLUEB, Bologna, pp. 211–228.

Troelstra, A.S. and Dalen, D. van (1988): *Constructivism in Mathematics. An Introduction*, 2 vols., North-Holland, Amsterdam.

Troelstra, A.S. and Schwichtenberg, H. (1996): *Basic Proof Theory*, CUP, Cambridge.

Whitehead, A.N. and Russell, B. (1910,1912,1913): *Principia Mathematica*, CUP, Cambridge I: 1910, II: 1912, III: 1913.

Wittgenstein, L. (1922): *Tractatus logico-philosophicus*, Routledge & Kegan Paul, London.

Writings by Per Martin-Löf Available to Me

(1968): *Notes on Constructive Mathematics*, Almqvist & Wiksell, Stockholm, 1970.
(1969): Infinite terms and a system of natural deduction, *Compositio Mathematica* 24 (1972) 93–103.
(1970): Hauptsatz for the intuitionistic theory of iterated inductive definitions, in: Fenstad, J.E. (ed.), *Proceedings of the Second Scandinavian Logic Symposium*, North-Holland, Amsterdam, 1971, pp. 179–216.
(1970A): Hauptsatz for the theory of species, in: Fenstad, J.E. (ed.), *Proceedings of the Second Scandinavian Logic Symposium*, North-Holland, Amsterdam, 1971, pp. 217–233.
(1970B): A construction of the provable wellorderings of the theory of species (unpublished typed paper).
(1971): Hauptsatz for intuitionistic simple type theory, in: Suppes, P. et al. (eds.), *Logic, Methodology and Philosophy of Science IV*, North-Holland, Amsterdam, 1973, pp. 279–290.
(1971A): A theory of types. Technical report 71-3, University of Stockholm, Stockholm, 1971.
(1971B): On the strength of intuitionistic reasoning (Contribution to the *Symposium on Perspectives in the Philosophy of Mathematics at the IVth International Congress for Logic, Methodology and Philosophy of Science*, Bucharest, August 29 – September 4 (1971) (unpublished typed paper).
(1972): An intuitionistic theory of types, Technical report, University of Stockholm, Stockholm, 1972.
(1972A): About models for intuitionistic type theories and the notion of definitional equality, in: Kanger, S. (ed.), *Proceedings of the Third Scandinavian Logic Symposium*, North-Holland, Amsterdam, 1975, pp. 81–109.
(1973): An intuitionistic theory of types: predicative part, in: Rose, H.E. and Shepherdson, J. (eds.), *Logic Colloquium '73*, North-Holland, Amsterdam, 1975, pp. 73–118.
(1975) together with Peter Hancock: Syntax and semantics of the language of primitive recursive functions, preprint No. 3, Department of Mathematics, University of Stockholm, Stockholm, 1975.
(1975A): Syntax and semantics of mathematical language (Notes written by Peter Hancock on half of a lecture series given by Per Martin-Löf at Oxford University in Michaelmas Term 1975) (unpublished typed lecture notes).
(1979): Constructive mathematics and computer programming, in: Cohen, L.J. et al. (eds.), *Logic, Methodology and Philosophy of Science VI*, North-Holland, Amsterdam, 1982, pp. 153–175.
(1980): *Intuitionistic Type Theory. Notes by Giovanni Sambin of a Series of Lectures given in Padua*, June 1980, Bibliopolis, Napoli, 1984.
(1983): Outline of the Siena lectures, March–April 1983 (unpublished handwritten lecture notes).
(1983A): On the meanings of the logical constants and the justifications of the logical laws, in: Bernardi, C. and Pagli, P. (eds.), *Atti degli incontri di logica matematica*, II, Scuola di Specializzazione in Logica Matematica, Dip. di Matematica, Università di Siena, Siena, 1984, pp. 203–281.
(1983B): Notes on the domain interpretation of type theory (unpublished handwritten notes).
(1983C): Unifying Scott's theory of domains for denotational semantics and intuitionistic type theory (Abstract), in: *Atti del Congresso* Logica e filosofia della scienza, oggi. *San Gimignano, 7–11 dicembre 1983*, I, CLUEB, Bologna, 1986.
(1985): Truth of a proposition, evidence of a judgement, validity of a proof, *Synthese* 73 (1987) 407–420.
(1986): The category structure of intuitionistic type theory (Talk given at the Société Belge de Logique et de Philosophie des Sciences, 26 avril 1986) (unpublished handwritten notes of a talk).
(1987): Philosophical implications of type theory (Series of lectures given at the Istituto di Filosofia, Facoltà di Lettere e Filosofia, Università degli Studi di Firenze, March–May 1987) (unpublished handwritten lecture notes).
(1988): Mathematics of infinity, in: Martin-Löf, P. and Mints, G. (eds.), *Colog 88, International Conference on Computer Logic*, Tallinn, December 1988, Springer, Berlin, 1990, pp. 146–197.
(1990): A path from logic to metaphysics, in: *Atti del Congresso* Nuovi problemi della logica e della filosofia della scienza. *Viareggio, 8–13 gennaio 1990*, II, CLUEB, Bologna, 1991, pp. 141–149.
(1990–91): Choice sequences in type theory (Notes written by Petri Mäenpää and Aarne Ranta on a lecture series given by Per Martin-Löf at Stockholm University in 1990–1991) (unpublished handwritten lecture notes).

(1992): On the relation between mathematics, logic and the theory of knowledge (Talk given in Paris, April 1992) (unpublished typed paper).

(1992A): Analytic and synthetic judgements in type theory (Talk given in Florence at the workshop on Kant and Contemporary Epistemology, 27–30 May 1992) (unpublished typed paper).

(1993): Philosophical aspects of intuitionistic type theory (Notes written by Giovanni Sommaruga–Rosolemos on a lecture series given by Per Martin-Löf at Leiden University, September–December 1993) (unpublished handwritten lecture notes) (Addendum: in the meantime there also exists a transcription of these lectures by M. Wijers).

(1994): Truth and knowability (Notes written by Göran Sundholm of a talk given at Stockholm University, 18 March 1994) (unpublished typed paper).

(1994A): Verificationism then and now (Talk given at the conference *The Foundational Debate: Complexity and Constructivity in Mathematics and Physics*, Vienna, 15–17 September 1994) (unpublished typed paper).

Note: the dates in brackets are the dates when the resp. writings were composed or called forth by Martin-Löf.

Version (1) of constructive type theory is (1971A).
Version (2) is (1972).
Version (3) is (1973).
Version (4) is (1975A).
Version (5) is (1979).
Version (6) is (1980).
Version (6a) is the Preface of (1980).
Version (7) is (1987).
Version (8) is (1993).

1. J. M. Bochénski, *A Precis of Mathematical Logic.* Translated from French and German by O. Bird. 1959 ISBN 90-277-0073-7
2. P. Guiraud, *Problèmes et méthodes de la statistique linguistique.* 1959 ISBN 90-277-0025-7
3. H. Freudenthal (ed.), *The Concept and the Role of the Model in Mathematics and Natural and Social Sciences.* 1961 ISBN 90-277-0017-6
4. E. W. Beth, *Formal Methods.* An Introduction to Symbolic Logic and to the Study of Effective Operations in Arithmetic and Logic. 1962 ISBN 90-277-0069-9
5. B. H. Kazemier and D. Vuysje (eds.), *Logic and Language.* Studies dedicated to Professor Rudolf Carnap on the Occasion of His 70th Birthday. 1962 ISBN 90-277-0019-2
6. M. W. Wartofsky (ed.), *Proceedings of the Boston Colloquium for the Philosophy of Science, 1961–1962.* [Boston Studies in the Philosophy of Science, Vol. I] 1963 ISBN 90-277-0021-4
7. A. A. Zinov'ev, *Philosophical Problems of Many-valued Logic.* A revised edition, edited and translated (from Russian) by G. Küng and D.D. Comey. 1963 ISBN 90-277-0091-5
8. G. Gurvitch, *The Spectrum of Social Time.* Translated from French and edited by M. Korenbaum and P. Bosserman. 1964 ISBN 90-277-0006-0
9. P. Lorenzen, *Formal Logic.* Translated from German by F.J. Crosson. 1965 ISBN 90-277-0080-X
10. R. S. Cohen and M. W. Wartofsky (eds.), *Proceedings of the Boston Colloquium for the Philosophy of Science, 1962–1964.* In Honor of Philipp Frank. [Boston Studies in the Philosophy of Science, Vol. II] 1965 ISBN 90-277-9004-0
11. E. W. Beth, *Mathematical Thought.* An Introduction to the Philosophy of Mathematics. 1965 ISBN 90-277-0070-2
12. E. W. Beth and J. Piaget, *Mathematical Epistemology and Psychology.* Translated from French by W. Mays. 1966 ISBN 90-277-0071-0
13. G. Küng, *Ontology and the Logistic Analysis of Language.* An Enquiry into the Contemporary Views on Universals. Revised ed., translated from German. 1967 ISBN 90-277-0028-1
14. R. S. Cohen and M. W. Wartofsky (eds.), *Proceedings of the Boston Colloquium for the Philosophy of Sciences, 1964–1966.* In Memory of Norwood Russell Hanson. [Boston Studies in the Philosophy of Science, Vol. III] 1967 ISBN 90-277-0013-3
15. C. D. Broad, *Induction, Probability, and Causation.* Selected Papers. 1968 ISBN 90-277-0012-5
16. G. Patzig, *Aristotle's Theory of the Syllogism.* A Logical-philosophical Study of *Book A* of the *Prior Analytics.* Translated from German by J. Barnes. 1968 ISBN 90-277-0030-3
17. N. Rescher, *Topics in Philosophical Logic.* 1968 ISBN 90-277-0084-2
18. R. S. Cohen and M. W. Wartofsky (eds.), *Proceedings of the Boston Colloquium for the Philosophy of Science, 1966–1968, Part I.* [Boston Studies in the Philosophy of Science, Vol. IV] 1969 ISBN 90-277-0014-1
19. R. S. Cohen and M. W. Wartofsky (eds.), *Proceedings of the Boston Colloquium for the Philosophy of Science, 1966–1968, Part II.* [Boston Studies in the Philosophy of Science, Vol. V] 1969 ISBN 90-277-0015-X
20. J. W. Davis, D. J. Hockney and W. K. Wilson (eds.), *Philosophical Logic.* 1969 ISBN 90-277-0075-3
21. D. Davidson and J. Hintikka (eds.), *Words and Objections.* Essays on the Work of W. V. Quine. 1969, rev. ed. 1975 ISBN 90-277-0074-5; Pb 90-277-0602-6
22. P. Suppes, *Studies in the Methodology and Foundations of Science. Selected Papers from 1951 to 1969.* 1969 ISBN 90-277-0020-6

23. J. Hintikka, *Models for Modalities.* Selected Essays. 1969
ISBN 90-277-0078-8; Pb 90-277-0598-4
24. N. Rescher *et al.* (eds.), *Essays in Honor of Carl G. Hempel.* A Tribute on the Occasion of His 65th Birthday. 1969 ISBN 90-277-0085-0
25. P. V. Tavanec (ed.), *Problems of the Logic of Scientific Knowledge.* Translated from Russian. 1970 ISBN 90-277-0087-7
26. M. Swain (ed.), *Induction, Acceptance, and Rational Belief.* 1970 ISBN 90-277-0086-9
27. R. S. Cohen and R. J. Seeger (eds.), *Ernst Mach: Physicist and Philosopher.* [Boston Studies in the Philosophy of Science, Vol. VI]. 1970 ISBN 90-277-0016-8
28. J. Hintikka and P. Suppes, *Information and Inference.* 1970 ISBN 90-277-0155-5
29. K. Lambert, *Philosophical Problems in Logic.* Some Recent Developments. 1970
ISBN 90-277-0079-6
30. R. A. Eberle, *Nominalistic Systems.* 1970 ISBN 90-277-0161-X
31. P. Weingartner and G. Zecha (eds.), *Induction, Physics, and Ethics.* 1970 ISBN 90-277-0158-X
32. E. W. Beth, *Aspects of Modern Logic.* Translated from Dutch. 1970 ISBN 90-277-0173-3
33. R. Hilpinen (ed.), *Deontic Logic.* Introductory and Systematic Readings. 1971
See also No. 152. ISBN Pb (1981 rev.) 90-277-1302-2
34. J.-L. Krivine, *Introduction to Axiomatic Set Theory.* Translated from French. 1971
ISBN 90-277-0169-5; Pb 90-277-0411-2
35. J. D. Sneed, *The Logical Structure of Mathematical Physics.* 2nd rev. ed., 1979
ISBN 90-277-1056-2; Pb 90-277-1059-7
36. C. R. Kordig, *The Justification of Scientific Change.* 1971
ISBN 90-277-0181-4; Pb 90-277-0475-9
37. M. Čapek, *Bergson and Modern Physics.* A Reinterpretation and Re-evaluation. [Boston Studies in the Philosophy of Science, Vol. VII] 1971 ISBN 90-277-0186-5
38. N. R. Hanson, *What I Do Not Believe, and Other Essays.* Ed. by S. Toulmin and H. Woolf. 1971 ISBN 90-277-0191-1
39. R. C. Buck and R. S. Cohen (eds.), *PSA 1970.* Proceedings of the Second Biennial Meeting of the Philosophy of Science Association, Boston, Fall 1970. In Memory of Rudolf Carnap. [Boston Studies in the Philosophy of Science, Vol. VIII] 1971
ISBN 90-277-0187-3; Pb 90-277-0309-4
40. D. Davidson and G. Harman (eds.), *Semantics of Natural Language.* 1972
ISBN 90-277-0304-3; Pb 90-277-0310-8
41. Y. Bar-Hillel (ed.), *Pragmatics of Natural Languages.* 1971
ISBN 90-277-0194-6; Pb 90-277-0599-2
42. S. Stenlund, *Combinators, γ Terms and Proof Theory.* 1972 ISBN 90-277-0305-1
43. M. Strauss, *Modern Physics and Its Philosophy.* Selected Paper in the Logic, History, and Philosophy of Science. 1972 ISBN 90-277-0230-6
44. M. Bunge, *Method, Model and Matter.* 1973 ISBN 90-277-0252-7
45. M. Bunge, *Philosophy of Physics.* 1973 ISBN 90-277-0253-5
46. A. A. Zinov'ev, *Foundations of the Logical Theory of Scientific Knowledge (Complex Logic).* Revised and enlarged English edition with an appendix by G. A. Smirnov, E. A. Sidorenka, A. M. Fedina and L. A. Bobrova. [Boston Studies in the Philosophy of Science, Vol. IX] 1973
ISBN 90-277-0193-8; Pb 90-277-0324-8
47. L. Tondl, *Scientific Procedures.* A Contribution concerning the Methodological Problems of Scientific Concepts and Scientific Explanation. Translated from Czech by D. Short. Edited by R.S. Cohen and M.W. Wartofsky. [Boston Studies in the Philosophy of Science, Vol. X] 1973
ISBN 90-277-0147-4; Pb 90-277-0323-X

48. N. R. Hanson, *Constellations and Conjectures.* 1973 ISBN 90-277-0192-X
49. K. J. J. Hintikka, J. M. E. Moravcsik and P. Suppes (eds.), *Approaches to Natural Language.* 1973 ISBN 90-277-0220-9; Pb 90-277-0233-0
50. M. Bunge (ed.), *Exact Philosophy.* Problems, Tools and Goals. 1973 ISBN 90-277-0251-9
51. R. J. Bogdan and I. Niiniluoto (eds.), *Logic, Language and Probability.* 1973 ISBN 90-277-0312-4
52. G. Pearce and P. Maynard (eds.), *Conceptual Change.* 1973 ISBN 90-277-0287-X; Pb 90-277-0339-6
53. I. Niiniluoto and R. Tuomela, *Theoretical Concepts and Hypothetico-inductive Inference.* 1973 ISBN 90-277-0343-4
54. R. Fraissé, *Course of Mathematical Logic* – Volume 1: *Relation and Logical Formula.* Translated from French. 1973 ISBN 90-277-0268-3; Pb 90-277-0403-1 (For *Volume 2* see under No. 69).
55. A. Grünbaum, *Philosophical Problems of Space and Time.* Edited by R.S. Cohen and M.W. Wartofsky. 2nd enlarged ed. [Boston Studies in the Philosophy of Science, Vol. XII] 1973 ISBN 90-277-0357-4; Pb 90-277-0358-2
56. P. Suppes (ed.), *Space, Time and Geometry.* 1973 ISBN 90-277-0386-8; Pb 90-277-0442-2
57. H. Kelsen, *Essays in Legal and Moral Philosophy*. Selected and introduced by O. Weinberger. Translated from German by P. Heath. 1973 ISBN 90-277-0388-4
58. R. J. Seeger and R. S. Cohen (eds.), *Philosophical Foundations of Science.* [Boston Studies in the Philosophy of Science, Vol. XI] 1974 ISBN 90-277-0390-6; Pb 90-277-0376-0
59. R. S. Cohen and M. W. Wartofsky (eds.), *Logical and Epistemological Studies in Contemporary Physics.* [Boston Studies in the Philosophy of Science, Vol. XIII] 1973 ISBN 90-277-0391-4; Pb 90-277-0377-9
60. R. S. Cohen and M. W. Wartofsky (eds.), *Methodological and Historical Essays in the Natural and Social Sciences. Proceedings of the Boston Colloquium for the Philosophy of Science, 1969–1972.* [Boston Studies in the Philosophy of Science, Vol. XIV] 1974 ISBN 90-277-0392-2; Pb 90-277-0378-7
61. R. S. Cohen, J. J. Stachel and M. W. Wartofsky (eds.), *For Dirk Struik. Scientific, Historical and Political Essays*. [Boston Studies in the Philosophy of Science, Vol. XV] 1974 ISBN 90-277-0393-0; Pb 90-277-0379-5
62. K. Ajdukiewicz, *Pragmatic Logic*. Translated from Polish by O. Wojtasiewicz. 1974 ISBN 90-277-0326-4
63. S. Stenlund (ed.), *Logical Theory and Semantic Analysis.* Essays dedicated to Stig Kanger on His 50th Birthday. 1974 ISBN 90-277-0438-4
64. K. F. Schaffner and R. S. Cohen (eds.), *PSA 1972. Proceedings of the Third Biennial Meeting of the Philosophy of Science Association.* [Boston Studies in the Philosophy of Science, Vol. XX] 1974 ISBN 90-277-0408-2; Pb 90-277-0409-0
65. H. E. Kyburg, Jr., *The Logical Foundations of Statistical Inference.* 1974 ISBN 90-277-0330-2; Pb 90-277-0430-9
66. M. Grene, *The Understanding of Nature.* Essays in the Philosophy of Biology. [Boston Studies in the Philosophy of Science, Vol. XXIII] 1974 ISBN 90-277-0462-7; Pb 90-277-0463-5
67. J. M. Broekman, *Structuralism: Moscow, Prague, Paris*. Translated from German. 1974 ISBN 90-277-0478-3
68. N. Geschwind, *Selected Papers on Language and the Brain.* [Boston Studies in the Philosophy of Science, Vol. XVI] 1974 ISBN 90-277-0262-4; Pb 90-277-0263-2

69. R. Fraissé, *Course of Mathematical Logic* – Volume 2: *Model Theory.* Translated from French. 1974 ISBN 90-277-0269-1; Pb 90-277-0510-0
(For *Volume 1* see under No. 54)
70. A. Grzegorczyk, *An Outline of Mathematical Logic.* Fundamental Results and Notions explained with all Details. Translated from Polish. 1974 ISBN 90-277-0359-0; Pb 90-277-0447-3
71. F. von Kutschera, *Philosophy of Language.* 1975 ISBN 90-277-0591-7
72. J. Manninen and R. Tuomela (eds.), *Essays on Explanation and Understanding.* Studies in the Foundations of Humanities and Social Sciences. 1976 ISBN 90-277-0592-5
73. J. Hintikka (ed.), *Rudolf Carnap, Logical Empiricist.* Materials and Perspectives. 1975 ISBN 90-277-0583-6
74. M. Čapek (ed.), *The Concepts of Space and Time.* Their Structure and Their Development. [Boston Studies in the Philosophy of Science, Vol. XXII] 1976 ISBN 90-277-0355-8; Pb 90-277-0375-2
75. J. Hintikka and U. Remes, *The Method of Analysis.* Its Geometrical Origin and Its General Significance. [Boston Studies in the Philosophy of Science, Vol. XXV] 1974 ISBN 90-277-0532-1; Pb 90-277-0543-7
76. J. E. Murdoch and E. D. Sylla (eds.), *The Cultural Context of Medieval Learning.* [Boston Studies in the Philosophy of Science, Vol. XXVI] 1975 ISBN 90-277-0560-7; Pb 90-277-0587-9
77. S. Amsterdamski, *Between Experience and Metaphysics.* Philosophical Problems of the Evolution of Science. [Boston Studies in the Philosophy of Science, Vol. XXXV] 1975 ISBN 90-277-0568-2; Pb 90-277-0580-1
78. P. Suppes (ed.), *Logic and Probability in Quantum Mechanics.* 1976 ISBN 90-277-0570-4; Pb 90-277-1200-X
79. H. von Helmholtz: *Epistemological Writings. The Paul Hertz / Moritz Schlick Centenary Edition of 1921 with Notes and Commentary by the Editors.* Newly translated from German by M. F. Lowe. Edited, with an Introduction and Bibliography, by R. S. Cohen and Y. Elkana. [Boston Studies in the Philosophy of Science, Vol. XXXVII] 1975 ISBN 90-277-0290-X; Pb 90-277-0582-8
80. J. Agassi, *Science in Flux.* [Boston Studies in the Philosophy of Science, Vol. XXVIII] 1975 ISBN 90-277-0584-4; Pb 90-277-0612-2
81. S. G. Harding (ed.), *Can Theories Be Refuted?* Essays on the Duhem-Quine Thesis. 1976 ISBN 90-277-0629-8; Pb 90-277-0630-1
82. S. Nowak, *Methodology of Sociological Research.* General Problems. 1977 ISBN 90-277-0486-4
83. J. Piaget, J.-B. Grize, A. Szeminśska and V. Bang, *Epistemology and Psychology of Functions.* Translated from French. 1977 ISBN 90-277-0804-5
84. M. Grene and E. Mendelsohn (eds.), *Topics in the Philosophy of Biology.* [Boston Studies in the Philosophy of Science, Vol. XXVII] 1976 ISBN 90-277-0595-X; Pb 90-277-0596-8
85. E. Fischbein, *The Intuitive Sources of Probabilistic Thinking in Children.* 1975 ISBN 90-277-0626-3; Pb 90-277-1190-9
86. E. W. Adams, *The Logic of Conditionals.* An Application of Probability to Deductive Logic. 1975 ISBN 90-277-0631-X
87. M. Przełęcki and R. Wójcicki (eds.), *Twenty-Five Years of Logical Methodology in Poland.* Translated from Polish. 1976 ISBN 90-277-0601-8
88. J. Topolski, *The Methodology of History.* Translated from Polish by O. Wojtasiewicz. 1976 ISBN 90-277-0550-X

89. A. Kasher (ed.), *Language in Focus: Foundations, Methods and Systems.* Essays dedicated to Yehoshua Bar-Hillel. [Boston Studies in the Philosophy of Science, Vol. XLIII] 1976 ISBN 90-277-0644-1; Pb 90-277-0645-X
90. J. Hintikka, *The Intentions of Intentionality and Other New Models for Modalities.* 1975 ISBN 90-277-0633-6; Pb 90-277-0634-4
91. W. Stegmüller, *Collected Papers on Epistemology, Philosophy of Science and History of Philosophy.* 2 Volumes. 1977 Set ISBN 90-277-0767-7
92. D. M. Gabbay, *Investigations in Modal and Tense Logics with Applications to Problems in Philosophy and Linguistics.* 1976 ISBN 90-277-0656-5
93. R. J. Bogdan, *Local Induction.* 1976 ISBN 90-277-0649-2
94. S. Nowak, *Understanding and Prediction.* Essays in the Methodology of Social and Behavioral Theories. 1976 ISBN 90-277-0558-5; Pb 90-277-1199-2
95. P. Mittelstaedt, *Philosophical Problems of Modern Physics.* [Boston Studies in the Philosophy of Science, Vol. XVIII] 1976 ISBN 90-277-0285-3; Pb 90-277-0506-2
96. G. Holton and W. A. Blanpied (eds.), *Science and Its Public: The Changing Relationship.* [Boston Studies in the Philosophy of Science, Vol. XXXIII] 1976 ISBN 90-277-0657-3; Pb 90-277-0658-1
97. M. Brand and D. Walton (eds.), *Action Theory.* 1976 ISBN 90-277-0671-9
98. P. Gochet, *Outline of a Nominalist Theory of Propositions.* An Essay in the Theory of Meaning and in the Philosophy of Logic. 1980 ISBN 90-277-1031-7
99. R. S. Cohen, P. K. Feyerabend, and M. W. Wartofsky (eds.), *Essays in Memory of Imre Lakatos.* [Boston Studies in the Philosophy of Science, Vol. XXXIX] 1976 ISBN 90-277-0654-9; Pb 90-277-0655-7
100. R. S. Cohen and J. J. Stachel (eds.), *Selected Papers of Léon Rosenfield.* [Boston Studies in the Philosophy of Science, Vol. XXI] 1979 ISBN 90-277-0651-4; Pb 90-277-0652-2
101. R. S. Cohen, C. A. Hooker, A. C. Michalos and J. W. van Evra (eds.), *PSA 1974. Proceedings of the 1974 Biennial Meeting of the Philosophy of Science Association.* [Boston Studies in the Philosophy of Science, Vol. XXXII] 1976 ISBN 90-277-0647-6; Pb 90-277-0648-4
102. Y. Fried and J. Agassi, *Paranoia.* A Study in Diagnosis. [Boston Studies in the Philosophy of Science, Vol. L] 1976 ISBN 90-277-0704-9; Pb 90-277-0705-7
103. M. Przełęcki, K. Szaniawski and R. Wójcicki (eds.), *Formal Methods in the Methodology of Empirical Sciences.* 1976 ISBN 90-277-0698-0
104. J. M. Vickers, *Belief and Probability.* 1976 ISBN 90-277-0744-8
105. K. H. Wolff, *Surrender and Catch.* Experience and Inquiry Today. [Boston Studies in the Philosophy of Science, Vol. LI] 1976 ISBN 90-277-0758-8; Pb 90-277-0765-0
106. K. Kosík, *Dialectics of the Concrete.* A Study on Problems of Man and World. [Boston Studies in the Philosophy of Science, Vol. LII] 1976 ISBN 90-277-0761-8; Pb 90-277-0764-2
107. N. Goodman, *The Structure of Appearance.* 3rd ed. with an Introduction by G. Hellman. [Boston Studies in the Philosophy of Science, Vol. LIII] 1977 ISBN 90-277-0773-1; Pb 90-277-0774-X
108. K. Ajdukiewicz, *The Scientific World-Perspective and Other Essays, 1931-1963.* Translated from Polish. Edited and with an Introduction by J. Giedymin. 1978 ISBN 90-277-0527-5
109. R. L. Causey, *Unity of Science.* 1977 ISBN 90-277-0779-0
110. R. E. Grandy, *Advanced Logic for Applications.* 1977 ISBN 90-277-0781-2
111. R. P. McArthur, *Tense Logic.* 1976 ISBN 90-277-0697-2
112. L. Lindahl, *Position and Change.* A Study in Law and Logic. Translated from Swedish by P. Needham. 1977 ISBN 90-277-0787-1
113. R. Tuomela, *Dispositions.* 1978 ISBN 90-277-0810-X

114. H. A. Simon, *Models of Discovery and Other Topics in the Methods of Science.* [Boston Studies in the Philosophy of Science, Vol. LIV] 1977 ISBN 90-277-0812-6; Pb 90-277-0858-4
115. R. D. Rosenkrantz, *Inference, Method and Decision.* Towards a Bayesian Philosophy of Science. 1977 ISBN 90-277-0817-7; Pb 90-277-0818-5
116. R. Tuomela, *Human Action and Its Explanation.* A Study on the Philosophical Foundations of Psychology. 1977 ISBN 90-277-0824-X
117. M. Lazerowitz, *The Language of Philosophy.* Freud and Wittgenstein. [Boston Studies in the Philosophy of Science, Vol. LV] 1977 ISBN 90-277-0826-6; Pb 90-277-0862-2
118. Not published 119. J. Pelc (ed.), *Semiotics in Poland, 1894–1969.* Translated from Polish. 1979 ISBN 90-277-0811-8
120. I. Pörn, *Action Theory and Social Science.* Some Formal Models. 1977 ISBN 90-277-0846-0
121. J. Margolis, *Persons and Mind.* The Prospects of Nonreductive Materialism. [Boston Studies in the Philosophy of Science, Vol. LVII] 1977 ISBN 90-277-0854-1; Pb 90-277-0863-0
122. J. Hintikka, I. Niiniluoto, and E. Saarinen (eds.), *Essays on Mathematical and Philosophical Logic.* 1979 ISBN 90-277-0879-7
123. T. A. F. Kuipers, *Studies in Inductive Probability and Rational Expectation.* 1978 ISBN 90-277-0882-7
124. E. Saarinen, R. Hilpinen, I. Niiniluoto and M. P. Hintikka (eds.), *Essays in Honour of Jaakko Hintikka on the Occasion of His 50th Birthday.* 1979 ISBN 90-277-0916-5
125. G. Radnitzky and G. Andersson (eds.), *Progress and Rationality in Science.* [Boston Studies in the Philosophy of Science, Vol. LVIII] 1978 ISBN 90-277-0921-1; Pb 90-277-0922-X
126. P. Mittelstaedt, *Quantum Logic.* 1978 ISBN 90-277-0925-4
127. K. A. Bowen, *Model Theory for Modal Logic.* Kripke Models for Modal Predicate Calculi. 1979 ISBN 90-277-0929-7
128. H. A. Bursen, *Dismantling the Memory Machine.* A Philosophical Investigation of Machine Theories of Memory. 1978 ISBN 90-277-0933-5
129. M. W. Wartofsky, *Models.* Representation and the Scientific Understanding. [Boston Studies in the Philosophy of Science, Vol. XLVIII] 1979 ISBN 90-277-0736-7; Pb 90-277-0947-5
130. D. Ihde, *Technics and Praxis.* A Philosophy of Technology. [Boston Studies in the Philosophy of Science, Vol. XXIV] 1979 ISBN 90-277-0953-X; Pb 90-277-0954-8
131. J. J. Wiatr (ed.), *Polish Essays in the Methodology of the Social Sciences.* [Boston Studies in the Philosophy of Science, Vol. XXIX] 1979 ISBN 90-277-0723-5; Pb 90-277-0956-4
132. W. C. Salmon (ed.), *Hans Reichenbach: Logical Empiricist.* 1979 ISBN 90-277-0958-0
133. P. Bieri, R.-P. Horstmann and L. Krüger (eds.), *Transcendental Arguments in Science.* Essays in Epistemology. 1979 ISBN 90-277-0963-7; Pb 90-277-0964-5
134. M. Marković and G. Petrović (eds.), *Praxis.* Yugoslav Essays in the Philosophy and Methodology of the Social Sciences. [Boston Studies in the Philosophy of Science, Vol. XXXVI] 1979 ISBN 90-277-0727-8; Pb 90-277-0968-8
135. R. Wójcicki, *Topics in the Formal Methodology of Empirical Sciences.* Translated from Polish. 1979 ISBN 90-277-1004-X
136. G. Radnitzky and G. Andersson (eds.), *The Structure and Development of Science.* [Boston Studies in the Philosophy of Science, Vol. LIX] 1979 ISBN 90-277-0994-7; Pb 90-277-0995-5
137. J. C. Webb, *Mechanism, Mentalism and Metamathematics.* An Essay on Finitism. 1980 ISBN 90-277-1046-5
138. D. F. Gustafson and B. L. Tapscott (eds.), *Body, Mind and Method.* Essays in Honor of Virgil C. Aldrich. 1979 ISBN 90-277-1013-9

139. L. Nowak, *The Structure of Idealization.* Towards a Systematic Interpretation of the Marxian Idea of Science. 1980 ISBN 90-277-1014-7
140. C. Perelman, *The New Rhetoric and the Humanities.* Essays on Rhetoric and Its Applications. Translated from French and German. With an Introduction by H. Zyskind. 1979 ISBN 90-277-1018-X; Pb 90-277-1019-8
141. W. Rabinowicz, *Universalizability.* A Study in Morals and Metaphysics. 1979 ISBN 90-277-1020-2
142. C. Perelman, *Justice, Law and Argument.* Essays on Moral and Legal Reasoning. Translated from French and German. With an Introduction by H.J. Berman. 1980 ISBN 90-277-1089-9; Pb 90-277-1090-2
143. S. Kanger and S. Öhman (eds.), *Philosophy and Grammar.* Papers on the Occasion of the Quincentennial of Uppsala University. 1981 ISBN 90-277-1091-0
144. T. Pawlowski, *Concept Formation in the Humanities and the Social Sciences.* 1980 ISBN 90-277-1096-1
145. J. Hintikka, D. Gruender and E. Agazzi (eds.), *Theory Change, Ancient Axiomatics and Galileo's Methodology.* Proceedings of the 1978 Pisa Conference on the History and Philosophy of Science, Volume I. 1981 ISBN 90-277-1126-7
146. J. Hintikka, D. Gruender and E. Agazzi (eds.), *Probabilistic Thinking, Thermodynamics, and the Interaction of the History and Philosophy of Science.* Proceedings of the 1978 Pisa Conference on the History and Philosophy of Science, Volume II. 1981 ISBN 90-277-1127-5
147. U. Mönnich (ed.), *Aspects of Philosophical Logic.* Some Logical Forays into Central Notions of Linguistics and Philosophy. 1981 ISBN 90-277-1201-8
148. D. M. Gabbay, *Semantical Investigations in Heyting's Intuitionistic Logic.* 1981 ISBN 90-277-1202-6
149. E. Agazzi (ed.), *Modern Logic – A Survey.* Historical, Philosophical, and Mathematical Aspects of Modern Logic and Its Applications. 1981 ISBN 90-277-1137-2
150. A. F. Parker-Rhodes, *The Theory of Indistinguishables.* A Search for Explanatory Principles below the Level of Physics. 1981 ISBN 90-277-1214-X
151. J. C. Pitt, *Pictures, Images, and Conceptual Change.* An Analysis of Wilfrid Sellars' Philosophy of Science. 1981 ISBN 90-277-1276-X; Pb 90-277-1277-8
152. R. Hilpinen (ed.), *New Studies in Deontic Logic.* Norms, Actions, and the Foundations of Ethics. 1981 ISBN 90-277-1278-6; Pb 90-277-1346-4
153. C. Dilworth, *Scientific Progress.* A Study Concerning the Nature of the Relation between Successive Scientific Theories. 3rd rev. ed., 1994 ISBN 0-7923-2487-0; Pb 0-7923-2488-9
154. D. Woodruff Smith and R. McIntyre, *Husserl and Intentionality.* A Study of Mind, Meaning, and Language. 1982 ISBN 90-277-1392-8; Pb 90-277-1730-3
155. R. J. Nelson, *The Logic of Mind.* 2nd. ed., 1989 ISBN 90-277-2819-4; Pb 90-277-2822-4
156. J. F. A. K. van Benthem, *The Logic of Time.* A Model-Theoretic Investigation into the Varieties of Temporal Ontology, and Temporal Discourse. 1983; 2nd ed., 1991 ISBN 0-7923-1081-0
157. R. Swinburne (ed.), *Space, Time and Causality.* 1983 ISBN 90-277-1437-1
158. E. T. Jaynes, *Papers on Probability, Statistics and Statistical Physics.* Ed. by R. D. Rozenkrantz. 1983 ISBN 90-277-1448-7; Pb (1989) 0-7923-0213-3
159. T. Chapman, *Time: A Philosophical Analysis.* 1982 ISBN 90-277-1465-7
160. E. N. Zalta, *Abstract Objects.* An Introduction to Axiomatic Metaphysics. 1983 ISBN 90-277-1474-6
161. S. Harding and M. B. Hintikka (eds.), *Discovering Reality.* Feminist Perspectives on Epistemology, Metaphysics, Methodology, and Philosophy of Science. 1983 ISBN 90-277-1496-7; Pb 90-277-1538-6

162. M. A. Stewart (ed.), *Law, Morality and Rights.* 1983 ISBN 90-277-1519-X
163. D. Mayr and G. Süssmann (eds.), *Space, Time, and Mechanics.* Basic Structures of a Physical Theory. 1983 ISBN 90-277-1525-4
164. D. Gabbay and F. Guenthner (eds.), *Handbook of Philosophical Logic.* Vol. I: Elements of Classical Logic. 1983 ISBN 90-277-1542-4
165. D. Gabbay and F. Guenthner (eds.), *Handbook of Philosophical Logic.* Vol. II: Extensions of Classical Logic. 1984 ISBN 90-277-1604-8
166. D. Gabbay and F. Guenthner (eds.), *Handbook of Philosophical Logic.* Vol. III: Alternative to Classical Logic. 1986 ISBN 90-277-1605-6
167. D. Gabbay and F. Guenthner (eds.), *Handbook of Philosophical Logic.* Vol. IV: Topics in the Philosophy of Language. 1989 ISBN 90-277-1606-4
168. A. J. I. Jones, *Communication and Meaning.* An Essay in Applied Modal Logic. 1983 ISBN 90-277-1543-2
169. M. Fitting, *Proof Methods for Modal and Intuitionistic Logics.* 1983 ISBN 90-277-1573-4
170. J. Margolis, *Culture and Cultural Entities.* Toward a New Unity of Science. 1984 ISBN 90-277-1574-2
171. R. Tuomela, *A Theory of Social Action.* 1984 ISBN 90-277-1703-6
172. J. J. E. Gracia, E. Rabossi, E. Villanueva and M. Dascal (eds.), *Philosophical Analysis in Latin America.* 1984 ISBN 90-277-1749-4
173. P. Ziff, *Epistemic Analysis.* A Coherence Theory of Knowledge. 1984 ISBN 90-277-1751-7
174. P. Ziff, *Antiaesthetics.* An Appreciation of the Cow with the Subtile Nose. 1984 ISBN 90-277-1773-7
175. W. Balzer, D. A. Pearce, and H.-J. Schmidt (eds.), *Reduction in Science.* Structure, Examples, Philosophical Problems. 1984 ISBN 90-277-1811-3
176. A. Peczenik, L. Lindahl and B. van Roermund (eds.), *Theory of Legal Science.* Proceedings of the Conference on Legal Theory and Philosophy of Science (Lund, Sweden, December 1983). 1984 ISBN 90-277-1834-2
177. I. Niiniluoto, *Is Science Progressive?* 1984 ISBN 90-277-1835-0
178. B. K. Matilal and J. L. Shaw (eds.), *Analytical Philosophy in Comparative Perspective.* Exploratory Essays in Current Theories and Classical Indian Theories of Meaning and Reference. 1985 ISBN 90-277-1870-9
179. P. Kroes, *Time: Its Structure and Role in Physical Theories.* 1985 ISBN 90-277-1894-6
180. J. H. Fetzer, *Sociobiology and Epistemology.* 1985 ISBN 90-277-2005-3; Pb 90-277-2006-1
181. L. Haaparanta and J. Hintikka (eds.), *Frege Synthesized.* Essays on the Philosophical and Foundational Work of Gottlob Frege. 1986 ISBN 90-277-2126-2
182. M. Detlefsen, *Hilbert's Program.* An Essay on Mathematical Instrumentalism. 1986 ISBN 90-277-2151-3
183. J. L. Golden and J. J. Pilotta (eds.), *Practical Reasoning in Human Affairs.* Studies in Honor of Chaim Perelman. 1986 ISBN 90-277-2255-2
184. H. Zandvoort, *Models of Scientific Development and the Case of Nuclear Magnetic Resonance.* 1986 ISBN 90-277-2351-6
185. I. Niiniluoto, *Truthlikeness.* 1987 ISBN 90-277-2354-0
186. W. Balzer, C. U. Moulines and J. D. Sneed, *An Architectonic for Science.* The Structuralist Program. 1987 ISBN 90-277-2403-2
187. D. Pearce, *Roads to Commensurability.* 1987 ISBN 90-277-2414-8
188. L. M. Vaina (ed.), *Matters of Intelligence.* Conceptual Structures in Cognitive Neuroscience. 1987 ISBN 90-277-2460-1

189. H. Siegel, *Relativism Refuted.* A Critique of Contemporary Epistemological Relativism. 1987 ISBN 90-277-2469-5
190. W. Callebaut and R. Pinxten, *Evolutionary Epistemology.* A Multiparadigm Program, with a Complete Evolutionary Epistemology Bibliograph. 1987 ISBN 90-277-2582-9
191. J. Kmita, *Problems in Historical Epistemology.* 1988 ISBN 90-277-2199-8
192. J. H. Fetzer (ed.), *Probability and Causality.* Essays in Honor of Wesley C. Salmon, with an Annotated Bibliography. 1988 ISBN 90-277-2607-8; Pb 1-5560-8052-2
193. A. Donovan, L. Laudan and R. Laudan (eds.), *Scrutinizing Science.* Empirical Studies of Scientific Change. 1988 ISBN 90-277-2608-6
194. H.R. Otto and J.A. Tuedio (eds.), *Perspectives on Mind.* 1988 ISBN 90-277-2640-X
195. D. Batens and J.P. van Bendegem (eds.), *Theory and Experiment.* Recent Insights and New Perspectives on Their Relation. 1988 ISBN 90-277-2645-0
196. J. Österberg, *Self and Others.* A Study of Ethical Egoism. 1988 ISBN 90-277-2648-5
197. D.H. Helman (ed.), *Analogical Reasoning.* Perspectives of Artificial Intelligence, Cognitive Science, and Philosophy. 1988 ISBN 90-277-2711-2
198. J. Woleński, *Logic and Philosophy in the Lvov-Warsaw School.* 1989 ISBN 90-277-2749-X
199. R. Wójcicki, *Theory of Logical Calculi.* Basic Theory of Consequence Operations. 1988 ISBN 90-277-2785-6
200. J. Hintikka and M.B. Hintikka, *The Logic of Epistemology and the Epistemology of Logic.* Selected Essays. 1989 ISBN 0-7923-0040-8; Pb 0-7923-0041-6
201. E. Agazzi (ed.), *Probability in the Sciences.* 1988 ISBN 90-277-2808-9
202. M. Meyer (ed.), *From Metaphysics to Rhetoric.* 1989 ISBN 90-277-2814-3
203. R.L. Tieszen, *Mathematical Intuition.* Phenomenology and Mathematical Knowledge. 1989 ISBN 0-7923-0131-5
204. A. Melnick, *Space, Time, and Thought in Kant.* 1989 ISBN 0-7923-0135-8
205. D.W. Smith, *The Circle of Acquaintance.* Perception, Consciousness, and Empathy. 1989 ISBN 0-7923-0252-4
206. M.H. Salmon (ed.), *The Philosophy of Logical Mechanism.* Essays in Honor of Arthur W. Burks. With his Responses, and with a Bibliography of Burk's Work. 1990 ISBN 0-7923-0325-3
207. M. Kusch, *Language as Calculus vs. Language as Universal Medium.* A Study in Husserl, Heidegger, and Gadamer. 1989 ISBN 0-7923-0333-4
208. T.C. Meyering, *Historical Roots of Cognitive Science.* The Rise of a Cognitive Theory of Perception from Antiquity to the Nineteenth Century. 1989 ISBN 0-7923-0349-0
209. P. Kosso, *Observability and Observation in Physical Science.* 1989 ISBN 0-7923-0389-X
210. J. Kmita, *Essays on the Theory of Scientific Cognition.* 1990 ISBN 0-7923-0441-1
211. W. Sieg (ed.), *Acting and Reflecting.* The Interdisciplinary Turn in Philosophy. 1990 ISBN 0-7923-0512-4
212. J. Karpiński, *Causality in Sociological Research.* 1990 ISBN 0-7923-0546-9
213. H.A. Lewis (ed.), *Peter Geach: Philosophical Encounters.* 1991 ISBN 0-7923-0823-9
214. M. Ter Hark, *Beyond the Inner and the Outer.* Wittgenstein's Philosophy of Psychology. 1990 ISBN 0-7923-0850-6
215. M. Gosselin, *Nominalism and Contemporary Nominalism.* Ontological and Epistemological Implications of the Work of W.V.O. Quine and of N. Goodman. 1990 ISBN 0-7923-0904-9
216. J.H. Fetzer, D. Shatz and G. Schlesinger (eds.), *Definitions and Definability.* Philosophical Perspectives. 1991 ISBN 0-7923-1046-2
217. E. Agazzi and A. Cordero (eds.), *Philosophy and the Origin and Evolution of the Universe.* 1991 ISBN 0-7923-1322-4

218. M. Kusch, *Foucault's Strata and Fields.* An Investigation into Archaeological and Genealogical Science Studies. 1991 ISBN 0-7923-1462-X
219. C.J. Posy, *Kant's Philosophy of Mathematics.* Modern Essays. 1992 ISBN 0-7923-1495-6
220. G. Van de Vijver, *New Perspectives on Cybernetics.* Self-Organization, Autonomy and Connectionism. 1992 ISBN 0-7923-1519-7
221. J.C. Nyíri, *Tradition and Individuality.* Essays. 1992 ISBN 0-7923-1566-9
222. R. Howell, *Kant's Transcendental Deduction.* An Analysis of Main Themes in His Critical Philosophy. 1992 ISBN 0-7923-1571-5
223. A. García de la Sienra, *The Logical Foundations of the Marxian Theory of Value.* 1992 ISBN 0-7923-1778-5
224. D.S. Shwayder, *Statement and Referent.* An Inquiry into the Foundations of Our Conceptual Order. 1992 ISBN 0-7923-1803-X
225. M. Rosen, *Problems of the Hegelian Dialectic.* Dialectic Reconstructed as a Logic of Human Reality. 1993 ISBN 0-7923-2047-6
226. P. Suppes, *Models and Methods in the Philosophy of Science: Selected Essays.* 1993 ISBN 0-7923-2211-8
227. R. M. Dancy (ed.), *Kant and Critique: New Essays in Honor of W. H. Werkmeister.* 1993 ISBN 0-7923-2244-4
228. J. Woleński (ed.), *Philosophical Logic in Poland.* 1993 ISBN 0-7923-2293-2
229. M. De Rijke (ed.), *Diamonds and Defaults.* Studies in Pure and Applied Intensional Logic. 1993 ISBN 0-7923-2342-4
230. B.K. Matilal and A. Chakrabarti (eds.), *Knowing from Words.* Western and Indian Philosophical Analysis of Understanding and Testimony. 1994 ISBN 0-7923-2345-9
231. S.A. Kleiner, *The Logic of Discovery.* A Theory of the Rationality of Scientific Research. 1993 ISBN 0-7923-2371-8
232. R. Festa, *Optimum Inductive Methods.* A Study in Inductive Probability, Bayesian Statistics, and Verisimilitude. 1993 ISBN 0-7923-2460-9
233. P. Humphreys (ed.), *Patrick Suppes: Scientific Philosopher.* Vol. 1: Probability and Probabilistic Causality. 1994 ISBN 0-7923-2552-4
234. P. Humphreys (ed.), *Patrick Suppes: Scientific Philosopher.* Vol. 2: Philosophy of Physics, Theory Structure, and Measurement Theory. 1994 ISBN 0-7923-2553-2
235. P. Humphreys (ed.), *Patrick Suppes: Scientific Philosopher.* Vol. 3: Language, Logic, and Psychology. 1994 ISBN 0-7923-2862-0
Set ISBN (Vols 233–235) 0-7923-2554-0
236. D. Prawitz and D. Westerståhl (eds.), *Logic and Philosophy of Science in Uppsala.* Papers from the 9th International Congress of Logic, Methodology, and Philosophy of Science. 1994 ISBN 0-7923-2702-0
237. L. Haaparanta (ed.), *Mind, Meaning and Mathematics.* Essays on the Philosophical Views of Husserl and Frege. 1994 ISBN 0-7923-2703-9
238. J. Hintikka (ed.), *Aspects of Metaphor.* 1994 ISBN 0-7923-2786-1
239. B. McGuinness and G. Oliveri (eds.), *The Philosophy of Michael Dummett.* With Replies from Michael Dummett. 1994 ISBN 0-7923-2804-3
240. D. Jamieson (ed.), *Language, Mind, and Art.* Essays in Appreciation and Analysis, In Honor of Paul Ziff. 1994 ISBN 0-7923-2810-8
241. G. Preyer, F. Siebelt and A. Ulfig (eds.), *Language, Mind and Epistemology.* On Donald Davidson's Philosophy. 1994 ISBN 0-7923-2811-6
242. P. Ehrlich (ed.), *Real Numbers, Generalizations of the Reals, and Theories of Continua.* 1994 ISBN 0-7923-2689-X

243. G. Debrock and M. Hulswit (eds.), *Living Doubt.* Essays concerning the epistemology of Charles Sanders Peirce. 1994 ISBN 0-7923-2898-1
244. J. Srzednicki, *To Know or Not to Know.* Beyond Realism and Anti-Realism. 1994 ISBN 0-7923-2909-0
245. R. Egidi (ed.), *Wittgenstein: Mind and Language.* 1995 ISBN 0-7923-3171-0
246. A. Hyslop, *Other Minds.* 1995 ISBN 0-7923-3245-8
247. L. Pólos and M. Masuch (eds.), *Applied Logic: How, What and Why.* Logical Approaches to Natural Language. 1995 ISBN 0-7923-3432-9
248. M. Krynicki, M. Mostowski and L.M. Szczerba (eds.), *Quantifiers: Logics, Models and Computation.* Volume One: Surveys. 1995 ISBN 0-7923-3448-5
249. M. Krynicki, M. Mostowski and L.M. Szczerba (eds.), *Quantifiers: Logics, Models and Computation.* Volume Two: Contributions. 1995 ISBN 0-7923-3449-3
Set ISBN (Vols 248 + 249) 0-7923-3450-7
250. R.A. Watson, *Representational Ideas from Plato to Patricia Churchland.* 1995 ISBN 0-7923-3453-1
251. J. Hintikka (ed.), *From Dedekind to Gödel.* Essays on the Development of the Foundations of Mathematics. 1995 ISBN 0-7923-3484-1
252. A. Wiśniewski, *The Posing of Questions.* Logical Foundations of Erotetic Inferences. 1995 ISBN 0-7923-3637-2
253. J. Peregrin, *Doing Worlds with Words.* Formal Semantics without Formal Metaphysics. 1995 ISBN 0-7923-3742-5
254. I.A. Kieseppä, *Truthlikeness for Multidimensional, Quantitative Cognitive Problems.* 1996 ISBN 0-7923-4005-1
255. P. Hugly and C. Sayward: *Intensionality and Truth.* An Essay on the Philosophy of A.N. Prior. 1996 ISBN 0-7923-4119-8
256. L. Hankinson Nelson and J. Nelson (eds.): *Feminism, Science, and the Philosophy of Science.* 1997 ISBN 0-7923-4162-7
257. P.I. Bystrov and V.N. Sadovsky (eds.): *Philosophical Logic and Logical Philosophy.* Essays in Honour of Vladimir A. Smirnov. 1996 ISBN 0-7923-4270-4
258. Å.E. Andersson and N-E. Sahlin (eds.): *The Complexity of Creativity.* 1996 ISBN 0-7923-4346-8
259. M.L. Dalla Chiara, K. Doets, D. Mundici and J. van Benthem (eds.): *Logic and Scientific Methods.* Volume One of the Tenth International Congress of Logic, Methodology and Philosophy of Science, Florence, August 1995. 1997 ISBN 0-7923-4383-2
260. M.L. Dalla Chiara, K. Doets, D. Mundici and J. van Benthem (eds.): *Structures and Norms in Science.* Volume Two of the Tenth International Congress of Logic, Methodology and Philosophy of Science, Florence, August 1995. 1997 ISBN 0-7923-4384-0
Set ISBN (Vols 259 + 260) 0-7923-4385-9
261. A. Chakrabarti: *Denying Existence.* The Logic, Epistemology and Pragmatics of Negative Existentials and Fictional Discourse. 1997 ISBN 0-7923-4388-3
262. A. Biletzki: *Talking Wolves.* Thomas Hobbes on the Language of Politics and the Politics of Language. 1997 ISBN 0-7923-4425-1
263. D. Nute (ed.): *Defeasible Deontic Logic.* 1997 ISBN 0-7923-4630-0
264. U. Meixner: *Axiomatic Formal Ontology.* 1997 ISBN 0-7923-4747-X
265. I. Brinck: *The Indexical 'I'.* The First Person in Thought and Language. 1997 ISBN 0-7923-4741-2
266. G. Hölmström-Hintikka and R. Tuomela (eds.): *Contemporary Action Theory.* Volume 1: Individual Action. 1997 ISBN 0-7923-4753-6; Set: 0-7923-4754-4

267. G. Hölmström-Hintikka and R. Tuomela (eds.): *Contemporary Action Theory.* Volume 2: Social Action. 1997 ISBN 0-7923-4752-8; Set: 0-7923-4754-4
268. B.-C. Park: *Phenomenological Aspects of Wittgenstein's Philosophy.* 1998 ISBN 0-7923-4813-3
269. J. Paśniczek: *The Logic of Intentional Objects.* A Meinongian Version of Classical Logic. 1998 Hb ISBN 0-7923-4880-X; Pb ISBN 0-7923-5578-4
270. P.W. Humphreys and J.H. Fetzer (eds.): *The New Theory of Reference.* Kripke, Marcus, and Its Origins. 1998 ISBN 0-7923-4898-2
271. K. Szaniawski, A. Chmielewski and J. Woleński (eds.): *On Science, Inference, Information and Decision Making.* Selected Essays in the Philosophy of Science. 1998 ISBN 0-7923-4922-9
272. G.H. von Wright: *In the Shadow of Descartes.* Essays in the Philosophy of Mind. 1998 ISBN 0-7923-4992-X
273. K. Kijania-Placek and J. Woleński (eds.): *The Lvov–Warsaw School and Contemporary Philosophy.* 1998 ISBN 0-7923-5105-3
274. D. Dedrick: *Naming the Rainbow.* Colour Language, Colour Science, and Culture. 1998 ISBN 0-7923-5239-4
275. L. Albertazzi (ed.): *Shapes of Forms.* From Gestalt Psychology and Phenomenology to Ontology and Mathematics. 1999 ISBN 0-7923-5246-7
276. P. Fletcher: *Truth, Proof and Infinity.* A Theory of Constructions and Constructive Reasoning. 1998 ISBN 0-7923-5262-9
277. M. Fitting and R.L. Mendelsohn (eds.): *First-Order Modal Logic.* 1998 Hb ISBN 0-7923-5334-X; Pb ISBN 0-7923-5335-8
278. J.N. Mohanty: *Logic, Truth and the Modalities from a Phenomenological Perspective.* 1999 ISBN 0-7923-5550-4
279. T. Placek: *Mathematical Intiutionism and Intersubjectivity.* A Critical Exposition of Arguments for Intuitionism. 1999 ISBN 0-7923-5630-6
280. A. Cantini, E. Casari and P. Minari (eds.): *Logic and Foundations of Mathematics.* 1999 ISBN 0-7923-5659-4 set ISBN 0-7923-5867-8
281. M.L. Dalla Chiara, R. Giuntini and F. Laudisa (eds.): *Language, Quantum, Music.* 1999 ISBN 0-7923-5727-2; set ISBN 0-7923-5867-8
282. R. Egidi (ed.): *In Search of a New Humanism.* The Philosophy of Georg Hendrik von Wright. 1999 ISBN 0-7923-5810-4
283. F. Vollmer: *Agent Causality.* 1999 ISBN 0-7923-5848-1
284. J. Peregrin (ed.): *Truth and Its Nature (if Any).* 1999 ISBN 0-7923-5865-1
285. M. De Caro (ed.): *Interpretations and Causes.* New Perspectives on Donald Davidson's Philosophy. 1999 ISBN 0-7923-5869-4
286. R. Murawski: *Recursive Functions and Metamathematics.* Problems of Completeness and Decidability, Gödel's Theorems. 1999 ISBN 0-7923-5904-6
287. T.A.F. Kuipers: *From Instrumentalism to Constructive Realism.* On Some Relations between Confirmation, Empirical Progress, and Truth Approximation. 2000 ISBN 0-7923-6086-9
288. G. Holmström-Hintikka (ed.): *Medieval Philosophy and Modern Times.* 2000 ISBN 0-7923-6102-4
289. E. Grosholz and H. Breger (eds.): *The Growth of Mathematical Knowledge.* 2000 ISBN 0-7923-6151-2

SYNTHESE LIBRARY

290. G. Sommaruga: *History and Philosophy of Constructive Type Theory*. 2000
ISBN 0-7923-6180-6

Previous volumes are still available.

KLUWER ACADEMIC PUBLISHERS – DORDRECHT / BOSTON / LONDON